Empirical Process Techniques for Dependent Data

Herold Dehling
Thomas Mikosch
Michael Sørensen
Editors

Birkhäuser
Boston • Basel • Berlin

Herold Dehling
Ruhr-Universität Bochum
Fakultät für Mathematik
44780 Bochum
Germany

Thomas Mikosch
University of Copenhagen
Laboratory of Actuarial Mathematics
DK-2100 Copenhagen
Denmark

Michael Sørensen
University of Copenhagen
Department of Statistics and
Operations Research
Copenhagen
DK-2100 Denmark

Library of Congress Cataloging-in-Publication Data

Empirical process techniques for dependent data / Herold Dehling, Thomas Mikosch, and Michael Sørensen, editors.
p. cm.
Includes bibliographical references.
ISBN 0-8176-4201-3 (alk. paper) – ISBN 3-7643-4201-3
1. Nonparametric statistics. 2. Estimation theory. 3. Limit theorems (Probability theory) I. Dehling, Herold. II. Mikosch, Thomas. III. Sørensen, Michael.

QA278.8.E47 2002
519.5–dc21 2002071106
CIP

AMS Subject Classifications: Primary: 60FXX, 60GXX, 60JXX, 62GXX, 62MXX

Printed on acid-free paper.

Birkhäuser

ISBN 0-8176-4201-3 SPIN 10772146
ISBN 3-7643-4201-3

Reformatted from editors' files by John Spiegelman, Philadelphia, PA.
Printed and bound by Hamilton Printing Company, Rensselaer, NY.
Printed in the United States of America.

9 8 7 6 5 4 3 2 1

Birkhäuser Boston • Basel • Berlin
A member of BertelsmannSpringer Science+Business Media GmbH

Contents

Preface

This volume aims to bring together empirical process theory with different concepts of dependence in probability theory and statistics. This convergence is needed to respond to situations in which increasingly complicated stochastic models are being used to capture the dependence structure of real-life data.

Data with complex dependencies are found in fields as diverse as geology, genomics, environmental sciences, finance, insurance, meteorology, physiology, and telecommunications, to name several. Present-day computers, with their high speed and large memory requirements, have made it possible to use models for dependence in statistical applications. Such advances in technology have brought the inference for stochastic processes into mainstream statistics.

As a matter of fact, the investigation of the properties of estimators and test statistics for stochastic processes is much more difficult than for models that assume observations to be independent. Here, another important development in statistical science comes in. Over the last few decades, it has become evident that empirical process techniques are extremely useful when it comes to studying the asymptotic properties of parametric as well as nonparametric statistical procedures. Given the large increase in the number of statistical problems that utilize concepts and methods from empirical process theory, it is not surprising that a theory of empirical processes for dependent data has been developed in recent years.

Our main goal is to give an introduction to empirical process techniques for dependent data and to survey recent developments that are widely scattered over the statistics and probability literature. It is worthwhile stressing that we do not follow standard patterns by sticking to the standard empirical process for stationary sequences. In our own research we have often experienced the fact that researchers with quite different interests and expertise have borrowed techniques from classical empirical process theory without being aware of parallel developments in other areas. Such areas include *spectral analysis of time series*, *bootstrap for dependent sequences*, *extreme value theory for dependent data*, and *empirical processes for mixing dependent observations*, including *processes with long-range dependence*. This volume is a unique collection of material from the diverse areas mentioned. The interested reader will easily find the common spirit throughout the various parts of this book and, hopefully, will be able to discover analogies with techniques used in other fields that are not covered in this text.

Earlier monographs on empirical processes such as *Empirical Processes with Applications to Statistics* by Shorack and Wellner, *Weak Convergence and Empirical Processes* by van der Vaart and Wellner, and *Convergence of Stochastic Processes* by Pollard cover the case of independent observations. We believe that this volume is a timely and indispensable supplement to these books. It can also be seen as a follow-up to the 1986 Birkhäuser volume *Dependence in Probability and Statistics*, edited by Eberlein and Taqqu, which served as a standard reference for a generation of researchers.

In Part I, Herold Dehling and Walter Philipp give a *tutorial on empirical process techniques for dependent data*. They present a gentle introduction to empirical processes and dependence from the early roots of Hermann Weyl up to recent developments via the classical contributions of Patrick Billingsley. Dehling and Philipp start with the basics of empirical process theory for iid data, including the Glivenko–Cantelli theorem and the Donsker empirical central limit theorem, after which they turn to weakly dependent sequences and sketch the technical arguments needed in the case of weak dependence. Different mixing properties are explained in detail and illustrated through various examples. Extensive overviews of covariance inequalities and their applications to moment bounds for sums of dependent variables are also given. All these tools, in conjunction with chaining arguments, are then used to derive a multitude of empirical central limit theorems for mixing sequences.

Dehling and Philipp also present a unique survey of recent developments on the empiricial process of U-statistics structure. One section is devoted to the limit theory for the empirical process of strongly dependent or long-memory stationary sequences. An accessible introduction is given to this field, stressing the crucial differences between the weakly and strongly dependent worlds of processes. This tutorial chapter, with a bibliography of some 150 references, allows the reader to obtain a thorough understanding of the basic theory of empirical processes and dependence as well as its historical development.

Techniques for the empirical process of stationary sequences are presented in Part II. Various authors give accounts of their research on the classical empirical process in the extended framework of dependent data. Patrick Ango Nze and Paul Doukhan introduce a new concept of weak dependence based on the covariances of functions of the stationary sequence. They apply this notion to derive weak limit theory for dependent sequences, including the empirical central limit theorem. Jérôme Dedecker and Sana Louhichi, in their presentation of a recent development concerning empirical central limit theorems, examine maximal inequalities which are crucial for proving tightness of the empirical process for mixing sequences. They work in the abstract context of the empirical process indexed by classes of functions, assuming entropy type conditions.

Sara van de Geer continues with a generalization of Hoeffding's inequality to the case of dependent variables and illustrates the theory with an application to a general nonlinear autoregressive process. Florence Merlevède and Magda Peligrad continue with the coupling of sequences of dependent, respectively independent, variables with the same marginals. They apply these results to obtain uniform strong laws of large numbers and central limit theorems for various classes of dependent sequences. István Berkes and Lajos Horváth review results on the behavior of empirical processes of residuals from ARMA, ARCH and GARCH time-series models that are widely used in practice, including financial econometrics.

The empirical process of long-range dependent data exhibits asymptotic behaviour which is very much different from weakly dependent data. Such empirical processes are treated in Part III, which is written by masters of their trade and accessible to the nonspecialist who cannot follow the branching specialized literature.

Hira L. Koul and Donatas Surgailis consider linear processes with long-range dependence which constitute a major class of time-series models that are important for statistical applications. They review asymptotic expansions for the empirical process of finite and infinite variance sequences with long-range dependence and discuss applications to goodness-of-fit testing for the marginal stationary error distribution in linear regression models and to M-estimation in the one sample location model.

Liudas Giraitis and Donatas Surgailis continue with an extension of the so-called reduction principle of Dehling and Taqqu (1989) which is closely related to the asymptotic expansions of the empirical process of long-range dependent stationary sequences. The latter principle is a fundamental result which allows one to derive the limiting structure of the empirical process for long-range dependent processes. The paper by Miguel A. Arcones provides a useful review of the literature on limit theorems for sums of nonlinear functions of a stationary sequence of

Gaussian random variables with particular emphasis on empirical processes.

In the classical work of Grenander, Rosenblatt, Bartlett and others from the 1950s it was realized that the integrated periodogram process, used in the spectral analysis of time series for goodness-of-fit tests, model fitting and change point analysis, can be considered as a quantity with properties analogous to the classical empirical process. These analogies motivated the adaptation of ideas and techniques from classical empirical process theory to the spectral analysis of time series. They are the goal of Part IV on *empirical spectral process techniques.*

The paper of Rainer Dahlhaus and Wolfgang Polonik makes the links between empirical process theory and spectral analysis transparent. They show in various examples how empirical process ideas come naturally into the picture by considering the integrated periodogram process indexed by appropriate classes of functions. Under entropy conditions the latter process weakly converges to the corresponding spectral distribution. This fact opens the door to goodness-of-fit tests in the spectral domain, nonparametric Whittle likelihood estimation for time series, change point analysis and many other statistical topics. They indicate how the theory has to be modified for the Gaussian and non-Gaussian stationary cases, and for locally nonstationary processes. They also show that the techniques used are similar to those in empirical process theory, including exponential inequalities, chaining arguments and others which originate from the latter theory.

The paper by Philippe Soulier stresses the analogies between the periodogram at the Fourier frequencies and a sequence of independent identically distributed random variables. Soulier points out how these similarities can be used to derive limit theory for functionals of the periodogram process. He obtains a functional central limit theorem for the empirical process of the periodogram process at the Fourier frequencies and shows how these results can be used in statistical applications to the estimation of the one-step-prediction error of a time series and goodness-of-fit testing. He also studies the behavior of the periodogram for fractional (long-range dependent) time series whose behavior is totally different from the weakly dependent case. Again, empirical process techniques play a major role for deriving functional central limit theorems and the resulting statistical applications.

Part V is devoted to the *tail empirical process.* The latter is perhaps less familiar; only recently it has become popular in the extreme value statistics community. The tail empirical process turns out to be a useful tool for parameter estimation for extremes of dependent data as well as for tail and high quantile estimation. Holger Drees gives a unique review on the limit theory for tail processes. It includes Rootzén's result for the uniform tail empirical process and generalizations thereof, and results for the uniform tail quantile process. The results are applied to derive the asymptotic normality of statistical tail functionals used to deduce asymptotic normality of various estimators in extreme value statistics under mixing conditions.

Over the last 15 years bootstrap techniques have been developed for dependent data as well. Since the bootstrap is based on the empirical distribution of the data, the link with the main ideas of this volume is quite natural. Therefore, Part VI is entirely devoted to *the bootstrap for dependent data.* Dragan Radulović gives an introduction to the basic ideas in this field and also surveys recent developments on the bootstrap for stationary sequences, including new results for Markov chains. Whereas he focuses on the bootstrap based on the empirical distribution of the sample, Efstathios Paparoditis considers nonparametric resampling methods based on the empirical distribution of the periodogram process. Both papers give a unique overview of bootstrap techniques for dependent data which are available in the literature.

This volume grew out of a MaPhySto Instructional Workshop on Empirical Process Techniques for Dependent Data organized by Thomas Mikosch, Søren Feodor Nielsen and Michael Sørensen in November 2000 at the University of Copenhagen. The workshop was made possible by generous support from the Centre for Mathematical Physics and Stochastics (MaPhySto) which is funded by the Danish National Research Foundation. Supplementary support was received from the research training network on Statistical Methods for Dynamical Stochastic Models (DYNSTOCH) under the EU programme Improving Human Potential and from the Centre for Analytical Finance (CAF) funded by the Danish Social Science Research Council. Many of the contributors to this volume taught at the instructional workshop. In particular, a main series of lectures was given by Herold Dehling and Walter Philip, which becames the starting point for this volume. We are delighted that the authors here, who are top specialists in their field of expertise, agreed to contribute to this book.

We take pleasure in thanking Daniel Straumann from the University of Copenhagen, who masterfully handled the difficult job of merging the different LaTeX styles of the authors. We would also like to thank Ann Kostant from Birkhäuser Boston who carefully supervised the editing of this volume.

We are convinced that this volume is a useful collection of material on dependent sequences and empirical processes. The mix of time-series analysis, statistics, empirical process theory, extreme value theory, and many other areas may be unusual. On the other hand, we feel that this book will contribute to a better understanding of the parallel developments in different areas and lead to a beneficial exchange of ideas and concepts. Researchers in statistics, time-series analysis, extreme value theory, probability theory will find a useful toolbox in this book. The nonspecialized reader or the interested graduate student will get access to the worlds of dependence and empirical measures.

Herold G. Dehling Bochum, 1 June, 2002

Thomas Mikosch Copenhagen, 1 June, 2002

Michael Sørensen Copenhagen, 1 June, 2002

Empirical Process Techniques for Dependent Data

Part I

A Tutorial on Empirical Process Techniques for Dependent Data

Empirical Process Techniques for Dependent Data

HEROLD DEHLING AND WALTER PHILIPP

ABSTRACT. Let $(X_k)_{k\geq 1}$ be a sequence of random variables with common distribution function $F(x) = P(X_1 \leq x)$. Define the empirical distribution function

$$F_n(x) = \frac{1}{n}\#\{1 \leq i \leq n : X_i \leq x\},$$

and the empirical process by $\sqrt{n}(F_n(x) - F(x))$. In this chapter we provide a survey of classical as well as modern techniques in the study of empirical processes of dependent data. We begin with a sketch of the early roots of the field in the theory of uniform distribution mod 1, of sequences defined by $X_k = \{n_k\omega\}$, $\omega \in [0, 1]$, dating back to Weyl's celebrated 1916 paper. In the second section we provide the essential tools of empirical process theory, and we prove Donsker's classical empirical process invariance principle for i.i.d. processes. The third section provides an introduction to the subject of weakly dependent random variables. We introduce a variety of mixing concepts, provide necessary technical tools like correlation and moment inequalities, and prove central limit theorems for partial sums. The empirical process of weakly dependent data is investigated in the fourth section, where we put special emphasis on almost sure approximation techniques. The fifth section is devoted to the empirical distribution of U-statistics, defined as

$$U_n(x) = \binom{n}{2}^{-1}\#\{1 \leq i < j \leq n : h(X_i, X_j) \leq x\}$$

for some symmetric kernel h. We give some applications, e.g., to dimension estimation in the analysis of time series, and prove weak convergence of the corresponding empirical process. Empirical processes of long-range dependent data are the topic of the sixth section. We give an introduction to the area of long-range dependent processes, provide important technical tools for the study of their partial sums and investigate the limit behavior of the empirical process. It turns out that the limit process is of a completely different type as in the case of independent or weakly dependent data, and that this has important consequences for various functionals of the empirical process. The final section is devoted to pair correlations, i.e., U-statistics empirical processes over short intervals associated with the kernel $h(x, y) = |x - y|$.

1 Introduction and historical remarks

1.1 Introduction. It may come as a surprise to many probabilists and statisticians that the beginnings of empirical process theory for dependent random variables go back to the seminal paper by Hermann Weyl (1916) on uniform distribution mod 1. In this paper Weyl streamlined the theory of uniform distribution mod 1, including the presentation of what is nowadays called the Weyl criterion. He also proved several metric theorems which are precursors of the subject of the present course.

There are three additional reasons why we present Weyl's results as well as several related ones at the outset of this course. First, these results can be formulated with a minimum amount of technical preliminaries. Second, they nicely reveal the demarcation line between results which can be considered direct generalizations of the i.i.d. case and results which are not. Third, these results are at the interface of several branches of mathematics: probability theory and mathematical statistics, analysis, in particular ergodic theory and number theory and thus demonstrate the interplay between these disciplines.

Let $(n_k)_{k=1}^{\infty}$ be an increasing sequence of integers and let $([0, 1), \mathcal{B}, P)$ denote the unit interval with Lebesgue measure. The fractional parts

$$X_k := \{n_k \omega\}, \quad \omega \in [0, 1), \quad k \geq 1, \tag{1.1}$$

can be considered as random variables, defined on the probability space $([0, 1), \mathcal{B}, P)$, which have uniform distribution in the probabilistic sense, i.e.,

$$P(X_k \leq x) = x, \quad 0 \leq x \leq 1, \quad k \geq 1.$$

The random variables X_k are dependent. If, for instance $n_1 = 3, n_2 = 6$, then $X_1(\omega) = \{3\omega\}$ and $X_2(\omega) = \{6\omega\}$ and thus if $X_1(\omega) = x$, then $X_2(\omega) = \{2x\}$. Thus the value of X_2 is completely determined by the value of X_1. Let

$$F_N(x) := \frac{1}{N} \#\{k \leq N : X_k \leq x\} \tag{1.2}$$

denote the empirical distribution function of the sequence $(X_k)_{k=1}^{\infty}$ at stage N and set

$$D_N = D_N(\omega) := \sup_{0 \leq x < 1} \mid F_N(x) - x \mid .$$

Weyl called this quantity the discrepancy of the sequence $(X_k)_{k=1}^{N}$. In his paper Weyl proved that

$$D_N(\omega) \to 0, \tag{1.3}$$

for almost all ω, i.e., for all $0 \leq \omega < 1$ except possibly on a set of Lebesgue measure zero. Of course this relation is just the Glivenko–Cantelli theorem for the sequence (X_k).

We remark in passing that (1.3) continues to hold for sequences $(n_k)_{k=1}^{\infty}$ of real numbers, not necessarily integers, satisfying Weyl's growth condition. A nondecreasing sequence of real numbers $(n_k)_{k=1}^{\infty}$ is said to satisfy Weyl's growth condition, if $n_k - n_j \geq 1$ whenever

$$k - j \geq \frac{j}{(\log j)^{1+\epsilon}},$$

for some positive ϵ.

1.2 The Glivenko–Cantelli theorem for stationary ergodic sequences. Let us recall the Glivenko–Cantelli theorem for stationary ergodic sequences, defined on a probability space $(\Omega, \mathcal{F}, P)$.

Theorem 1.1. *Let (X_k) be a stationary ergodic sequence with common distribution function F and empirical distribution function F_N at stage N. Then*

$$D_N := \sup_{x \in \mathbb{R}} \mid F_N(x) - F(x) \mid \to 0, \tag{1.4}$$

with probability 1.

This is a direct generalization of the Glivenko–Cantelli theorem in the i.i.d. case since a sequence of i.i.d. random variables is stationary ergodic (the relevant definitions follow in the next two paragraphs). Theorem 1.1 demonstrates the centrality of results of this type in mathematical statistics. It says that the possibly unknown distribution function $F(x)$ can be recovered with probability 1 uniformly over the reals x by taking the empirical distribution function of a sample of size N from that distribution and letting $N \to \infty$. In statistical terminology, the empirical distribution function is a strongly consistent estimator of the unknown true distribution function F. The quantity D_N plays an important role in testing the hypothesis that F is the underlying distribution of the observations $X_1, \ldots, X_N$. The resulting test is called the Kolmogorov–Smirnov test, and D_N is called the Kolmogorov–Smirnov statistic.

Before we sketch a proof of Theorem 1.1, let us recall some of the definitions. A sequence $(X_k)_{k\geq 0}$ is called stationary if for each $k \geq 0$ and each $n \geq 1$ the distributions of $(X_{k+1}, \ldots, X_{k+n})$ and $(X_1, \ldots, X_n)$ are the same. Consider a measurable map T of Ω into itself. T is called measure preserving if $P(T^{-1}A) = P(A)$ for all $A \in \mathcal{F}$. If X is a random variable, defined on $(\Omega, \mathcal{F}, P)$, then the sequence $X_n(\omega) = X(T^{n-1}\omega)$, $n \geq 1$ is stationary. Here $T^j = T(T^{j-1})$ is the j-th iteration of T. The converse is almost true. For details see Breiman (1968, p. 107). A set $A \in \mathcal{F}$ is called invariant if $T^{-1}A = A$. The invariant sets form a sub-σ-field $\mathcal{J} \subset \mathcal{F}$. A measure-preserving map T is called ergodic if $P(A) = 0$ or 1 for every $A \in \mathcal{J}$.

Now Birkhoff's ergodic theorem (see Breiman (1968, p. 113)) states that if T is a measure-preserving map on $(\Omega, \mathcal{F}, P)$ and if X is a random variable with $E|X| < \infty$ then with probability 1

$$\lim_{n\to\infty} \frac{1}{N} \sum_{k=0}^{N-1} X(T^k\omega) = E(X \mid \mathcal{J}). \tag{1.5}$$

In addition, if T is ergodic, then $E(X|\mathcal{J}) = EX$ a.s., and in that case the r.h.s. can be replaced by EX. This special case of the ergodic theorem is often phrased by saying that the time average of the X-values along an orbit $(T^k\omega)_{k=1}^{\infty}$ converges to the spatial average EX.

Proof of Theorem 1.1. We first note that by (1.5) for fixed $x \in \mathbb{R}$,

$$F_N(x;\omega) = \frac{1}{N} \sum_{k=0}^{N-1} 1_{\{X(T^k\omega)\leq x\}} \to E1_{\{X\leq x\}} = F(x) \tag{1.6}$$

except on a set A_x with $P(A_x) = 0$.

In case that F is strictly increasing and continuous, the proof of (1.4) can be carried out by introducing the probability transformation $X \to F(X)$. As is well known $U := F(X)$ has uniform distribution over [0,1] and since $F(x) \leq u$ if and only if $x \leq F^{-1}(u)$ the proof of (1.4) can be reduced to the case of stationary ergodic sequences having uniform distribution over [0,1]. Using the monotonicity of F_N the proof of (1.4) can be completed by applying relation (1.6) at the points $x = \frac{k}{m}$, $k = 1, \ldots, m-1$, where m is arbitrarily large, but fixed. The case of general F can be completed as in Billingsley (1995, p. 269). □

As an example consider the map $T\omega = \{2\omega\}$, $\omega \in [0, 1)$ on the unit interval with Lebesgue measure. Then $T^k\omega = \{2^k\omega\} = X_k$ in (1.1) with $n_k = 2^k$. Thus the above theorem yields Weyl's result in this particular case.

1.3 Results from uniform distribution mod 1. Hermann Weyl proved his result by using what has been later called the Weyl criterion. Let $0 \leq x_k < 1$, $k = 1, 2, \ldots,$ and let D_N be the discrepancy of the first N of them, i.e.,

$$D_N = \sup_{0 \le x \le 1} \left| \frac{1}{N} \#\{i \le N : x_i \le x\} - x \right|.$$

Of course, in view of the definitions of Section 1.2, D_N is the Kolmogorov–Smirnov statistic of the constant random variables $X_k = x_k$ at stage N.

Theorem 1.2 (The Weyl criterion). *$D_N \to 0$ if and only if*

$$\lim_{N\to\infty} \frac{1}{N} \sum_{k \le N} e^{2\pi i h x_k} = 0 \tag{1.7}$$

for all integers $h \neq 0$.

Proof. Let F_N denote the (nonrandom) empirical distribution function of the first N points x_k, $k = 1, \ldots, N$. Convergence in distribution of F_N to a uniform distribution on $[0, 1)$ is equivalent to convergence of the corresponding Fourier coefficients. This is exactly the requirement in (1.7); see Billingsley (1995, p. 352). □

For further details, see, e.g., Kuipers and Niederreiter (1974) and Drmota and Tichy (1997).

A quantitative version of Weyl's theorem was given by Erdős and Turán (1948). Let H be a positive integer; then

$$D_N \le \frac{4}{H} + 4 \sum_{h=1}^{H} \frac{1}{h} \frac{1}{N} \left| \sum_{k=1}^{N} e^{2\pi i h x_k} \right|. \tag{1.8}$$

We remark in passing that the Erdős–Turán inequality was later generalized by several authors to obtain a full analogue to the classical Berry–Esséen inequality used to establish error terms in the central limit theorem for sums of random variables: let $F(x)$ be nondecreasing on $[0, 1]$ and let $G(x)$ satisfy a Lipschitz condition on $[0, 1]$, i.e.,

$$|G(x) - G(y)| \le M\,|x - y|, \quad 0 \le x, y \le 1.$$

Suppose that $F(0) = G(0) = 0$ and $F(1) = G(1) = 1$. Then for any positive integer H,

$$\sup_{0 \le x \le 1} |F(x) - G(x)| \le \frac{4M}{H+1} + \frac{4}{\pi} \sum_{h=1}^{H} \left(\frac{1}{h} - \frac{1}{H+1} \right) |\hat{F}(h) - \hat{G}(h)|$$

where

$$\hat{F}(h) = \int_0^1 e^{2\pi i h x}\, dF(x), \quad \hat{G}(h) = \int_0^1 e^{2\pi i h x}\, dG(x).$$

Thus if we set $F(x) = F_N(x)$ and $G(x) = x$, we get (1.8) as a special case.

We quote the following companion result to (1.8). For each integer $h \ge 1$, we have

$$\frac{1}{N} \left| \sum_{n \le N} e^{2\pi i h x_n} \right| \le 4h D_N. \tag{1.9}$$

To prove this, we use Koksma's inequality (see Kuipers and Niederreiter (1974) or Drmota and Tichy (1997)), which states that, given any function $f : [0, 1) \to \mathbb{R}$ with finite variation $V(f)$ and any set of numbers $0 \leq x_n < 1$, we have

$$\left| \frac{1}{N} \sum_{n \leq N} f(x_n) - \int_0^1 f(x)dx \right| \leq V(f)\, D_N. \tag{1.10}$$

The proof of (1.10) is easy. We define

$$H_N(x) = \frac{1}{N} \#\{i \leq N : x_i \leq x\} - x.$$

Then the left-hand side of (1.10) equals

$$\left| \int_0^1 f(x)\, dH_N(x) \right| = \left| \int_0^1 H_N(x) f'(x)\, dx \right| \leq \int_0^1 \sup_x |H_N(x)|\, |f'(x)|\, dx = D_N\, V(f),$$

using integration by parts and the fact that $H_N(0) = H_N(1) = 0$.

We apply Koksma's inequality to $f(\omega) = e^{2\pi i h \omega}$. Separating real and imaginary parts, we get $V(e^{2\pi i h \cdot}) \leq 4h\sqrt{2}$, and thus (1.9) with 4 replaced by $4\sqrt{2}$. However, an ad hoc argument yields the sharper inequality (1.9) (see Kuipers and Niederreiter (1974, Corollary 5.1)).

1.4 Proof of Weyl's version of the Glivenko–Cantelli theorem. To complete the proof of (1.3), we observe that by the Rademacher–Menshov law of large numbers for orthogonal sequences (see Stout (1974)), we have for almost every $\omega \in [0, 1)$, for each integer $h \neq 0$,

$$\lim_{N \to \infty} \frac{1}{N} \sum_{k \leq N} e^{2\pi i h n_k \omega} = 0,$$

and thus, in view of (1.8), we have for any $H \geq 1$,

$$\limsup_{N \to \infty} D_N(\omega) \leq \frac{4}{H} \quad \text{for almost every } \omega.$$

In fact, using this approach coupled with standard techniques in the theory of orthogonal functions, Cassels (1950) and Erdős and Koksma (1949) proved independently that for any $\epsilon > 0$,

$$N D_N \ll N^{\frac{1}{2}} (\log N)^{\frac{5}{2} + \epsilon} \quad \text{almost everywhere.} \tag{1.11}$$

Here, and in what follows, we frequently use the Vinogradov symbol "$\ll$" instead of the Landau symbol "O." It took some 30 years to improve upon (1.11). Roger Baker (1981) applied Hunt's (1968) deep maximal inequality for Fourier series, the inequality Hunt used to prove Carleson's theorem on the almost sure convergence of Fourier series. In this way Baker was able to reduce the $\frac{5}{2}$ in the exponent to $\frac{3}{2}$.

Returning to the question of improving (1.11) we quote a result of Walfisz (1930) who proved

$$\limsup_{N \to \infty} \frac{\left| \sum_{k \leq N} e^{2\pi i k^2 \omega} \right|}{N^{\frac{1}{2}} (\log N)^{\frac{1}{4} + \epsilon}} = 0 \quad \text{or} \quad \infty \quad \text{almost everywhere,} \tag{1.12}$$

according as $\epsilon > 0$ or $\epsilon = 0$. This combined with (1.9) shows that for the particular sequence $n_k = k^2$,

$$\limsup_{N\to\infty} \frac{N D_N(\omega)}{N^{\frac{1}{2}}(\log N)^{\frac{1}{4}}} = \infty \quad \text{almost everywhere.} \tag{1.13}$$

There is an interesting sideline to this discussion, namely the order of magnitude of trigonometric sums. We shall discuss this, including relation (1.12), in an appendix at the end of this section. Here we only report that Berkes and Philipp (1994) constructed a sequence $(n_k)_{k=1}^{\infty}$ of integers such that (1.13) holds with $\frac{1}{4}$ replaced by $\frac{1}{2}$ in the exponent of $\log N$.

This raises two questions, both of them still unsolved and difficult. The first is: what is the exact exponent on $\log N$ for general sequences $(n_k)_{k=1}^{\infty}$? The second question is: in analogy to the Chung–Smirnov law of the iterated logarithm (LIL), for which sequences $(n_k)_{k=1}^{\infty}$ does a bounded Chung–Smirnov LIL hold, i.e.,

$$0 < c \le \limsup_{N\to\infty} \frac{N D_N(\omega)}{\sqrt{N \log\log N}} \le C < \infty \quad \text{almost everywhere?} \tag{1.14}$$

In the absence of a Kolmogorov zero-one law it is not obvious that $c = C$. If the sequence $(n_k)_{k=1}^{\infty}$ grows exponentially fast, i.e., if $(n_k)_{k=1}^{\infty}$ satisfies a Hadamard gap condition

$$\frac{n_{k+1}}{n_k} \ge 1 + \rho \quad \text{for} \quad k \ge 1 \tag{1.15}$$

for some $\rho > 0$, then (1.14) holds. In view of (1.9), the lower bound follows with $c \ge \frac{1}{4}$ from a result of Erdős and Gál (1955) who proved an LIL for trigonometric sums in this particular case. The upper bound was known since 1955 as the Erdős–Gál conjecture and was established by Philipp (1974 and 1977) with $C \le \text{const} \cdot \rho^{-1}$.

Relation (1.14) was also established for Hardy–Littlewood–Fejér sequences n_k which are defined as follows: let $q_1, \ldots, q_\tau$ be a sequence of coprime integers (such as the first τ primes) and consider the multiplicative semigroup generated by them, i.e., the numbers $q_1^{\alpha_1} \cdots q_\tau^{\alpha_\tau}$, arranged in increasing order, where α_i, $1 \le i \le \tau$, are nonnegative integers. Then (1.14) holds with $c > \frac{1}{4}$ and $C < \infty$, depending only on the total number of primes involved in the factorization of $q_1, \ldots, q_\tau$ (Philipp (1994)). This established a conjecture of Roger Baker of 1979.

For sequences $(n_k)_{k=1}^{\infty}$ growing slower than (1.15), i.e., subexponentially, relation (1.14) no longer needs to hold. Berkes and Philipp (1994) constructed a class of sequences $(n_k)_{k=1}^{\infty}$ for which

$$\frac{n_{k+1}}{n_k} \ge 1 + \rho(k)$$

and

$$\limsup_{N\to\infty} \frac{N D_N(\omega)}{\log\log \frac{1}{\rho(N)} \sqrt{N \log\log N}} \ge c > 0 \quad \text{almost everywhere,}$$

for some absolute constant c. Here $\rho(k)$ is an arbitrary nonincreasing sequence of positive numbers with

$$\sup_{k\ge 1} \frac{\rho(k)}{\rho(k^2)} < \infty.$$

For instance, $\rho(k) = (\log k)^{-1}$ would do. In particular, this shows that there is no universal constant C such that (1.14) holds for all sequences $(n_k)_{k=1}^{\infty}$ satisfying the Hadamard gap condition (1.15). The constant $C = C(\rho)$ will depend on ρ.

A third question which could be asked in connection with the Chung–Smirnov LIL is the following: for which sequences $(n_k)_{k=1}^{\infty}$ of integers does the properly normalized Kolmogorov–Smirnov statistic converge in distribution, or when does the empirical process even converge in distribution?

These results demonstrate nicely the scope and extent to which limit theorems for sums and for empirical processes for independent random variables can be generalized in the dependent case. The following appendix to this section is designed to round out the general picture, though it does not directly relate to empirical process theory.

1.5 Appendix. Let $(n_k)_{k=1}^{\infty}$ be an increasing sequence of integers, and set

$$S_N(\omega) = \sum_{k \le N} e^{2\pi i n_k \omega}.$$

We will address the question of the size of S_N. The classical Rademacher–Menshov theory of orthogonal functions (see Stout (1974)) immediately yields, for each $\epsilon > 0$,

$$S_N(\omega) \ll N^{\frac{1}{2}} (\log N)^{\frac{3}{2}+\epsilon} \text{ almost everywhere.} \tag{1.16}$$

This was improved upon in 1926 by Kolmogorov and Seliverstoff and independently and simultaneously by Plessner who replaced the exponent $\frac{3}{2} + \epsilon$ by $1 + \epsilon$. This estimate can be further improved to $\frac{1}{2} + \epsilon$ by a straightforward application of Carleson's theorem (see Carleson (1966)). In 1994 Berkes and Philipp showed that the exponent $\frac{1}{2}$ on the logarithm cannot be replaced by a smaller one. They proved the following result.

Theorem 1.3. *Let $f \ge 0$ be a nondecreasing function with*

$$\sup_{k \ge 1} \frac{f(k^2)}{f(k)} < \infty. \tag{1.17}$$

Then

$$\sum_{k \le N} e^{2\pi i n_k \omega} = o(N^{\frac{1}{2}} f(N)) \quad a.e. \tag{1.18}$$

for all increasing sequences $(n_k)_{k=1}^{\infty}$ of integers if and only if

$$\sum_{k \ge 1} \frac{1}{k\,(f(k))^2} < \infty. \tag{1.19}$$

Remark 1

Condition (1.17) is certainly satisfied for powers of the logarithm which provide the watershed between convergence and divergence of the series (1.19).

Remark 2

If the series (1.19) converges, then condition (1.17) is not needed. In fact this is the part of the theorem which follows directly from Carleson's theorem. Indeed, (1.19) implies that the series

$$\sum_{k \ge 1} \frac{1}{k^{\frac{1}{2}} f(k)} e^{2\pi i n_k \omega}$$

converges a.e. and thus (1.18) follows from the Kronecker lemma (see, e.g., Révész (1968, p. 35)).

Remark 3

The sequence $(n_k)_{k=1}^{\infty}$, constructed to treat the divergence part of the theorem, as well as the sequences $(n_k)_{k=1}^{\infty}$ with "bad" behavior in similar contexts, mostly follow the same pattern. We will try to show the probabilistic ideas behind these constructions. Set $m_k = [kf(k)^2]$. As the series in (1.19) is assumed to diverge, $\sum m_k^{-1} = \infty$. The real part of the sum

$$\sum_{j \le m_k} e^{2\pi i j \omega} \tag{1.20}$$

is in essence the Dirichlet kernel $\Delta(m_k)$ (see, e.g., Rudin (1976, p. 189)). Now $\Delta(m_k)$ has Cauchy-like tails; that is,

$$P(\Delta(m_k) \ge c m_k) \ge c m_k^{-1} \tag{1.21}$$

for some positive constant c. Next we multiply the arguments in (1.20) by the same huge integer a_k. For instance $a_k = 2^{k^2}$ will do. This guarantees that the random variables

$$Z_k = Z_k(\omega) := \sum_{j \le m_k} \cos 2\pi j a_k \omega, \quad k \ge 1 \tag{1.22}$$

are asymptotically independent. This is the same phenomenon as dealt with in (1.15), only that the gap is much bigger than the Hadamard gap condition would permit. Moreover, each Z_k has the same distribution as the real part of (1.20) since the map $T\omega = a_k \omega \bmod 1$ preserves Lebesgue measure. Thus the Z_k, $k \ge 1$, constitute the steps of an approximate Cauchy random walk. But for a Cauchy process, the largest jump up to time t has about the same size as the process itself up to this jump, that is $\max_{k \le M} Z_k$ is comparable to $\sum_{k \le M} Z_k$. Since the random variables Z_k are asymptotically independent, relation (1.21), the assumed divergence of the series $\sum m_k^{-1}$ and the divergence part of the Borel–Cantelli lemma imply that, with probability 1, infinitely many of the events $\{Z_k \ge c m_k\}$ happen; and so do, by the above remark, infinitely many of the events $\{\sum_{k \le M} Z_k \ge c m_M\}$.

Finally, a little calculus shows that the sequence

$$(n_k) := \left\{ \bigcup_{k \ge 1} \bigcup_{j \le m_k} \{j a_k\} \right\},$$

which is already arranged in increasing order, will satisfy

$$\limsup_{N \to \infty} \frac{|\sum_{k \le N} e^{2\pi i n_k \omega}|}{N^{\frac{1}{2}} f(N)} = \infty \quad a.e.,$$

where f is the function of Theorem 1.3. For the details see Berkes and Philipp (1994).

In a similar direction Fiedler, Jurkat and Körner (1977) proved the following profound result for the sequence $n_k = k^2$.

Theorem 1.4. *Let $g \ge 0$ be a nondecreasing function. Then*

$$\sum_{k \le N} e^{2\pi i k^2 \omega} = o(N^{\frac{1}{2}} g(N)) \quad a.e.$$

if and only if

$$\sum_{k\geq 1} \frac{1}{k\,(g(k))^4} < \infty.$$

Remark 4

In particular if $g(k) = (\log k)^{\frac{1}{4}+\epsilon}$, then

$$\limsup_{N\to\infty} \frac{|\sum_{k\leq N} e^{2\pi i k^2 \omega}|}{N^{\frac{1}{2}} (\log k)^{\frac{1}{4}+\epsilon}} = \infty \quad \text{or} \quad 0 \quad \text{a.e.,}$$

according as $\epsilon = 0$ or $\epsilon > 0$. As was noted in Fiedler, Jurkat and Körner (1977, p. 131) this last result is implicitly contained in Walfisz (1930, pp. 383–384).

2 Invariance principles for i.i.d. observations

The aim of this section is twofold. First we will survey some major results concerning the empirical process of i.i.d. variables, which will serve as motivation for the dependent case. At the same time, we will introduce important technical tools that will also be used in the study of dependent empirical processes.

2.1 Weak convergence in metric spaces. In this section we will introduce concepts for the study of weak convergence of measures in infinite dimensional spaces. Historically, the interest in this subject arose in the investigation of the limit behavior of the partial sum process and of the empirical process. Given a sequence of i.i.d. random variables $X_1, X_2, \ldots$ with finite variance σ^2, we introduce the partial sum process $S_n(t)$, $0 \leq t \leq 1$, by

$$S_n(t) = \begin{cases} \frac{1}{\sigma\sqrt{n}} \sum_{i=1}^{k} (X_i - EX_i) & \text{if } t = k/n \\ \text{linearly interpolated} & \text{elsewhere} \end{cases}$$

and the empirical process $Y_n(t)$, $t \in \mathbb{R}$, by

$$Y_n(t) = \sqrt{n}\,(F_n(t) - F(t))\,.$$

Both processes define random variables on the original probability space (Ω, F, P) with values in some space of functions on $[0, 1]$ and on $\mathbb{R}$, respectively. Thus their asymptotic analysis directly leads to questions of weak convergence of measures on infinite dimensional spaces.

Historically, the material of this section was developed in the 1950s, mainly by Prohorov, Skorohod and Donsker. Important motivation had been provided in the 1940s by Erdős and Kac (1946) who observed that many functionals of a partial sum process have a limit distribution that does not depend on the distribution of the underlying process. Examples of such functionals are the maximum of the centered partial sums

$$M_n = \frac{1}{\sigma\sqrt{n}} \max_{1\leq k\leq n} \sum_{i=1}^{k} (X_i - EX_i)$$

or the area under the graph of $S_n(t)$,

$$I = \int_0^1 |S_n(t)|\,dt.$$

Erdős and Kac (1946) coined the term 'invariance principle' for their result as they established invariance of the limit distribution of the underlying distribution of the sequence $(X_i)_{i\geq 1}$. Donsker (1951) then showed that by introducing the partial sum process and by analyzing its limit distribution one could elegantly formalize the invariance principle.

For the empirical process it was Doob (1949) who noted that for a large class of functionals the limit distribution could be computed by evaluating the distribution of the corresponding functional of the Brownian bridge process. Again, Donsker (1952) formalized this idea by proving weak convergence of the empirical process.

Definition 2.1. *Let S be a separable metric space with Borel σ-field $\mathcal{B}$. A sequence of probability measures $(\mu_n)_{n\geq 1}$ on S is said to converge weakly to μ if*

$$\int_S f\,d\mu_n \to \int_S f\,d\mu \tag{2.1}$$

for all bounded continuous functions $f : S \to \mathbb{R}$. A sequence of S-valued random variables is said to converge in distribution if their distributions converge weakly.

For the partial sum process, the natural metric space is $C[0, 1]$, the space of all continuous functions $f : [0, 1] \to \mathbb{R}$, equipped with the sup-norm. $C[0, 1]$ is a separable space and thus the above definition of weak convergence applies.

Definition 2.2. *Brownian motion on $[0, 1]$ is a stochastic process $\{W_t; 0 \leq t \leq 1\}$ with the following properties:*

1. $W_0 = 0$,
2. *The sample paths $t \longmapsto W_t$ are continuous,*
3. *For any $0 \leq t_0 < t_1 < \cdots < t_k \leq 1$, the increments $W(t_1) - W(t_0), \ldots, W(t_k) - W(t_{k-1})$ are independent and normally distributed with mean zero and variances $\sigma^2(t_i - t_{i-1})$, $i = 1, \ldots, k$.*

If $\sigma^2 = 1$, we call $\{W_t, 0 \leq t \leq 1\}$ standard Brownian motion.

Theorem 2.3. *The partial sum process $(S_n)_{n\geq 1}$ converges in distribution in $C[0, 1]$ to standard Brownian motion.*

In most applications, one is not so much interested in the entire partial sum process but rather in some functional $h(S_n)$. To treat such situations, the so-called continuous mapping theorem is needed.

Lemma 2.4. *Let $(Y_n)_{n\geq 1}$ be a sequence of S-valued random variables converging in distribution to Y. If S' is another separable metric space and if $h : S \to S'$ is a continuous map, then $(h(Y_n))_{n\geq 1}$ converges in distribution to $h(Y)$.*

Example 2.5. The functionals of the partial sum process $(S_n(t), 0 \leq t \leq 1)$ discussed earlier, are obtained by applying the continuous maps M, respectively, $I : C[0, 1] \to \mathbb{R}$, defined by

$$\begin{aligned} M(f) &= \max_{0\leq t\leq 1} |f(t)|, \\ I(f) &= \int_0^1 |f(t)|\,dt. \end{aligned}$$

Whereas the choice of the function space was completely natural for the partial sum process, the situation is much less clear for the empirical process. Let us first consider the special case where the underlying random variables have a uniform distribution on [0, 1]. We will later show that the general case can be reduced to this situation. We call the associated empirical process, $U_n(t) := \sqrt{n}(F_n(t)-t), 0 \le t \le 1$, the uniform empirical process. The sample path $t \longmapsto U_n(t)$ has discontinuities of the first kind and thus the natural function space is $D[0, 1]$, the space of all real-valued functions on [0, 1] that are right-continuous with left limits—also known by the acronym càdlàg-functions for the French "continu à droite, limite à gauche."

It might seem natural to equip $D[0, 1]$ with the sup-norm $\|\cdot\|_\infty$, but then several difficulties arise. To begin with, $(D[0, 1], \|\cdot\|_\infty)$ is a nonseparable space, as can be seen from the fact that the functions $\zeta_t(x) := 1_{[0,t]}(x)$, $0 \le t \le 1$, have mutual distance 1. As a result of nonseparability, there are several candidates for a σ-field on $D[0, 1]$. The Borel-σ-field, which is generated by all open sets, is different, e.g., from the σ-field generated by the open balls. The most natural of these σ-fields is the Borel-σ-field. However, the empirical process Y_n, defined in Section 2.1 above, is nonmeasurable with respect to this σ-field (see Billingsley (1968)).

There are at least four different ways to cope with the measurability and nonseparability problems, and to define weak convergence suitable for empirical process theory. Historically, the first approach is due to Skorohod who suggested introducing a new metric that makes $D[0, 1]$ a separable metric space.

Definition 2.6. *Let f, g be elements of $D[0, 1]$. Then we define their Skorohod distance by*

$$d(f, g) := \inf \left\{ \sup_t(| f(t) - g(\lambda(t)) | + | \lambda(t) - t |) : \lambda \textit{ is a homeomorphism of } [0, 1] \right\}.$$

The topology generated by d on $D[0, 1]$ is called the Skorohod topology and the space $D[0, 1]$ endowed with this topology is called Skorohod space.

There is a heuristic justification for this choice of metric as it makes functions f, g close if one of them can be deformed into the other by horizontal or vertical moves. In this way, functions with nearby jumps will be close in Skorohod distance, but not in sup-norm. In addition to empirical process theory, the Skorohod distance plays an important role in the study of any kind of stochastic processes that possibly have jumps, like stable processes. The main drawback of Skorohod's approach to weak convergence on $D[0, 1]$ is the lack of a natural generalization to random variables with values in higher dimensions.

There are three approaches to weak convergence of empirical processes that use the uniform topology on $D[0, 1]$. The first of these, due to Dudley, requires convergence of the integrals in (2.1) for the smaller class of functions that are continuous w.r.t. the uniform topology and measurable w.r.t. the σ-field generated by the open balls. The two approaches that are presently most accepted are due to Dudley and Philipp (1983) and to Hoffmann-Jørgensen (1991). In both approaches, one works with the uniform topology and its Borel-σ-field. Nonmeasurability of the empirical process is accepted, but handled in different ways. Dudley and Philipp formulate their results as pointwise approximation theorems, much in the spirit of Kiefer's (1972) almost sure invariance principle, but employing outer measure and the notion of measurable cover functions. Hoffmann-Jørgensen replaces integrals in (2.1) by outer integrals.

When treating empirical processes indexed by intervals of the type $[0, t]$, the approach based on Skorohod space is still the least complicated as one may use standard weak convergence notions.

2.2 Weak convergence in $D[0, 1]$. When working in Skorohod space, the definition of weak convergence is just the general one given earlier for separable metric spaces. That is, a sequence of $D[0, 1]$-valued random variables $(Y_n)_{n \geq 1}$ converges in distribution to the random variable Y, if $Ef(Y_n) \to Ef(Y)$ for any bounded continuous maps $f : D[0, 1] \to \mathbb{R}$. Note that $D[0, 1]$-valued random variables are simply stochastic processes $Y(t), 0 \leq t \leq 1$. In order to prove weak convergence of $(Y_n)_{n \geq 1}$ to X, roughly speaking one has to verify two properties.

(i) weak convergence of all finite-dimensional projections $(Y_n(t_0), \ldots Y_n(t_k))$ for any $0 \leq t_0 < t_1 < \cdots < t_k \leq 1$,

(ii) uniform closeness of the processes $(Y_n(t), 0 \leq t \leq 1)$ to their finite-dimensional projections $(Y_n(t_0), \ldots, Y_n(t_k))$ for a sufficiently fine grid of points $0 \leq t_0 < t_1 < \cdots < t_k \leq 1$. (The mathematically precise condition is called 'tightness'.)

This intuitive idea is formalized in the following theorem.

Theorem 2.7. *Let $Y, Y_1, Y_2, \ldots$ be random variables with values in $D[0, 1]$ and suppose that $P(Y \in C) = 1$ for some separable subset $C \subset D[0, 1]$. Then $(Y_n)_{n \geq 1}$ converges in distribution to Y if the following two conditions hold:*

(i) *the finite-dimensional distributions of Y_n converge to the finite-dimensional distribution of Y;*

(ii) *for each $\epsilon > 0$, $\delta > 0$, one can find a grid $0 = t_0 < t_1 < \cdots < t_m = 1$ such that*

$$\limsup_{n \to \infty} P\left\{ \max_{1 \leq i \leq m} \sup_{t_i \leq t < t_{i+1}} |Y_n(t) - Y_n(t_i)| > \delta \right\} < \epsilon.$$

Proof. See Pollard (1984). □

The first condition for weak convergence of stochastic processes is often easily established. For empirical processes, (i) is simply a consequence of the multivariate central limit theorem. To see this, recall from Section 2.1 above, that

$$Y_n(t) = \frac{1}{\sqrt{n}} \sum_{i=1}^{n} \left[1_{\{X_i \leq t\}} - P(X_i \leq t) \right],$$

so that

$$(Y_n(t_0), \ldots, Y_n(t_k)) = \frac{1}{\sqrt{n}} \sum_{i=1}^{n} \left(1_{\{X_i \leq t_0\}} - P(X_i \leq t_0), \ldots, 1_{\{X_i \leq t_k\}} - P(X_i \leq t_k) \right),$$

which is simply a normalized partial sum of the random vectors

$$(1_{\{X_i \leq t_0\}} - P(X_i \leq t_0), \ldots, 1_{\{X_i \leq t_k\}} - P(X_i \leq t_k)).$$

If the underlying observations X_i are i.i.d., one can apply the classical multivariate CLT to show that $(Y_n(t_0), \ldots, Y_n(t_k))$ is asymptotically normal with mean zero and covariance matrix $\Sigma = (\sigma(t_i, t_j))_{0 \leq i, j \leq k}$, whose entries are given by

$$\sigma(t_i, t_j) = \mathrm{Cov}(1_{\{X_1 \leq t_i\}}, 1_{\{X_1 \leq t_j\}}) = F(t_i \wedge t_j) - F(t_i) F(t_j).$$

Verification of the second condition for weak convergence is generally the more critical issue, and requires creative ideas for each case. Chaining is the crucial technique used in a more or less hidden form in all proofs. In order to bound

$$P(\sup_{t_i \leq t < t_{i+1}} |Y_n(t) - Y_n(t_i)| > \delta),$$

one introduces a refining sequence of grids on the interval $[t_i, t_{i+1}]$:

$$t_i = s_0(k) < s_1(k) < \cdots < s_{j_k}(k) = t_{i+1}$$

that is nested in the sense that $\{s_j(k) : j = 0, \ldots, j_k\} \subset \{s_j(k+1) : j = 0, \ldots, j_{k+1}\}$. The technique borrows its name from the fact that given any $t \in [t_i, t_{i+1}]$, one establishes a chain $t_i = s_0(1) < s_{j_1(t)}(1) < \cdots < s_{j_K(t)}(K) \le t \le s_{j_K(t)+1}(K)$, linking t to the left endpoint of the interval $[t_i, t_{i+1}]$. In this way, we can bound $|Y_n(t) - Y_n(t_i)|$ by

$$|Y_n(t) - Y_n(t_i)| < \sum_{k=1}^{K} |Y_n(s_{j_k(t)}(k)) - Y_n(s_{j_{k-1}(t)}(k-1))| + |Y_n(t) - Y_n(s_{j_K(t)}(K))|. \quad (2.2)$$

There are only finitely many terms of the type $|Y_n(s_{j_k(t)}(k)) - Y_n(s_{j_{k-1}(t)}(k-1))|$ occurring in the sum on the r.h.s. of (2.2), and these can be bound by a Bonferroni-type inequality. The final term, $|Y_n(t) - Y_n(s_{j_K(t)}(K))|$, still has an uncountable index set. In order to control this term, we exploit the special structure of the empirical process, and especially monotonicity of the empirical distribution function $F_n(t)$. As $s_{j_K(t)}(K) \le t \le s_{j_K(t)+1}(K)$, we have

$$\begin{aligned}
&|Y_n(t) - Y_n(s_{j_K(t)}(K))| \\
&= \sqrt{n}\left|[F_n(t) - F_n(s_{j_K(t)}(K))] - [F(t) - F(s_{j_K(t)}(K))]\right| \\
&\le \sqrt{n}\left[F_n(t) - F_n(s_{j_K(t)}(K))\right] + \sqrt{n}\left[F(t) - F(s_{j_K(t)}(K))\right] \\
&\le \sqrt{n}\left[F_n(s_{j_K(t)+1}(K)) - F_n(s_{j_K(t)}(K))\right] \\
&\quad + \sqrt{n}\left[F(s_{j_K(t)+1}(K)) - F(s_{j_K(t)}(K))\right] \\
&\le \sqrt{n}\left|[F_n(s_{j_K(t)+1}(K)) - F_n(s_{j_K(t)}(K))] - [F(s_{j_K(t)+1}(K)) - F(s_{j_K(t)}(K))]\right| \\
&\quad + 2\sqrt{n}\left[F(s_{j_K(t)+1}(K)) - F(s_{j_K(t)}(K))\right] \\
&= \left|Y_n(s_{j_{K(t)+1}}(K)) - Y_n(s_{j_K(t)}(K))\right| + 2\sqrt{n}\left[F(s_{j_K(t)+1}(K)) - F(s_{j_K(t)}(K))\right].
\end{aligned}$$

All together, we have thus found an upper bound for the uncountably many terms $Y_n(t) - Y_n(t_i)$ by a sum of just a finite number of terms.

2.3 Donsker's empirical process invariance principle. As an illustration of empirical process techniques, we will prove in this section Donsker's classical invariance principle for the empirical process of i.i.d. random variables. Let $X_i, i = 1, 2, \ldots$ be independent random variables with common distribution function $F(x) = P(X_i \le x)$. We will first consider the special case that F is the uniform distribution on (0, 1), i.e., $F(x) = x$ for $x \in [0, 1]$, $F(x) \equiv 0$ on $(-\infty, 0]$ and $F(x) \equiv 1$ on $[1, \infty]$. The general case can be reduced to this special case by use of the so-called quantile transform. Defining the quantile function $F^{-1} : (0, 1) \to \mathbb{R}$ by

$$F^{-1}(t) = \inf\{x : F(x) \ge t\},$$

one can show that $F^{-1}(U)$ has distribution F if U is a uniform (0, 1) random variable. We may thus assume without loss of generality that $X_i = F^{-1}(U_i)$ for an i.i.d. process $\{U_i, i \ge 1\}$, of uniform (0, 1) random variables. The empirical distribution functions $F_n^X(x)$ and $F_n^U(u)$ of the two processes are linked by the identity

$$F_n^X(x) = \frac{1}{n}\sum_{i=1}^{n} 1_{\{X_i \le x\}} = \frac{1}{n}\sum_{i=1}^{n} 1_{\{U_i \le F(x)\}} = F_n^U(F(x)),$$

and thus

$$\sqrt{n}\,(F_n^X(x) - F(x)) = \sqrt{n}\,(F_n^U(F(x)) - F(x)).$$

This identity enables us to translate any limit theorem for the uniform empirical process into a limit theorem for a general empirical process.

Definition 2.8. *A mean-zero Gaussian process $W^0(t)$, $0 \le t \le 1$, with continuous sample paths, and covariance function*

$$\mathrm{Cov}(W^0(s), W^0(t)) = s \wedge t - st, \quad 0 \le s, t \le 1,$$

is called a standard Brownian bridge.

One can show that a Brownian bridge can be obtained from a Brownian motion $W_t, 0 \le t \le 1$, via the transformation $W_t^0 = W_t - tW_1$. Thus the existence of a Brownian bridge can be reduced to the existence of Brownian motion.

Theorem 2.9. *The uniform empirical process $U_n(t) := \sqrt{n}\,(F_n^U(t) - t)$, $0 \le t \le 1$, converges in distribution to the Brownian bridge $W^0(t)$.*

For the proof of this theorem we need the following lemma (see Billingsley (1968), Theorem 12.2).

Lemma 2.10. *Let $\zeta_1, \ldots, \zeta_m$ be real-valued random variables, and define*

$$\begin{aligned} S_k &= \zeta_1 + \cdots + \zeta_k, \quad k \ge 0, \\ M_m &= \max_{0 \le k \le m} |S_k|. \end{aligned}$$

Suppose that there are nonnegative numbers $u_1, \ldots, u_m$ and $\gamma > 0, \alpha > 1$ such that

$$E|S_j - S_i|^\gamma \le \Big(\sum_{i<l\le j} u_l\Big)^\alpha.$$

Then there exists a constant $K \ge 0$ such that

$$P(M_m \ge \lambda) \le \frac{K}{\lambda^\gamma}(u_1 + \cdots + u_m)^\alpha$$

holds for all $\lambda > 0$.

Proposition 2.11. *Let $Y_n(t) = \sqrt{n}\,(F_n^U(t) - t)$, $0 \le t \le 1$, be the uniform empirical process, and suppose that for real numbers α, β, γ, q with $\gamma, q > 0$, $\alpha > 1$ and $q > (\alpha - \beta)/2$,*

$$E|Y_n(t) - Y_n(s)|^\gamma \le C\left(|t-s|^\alpha + \frac{|t-s|^\beta}{n^q}\right) \tag{2.3}$$

holds for all $s, t \in [0, 1]$. Then, given $\epsilon, \delta > 0$ there exists a grid $0 = t_0 < t_1 < \cdots < t_k = 1$ such that

$$P\Big(\max_{i=1,\ldots,k} \sup_{t_{i-1} \le t < t_i} |Y_n(t) - Y_n(t_{i-1})| \ge \varepsilon\Big) \le \delta$$

for all n sufficiently large.

Proof. Note that w.l.o.g. $\alpha > \beta$, as otherwise we can replace the right-hand side of (2.3) by

$$2C\,|t-s|^\alpha \le 2C\left(|t-s|^\alpha + \frac{1}{n^{3\alpha/4}}\right),$$

and thus (2.3) holds with $\beta = 0$ and $q = 3\alpha/4 > (\alpha - \beta)/2$. Define $r := q/(\alpha - \beta)$ and note that $r > 1/2$ and also that if $|t - s| \geq n^{-r}$, we get $|t - s|^{\beta}/n^q \leq |t - s|^{\alpha}$ and thus

$$E|Y_n(t) - Y_n(s)|^{\gamma} \leq 2C\,|t - s|^{\alpha}.$$

We shall show that any grid $0 = t_0 < t_1 < \cdots < t_k = 1$ with

$$\max_i |t_i - t_{i-1}| < \left(\delta \left(\frac{\epsilon}{2}\right)^{\gamma} / K\right)^{1/(\alpha-1)}$$

will work. Here K is the constant supplied by Lemma 2.10. Choose such a grid. Set $\eta := \epsilon n^{-\frac{1}{2}}/2 > n^{-r}$ if $n \geq n_0$ as $r > 1/2$. Introduce the points $s_j = t_{i-1} + j\eta$, $j = 0, \ldots, m$, where $m := \left[\frac{t_i - t_{i-1}}{\eta}\right]$ and $s_m = t_i$.

Now apply Lemma 2.10 to $\zeta_j = Y_n(s_j) - Y_n(s_{j-1})$, $j = 1, \ldots, m$. By the above remark we get, for all $\lambda > 0$,

$$P(\max_{j=1,\ldots,m} |Y_n(s_j) - Y_n(t_{i-1})| \geq \lambda) \leq \frac{K}{\lambda^{\gamma}}(t_i - t_{i-1})^{\alpha}.$$

For $s_{j-1} \leq t < s_j$ we have

$$\begin{aligned} Y_n(t) - Y_n(t_{i-1}) &= \sqrt{n}\,(F_n(t) - F_n(t_{i-1}) + t_{i-1} - t) \\ &\leq \sqrt{n}\left(F_n(s_j) - F_n(t_{i-1}) + t_{i-1} - s_j\right) \\ &\quad + \sqrt{n}\left(s_j - s_{j-1}\right) \\ &\leq \left(Y_n(s_j) - Y_n(t_{i-1})\right) + \frac{1}{2}\epsilon \end{aligned}$$

and, in the same way,

$$Y_n(t) - Y_n(t_{i-1}) \geq \left(Y_n(s_{j-1}) - Y_n(t_{i-1})\right) - \frac{1}{2}\epsilon$$

so that

$$\sup_{t_{i-1} \leq t < t_i} |Y_n(t) - Y_n(t_{i-1})| \leq \max_{j=1,\ldots,m} |Y_n(s_j) - Y_n(t_{i-1})| + \frac{1}{2}\epsilon.$$

Then

$$\begin{aligned} P\left(\sup_{t_{i-1} \leq t < t_i} |Y_n(t) - Y_n(t_{i-1})| \geq \varepsilon\right) &\leq P\left(\max_{j=1,\ldots,m} |Y_n(s_j) - Y_n(t_{i-1})| \geq \frac{\varepsilon}{2}\right) \\ &\leq \frac{2^{\gamma} K}{\varepsilon^{\gamma}}(t_i - t_{i-1})^{\alpha}, \end{aligned}$$

and thus

$$\begin{aligned} P\left(\max_{1 \leq i \leq k} \sup_{t_{i-1} \leq t < t_i} |Y_n(t) - Y_n(t_{i-1})| \geq \varepsilon\right) &\leq \frac{2^{\gamma} K}{\varepsilon^{\gamma}} \sum_{i=1}^{k} (t_i - t_{i-1})^{\alpha} \\ &\leq \frac{2^{\gamma} K}{\varepsilon^{\gamma}} \max_i (t_i - t_{i-1})^{\alpha - 1} \sum_{i=1}^{k} (t_i - t_{i-1}) \\ &= \frac{2^{\gamma} K}{\varepsilon^{\gamma}} \max_i |t_i - t_{i-1}|^{\alpha - 1} < \delta. \end{aligned}$$

□

Proof of Donsker's empirical process invariance principle. As pointed out earlier, convergence of the finite-dimensional distributions is a consequence of the multivariate CLT for sums of i.i.d. vectors, and hence only (ii) remains to be verified. In view of Proposition 2.11 it is enough to establish (2.3). Note that for $0 \leq s < t \leq 1$,

$$\begin{aligned} Y_n(t) - Y_n(s) &= \frac{1}{\sqrt{n}} \sum_{i=1}^{n} \left[1_{\{s < U_i \leq t\}} - (t-s) \right] \\ &= \frac{1}{\sqrt{n}} \sum_{i=1}^{n} Z_i, \quad \text{say.} \end{aligned}$$

Hence, by independence and as $EZ_i = 0$,

$$\begin{aligned} E|Y_n(t) - Y_n(s)|^4 &= \frac{1}{n^2} \sum_{1 \leq i,j,k,l \leq n} EZ_i Z_j Z_k Z_l \\ &= \frac{1}{n^2} \left[nEZ_1^4 + 6n^2 (EZ_1^2)^2 \right]. \end{aligned}$$

Note that $EZ_1^2 = (t-s)(1-(t-s)) \leq t-s$ and $EZ_1^4 \leq EZ_1^2 \leq t-s$, so that we finally get

$$E|Y_n(t) - Y_n(s)|^4 \leq 6\,|t-s|^2 + \frac{|t-s|}{n}.$$

Hence (2.3) is satisfied with $\alpha = 2$ and $\beta = q = 1$. □

3 Weakly dependent random variables

Let $(\xi_j, j \geq 1)$ be a sequence of random variables on the probability space $(\Omega, \mathcal{F}, P)$, not necessarily stationary. For the moment, we might want to call the sequence $(\xi_j, j \geq 1)$ weakly dependent if, for any $k \geq 1$,

$$E\,|E\,(\xi_{k+n} \mid \xi_1, \ldots, \xi_k)| \to 0, \text{ as } n \to \infty. \tag{3.1}$$

Unfortunately, this definition does not encompass all sequences of random variables which have been called weakly dependent in the literature, but this definition serves its purpose. Of course, martingale differences satisfy (3.1) and so do mixingale difference sequences. Gaussian sequences with pairwise correlation coefficients tending to zero, lacunary random variables, and many others all satisfy (3.1) either directly or in a slightly modified form.

This definition works reasonably well if one is interested in sums of the random variables ξ. However, for empirical processes this definition is lacking. To see why, let $\{X_j, j \geq 1\}$ be a stationary sequence of real-valued random variables with common distribution function $F(x)$. The building blocks for the empirical process are the discontinuous random functions

$$\xi_j = \xi_j(x, \omega) = 1(X_j(\omega) \leq x) - F(x), \quad x \in \mathbb{R}.$$

So even if the sequence $\{X_j, j \geq 1\}$ satisfies relation (3.1), the functions $\{\xi_j, j \geq 1\}$ need not, even when (3.1) is appropriately adapted to accommodate random functions.

Thus we will abandon the definition (3.1) and we will concentrate on weak dependence relations which are meaningful in connection with empirical processes.

3.1 Mixing conditions. In this section we will present the main mixing conditions studied in the literature. In contrast to (3.1) most of these do not presuppose the existence of a finite expectation. Let $\mathcal{M}_a^b$ denote the σ-field generated by the random variables $\xi_a, \ldots, \xi_b$.

Definition 3.1. *The sequence* $\{\xi_j,\ j \geq 1\}$ *is called* strongly mixing *if*

$$\alpha(n) := \sup\left\{|P(A \cap B) - P(A)P(B)| : A \in \mathcal{M}_1^k,\ B \in \mathcal{M}_{k+n}^\infty,\ k \geq 1\right\} \to 0. \tag{3.2}$$

The sequence is called absolutely regular *if*

$$\beta(n) := E \sup\left\{\left|P(B|\mathcal{M}_1^k) - P(B)\right| : B \in \mathcal{M}_{k+n}^\infty,\ k \geq 1\right\} \to 0. \tag{3.3}$$

(The supremum is measurable since in a Polish space it is sufficient to extend the supremum only over countably many sets B.) The sequence is called ρ*-mixing if the maximal correlation coefficient*

$$\rho(n) := \sup\left\{\frac{|EXY|}{\|X\|_2\|Y\|_2} : X \in L^2(\mathcal{M}_1^k),\ Y \in L^2(\mathcal{M}_{k+n}^\infty),\ EX = EY = 0,\ k \geq 1\right\} \to 0. \tag{3.4}$$

The sequence is called ϕ-mixing *or* uniformly mixing *if*

$$\phi(n) := \sup\left\{\frac{|P(A \cap B) - P(A)P(B)|}{P(A)} : A \in \mathcal{M}_1^k,\ P(A) > 0,\ B \in \mathcal{M}_{k+n}^\infty,\ k \geq 1\right\} \to 0. \tag{3.5}$$

It is called ψ-mixing *if*

$$\psi(n) := \sup\left\{\frac{|P(A \cap B) - P(A)P(B)|}{P(A)P(B)} : A \in \mathcal{M}_1^k,\ B \in \mathcal{M}_{k+n}^\infty, \right.$$
$$\left. P(A), P(B) > 0,\ k \geq 1\right\} \to 0. \tag{3.6}$$

and it is called $*$-mixing *if in the last supremum* $\mathcal{M}_{k+n}^\infty$ *is replaced by* $\mathcal{M}_{k+n}^{k+n}$.

Notice that all mixing coefficients, considered as functions of n, are nonincreasing, except the one associated with $*$-mixing.

The strong mixing condition was introduced by Rosenblatt (1956), the ϕ-mixing condition by Ibragimov (1959), the $*$-mixing condition by Blum, Hanson and Koopmans (1963), the ψ-mixing condition in its present form was introduced by Philipp (1969). The ρ-mixing condition was introduced by Kolmogorov and Rozanov (1960) and the absolute regularity condition by Volkonskii and Rozanov (1959). Here we shall present only a few basic facts about mixing conditions, referring the reader for details and for more information to the soon to be published authoritative three volume treatise by Rick Bradley (2001).

Both for the strong mixing as well as for the uniform mixing condition, there are equivalent formulations in terms of the difference between the conditional probability $P(B|\mathcal{M}_1^k)$ and the total probability $P(B)$. We define

$$\alpha'(n) := \sup\left\{E\left|P(B|\mathcal{M}_1^k) - P(B)\right| : B \in \mathcal{M}_{k+n}^\infty,\ k \geq 1\right\}, \tag{3.7}$$

$$\phi'(n) := \sup\left\{\|P(B|\mathcal{M}_1^k) - P(B)\|_\infty : B \in \mathcal{M}_{k+n}^\infty,\ k \geq 1\right\}. \tag{3.8}$$

Then strong mixing is equivalent to $\alpha'(n) \to 0$ as $n \to \infty$ and uniform mixing is equivalent to $\phi'(n) \to 0$ a.s. as $n \to \infty$. The first claim will be established later in connection with correlation inequalities (see remarks after Corollary 3.10). Concerning the 2nd equivalence, we show that $\phi(n) = \phi'(n)$ almost surely. Indeed, since

$$\int_A P(B|\mathcal{M}_1^k)\,dP = P(A \cap B) \quad \text{for all } A \in \mathcal{M}_1^k$$

we have for all $B \in \mathcal{M}_{k+n}^{\infty}$,

$$|P(A \cap B) - P(A)P(B)| \leq \int_A |P(B|\mathcal{M}_1^k) - P(B)|\,dP \leq \phi'(n)\,P(A)$$

and thus $\phi(n) \leq \phi'(n)$ almost surely. Conversely, let

$$A^+ = \{P(B|\mathcal{M}_1^k) - P(B) > \phi(n)\} \in \mathcal{M}_1^k.$$

Then by (3.5)

$$\begin{aligned}
\phi(n)\,P(A^+) &\geq \left|P(A^+ \cap B) - P(A^+)P(B)\right| \\
&= \left|\int_{A^+} (P(B|\mathcal{M}_1^k) - P(B))\,dP\right| > \phi(n)\,P(A^+)
\end{aligned}$$

unless $P(A^+) = 0$. Similarly, $P(A^-) = 0$ with

$$A^- = \left\{P(B|\mathcal{M}_1^k) - P(B) < -\phi(n)\right\}$$

and thus we get that $|P(B|\mathcal{M}_1^k) - P(B)| \leq \phi(n)$ a.s. for any B. Using the fact that $\mathcal{M}_1^k$ is countably generated, this implies that $\phi'(n) \leq \phi(n)$ a.s.; see Bradley (2001, Vol. 1, Chapter 3) for details.

It is obvious that $\alpha'(n) \leq \beta(n) \leq \phi'(n) \leq \psi(n)$ a.s. so that a ψ-mixing sequence is ϕ-mixing which in turn implies absolute regularity and this in turn implies strong mixing. There are well-known counterexamples to the reverse implications.

We mention two more weak dependence structures which have played a role in empirical process theory. The first one was extensively treated in Section 1. Let $\{n_k, k \geq 1\}$ be a sequence of positive numbers satisfying a Hadamard gap condition

$$\frac{n_{k+1}}{n_k} \geq q > 1, \quad k \geq 1. \tag{3.9}$$

Let ω be a random variable uniformly distributed over $[0, 1]$ and let $\{x\} := x - [x]$ denote the fractional part of a real number x. The sequence $\{\{n_k\omega\}, k \geq 1\}$ is called a lacunary sequence of random variables. We call such a sequence quasi-lacunary if the corresponding sequence $\{n_k, k \geq 1\}$ satisfies an Erdős gap condition

$$\frac{n_{k+1}}{n_k} \geq 1 + \gamma(k)k^{-\frac{1}{2}}, \quad k \geq 1, \tag{3.10}$$

for some nondecreasing sequence $\gamma(k) = o(k^{\frac{1}{2}})$.

A sequence $(Y_j, j \geq 1)$ of real-valued random variables is called associated if

$$\operatorname{Cov}(f(Y_1, \ldots, Y_n), g(Y_1, \ldots, Y_n)) \geq 0$$

holds for any $n \geq 1$ and any coordinate-wise nondecreasing functions $f, g : \mathbb{R}^n \to \mathbb{R}$ with $\operatorname{Var}(f(Y_1, \ldots, Y_n)) < \infty$ and $\operatorname{Var}(g(Y_1, \ldots, Y_n)) < \infty$. Association was introduced by Esary,

Proschan and Walkup (1967), in the context of reliability theory. This condition also plays an important role in statistical mechanics.

In many important situations, a given stochastic process does not satisfy a weak dependence condition, but it can be written as a functional of another process that is weakly dependent. This situation is covered by the following definition.

Definition 3.2. *Let $(\Omega, \mathcal{F}, P)$ be a probability space and let $(Z_n)_{n\in\mathbb{Z}}$ be a stationary stochastic process. We call a process $(X_n)_{n\in\mathbb{Z}}$ a functional of $(Z_n)_{n\in\mathbb{Z}}$ if there exists a measurable function $f : \mathbb{R}^{\mathbb{Z}} \to \mathbb{R}$ such that*

$$X_n = f((Z_{n+k})_{k\in\mathbb{Z}}) \tag{3.11}$$

holds for all $n \in \mathbb{Z}$.

More precisely, a process $(X_n)_{n\in\mathbb{Z}}$ that can be expressed as (3.11) is called a two-sided functional. For processes indexed by the nonnegative integers, one-sided functionals defined by $X_n = f((Z_{n+k})_{k\geq 1})$ are more appropriate. For simplicity, we will concentrate in these lecture notes on two-sided functionals. Functionals arise naturally in a variety of applications, e.g., in statistics and in the theory of dynamical systems. Of course, the usefulness of functionals depends heavily on the mixing properties of the underlying process $(Z_n)_{n\in\mathbb{Z}}$ as well as on the continuity of the functional. We will return to these issues later on.

3.2 Examples.. Here we only give a few standard examples of weakly dependent processes, referring the reader to the work of Bradley (2001).

Example 3.3. A sequence $(X_k)_{k\geq 0}$ is called m-dependent if the σ-fields $\mathcal{M}_1^k$ and $\mathcal{M}_{k+m}^{\infty}$ are independent for each $k \geq 1$. Obviously, m-dependent sequences are ψ-mixing with

$$\psi(k) = \begin{cases} \infty & \text{for } k < m, \\ 0 & \text{for } k \geq m. \end{cases}$$

Example 3.4. Let $(X_k)_{k\geq 0}$ be a strictly stationary Markov chain with an at most countable state space. If $(X_k)_{k\geq 0}$ is irreducible and aperiodic, then $\beta(n) \to 0$. If the state space is finite, then even $\phi(n) \to 0$ exponentially fast; see Billingsley (1999, p. 201).

Example 3.5. Each irrational number $\omega \in (0, 1)$ can be written as an infinite continued fraction

$$\omega = \cfrac{1}{a_1(\omega) + \cfrac{1}{a_2(\omega) + \cfrac{1}{a_3(\omega) + \dots}}},$$

where

$$a_1(\omega) = \left[\frac{1}{\omega}\right]$$

and $a_{n+1}(\omega) = a_n(T\omega)$ with

$$T\omega = \left\{\frac{1}{\omega}\right\} = \frac{1}{\omega} - \left[\frac{1}{\omega}\right] = \frac{1}{\omega} \bmod 1.$$

Here $[x]$ denotes the largest integer not exceeding x. The a_n are positive integers.

Introduce on the set of irrational numbers in (0, 1) a probability measure P, the so-called Gauss measure, by

$$P(d\omega) = \frac{1}{\log 2}\frac{d\omega}{1+\omega}.$$

Then P is invariant under T (in the ergodic theoretic sense). Equivalently, T preserves P or $(a_n)_{n\geq 0}$ is a stationary sequence of random variables. Indeed, for any interval $(0, \alpha)$ with $0 < \alpha \leq 1$,

$$\begin{aligned} P(T^{-1}(0,\alpha)) = P(\omega : T\omega < \alpha) &= P\left(\bigcup_{k=1}^{\infty}\left\{\omega : 0 < \frac{1}{\omega} - k < \alpha\right\}\right) \\ &= P\left(\bigcup_{k=1}^{\infty}\left\{\omega : \frac{1}{k+\alpha} < \omega < \frac{1}{k}\right\}\right) \\ &= \sum_{k=1}^{\infty}\int_{1/(k+\alpha)}^{1/k}\frac{d\omega}{\log 2(1+\omega)} \\ &= \int_0^{\alpha}\frac{d\omega}{\log 2(1+\omega)} = P(0,\alpha). \end{aligned}$$

The next to last equality is established by performing the integration and by observing that the partial sums of the resulting series are telescoping. Moreover it can be shown that the sequence $(a_n)_n$ is ψ-mixing with mixing coefficients satisfying

$$\psi(n) \leq 2^{-n+1}\log 2;$$

see Iosifescu (1992).

Example 3.6. Markov chains satisfying the Doeblin condition are ϕ-mixing with $\phi(n) \to 0$ exponentially fast. See Doob (1953, p. 221) and Bradley (2001, vol. I, Chapter 7).

Example 3.7. Autoregressive processes are popular models in the analysis of time series. An autoregressive process of order p is defined as the solution of the stochastic difference equation

$$X_n = a_1 X_{n-1} + a_2 X_{n-2} + \cdots + a_p X_{n-p} + Z_n, \tag{3.12}$$

where $a_1, \ldots, a_p$ are fixed real numbers and $(Z_n)_{n\geq 0}$ is an i.i.d. process. If all roots of the characteristic polynomial $a(z) = a_p z^p + \cdots + a_1 z - 1$ lie outside the unit circle, (3.12) may be iterated backwards to obtain

$$X_n = Z_n + \sum_{k=1}^{\infty} b_k Z_{n-k},$$

for a sequence (b_k) of real numbers. Thus, $(X_n)_{n\geq 1}$ is a functional of the i.i.d. process $(Z_n)_{n\in\mathbb{Z}}$.

Example 3.8. In dynamical systems, functionals arise in connection with certain partitions of the state space. Let $(\mathcal{X}, \mathcal{A}, \mu, T)$ be a measure-preserving dynamical system, $g : \mathcal{X} \to \mathbb{R}$ a measurable function and

$$X_n = g(T^n), \quad n \geq 0,$$

a stochastic process, defined on the probability space $(\mathcal{X}, \mathcal{A}, \mu)$. Given a partition $\mathcal{A} = \{A_i, i \in I\}$ of $\mathcal{X}$, we define the label process $(Z_n)_{n\geq 0}$ by $Z_n = i$ if $T^n \in A_i$. If $\mathcal{A}$ is a generator (for this and other notions from ergodic theory, see Denker, Grillenberger, Siegmund (1976)), we can invert this relation and write

$$X_n = f((Z_{n+k})_{k\geq 0}).$$

If $\mathcal{A}$ is a weak Bernoulli partition, the label process $(Z_n)_{n\geq 0}$ is absolutely regular, and it is this example that motivates the study of functionals of absolutely regular processes. For more examples, see Borovkova, Burton, and Dehling (2001).

3.3 Covariance inequalities. For two σ-fields $\mathcal{F}$ and $\mathcal{G}$ we define

$$\alpha(\mathcal{F}, \mathcal{G}) := \sup\{|P(A\cap B) - P(A)P(B)| : A \in \mathcal{F}, B \in \mathcal{G}\}.$$

Observe that $\alpha(\mathcal{F}, \mathcal{G})$ measures the degree of dependence of the σ-fields $\mathcal{F}$, $\mathcal{G}$. Indeed, for independent σ-fields, $\alpha(\mathcal{F}, \mathcal{G}) = 0$. Also note that the mixing coefficients $\alpha(n)$ can be expressed as $\alpha(n) = \sup_k \alpha(\mathcal{M}_1^k, \mathcal{M}_{k+n}^\infty)$. For random variables ξ and η let $\sigma(\xi)$ and $\sigma(\eta)$ be the σ-fields generated by them.

Lemma 3.9. *Let ξ and η be real-valued bounded random variables. Then*

$$|E\xi\eta - E\xi E\eta| \leq 4\alpha(\sigma(\xi), \sigma(\eta))\, \|\xi\|_\infty\, \|\eta\|_\infty. \tag{3.13}$$

Proof. Without loss of generality we may assume that $\|\xi\|_\infty = \|\eta\|_\infty = 1$. Set $\mathcal{F} = \sigma(\xi)$ and $\mathcal{G} = \sigma(\eta)$, and observe that

$$\begin{aligned} |E\xi\eta - E\xi E\eta| = |E\{\xi\,(E(\eta|\mathcal{F}) - E\eta)\}| &\leq E|E(\eta|\mathcal{F}) - E\eta| \\ &= E(\xi_1\,(E(\eta|\mathcal{F}) - E\eta)) \\ &= E\xi_1\eta - E\xi_1 E\eta, \end{aligned}$$

where $\xi_1 := \mathrm{sign}\{E(\eta|\mathcal{F}) - E\eta\}$. The ξ_1 is $\mathcal{F}$-measurable and we can apply the same argument again, now with the roles of ξ_1 and η interchanged. In this way we then get

$$|E(\xi_1\eta) - E\xi_1 E\eta| \leq |E(\xi_1\eta_1) - E\xi_1 E\eta_1|,$$

where $\eta_1 = \mathrm{sign}\{E(\xi_1|\mathcal{G}) - E\xi_1\}$. Now set $A = \{\xi_1 = 1\}$ and $B = \{\eta_1 = 1\}$. Then

$$\begin{aligned} |E\xi_1\eta_1 - E\xi_1 E\eta_1| &= |P(A\cap B) - P(A^c\cap B) - P(A\cap B^c) + P(A^c\cap B^c) \\ &\quad -P(A)P(B) + P(A^c)P(B) + P(A)P(B^c) - P(A^c)P(B^c)| \\ &\leq 4\alpha(\mathcal{F}, \mathcal{G}). \end{aligned}$$

The three inequalities together establish (3.13). □

Corollary 3.10. *Let η be a real-valued bounded random variable and let $\mathcal{F}$ be a σ-field. Then*

$$E|E(\eta|\mathcal{F}) - E\eta| \leq 4\alpha(\mathcal{F}, \sigma(\eta))\, \|\eta\|_\infty. \tag{3.14}$$

Remarks. Lemma 3.9 is essentially due to Volkonskii and Rozanov (1959). Both Lemma 3.9 and the corollary remain valid for complex valued random variables, in which case the constant 4 needs to be replaced by no more than 2π (Dvoretzky (1970)).

We can apply (3.14) to show that strong mixing is equivalent with "ϕ-mixing on average." Indeed, if $(\xi_k)_{k\geq 0}$ is a sequence of random variables satisfying a strong mixing condition, then setting $\eta = 1_B$ in (3.14) for any $B \in \mathcal{M}_{k+n}^\infty$, we obtain

$$\sup_{B\in\mathcal{M}_{k+n}^\infty} E|P(B|\mathcal{M}_1^k) - P(B)| \leq 4\alpha(n). \tag{3.15}$$

Thus $\alpha'(n) \leq 4\alpha(n)$. On the other hand, the third line in the proof of Lemma 3.9 with $\xi = 1_A$, $\eta = 1_B$, $\mathcal{F} = \mathcal{M}_1^k$ implies $\alpha(n) \leq \alpha'(n)$. This concludes the proof of the equivalence of conditions (3.2) and $\alpha'(n) \to 0$.

The following lemma is due to Davydov (1970), the special case $r = s$ to Ibragimov (1962).

Lemma 3.11. *Let $1 \leq r, s, t \leq \infty$ satisfy $r^{-1} + s^{-1} + t^{-1} = 1$, and let ξ and η be real-valued random variables. Then*

$$|E(\xi\eta) - E\xi E\eta| \leq 10 \|\xi\|_r \|\eta\|_s (\alpha(\sigma(\xi), \sigma(\eta)))^{1/t}, \tag{3.16}$$

provided that $\|\xi\|_r < \infty$ and $\|\eta\|_s < \infty$.

The lemma easily follows from Lemma 3.9 and a truncation argument.

For ϕ-mixing sequences slightly stronger inequalities hold. For instance, the inequality (3.16) can be improved to

$$|E(\xi\eta) - E\xi E\eta| \leq 2 \|\xi\|_r \|\eta\|_s (\phi(\sigma(\xi), \sigma(\eta)))^{1/r}, \tag{3.17}$$

where $1 \leq r, s \leq \infty$ satisfy $r^{-1} + s^{-1} = 1$ and where for two σ-fields $\mathcal{F}$ and $\mathcal{G}$ we define

$$\phi(\mathcal{F}, \mathcal{G}) = \sup\{|P(B|A) - P(B)| : A \in \mathcal{F},\ B \in \mathcal{G}\}.$$

The inequality (3.17) easily extends to Hilbert space valued random variables. The classical proof due to Ibragimov (1962) (see also Billingsley (1968)) works in this setting as well. In contrast, the extension of Lemmas 3.9 and 3.11 to Hilbert space valued random variables, due to Dehling and Philipp (1982), is far from being trivial and requires the use of Grothendieck's inequality. For Hilbert space valued random variables, slightly larger constants appear in the bounds on the r.h.s. of (3.13) and (3.16). Specifically, we get

$$|E(\xi, \eta) - (E\xi, E\eta)| \leq 10 \|\xi\|_\infty \|\eta\|_\infty \alpha(\sigma(\xi), \sigma(\eta)) \tag{3.18}$$

for bounded random variables, and

$$|E(\xi, \eta) - (E\xi, E\eta)| \leq 15 \|\xi\|_r \|\eta\|_s (\alpha(\sigma(\xi), \sigma(\eta)))^{1/t} \tag{3.19}$$

if $\|\xi\|_r < \infty$, $\|\eta\|_s < \infty$ and $r^{-1} + s^{-1} + t^{-1} = 1$.

For most applications the above inequalities are adequate. However, when pushing for results which are best possible or almost best possible, sharper versions than Lemma 3.11 are needed. One such covariance inequality is due to Rio (1993). Let X be a real-valued random variable with distribution function F_X, i.e.,

$$F_X(x) := P(X \leq x).$$

Denote by F_X^{-1} its generalized inverse or quantile function. Define the upper tail quantile function $Q_X : (0, 1) \to [0, \infty)$ by

$$Q_X(u) = F_X^{-1}(1 - u).$$

Lemma 3.12. *Let X and Y be real-valued random variables with finite expectation. Suppose that*

$$\int_0^1 Q_{|X|}(u)\, Q_{|Y|}(u)\, du < \infty.$$

Then $E|XY| < \infty$ and

$$|EXY - EXEY| \leq 4 \int_0^{\alpha} Q_{|X|}(u)\, Q_{|Y|}(u)\, du,$$

where $\alpha = \alpha(\sigma(X), \sigma(Y))$.

In the remainder of this subsection we study *covariance inequalities for functionals*, as defined by (3.11). As pointed out earlier, certain conditions have to be imposed on the underlying process (Z_n) as well as on the functional f in order to obtain meaningful results. From now on, we assume that (Z_n) is a stationary absolutely regular process. Concerning f, a certain type of continuity will be demanded, which basically guarantees that f can be approximated by a finite horizon functional.

Definition 3.13. *Let $r \geq 1$ and define the σ-fields $\mathcal{A}_{-l}^{l} = \sigma(Z_{-l}, \ldots, Z_l)$. Suppose $(a_l)_{l\geq 0}$ are nonnegative constants with $a_l \to 0$ as $l \to \infty$. We say that $(X_n)_{n\in\mathbb{Z}}$ is an r-approximating functional of $(Z_n)_{n\in\mathbb{Z}}$ with constants $(a_l)_{l\geq 0}$ if*

$$E\left|X_0 - E\left(X_0|\mathcal{A}_{-l}^{l}\right)\right|^r \leq a_l \tag{3.20}$$

holds for any $l \geq 0$.

In many applications, a constant that is derived from the r-approximation constants enters. For the sake of brevity of future formulas, we introduce

$$\bar{a}_k := \left(2\sum_{i=k}^{\infty} a_i\right)^{1/2}. \tag{3.21}$$

For 1-approximating functionals of absolutely regular processes, correlation inequalities can be established that are analogous to the inequalities obtained earlier in this section for mixing processes. Here we quote the most relevant results, referring for proofs and further inequalities to Borovkova, Burton, and Dehling (2001).

Lemma 3.14. *Let $(X_n)_{n\in\mathbb{N}}$ be a 1-approximating functional with constants $(a_k)_{k\geq 0}$ of an absolutely regular process with mixing coefficients $(\beta_k)_{k\geq 0}$. Then*
(i) *if X_0 is a.s. bounded, we have, for all $i, k \geq 0$,*

$$|E(X_i X_{i+k}) - EX_i EX_{i+k}| \leq 4\,\|X_0\|_\infty\, a_{\lfloor \frac{k}{3} \rfloor} + 2\,\|X_0\|_\infty^2\, \beta_{\lfloor \frac{k}{3} \rfloor}; \tag{3.22}$$

(ii) *if $\|X_0\|_{2+\delta} < \infty$ for some $\delta > 0$, we have, for all $i, k \geq 0$,*

$$\begin{aligned} &|E(X_i X_{i+k}) - EX_i EX_{i+k}| \\ &\quad \leq 4\,\|X_0\|_{2+\delta}^{(2+\delta)/(1+\delta)}, \left(a_{\lfloor \frac{k}{3} \rfloor}\right)^{\delta/(1+\delta)} + 2\,\|X_0\|_{2+\delta}^2 \left(\beta_{\lfloor \frac{k}{3} \rfloor}\right)^{\delta/(2+\delta)}; \end{aligned} \tag{3.23}$$

(iii) *if X_0 is a.s. bounded, we have, for any $\delta > 0$ and for all nonnegative integers $i < j < k < l$,*

$$\begin{aligned} &|E(X_i X_j X_k X_l) - EX_i E(X_j X_k X_l)| \\ &\leq \left\{6\left(\beta_{\lfloor \frac{j-i}{3} \rfloor}\right)^{\delta/(2+\delta)} \|X_0\|_{2+\delta}^2 + 8\left(a_{\lfloor \frac{j-i}{3} \rfloor}\right)^{\delta/(1+\delta)} \|X_0\|_{2+\delta}^{(2+\delta)/(1+\delta)}\right\} \|X_0\|_\infty^2, \end{aligned} \tag{3.24}$$

$$\begin{aligned} &|E(X_i X_j X_k X_l) - E(X_i X_j) E(X_k X_l)| \\ &\leq \left\{4\left(\beta_{\lfloor \frac{k-j}{3} \rfloor}\right)^{\delta/(2+\delta)} \|X_0\|_{2+\delta}^2 + 8\left(a_{\lfloor \frac{k-j}{3} \rfloor}\right)^{\delta/(1+\delta)} \|X_0\|_{2+\delta}^{(2+\delta)/(1+\delta)}\right\} \|X_0\|_\infty^2; \end{aligned} \tag{3.25}$$

(iv) *if* $\|X_0\|_{4+\delta} < \infty$ *for some* $\delta > 0$*, we have, for all nonnegative integers* $i < j < k < l$,

$$
\begin{aligned}
&|E(X_i X_j X_k X_l) - EX_i E(X_j X_k X_l)| \\
&\quad \le 6\left(\beta_{\lfloor \frac{j-i}{3} \rfloor}\right)^{\delta/(4+\delta)} \|X_0\|_{4+\delta}^4 + 8\left(a_{\lfloor \frac{j-i}{3} \rfloor}\right)^{\delta/(3+\delta)} \|X_0\|_{4+\delta}^{(12+3\delta)/(3+\delta)}, \qquad (3.26)
\end{aligned}
$$

$$
\begin{aligned}
&|E(X_i X_j X_k X_l) - E(X_i X_j) E(X_k X_l)| \\
&\quad \le 4\left(\beta_{\lfloor \frac{k-j}{3} \rfloor}\right)^{\delta/(4+\delta)} \|X_0\|_{4+\delta}^4 + 8\left(a_{\lfloor \frac{k-j}{3} \rfloor}\right)^{\delta/(3+\delta)} \|X_0\|_{4+\delta}^{(12+3\delta)/(3+\delta)}. \qquad (3.27)
\end{aligned}
$$

In connection with U-statistics, we will have to study correlations of random variables $h(X_i, X_j)$ and $h(X_{i'}, X_{j'})$, for some measurable functions h. This, however, requires extra continuity assumptions that have to be imposed on h. As the course of our investigations will lead us to more complex functions of the type $h(X_{i_1}, X_{j_1})\,h(X_{i_2}, X_{j_2})$, we will formulate the assumptions in this more general context. Let $(\xi_k)_{k\in\mathbb{Z}}$ be a stochastic process with the same 1-dimensional marginal distribution as $(X_n)_{n\in\mathbb{Z}}$. For a finite index set $I = \{i_1, \ldots, i_k\} \subset \mathbb{Z}$ we denote by ξ_I the vector of random variables $(\xi_{i_1}, \ldots, \xi_{i_k})$.

Definition 3.15. *Let* $I \subset \mathbb{Z}$ *be a finite set of integers. A measurable function* $g : \mathbb{R}^I \to \mathbb{R}$ *is called p-continuous if there exists a function* $\gamma : (0, \infty) \to (0, \infty)$ *with* $\gamma(\epsilon) \to 0$ *as* $\epsilon \to 0$ *such that*

$$
E|g(\xi_{I_1}, \xi_{I_2}) - g(\xi_{I_1}, \xi'_{I_2})|^p 1_{\{|\xi_{I_2} - \xi_{I'_2}| \le \epsilon\}} \le \gamma(\epsilon)
$$

holds for all disjoint index sets I_1 *and* I_2 *with* $I_1 \cup I_2 = I$ *and for all random vectors* $\xi_{I_1}, \xi_{I_2}, \xi'_{I_2}$ *such that* (ξ_{I_1}, ξ_{I_2}) *has distribution* $P_{X_{I_1}, X_{I_2}}$ *or* $P_{X_{I_1}} \times P_{X_{I_2}}$*, and* ξ'_{I_2} *has the same distribution as* X_{I_2}*, i.e.,* $P_{X_{I_2}}$.

Example 3.16. In the analysis of the Grassberger–Procaccia dimension estimator (Grassberger, Procaccia (1983)), the kernel $h_t(x, y) = 1_{\{|x-y|\le t\}}$ plays an important role. This kernel is p-continuous for any $p > 0$ provided

- the family of distribution functions of $|X_i - X_j|$, $i, j \in \mathbb{N}_0$, is equicontinuous,
- $\int\int h_t(x, y)\, dF(x)\, dF(y)$ is continuous at t, where F denotes the marginal distribution of X_i.

Indeed, $|X_0 - X_k| \le t - \epsilon$ and $|X_0 - X'_0| \le \epsilon$ imply that $|X'_0 - X_k| \le t$ and, in the same way, $|X_0 - X_k| > t + \epsilon$ and $|X_0 - X'_0| \le \epsilon$ imply $|X'_0 - X_k| > t$. Hence

$$
\left| 1_{\{|X_0 - X_k| \le t\}} - 1_{\{|X'_0 - X_k| \le t\}} \right| 1_{\{|X_0 - X'_0| \le \epsilon\}} \le 1_{\{t-\epsilon \le |X_0 - X_k| \le t+\epsilon\}}
$$

and thus we get, for any $p > 0$,

$$
\begin{aligned}
E\left| 1_{\{|X_0 - X_k| \le t\}} - 1_{\{|X'_0 - X_k| \le t\}} \right|^p 1_{\{|X_0 - X'_0| \le \epsilon\}} &\le P(t - \epsilon \le |X_0 - X_k| \le t + \epsilon) \\
&\le \gamma(\epsilon)
\end{aligned}
$$

where $\gamma(\epsilon) = \max(\sup_k P(t - \epsilon \le |X_0 - X_k| \le t + \epsilon), P(t - \epsilon \le |X - Y| \le t + \epsilon)$, with X, Y independent and with distribution function F.

The following two lemmas are from Borovkova, Burton, and Dehling (2001).

Lemma 3.17. *Let $h : \mathbb{R}^2 \to \mathbb{R}$ be a symmetric p-continuous kernel and define the function $g : \mathbb{R}^4 \to \mathbb{R}$ by*

$$g(x_{i_1}, x_{i_2}, x_{i_3}, x_{i_4}) := h(x_{i_1}, x_{i_2}) h(x_{i_3}, x_{i_4}).$$

Then the following hold:

(i) *If h is bounded, g is also p-continuous.*

(ii) *If $E|h(X_{i_1}, X_{i_2})|^p < \infty$, $E|h(X_{i_3}, X_{i_4})|^p < \infty$ and $E_{X_0} E_{X_k} |h(X_0, X_k)|^p < \infty$, then g is $p/2$-continuous.*

Here and in what follows, E_{X_0} or more generally $E_{X_{i_1}, \ldots, X_{i_j}}$ denotes the conditional expectation, given the remaining variables. The essential correlation inequality for U-statistics is formulated in the following lemma.

Lemma 3.18. *Let $(X_n)_{n \in \mathbb{Z}}$ be a 1-approximating functional with constants $(a_k)_{k \geq 0}$ of an absolutely regular process with mixing coefficients $(\beta_k)_{k \geq 0}$ and let $g : \mathbb{R}^l \to \mathbb{R}$ be 1-continuous. Moreover, let $1 \leq i_1 < i_2 < \cdots < i_l$ be integers and assume that for some $M < \infty$ and $j \in \{1, 2, \ldots, l-1\}$,*

$$\begin{aligned} \left(E\left|g(X_{i_1}, \ldots, X_{i_l})\right|^r\right)^{1/r} &\leq M, \\ \left(E_{X_{i_1}, \cdots, X_{i_j}} E_{X_{i_{j+1}}, \ldots, X_{i_l}} \left|g(X_{i_1}, \cdots, X_{i_l})\right|^r\right)^{1/r} &\leq M. \end{aligned}$$

Then

$$\begin{aligned} &\left| Eg(X_{i_1}, \ldots, X_{i_l}) - E_{X_{i_1}, \ldots, X_{i_j}} E_{X_{i_{j+1}}, \ldots, X_{i_l}} g(X_{i_1}, \ldots, X_{i_l}) \right| \\ &\quad \leq 4M \left(\beta_{\lfloor k/3 \rfloor} + \bar{a}_{\lfloor k/3 \rfloor}\right)^{1/s} + 2\gamma(\bar{a}_{\lfloor k/3 \rfloor}) \end{aligned} \tag{3.28}$$

where $k = |i_{j+1} - i_j|$ and where r, s are nonnegative numbers satisfying $r^{-1} + s^{-1} = 1$.

Lemma 3.18 generalizes a result by Yoshihara (1976), established for absolutely regular processes, rather than for functionals of such processes.

3.4 Moment inequalities for sums. In this section we will investigate bounds on the moments of partial sums $\sum_{j \leq n} \xi_j$ of weakly dependent random variables. These results are important technical tools in later proofs of central limit theorems and of empirical process invariance principles. We will formulate and prove these results for random variables with values in a separable Hilbert space H or Banach space B.

A sequence of H-valued random variables is called weakly stationary, if $E\xi_j$ is independent of j and if $E(\xi_i, \xi_j)$ only depends on $|j - i|$, for any $i, j \geq 1$. Note that a (strictly) stationary sequence with finite 2nd moments is automatically weakly stationary.

Lemma 3.19. *Let $\{\xi_j, j \geq 1\}$ be a weakly stationary sequence of H-valued random variables with mean 0 and with $(2+\delta)$-th moments bounded. If $\{\xi_j;\ j \geq 1\}$ is strongly mixing with rate $\alpha(n)$ satisfying*

$$\sum_{n \geq 1} (\alpha(n))^{\delta/(2+\delta)} < \infty,$$

then

$$\frac{1}{n} E \| \sum_{j \leq n} \xi_j \|^2 \leq 30 \left(1 + \sum_{j \geq 1} (\alpha(j))^{\delta/(2+\delta)}\right) \sup_{j \geq 1} \|\xi_j\|_{2+\delta}^2, \tag{3.29}$$

and

$$E\|\sum_{j\le n}\xi_j\|^2 = \sigma^2 n + o(n)$$

where $\sigma^2 = E\|\xi_1\|^2 + 2\sum_{j\ge 2} E\{(\xi_1,\xi_j)\}$.

Proof. By (3.19) with $r = s = 2+\delta$ and $t = (2+\delta)/\delta$, we have

$$|E(\xi_1,\xi_{n+1})| \le 15(\alpha(n))^{\delta/(2+\delta)} \sup_j \|\xi_j\|_{2+\delta}^2. \tag{3.30}$$

Thus, the series defining σ^2 converges. Moreover,

$$\begin{aligned} E\|\sum_{j\le n}\xi_j\|^2 &= \sum_{1\le k,l\le n} E(\xi_k,\xi_l) \\ &= n\,E\|\xi_1\|^2 + 2\sum_{j=2}^{n}(n-j+1)\,E\{(\xi_1,\xi_j)\}. \end{aligned}$$

From this identity and (3.30) we immediately get the inequality (3.29). Moreover,

$$\begin{aligned} E\|\sum_{j\le n}\xi_j\|^2 &= n\left(E\|\xi_1\|^2 + 2\sum_{j=2}^{n}E(\xi_1,\xi_j)\right) + 2n\sum_{j=2}^{n}\frac{j-1}{n}E(\xi_1,\xi_j) \quad (3.31)\\ &= \sigma^2 n + o(n), \end{aligned}$$

as $\sum_{j=2}^{n}\frac{j-1}{n}E(\xi_1,\xi_j) = o(1)$, by (3.30) and because $j(\alpha(j))^{\delta/(2+\delta)} \to 0$ by the monotonicity of α and by the assumption on the mixing rate. □

The following theorem is due to Sotres and Ghosh (1977), for the case of real-valued random variables. However, their proof also works for random variables with values in a separable Banach space B, as was observed by Dehling and Philipp (1982).

Theorem 3.20. *Let* $0 < \delta \le 1$ *and let* $\{\xi_j, j \ge 1\}$ *be a sequence of* B*-valued random variables, strongly mixing with rate*

$$\alpha(n) \ll n^{-(1+\rho)(1+2/\delta)}, \quad 0 < \rho \le 1.$$

Suppose that $\sup_{j\ge 1} E\|\xi_j\|^{2+\delta} \le b < \infty$, *for some* $b \ge 1$ *and that for some* $A < \infty$

$$E\|\sum_{j=a+1}^{a+n}\xi_j\|^2 \le A\,n \quad \textit{for all } n \ge 1, a \ge 0. \tag{3.32}$$

Then

$$E\|\sum_{j\le n}\xi_j\|^{2+\epsilon} \ll n^{1+\frac{1}{2}\epsilon}(A^{1+\frac{1}{2}\epsilon} + b)$$

for any ϵ *with* $0 \le \epsilon \le \rho\delta/7$.

Under a stronger mixing condition, one can even provide estimates for third moments of sums of strongly mixing processes. The following result is due to Dehling (1983).

Lemma 3.21. *Let $\{\xi_j, j \geq 1\}$ be a strongly mixing sequence of B-valued random variables with mean zero and with $(3+\delta)$-th moments uniformly bounded by $\rho_{3+\delta}$ for some $\delta > 0$. Suppose that the mixing coefficients $\alpha(k)$ satisfy*

$$\alpha(k) \ll k^{-(1+\epsilon)(3+9/\delta)} \tag{3.33}$$

for some $\epsilon > 0$. If

$$\sigma^2 := \sup_n \frac{1}{n} E\|\sum_{\nu \leq n} \xi_\nu\|^2,$$

then we have

$$E\|\sum_{\nu=1}^{n} \xi_\nu\|^3 \leq Cn^{3/2}\left(\sigma^3 + \rho_{3+\delta}^{3/(3+\delta)}\right), \tag{3.34}$$

where C is a constant depending only on ϵ, δ and the constant implied by $\ll$ in (3.33).

The following lemma, in conjunction with Proposition 2.11, serves the same purpose as in Section 2, namely to establish tightness of the empirical process sequence. For ϕ-mixing sequences the lemma and its proof can be found in Billingsley (1968, pp. 195–196). Its extension to strongly mixing sequences sequences is due to Deo (1973).

Lemma 3.22. *Let $(\xi_n)_{n\geq 0}$ be a stationary sequence of real-valued random variables with $E\xi_0 = 0$ and $\|\xi_0\|_\infty \leq 1$. Suppose $(\xi_n)_{n\geq 0}$ satisfies a strong mixing condition with*

$$\alpha(n) \leq Cn^{-6-\epsilon}$$

for some $C \geq 1$ and $0 < \epsilon < 1$. Then, with $\alpha = 1 + \frac{\epsilon}{10}$,

$$E\left(\sum_{i=1}^{n} \xi_i\right)^4 \leq 10^4 \frac{C}{\epsilon}\left(n^2(E\xi_0^2)^\alpha + n(E\xi_0^2)^{\alpha/2}\right). \tag{3.35}$$

Proof. We can rewrite and bound $E(\sum_{i=1}^n \xi_i)^4$ as follows:

$$E\left(\sum_{i=1}^{n} \xi_i\right)^4 = \sum_{1\leq i_1,i_2,i_3,i_4\leq n} E\left(\xi_{i_1}\xi_{i_2}\xi_{i_3}\xi_{i_4}\right) \leq 24n \sum_{0\leq i,j,k\leq n-1} \left|E(\xi_0\xi_i\xi_{i+j}\xi_{i+j+k})\right| \tag{3.36}$$

where the sum is extended over all triplets (i, j, k) with $i, j, k \geq 0$ and $i + j + k < n$. We set $r = 2 + \frac{1}{4}\epsilon$ and $s = (4 + \frac{1}{2}\epsilon)/(1 + \frac{1}{4}\epsilon)$ so that $\frac{1}{r} + \frac{2}{s} = 1$ and $s < 4$. Then by Lemma 3.11 with $t = s$, we obtain

$$\begin{aligned}
\left|E\left(\xi_0(\xi_i\xi_{i+j}\xi_{i+j+k})\right)\right| &\leq 10\,\alpha(i)^{\frac{1}{r}}\|\xi_0\|_s\|\xi_i\xi_{i+j}\xi_{i+j+k}\|_s \\
&\leq 10\,\alpha(i)^{\frac{1}{r}}\|\xi_0\|_s^2, \\
\left|E\left((\xi_0\xi_i\xi_{i+j})\xi_{i+j+k}\right)\right| &\leq 10\,\alpha(k)^{\frac{1}{r}}\|\xi_0\|_s^2, \\
\left|E\left((\xi_0\xi_i)(\xi_{i+j}\xi_{i+j+k})\right)\right| &\leq |E(\xi_0\xi_i)|\,|E(\xi_0\xi_k)| + 10\,\alpha(j)^{\frac{1}{r}}\|\xi_0\xi_i\|_s\|\xi_0\xi_k\|_s \\
&\leq 100\,\alpha(i)^{\frac{1}{r}}\alpha(k)^{\frac{1}{r}}\|\xi_0\|_s^4 + 10\,\alpha(j)^{\frac{1}{r}}\|\xi_0\|_s^2.
\end{aligned}$$

We now estimate each term on the r.h.s. of (3.36) using one of the above three inequalities, according to which of the indices i, j, k, is the largest. Then

$$\begin{aligned} E\left(\sum_{i=1}^{n} \xi_i\right)^4 &\leq 2400\, n \left(\|\xi_0\|_s^4 \sum_{j=0}^{n-1} \sum_{i,k\leq j} \alpha(i)^{\frac{1}{r}} \alpha(k)^{\frac{1}{r}} + \frac{1}{10} \|\xi_0\|_s^2 \sum_{i=0}^{n-1} \sum_{j,k\leq i} \alpha(i)^{\frac{1}{r}} \right) \\ &\leq 2400\, C\, n \left(\|\xi_0\|_s^4\, n\, 4 + \frac{1}{10} \|\xi_0\|_s^2 \sum_{i=0}^{\infty} i^2 \alpha(i)^{\frac{1}{r}} \right) \\ &\leq 10^4\, \frac{C}{\epsilon} \left(n^2 (E\xi_0^2)^{4/s} + n \left(E\xi_0^2\right)^{2/s} \right). \end{aligned}$$

From here (3.35) follows, since $\frac{4}{s} = \frac{4+\epsilon}{4+\epsilon/2} \leq 1 + \epsilon/10$, as a quick calculation shows. □

3.5 A central limit theorem. For purposes which will become clear later, we need a multivariate central limit theorem where the dimension d and the bound on the $(2+\delta)$-th moment of the random vectors ξ_n $(1 \leq n \leq N)$ may increase with N. For that purpose we fix $N \geq 1$ and assume that $\{\xi_n, 1 \leq n \leq N\}$ is a sequence of random vectors of dimension d. Let f_N be the characteristic function of $N^{-\frac{1}{2}} \sum_{n\leq N} \xi_n$. Note that the constant implied by $\ll$ in the estimate of f_N below does not depend on d and b.

Theorem 3.23. *Let $\{\xi_n,\ 1 \leq n \leq N\}$ be a weakly stationary sequence of random vectors, $\xi_n = (\xi_{ni},\ 1 \leq i \leq d)$, with values in $\mathbb{R}^d$, centered at expectations and with $(2+\delta)$-th moments with $0 < \delta \leq 1$ bounded by $b \geq 1$. Suppose that the sequence satisfies the strong mixing condition* (3.2) *with rate*

$$\alpha(n) \ll n^{-(1+\rho)(1+2/\delta)}, \quad 0 < \rho < \frac{1}{4}. \tag{3.37}$$

Then the two series in

$$\gamma_{ij} = E\xi_{1i}\xi_{1j} + \sum_{k\geq 2} E\xi_{1i}\xi_{kj} + \sum_{k\geq 2} E\xi_{ki}\xi_{1j}$$

converge absolutely. Denote the matrix $(\gamma_{ij})_{1\leq i,j\leq d}$ by Γ. Then, as $N \to \infty$,

$$f_N(u) - \exp(-\frac{1}{2}(u, \Gamma u)) \ll b(Td)^{\frac{9}{2}} N^{-\frac{3}{16}} + N^{-\frac{1}{56}\rho\delta} b(Td)^{9/4}$$

for all $u \in \mathbb{R}^d$ with $|u| \leq T$. Here the constant implied by $\ll$ only depends on ρ, δ, and the constant implied by $\ll$ in (3.37).

Let $u = (u_1, \ldots, u_d) \in \mathbb{R}^d$ and let $\{\xi_n, n \geq 1\}$ be the sequence of Theorem 3.23. We write, for $n \geq 1$,

$$\zeta_n = (u, \xi_n) = \sum_{j\leq d} u_j \xi_{nj}. \tag{3.38}$$

Then $\{\zeta_n, n \geq 1\}$ satisfies the same mixing condition as $\{\xi_n, n \geq 1\}$.

Lemma 3.24. *We have, for the variance σ^2 of the sequence $\{\zeta_n, n \geq 1\}$,*

$$\sigma^2 = (u, \Gamma u) \leq C\, b\, d\, |u|^2$$

where the constant C only depends on ρ, δ and the constant implied by $\ll$ in (3.37), *and not on the dimension d, nor on the bound b.*

Proof. This follows at once from Lemma 3.19. □

One of the main ingredients of the proof of Theorem 3.23 and, at the same time, part of the standard techniques in the proofs of limit theorems for weakly dependent random variables is the so-called blocking technique. This technique can be briefly described as follows. We start with the sum of the first p random variables ζ_ν, say p is given below by (3.39). This is the first "large" block. The sum of the next q random variables ζ_ν is the first "small" block. Since q is much smaller than p this notation is self-evident. The second large block is the sum of the next p random variables, followed by the second small block of length q. We continue this alternating pattern until we run out of random variables, i.e., when we reach ζ_N. What is the purpose of this game? Since by Lemma 3.19 the variances of the blocks grow linearly with their lengths, the large blocks have variance about $\sigma^2 p$, whereas the small blocks have variance of only about $\sigma^2 q$. Thus the contribution of the small blocks is negligible, and they can simply be thrown out. Why did we introduce them in the first place? The small blocks are long enough to provide good separation between the large blocks in the sense that $\alpha(q)$ tends to zero. As a consequence the large blocks, considered as random vectors, are almost independent and thus the standard techniques for independent random variables can be easily modified to take care of them.

In the following this program will be formalized. Since we expand the characteristic functions using a Taylor expansion up to the $(2+\delta)$-th moments, we can assume without loss of generality that $b = 1$ by replacing ξ_n by $\xi_n b^{-1/(2+\delta)}$. Let

$$p = \left[N^{3/4}\right], \quad q = \left[N^{(1+\epsilon)/4}\right], \quad l = [N/(p+q)]. \tag{3.39}$$

We remark that the specific choice of p and q is somewhat arbitrary, and can be modified, e.g., to achieve better rates of approximation. Then

$$l = N^{1/4}(1 + O(N^{-1/4})). \tag{3.40}$$

We introduce the intervals

$$\begin{aligned} H_j &= ((j-1)(p+q), jp + (j-1)q], \\ I_j &= (jp + (j-1)q, j(p+q)] \end{aligned}$$

for $1 \le j \le l$ and $I_{l+1} = (l(p+q), N]$, and define the random variables y_j $(1 \le j \le l)$ and z_j $(1 \le j \le l+1)$ by

$$y_j = \sum_{\nu \in H_j} \zeta_\nu, \quad z_j = \sum_{\nu \in I_j} \zeta_\nu.$$

The y_j's are the large block sums and the z_j's the small block sums. The next lemma shows that the contribution of the small blocks is negligible, since their sum normalized by $N^{-\frac{1}{2}}$ tends to zero in probability.

Lemma 3.25. *Under the hypotheses of Theorem 3.23,*

$$E\left|\sum_{j \le l+1} z_j\right|^2 \ll bT^2 d^2 N^{13/16}.$$

Proof. By (3.38) and Cauchy's inequality,

$$\left|\sum_{\nu=a+1}^{a+n} \zeta_\nu\right| \le |u| \left|\sum_{\nu=a+1}^{a+n} \xi_\nu\right| \le T \sum_{j \le d} \left|\sum_{\nu=a+1}^{a+n} \xi_{\nu j}\right|$$

for all $a \geq 0$, $n \geq 1$ and $|u| \leq T$. Let $\epsilon = \frac{1}{7}\rho\delta$. We apply Theorem 3.20 to each sequence $\{\xi_{\nu j}, \nu \geq 1\}$ $(1 \leq j \leq d)$ and obtain

$$E\left|\sum_{\nu=a+1}^{a+n} \xi_{\nu j}\right|^{2+\gamma} \ll n^{1+\frac{\gamma}{2}}$$

uniformly for $a \geq 0$, $1 \leq j \leq d$ and $0 \leq \gamma \leq \epsilon$. More precisely, the constant implied by $\ll$ only depends on ρ, δ and the constant implied by $\ll$ in (3.37). Consequently,

$$E\left|\sum_{\nu=a+1}^{a+n} \zeta_{\nu}\right|^{2+\gamma} \ll n^{1+\frac{\gamma}{2}}(Td)^{2+\gamma} \tag{3.41}$$

with the same provision for the constant implied by $\ll$ as before. Thus

$$Ez_j^2 \ll N^{\frac{1}{4}(1+\epsilon)}T^2d^2 \tag{3.42}$$

and

$$Ez_{l+1}^2 \ll N^{3/4}T^2d^2. \tag{3.43}$$

Hence by Minkowski's inequality, (3.40), (3.42), (3.43) and since $\epsilon \leq 1/4$ by (3.37) and Theorem 3.20,

$$\left\|\sum_{j \leq l+1} z_j\right\|_2 \ll \left(lN^{(1+\epsilon)/8} + N^{3/8}\right) Td \ll T\, d\, N^{13/32}.$$

Lemma 3.26. *Under the hypotheses of Theorem 3.23,*

$$\left|\prod_{j \leq l} E\left\{\exp(iy_jN^{-\frac{1}{2}})\right\} - \exp(-\frac{1}{2}(u, \Gamma u))\right| \ll N^{-\frac{1}{4}}b\,(Td)^{\frac{9}{2}} + N^{-\epsilon/8}b\,(Td)^{9/4}.$$

Proof. The proof is a routine modification of the standard proof of the central limit theorem for independent random variables. Again we assume $b = 1$. We first observe that by Lemma 3.24,

$$\sigma^2 \leq CT^2d. \tag{3.44}$$

Thus by (3.31) and (3.30),

$$Ey_j^2 = \sigma^2N^{3/4} + O(N^{3(1-\epsilon)/4}T^2), \tag{3.45}$$

since $\epsilon \leq \rho$ and since $\zeta_\nu T^{-1}$ has $(2+\delta)$-th moment bounded by 1. Moreover, by (3.41),

$$E|y_j|^{2+\epsilon} \ll \left(N^{3/4}T^2d^2\right)^{1+\frac{1}{2}\epsilon}. \tag{3.46}$$

Thus for some θ with $|\theta| \leq 1$,

$$E\left\{\exp\left(iy_jN^{-\frac{1}{2}}\right)\right\} = 1 - \frac{1}{2}N^{-1}Ey_j^2 + \theta N^{-1-\epsilon/2}E|y_j|^{2+\epsilon} = 1 - r_j, \quad \text{say.} \tag{3.47}$$

Now by (3.45) and (3.46), uniformly in $j \geq 1$,

$$r_j = \frac{1}{2}N^{-\frac{1}{4}}\sigma^2 + O\left((N^{-\frac{1}{4}}T^2d^2)^{1+\frac{1}{2}\epsilon}\right).$$

Thus by (3.44),

$$r_j^2 \ll T^4d^2N^{-\frac{1}{2}} + \left(N^{-\frac{1}{2}}T^4d^4\right)^{1+\frac{1}{2}\epsilon}.$$

Consequently,

$$\log\prod_{j\leq l} E\left\{\exp(iy_jN^{-\frac{1}{2}})\right\} = -\frac{1}{2}\sigma^2 + O\left(N^{-\frac{1}{4}}(Td)^{9/2}\right) + O\left((T\,d)^{9/4}N^{-\epsilon/8}\right)$$

by (3.40) and (3.44). Hence

$$\prod_{j\leq l} E\{\exp(iy_jN^{-\frac{1}{2}})\} = e^{-\sigma^2/2}\left(1 + O(N^{-\frac{1}{4}}(Td)^{9/2}) + O((Td)^{9/4}\,N^{-\epsilon/8}\right).$$

The lemma now follows from Lemma 3.24. □

We can now finish the proof of Theorem 3.23 in the usual fashion. We have

$$f_N(u) - \exp(-\frac{1}{2}(u,\Gamma u)) = I + II + III, \quad \text{say}, \tag{3.48}$$

where

$$\begin{aligned} I &= E\left\{\exp(iN^{-\frac{1}{2}}\left(\sum_{j\leq l} y_j + \sum_{j\leq l+1} z_j\right)\right\} - E\left\{\exp\left(iN^{-\frac{1}{2}}\sum_{j\leq l} y_j\right)\right\}, \\ II &= E\left\{\exp(iN^{-\frac{1}{2}}\left(\sum_{j\leq l} y_j\right)\right\} - \prod_{j\leq l} E\left\{\exp(iN^{-\frac{1}{2}}y_j)\right\}, \\ III &= \prod_{j\leq l} E\left\{\exp(iN^{-\frac{1}{2}}y_j)\right\} - \exp(-\frac{1}{2}(u,\Gamma u)). \end{aligned}$$

Since for real x and y,

$$|e^{i(x+y)} - e^{iy}| = |e^{ix} - 1| \leq |x|$$

we have by Lemma 3.25 and Cauchy's inequality

$$I \ll N^{-\frac{1}{2}}E|\sum_{j\leq l+1} z_j| \ll T\,d\,b^{\frac{1}{2}}\,N^{-3/16}.$$

We apply Corollary 3.10 l times and obtain by (3.40), (3.39), and (3.37),

$$II \ll l\,\alpha(q) \ll N^{\frac{1}{4}}N^{-(1+\epsilon)/4} = N^{-\epsilon/4}.$$

Since III is estimated in Lemma 3.26 we obtain Theorem 3.23 from (3.48) and these estimates. □

3.6 Berry–Esseén inequality for strongly mixing processes. In this section we will present a central limit theorem for partial sums of strongly mixing random vectors, together with an estimate for the speed of convergence to a normal law. Our proof, taken from Dehling's 1981 thesis (see also Dehling (1983)) uses Lindeberg's method of substitution, also known as Bernstein's method. In a similar context, this technique was also used by Doukhan and Portal (1983 and 1987).

We first recall the definition of the Prohorov distance $\pi(\mu, \nu)$ of two probability measures on some separable metric space (S, d), given by

$$\pi(\mu, \nu) = \inf\{\epsilon > 0 : \mu(A) \leq \nu(A^\epsilon) + \epsilon\},$$

where $A^\epsilon = \{x \in S : d(x, A) \leq \epsilon\}$. If X and Y are S-valued random elements with distributions μ and ν, respectively, and $P(d(X, Y) \geq \epsilon) \leq \epsilon$, then

$$\pi(\mu, \nu) \leq \epsilon, \tag{3.49}$$

as is easily seen. The converse also holds, in the following sense: given two probability measures μ and ν on S satisfying (3.49), there exist random variables X and Y on some common probability space with values in S and with distributions μ and ν, respectively, such that $P(d(X, Y) \geq \epsilon) \leq \epsilon$. This is the statement of the celebrated Strassen–Dudley theorem.

Using Markov's inequality, we can also give an upper bound of the Prohorov distance of the distributions of two random vectors X and Y in terms of their L_2-distance. We have

$$\begin{aligned} P\left(\|X - Y\| > (E(\|X - Y\|^2))^{1/3}\right) &= P\left(\|X - Y\|^2 > (E(\|X - Y\|^2))^{2/3}\right) \\ &\leq \left(E(\|X - Y\|^2)\right)^{1/3} \end{aligned}$$

and thus by (3.49),

$$\pi(\mathcal{L}(X), \mathcal{L}(Y)) \leq \left(E(\|X - Y\|)^2\right)^{1/3}. \tag{3.50}$$

When combined with Lemma 4.15 below, Theorem 3.23 yields an estimate of the Prohorov distance of the law of $N^{-1/2} \sum_{n \leq N} \xi_n$ and the Gaussian law with mean 0 and covariance matrix Γ. The resulting estimate $O(N^{-\rho\delta/(112(d+3))})$ is rather weak, in particular, when its dependence on the dimension d is concerned. However, the estimate is still sufficient to prove many of the classical limit laws, such as the law of the iterated logarithm. Under somewhat more stringent hypotheses, the following theorem provides a much sharper estimate.

Theorem 3.27. *Let $\{\xi_\nu, \nu \geq 1\}$ be a strongly mixing sequence of $\mathbb{R}^d$-valued random vectors with mean zero and $E\|\xi_\nu\|^{3+\delta} \leq \rho_{3+\delta} < \infty$ for some $\delta > 0$. Suppose that the mixing coefficients satisfy*

$$\alpha(k) \ll k^{-(1+\epsilon)(3+9/\delta)} \tag{3.51}$$

for some $\epsilon > 0$. If Σ_n denotes the covariance matrix of $n^{-1/2} \sum_{\nu \leq n} \xi_\nu$, then

$$\pi(\mathcal{L}(n^{-1/2} \sum_{\nu \leq n} \xi_\nu; N(0, \Sigma_n)) \ll n^{-1/20} d^{3/2} (1 + \rho_{3+\delta}^{3/(3+\delta)}), \tag{3.52}$$

where the constant implied by $\ll$ only depends on ϵ, δ, and the constant implied by $\ll$ in (3.51).

We remark that the exponents and the constants in (3.51) and (3.52) are far from being sharp. Still, (3.52) is sufficient to obtain a law of the iterated logarithm and even strong approximation results, which was actually the main goal of Dehling (1981, 1983). Later, Rio (1995) applied Lindeberg's method in the case of real-valued random variables in a more subtle way to get much sharper results.

The proof of Theorem 3.27 uses again the blocking technique introduced earlier in connection with the proof of Theorem 3.23. Given an integer n, we subdivide the set $\{1, 2, \ldots, n\}$ into long blocks H_j of consecutive integers, separated by small blocks I_j of consecutive integers, with

$$\text{card } H_j = p = \lfloor n^{1/2} \rfloor \quad \text{and} \quad \text{card } I_j = q = [n^{1/4}]$$

for $j = 1, \ldots, l$, where $l = [n/(p+q)]$. Observe that $n^{1/2} \ll l \ll n^{1/2}$. In addition we define a last block I_{l+1} that contains the remaining integers so that

$$\text{card } I_{l+1} = n - l(p+q).$$

For fixed n we define the normalized block sums

$$X_j = p^{-1/2} \sum_{\nu \in H_j} \xi_\nu, \text{ for } 1 \le j \le l.$$

Using the moment estimate in Lemma 3.21 we get $E\|X_j\|^3 \le Cb^3$, where $b = \rho_{3+\delta}^{1/(3+\delta)}$, for some constant C, whose value depends only on ϵ, δ and the constant implied by $\ll$ in (3.51)

The ξ_ν-variables in the block sum defining X_j are separated by the short blocks of length q. In this situation we can apply the following auxiliary result, whose proof contains the application of the Lindeberg method.

Proposition 3.28. *Let $X_1, \ldots, X_n$ be a set of $\mathbb{R}^d$-valued random variables with mean zero, covariance matrix $\text{Cov}(X_i)$ and finite third absolute moments $\rho_3 := \sup E\|X_i\|^3$. Let $\mathcal{M}_a^b$ denote the σ-field generated by the random variables $X_a, \ldots, X_b$ and define the mixing constant $\alpha := \sup_{1\le k\le n-1} \alpha(\mathcal{M}_1^k, \mathcal{M}_{k+1}^n)$. Let $Z_1, \ldots, Z_n$ be a set of independent $N(0, \text{Cov}(X_i))$-distributed random variables. If $f : \mathbb{R}^d \to \mathbb{R}$ is three times differentiable, then*

$$\begin{aligned} &\left| Ef(n^{-1/2}(X_1 + \cdots + X_n)) - Ef(n^{-1/2}(Z_1 + \cdots + Z_n)) \right| \qquad (3.53) \\ &\quad \le C \left(\alpha^{2/3} n^{1/2} \sup_x \|D_x f\| \, \rho_3^{1/3} + d^2 \alpha^{1/3} \sup_x \|D_x^2 f\| \, \rho_3^{2/3} + n^{-1/2} \sup_x \|D_x^3 f\| \, \rho_3 \right), \end{aligned}$$

where C is an absolute constant.

Proof. Without loss of generality, we may assume that $Z_1, \ldots, Z_n$ is independent of $X_1, \ldots, X_n$. Then we can bound the term on the r.h.s. of (3.53) as follows:

$$\begin{aligned} &\left| Ef(n^{-1/2}(X_1 + \cdots + X_n)) - Ef(n^{-1/2}(Z_1 + \cdots + Z_n)) \right| \\ &\le \sum_{i=1}^n \left| Ef(n^{-1/2}(X_1 + \cdots + X_i + Z_{i+1} + \cdots + Z_n)) \right. \\ &\qquad \left. - Ef(n^{-1/2}(X_1 + \cdots + X_{i-1} + Z_i + \cdots + Z_n)) \right|. \end{aligned}$$

We estimate each of the n summands by using Taylor expansion of f about $W_i := n^{-1/2}(X_1 + \cdots + X_{i-1} + Z_{i+1} + \cdots + Z_n)$. In this way, we obtain

$$\begin{aligned}
&\Big| Ef(n^{-1/2}(X_1 + \cdots + X_i + Z_{i+1} + \cdots + Z_n)) \\
&\quad - Ef(n^{-1/2}(X_1 + \cdots + X_{i-1} + Z_i + \cdots + Z_n))\Big| \\
&\leq |E(D_{W_i} f(n^{-1/2}X_i))| + |E(D_{W_i} f(n^{-1/2}Z_i))| \\
&\quad + |E(D^2_{W_i} f(n^{-1/2}X_i)) - E(D^2_{W_i} f(n^{-1/2}Z_i))| \\
&\quad + \frac{1}{6} \sup_x \|D^3_x f\| \, E\left(\|\frac{1}{\sqrt{n}} X_i\|^3 + \|\frac{1}{\sqrt{n}} Z_i\|^3 \right).
\end{aligned} \tag{3.54}$$

Again, we estimate the three summands on the r.h.s. separately. The integrand in the first term can be written as

$$D_{W_i} f(n^{-1/2}X_i) = n^{-1/2}(\text{grad} f(W_i), X_i).$$

If $\mathcal{Z}$ denotes the σ-field generated by $Z_1, \ldots, Z_n$, then $\mathcal{Z}$ is independent of $\mathcal{M}_1^n$ and thus

$$\alpha(\mathcal{M}_1^{i-1} \vee \mathcal{Z}, \mathcal{M}_i^i) \leq \alpha.$$

Now, W_i is $\mathcal{M}_1^{i-1} \vee \mathcal{Z}$-measurable, and hence using the correlation inequality (3.19) with $r = 3$, $s = \infty$ and $t = 3/2$, we obtain

$$|ED_{W_i} f(n^{-1/2}X_i)| \leq 15\, \alpha^{2/3}\, n^{-1/2} \sup_x \|D_x f\| \, \|X_1\|_3.$$

Regarding the 2nd term we get $ED_{W_i} f(n^{-1/2}Z_i) = n^{-1/2}E(\text{grad} f(W_i), Z_i) = 0$, since Z_i is independent of W_i. The 3rd term can be rewritten as

$$|ED^2_{W_i} f(n^{-1/2}X_i) - ED^2_{W_i} f(n^{-1/2}Z_i)| = \frac{1}{n} |EX_i A(W_i) X_i^T - EZ_i A(W_i) Z_i^T|$$

where $A = A(W_i) = (a_{kl})_{1 \leq k,l \leq d}$ is the symmetric $d \times d$-matrix of 2nd order partial derivatives of f. Its entries are real-valued random variables, measurable with respect to $\sigma(W_i)$. Denoting by $X_i^{(1)}, \ldots, X_i^{(d)}$ and by $Z_i^{(1)}, \ldots, Z_i^{(d)}$ the elements of the random vector X_i and Z_i, we get

$$\begin{aligned}
|EX_i A(W_i) X_i^T - EZ_i A(W_i) Z_i^T| &\leq \sum_{k,l} |E(X_i^{(k)} X_i^{(l)} a_{kl}) - E(Z_i^{(k)} Z_i^{(l)} a_{kl})| \\
&\leq \sum_{k,l} |E(X_i^{(k)} X_i^{(l)} a_{kl}) - E(X_i^{(k)} X_i^{(l)}) E(a_{kl})| \\
&\quad + \sum_{k,l} |E(Z_i^{(k)} Z_i^{(l)} a_{kl}) - E(Z_i^{(k)} Z_i^{(l)}) E(a_{kl})|,
\end{aligned}$$

because X_i and Z_i have the same covariance matrix. The terms inside the second sum vanish again, because Z_i is independent of W_i. We apply the correlation inequality (3.16) with $r = 3/2$, $s = \infty$ and $t = 3$ to each term in the first sum, and get

$$|EX_i A(W_i) X_i^T - EZ_i A(W_i) Z_i^T| \leq d^2\, 10\, \alpha^{1/3} \|X_i\|_3^2 \sup_{k,l} \|a_{kl}\|_\infty$$

and hence

$$|ED^2_{W_i} f(n^{-1/2}X_i) - ED^2_{W_i} f(n^{-1/2}Z_i)| \le 10\,n^{-1}\,d^2\,\alpha^{1/3}\,\|X_i\|_3^2 \sup_x \|D^2_x f\|.$$

In order to bound the last term on the r.h.s. of (3.54), we use the fact that there exists a universal constant C such that $E\|Z_i\|^3 \le C(E\|Z_i\|^2)^{3/2}$, as the Z_i are Gaussian. Since $E\|Z_i\|^2 = E\|X_i\|^2 \le \|X_i\|_3^{2/3}$, we finally get

$$E(\|n^{-1/2}X_i\|^3 + \|n^{-1/2}Z_i\|^3) \le C\,n^{-3/2}\|X_i\|_3^3.$$

Combining the above estimates, we obtain the desired result. □

In order to obtain estimates for the Prohorov distance of the two measures involved in Proposition 3.28, we apply (3.53) to smooth functions that approximate the indicator of a given set A sufficiently well. A result of Yurinskii (1977), with some improvements by Dehling (1981, 1983) shows that, given any closed set $A \subset \mathbb{R}^d$, there exists for any $\epsilon \in (0, 1)$ a three times continuously differentiable function $\phi : \mathbb{R}^d \to [0, 1]$ such that

$$\begin{aligned} \phi(x) &\ge 1 - 2\epsilon \qquad \text{for } x \in A, \\ \phi(x) &\le 2\epsilon \qquad \text{for } x \in \mathbb{R}^d \setminus A^{\epsilon_1}, \end{aligned}$$

where $\epsilon_1 = 4\epsilon(\log(1/\epsilon))^{1/2}d^{1/2} + 4\epsilon d^{1/2}$. Here A^ϵ is defined as above, i.e., $A^\epsilon = \{x \in S : d(x, A) \le \epsilon\}$. Moreover, the first three derivatives of ϕ can be bounded as follows:

$$\|D^3_x\phi\| \le 2\epsilon^{-3}d^{-1/2}, \quad \|D^2_x\phi\| \le \epsilon^{-2}d^{-1/2}, \quad \|D_x\phi\| \le \epsilon^{-1}d^{-1/2}$$

(see Dehling (1983) for details of the construction). Observe that, for any measurable set $A \subset \mathbb{R}^d$ and any measurable function $\phi : \mathbb{R}^d \to [0, 1]$ satisfying the above conditions, we get

$$\mu(A) \le \nu(A^{\epsilon_1}) + 4\epsilon + \left|\int \phi(x)d(\mu - \nu)(x)\right|,$$

for any two probability measures μ, ν on $\mathbb{R}^d$. Applying this to (3.53), we get the following bound on the Prohorov distance of the measures involved in Proposition 3.28:

$$\begin{aligned} &\pi\left(\mathcal{L}(n^{-1/2}(X_1 + \cdots + X_n)), N(0, n^{-1}\sum_{i=1}^n \mathrm{Cov}(X_i))\right) \qquad (3.55) \\ &\le C\left(\alpha^{2/3}n^{1/2}\epsilon^{-1}d^{-1/2}\rho_3^{1/3} + d^2\alpha^{1/3}\rho_3^{2/3}\epsilon^{-2}d^{-1/2} + 2n^{-1/2}\rho_3\epsilon^{-3}d^{-1/2}\right) \\ &\quad +4\epsilon + 4\epsilon\left(\log\frac{1}{\epsilon}\right)^{1/2} d^{1/2} + 4\epsilon d^{1/2}. \end{aligned}$$

We now return to the block sums $X_j = p^{-1/2}\sum_{\nu\in H_j}\xi_\nu$, $1 \le j \le l$. In order to estimate the Prohorov distance of the distribution of the normalized partial sum $l^{-1/2}\sum_{j\le l} X_j$ to the normal distribution $N(0, l^{-1}\sum_{j\le l}\mathrm{Cov}(X_j))$, we apply (3.55). As the random variables X_i are defined in terms of blocks that are separated by small blocks of length q, and using $\alpha(k) \ll k^{-3}$ we obtain $\alpha = \alpha(q) \ll n^{-3/4}$. If we put $\epsilon = n^{-1/15}$, we get

$$\begin{aligned} &\pi(\mathcal{L}(l^{-1/2}\sum_{j\le l} X_j), N(0, l^{-1}\sum_{j\le l}\mathrm{Cov}X_j)) \\ &\ll n^{-1/2}n^{1/4}n^{1/15}d^{-1/2}b + n^{-1/4}n^{2/15}d^{3/2}b^2 + n^{-1/4}n^{3/15}d^{-1/2}b^3 + n^{-1/15} \\ &\quad +n^{-1/15}(\log n^{1/15})^{1/2}d^{1/2} + n^{-1/15}d^{1/2} \\ &\ll n^{-1/20}d^{3/2}(1 + b^3), \end{aligned}$$

where, as above $b = \sup_\nu \|\xi_\nu\|_3$.

To complete the proof of Theorem 3.27, two steps remain to be carried out. First we need an estimate for the difference between the normalized partial sum $\frac{1}{\sqrt{n}}\sum_{\nu=1}^n \xi_\nu$ and the corresponding normalized long block sum $\frac{1}{\sqrt{l}}\sum_{j=1}^l X_j$. Secondly, we need an estimate for the distance between the two normal laws with mean zero, but with different covariance matrices, namely $\mathrm{Cov}(l^{-1/2}\sum_{j\le l} X_j)$ and $l^{-1}\sum_{j\le l}\mathrm{Cov}(X_j)$. Regarding the first issue, we can assess the effect of deleting small blocks using the moment estimate (3.34) together with Minkowski's inequality. We get, using Lyapunov's inequality,

$$\begin{aligned}
\left\| \frac{1}{\sqrt{l}}\sum_{j=1}^l X_j - \frac{1}{\sqrt{n}}\sum_{\nu=1}^n \xi_\nu \right\|_2 &\le \left\| (lp)^{-1/2}\sum_{j=1}^l \sum_{\nu\in H_j} \xi_\nu - n^{-1/2}\sum_{\nu=1}^n \xi_\nu \right\|_2 \qquad (3.56)\\
&\le \frac{1}{\sqrt{n}}\left\| \sum_{j=1}^{l+1}\sum_{\nu\in I_j}\xi_\nu \right\|_2 + ((lp)^{-1/2} - n^{-1/2})\left\| \sum_{j=1}^{l}\sum_{\nu\in H_j} \right\|_2\\
&\ll [n^{-1/2}(lq+p+q)^{1/2} + (lp)^{1/2}((lp)^{-1/2} - n^{-1/2})]b\\
&\ll [n^{-1/2}(n^{3/4}+n^{1/2})^{1/2} + n^{-1}(n-lp)]b\\
&\ll [n^{-1/8} + n^{-1}(n^{3/4}+n^{1/2})]b\\
&\ll n^{-1/8}b,
\end{aligned}$$

and hence by (3.50),

$$\pi\left(\mathcal{L}(l^{-1/2}\sum_{j=1}^l X_j), \mathcal{L}(n^{-1/2}\sum_{\nu=1}^n \xi_\nu)\right) \ll n^{-1/12}b^{2/3}.$$

What remains is to estimate the Prohorov distance of the mean-zero Gaussian measures on $\mathbb{R}^d$ with covariance matrix Σ_n and $\Sigma_{n,1} := l^{-1}\sum_{i=1}^l \mathrm{Cov}(X_i)$. In addition, we introduce the covariance matrix $\Sigma_{n,2} := \mathrm{Cov}(l^{-1/2}\sum_{j=1}^l X_j)$. At this point, we need two further lemmas from Dehling (1983) which we state without proof.

Lemma 3.29. *Let X and Y be $\mathbb{R}^d$-valued square integrable random variables with mean zero. Then we have the following estimate for the Prohorov distance of the corresponding Gaussian distributions:*

$$\pi(N(0,\mathrm{Cov}(X)), N(0,\mathrm{Cov}(Y))) \le (E\|X-Y\|^2)^{1/3}. \qquad (3.57)$$

Lemma 3.30. *Let μ and ν be two Gaussian measures on $\mathbb{R}^d$ with mean zero and covariance operators Σ and Γ. Then the following estimate for the Prohorov distance holds:*

$$\pi(\mu,\nu) \le C\|\Sigma-\Gamma\|_1^{1/3} d^{1/6}\left(1 + |\log\left(\|\Sigma-\Gamma\|_1^{-1}\right) d|^{1/2}\right) \qquad (3.58)$$

where C is some absolute constant and where $\|\cdot\|_1$ denotes the trace class norm of a matrix, i.e., the sum of the absolute eigenvalues.

By Lemma 3.29 and (3.56) we get

$$\pi\left(N(0,\Sigma_n), N(0,\Sigma_{n,2})\right) \ll n^{-1/12}b^{2/3}.$$

To obtain a bound for the Prohorov distance of $N(0, \Sigma_{n,2})$ and $N(0, \Sigma_{n,1})$, we want to apply Lemma 3.30. Thus we need a bound on the trace class norm $\|\Sigma_{n,1} - \Sigma_{n,2}\|_1$. Let $e_1, \dots, e_d$ denote any orthonormal basis for $\mathbb{R}^d$. Then

$$\begin{aligned}(\Sigma_{n,2}e_k, e_k) &= E\left(\frac{1}{\sqrt{l}}\sum_{j\le l} X_j, e_k\right)^2 = \frac{1}{l}\sum_{j\le l} E(X_j, e_k)^2 + \frac{1}{l}\sum_{i\ne j} E(X_i, e_k)(X_i, e_k) \\ (\Sigma_{n,1}e_k, e_k) &= \frac{1}{l}\sum_{j\le l} E(X_j, e_k)^2,\end{aligned}$$

and thus we get $|((\Sigma_{n,2} - \Sigma_{n,1})e_k, e_k)| \le 2l^{-1}\sum_{1\le i<j\le l} |E(X_i, e_k)(X_j, e_k)|$. Applying (3.16) with $r = s = t = 3$ to the expectation on the r.h.s., we finally obtain

$$\begin{aligned}|((\Sigma_{n,2} - \Sigma_{n,1})e_k, e_k)| &\ll \frac{1}{l}\sum_{1\le i<j\le l} \alpha^{1/3}\,((j-i-1)\,p)\,b^2 \\ &= \frac{1}{l}\sum_{i=1}^{l} \alpha^{1/3}\,((i-1)p)\,(l-i)\,b^2,\end{aligned}$$

as the block sums X_i and X_j are separated by at least $(|j-i|-1)$ blocks of length p, and thus by $(j-i-1)\,p$ of the original random variables. Hence

$$|((\Sigma_{n,2} - \Sigma_{n,1})e_k, e_k)| \ll \frac{1}{l}n^{-1/4}lb^2 + l^{-1}\sum_{i=2}^{l} i^{-1}n^{-1/2}(l-i)b^2 \ll \frac{1}{l}n^{-1/4}b^2$$

and therefore $\|\Sigma_{n,2} - \Sigma_{n,1}\|_1 \ll dn^{-1/4}b^2$. Now, we apply Lemma 3.30 to get

$$\pi\left(N\left(0, \operatorname{Cov}\left(\frac{1}{\sqrt{l}}\sum_{j\le l} X_j\right)\right), N\left(0, \frac{1}{l}\sum_{j\le l}\operatorname{Cov} X_j\right)\right) \ll d^{1/4}n^{-1/16}b^{1/2},$$

thus finishing the proof of Theorem 3.27. □

3.7 A central limit theorem for functionals. Here we quote, without proof, a central limit theorem for functionals of absolutely regular processes, as stated in Borovkova, Burton and Dehling (2001).

Theorem 3.31. *Let $(X_k)_{k\in\mathbb{Z}}$ be a 1-approximating functional with constants $(a_k)_{k\ge0}$ of an absolutely regular process with mixing coefficients $(\beta_k)_{k\ge0}$. Suppose moreover that $EX_i = 0$, $E|X_0|^{4+\delta} < \infty$ and that*

$$\sum_{k=1}^{\infty} k^2\left(a_k^{\delta/(3+\delta)} + \beta_k^{\delta/(4+\delta)}\right) < \infty, \tag{3.59}$$

for some $\delta > 0$. Then, as $n \to \infty$,

$$\frac{1}{\sqrt{n}}\sum_{i=1}^{n} X_i \xrightarrow{d} \mathcal{N}(0, \sigma^2),$$

where $\sigma^2 = EX_0^2 + 2\sum_{k=1}^{\infty} E(X_0 X_k)$. *In the case* $\sigma^2 = 0$, *we adopt the convention that* $\mathcal{N}(0, 0)$ *denotes the point mass at the origin. If* X_0 *is bounded, the CLT continues to hold if (3.59) is replaced by the condition*

$$\sum_{k=1}^{\infty} k^2 (a_k + \beta_k) < \infty. \tag{3.60}$$

4 Empirical processes for mixing random variables

There is a vast literature on this subject, starting with Billingsley (1968, Section 22), where he proved weak convergence of the empirical process for functions of a stationary sequence of ϕ-mixing random variables to a tied-down Gaussian process with a suitable covariance function. Billingsley's result generalizes Donsker's theorem (1952) who proved weak convergence of the empirical process of i.i.d. random variables with uniform distribution over (0, 1) to the standard Brownian bridge. Billingsley's theorem was extended in turn by Sen (1971) and by Deo (1973). Deo weakened the mixing condition to strong mixing.

4.1 Weak convergence for empirical processes of strongly mixing random variables. As a warm-up we present a generalization of Donsker's invariance principle for empirical processes (Theorem 2.9) to sequences of strongly mixing random variables, due to Deo (1973).

Theorem 4.1. *Let* $\{\xi_j, j \geq 1\}$ *be a strictly stationary sequence of random variables satisfying the strong mixing condition (3.2) with rate*

$$a(n) \ll n^{-6-\epsilon}, \tag{4.1}$$

for some $0 < \epsilon < 1$. *Let* F *denote the common distribution function of the* ξ_j*'s. For* $s \in \mathbb{R}$ *write* $g_n(s) := 1_{\{\xi_n \leq s\}} - F(s)$. *Then the two series below defining the covariance function*

$$\Gamma(s, s') := Eg_1(s)g_1(s') + \sum_{n \geq 2} Eg_1(s)g_n(s') + \sum_{n \geq 2} Eg_n(s)g_1(s') \tag{4.2}$$

converge absolutely for $s, s' \in \mathbb{R}$. *Moreover, if* F_n *denotes the empirical distribution function, we get*

$$\sqrt{n}\,(F_n - F) \to \zeta \tag{4.3}$$

in distribution, where ζ *is a mean-zero Gaussian process on* $\mathbb{R}$ *with covariance function*

$$E\zeta(s)\zeta(s') = \Gamma(s, s'), \quad s, s' \in \mathbb{R}.$$

Again, as in the case of independent processes, we may assume without loss of generality that the underlying process $(\xi_j)_{j \geq 1}$ has a uniform distribution on [0, 1] as its 1-dimensional marginals. If the ξ_j have a continuous distribution, this follows immediately from the representation $\xi_j = F^{-1}(F(\xi_j))$, together with the observation that $F(\xi_j)$ has a uniform distribution on [0, 1], whenever F is continuous. For possibly noncontinuous marginal distributions, we can apply the following lemma.

Lemma 4.2. *Let* $(\xi_j)_{j \geq 1}$ *be a stationary sequence of random variables satisfying any of the above mixing conditions with rate* $\gamma(n)$. *Then there exists a stationary sequence* $(\eta_j)_{j \geq 1}$ *of* *random variables satisfying the same mixing condition with the same rate* $\gamma(n)$ *such that each* η_j *has uniform distribution on* $[0, 1]$ *and*

$$\xi_j = F^{-1}(\eta_j), \quad j = 1, 2, \ldots.$$

Here F *is the common distribution function of the* ξ_j*'s.*

Proof. Let $(U_j)_{j\geq 1}$ be a sequence of independent random variables uniformly distributed on $[0, 1]$ and independent of $(\xi_j)_{j\geq 1}$. Now define η_j in such a way that conditionally given $\xi_j = x$, η_j is uniformly distributed on $(F(x-), F(x)]$. Specifically define

$$h(x, u) = F(x-) + u\{F(x) - F(x-)\},$$

and define $\eta_j = h(\xi_j, U_j)$. It is easily verified that η_j is indeed uniformly distributed on $[0, 1]$. Moreover, the construction guarantees that the η_j have the same mixing properties as the ξ_j. Finally, we have that $\xi_j = F^{-1}(\eta_j)$, for $j = 1, 2, \ldots$. □

Proof of Theorem 4.1. In view of Lemma 4.2 we may assume without loss of generality that $F(s) = s, 0 \leq s \leq 1$. We apply Theorem 2.7 much in the same way as in Theorem 2.9, the case of independent random variables. First we observe that the finite dimensional distributions converge properly as a consequence of Theorem 3.23. Second to check condition (ii) of Theorem 2.7 we proceed as in the proof of Theorem 2.9. But instead of the ad hoc estimate of the fourth moments, we apply Lemma 3.22 instead. We obtain, with $\alpha = 1 + \epsilon/10 > 1$,

$$E|Y_n(t) - Y_n(s)|^4 \leq \frac{10^3 C}{\epsilon}\left((t - s)^{\alpha} + \frac{1}{n}(t - s)^{\alpha/2}\right).$$

Hence (2.3) holds with $\gamma = 4$, $\beta = \frac{1}{2}\alpha$ and $q = 1 > \frac{1}{4}\alpha = \frac{1}{2}(\alpha - \beta)$.

4.2 Strong approximation. A strong approximation theorem for the empirical process for functionals of a stationary sequence of strongly mixing random variables was established by Berkes and Philipp (1977). Except possibly for the mixing rate this result improved upon the previously known ones as well as on a functional LIL of Philipp (1977).

We present here a version for a strictly stationary strongly mixing sequence of d-dimensional random vectors with an arbitrary common distribution. For two vectors $x = (x_1, \ldots, x_d)$ and $y = (y_1, \ldots, y_d)$ we write $x \leq y$ if $x_i \leq y_i$, $1 \leq i \leq d$.

Theorem 4.3. *Let* $\{\xi_j, j \geq 1\}$ *be a strictly stationary sequence of* d*-dimensional random vectors satisfying the strong mixing condition (3.2) with rate*

$$a(n) \ll n^{-4-2d}. \tag{4.4}$$

Suppose there exists a random variable U*, uniformly distributed on* $(0, 1)$*, and independent of the sequence* (ξ_j)*. Let* F *denote the common distribution function of the* ξ_j*'s. For* $s \in \mathbb{R}^d$ *write* $g_n(s) := 1_{\{\xi_n \leq s\}} - F(s)$*. Then the two series below defining the covariance function*

$$\Gamma(s, s') := Eg_1(s)g_1(s') + \sum_{n\geq 2} Eg_1(s)g_n(s') + \sum_{n\geq 2} Eg_n(s)g_1(s') \tag{4.5}$$

converge absolutely for $s, s' \in \mathbb{R}^d$*. Moreover, there exists a sequence* $\{\zeta_j, j \geq 1\}$ *of independent identically distributed Gaussian processes, defined on the same probability space* $(\Omega, \mathcal{F}, P)$ *as* $(\xi_j)_{j\geq 1}$*, indexed by* $s \in \mathbb{R}^d$*, with*

$$E\zeta_1(s) = 0, \quad E\zeta_1(s)\zeta_1(s') = \Gamma(s, s'), \quad s, s' \in \mathbb{R}^d,$$

and a positive constant λ depending on d only such that

$$\sup_{s\in\mathbb{R}^d}\left|\sum_{j\le n}\left[1_{\{\xi_j\le s\}}-F(s)-\zeta_j(s)\right]\right| \ll n^{1/2}(\log n)^{-\lambda}, \tag{4.6}$$

with probability 1.

This result is due to Philipp (1982). In that paper it was derived as a corollary to a general approximation theorem for Banach space valued random elements, which are not necessarily measurable and where the Banach space need not be separable.

We shall present a sketch of the proof of Theorem 4.3 for the special case of real-valued random variables, i.e., $d = 1$. We first observe that, as a consequence of Lemma 4.2, there is no loss of generality to assume that the ξ_j's have uniform distribution on $[0, 1]$. Indeed, suppose we have proved the conclusion of Theorem 4.3 for the special case of uniformly distributed random variables $(\eta_j)_{j\ge1}$, i.e.,

$$\sup_{0\le s\le 1}\left|\sum_{j\le n}\left[1_{\{\eta_j\le s\}}-s-Z_j(s)\right]\right| \ll n^{\frac{1}{2}}(\log n)^{-\lambda} \quad \text{a.s.}$$

where $(Z_j(s))_{j\ge0}$ is a sequence of independent Gaussian processes with covariance function defined by (4.5), but with

$$g_n(s) := 1_{\{\eta_j\le s\}} - s.$$

Then we obtain the general result by replacing s by $F(s)$.

We use this opportunity to introduce the concept of a Kiefer process. Let $\{\eta_n, n \ge 1\}$ be a sequence of random variables uniformly distributed on $[0, 1]$. The empirical process of $\{\eta_n, n \ge 1\}$ is defined by

$$R(s,t) = \sum_{j\le t}\left(1_{\{\eta_j\le s\}} - s\right). \tag{4.7}$$

A separable Gaussian process $K(s, t)$ on $[0, 1] \times [0, \infty)$ is called a Kiefer process if $K(0, t) = K(1, t) = K(s, 0) = 0$ for all $0 \le s \le 1, t \ge 0$,

$$EK(s,t) = 0 \tag{4.8}$$

and

$$E\{K(s,t)K(s',t')\} = \min(t,t')\,\Gamma(s,s') \tag{4.9}$$

where $\Gamma(s, s')$ is the covariance function of a separable Gaussian process $G(s)$ on $[0, 1]$ with $G(0) = G(1) = 0$. Thus in this special case (4.6) can be reformulated as follows: there exists a Kiefer process $K(s, t)$ such that

$$\sup_{0\le s\le 1}\sup_{0\le t\le T}|R(s,t)-K(s,t)| \ll T^{\frac{1}{2}}(\log T)^{-\lambda} \quad \text{a.s.} \tag{4.10}$$

for some $\lambda > 0$.

For independent random variables η_n this result, with a much sharper error term, is due to Kiefer (1972). Indeed, it is easy to see that if $\xi_n = \eta_n$ are independent random variables uniformly distributed on [0, 1], then $\Gamma(s, s')$ in (4.5) reduces to $\Gamma(s, s') = s(1-s')$ for $0 \le s \le s' \le 1$. Hence in this case (4.10) holds, with the corresponding Kiefer process $K(s, t)$ reducing to a Brownian motion along the t-axis and to a Brownian bridge on [0, 1] along the s-axis. Kiefer's proof is based on a generalization of the Skorohod embedding theorem. The same result, with an error term slightly weaker than Kiefer's, but still much sharper than ours, has been established by Csörgő and Revész (1975) using a totally different method. Their method, based on the so-called quantile transform, was refined by Komlós, Major and Tusnády (1975) who obtained an error term which is nearly best possible.

An immediate corollary of an almost sure invariance principle of the form (4.10) is a functional law of the iterated logarithm. For functionals of mixing sequences such a result was proved by Philipp (1977) in Theorem 3.2. Here we observe that (4.10) even implies a two-dimensional functional law of the iterated logarithm.

Theorem 4.4. *Let $\{\eta_n, n \ge 1\}$ be a sequence of uniformly distributed random variables whose empirical process satisfies* (4.10). *Then the sequence*

$$\{(2N \log\log N)^{-1/2} R(s, Nt), N \ge 1\}$$

of functions on $[0, 1]^2$ *is with probability* 1 *relatively compact in the supremum norm and has the unit ball B in the reproducing kernel Hilbert space $H(\Gamma^*)$ as its set of limits. Here* $\Gamma^*(s, s', t, t') = \min(t, t')\Gamma(s, s')$.

Theorem 4.4 follows at once from (4.10) and Theorem 2 of Lai (1974; see the remark on p. 19) which implies that the conclusion of Theorem 4.4 holds for the sequence $\{(2N \log\log N)^{-1/2} K(s, Nt), N \ge 1\}$.

We obtain an interesting special case of Theorem 4.4, also due to Kiefer (1972, p. 32) when we apply it to a sequence $\{\eta_n, n \ge 1\}$ of independent uniformly distributed random variables. In this case the unit ball B in the reproducing kernel Hilbert space reduces to the class of functions h on $[0, 1]^2$ with

$$h(t_1, t_2) = \int_0^{t_1} \int_0^{t_2} f(u, v)\, du dv,$$

where

$$\int_0^1 \int_0^1 f^2(u, v)\, du dv \le 1;$$

see Lai (1974, p. 19).

It is perhaps worth mentioning that Theorem 4.4 reduces to the standard functional law of the iterated logarithm if we put $t = 1$ in Theorem 4.4. For independent random variables this result is due to Finkelstein (1971).

Another obvious consequence of (4.10) is a functional central limit theorem. Again we obtain a two-dimensional version.

Theorem 4.5. *Let $\{\eta_n, n \ge 1\}$ be a sequence of random variables as in Theorem* 4.4. *Then the sequence $\{N^{-\frac{1}{2}} R(s, Nt), N \ge 1\}$ of functions on* $[0, 1]^2$ *converges in distribution to a Gaussian process $X(s, t)$ which on* $[0, 1]^2$ *has the same distribution as $K(s, t)$.*

Theorem 4.5 follows at once from (4.10) since

$$\sup_{0 \le s, t \le 1} |N^{-\frac{1}{2}} R(s, Nt) - N^{-\frac{1}{2}} K(s, Nt)| \ll (\log N)^{-\lambda}$$

and since $X(s, t) = N^{-\frac{1}{2}} K(s, Nt)$ has on $[0, 1]^2$ the same covariance as $K(s, t)$.

4.3 Outline of proof of Theorem 4.3. One of the main tools used to prove Theorem 4.3 is the following result which is proved in Berkes and Philipp (1979). Denote by $\langle a, b\rangle$ the inner product of the vectors a, b.

Theorem 4.6. *Let $(X_k)_{k\geq 1}$ be a sequence of random variables with $X_k \in \mathbb{R}^{d_k}$, $d_k \geq 1$, and let $\{\mathcal{F}_k, k \geq 1\}$ be a nondecreasing sequence of σ-fields such that X_k is $\mathcal{F}_k$-measurable. Finally, let $\{G_k, k \geq 1\}$ be a sequence of probability distributions on $\mathbb{R}^{d_k}$ with characteristic functions $g_k(u)$, $u \in \mathbb{R}^{d_k}$, respectively. Suppose that for some nonnegative numbers λ_k, δ_k, and $T_k \geq 10^8 d_k$,*

$$E|E\{\exp(i\langle u, X_k\rangle)|\mathcal{F}_{k-1}\} - g_k(u)| \leq \lambda_k \tag{4.11}$$

for all u with $|u| \leq T_k$, and that

$$G_k\left\{u : |u| > \frac{T_k}{4}\right\} \leq \delta_k. \tag{4.12}$$

Suppose that there exists a random variable U, uniformly distributed on $[0, 1]$ and independent of the sequence $(X_k)_{k\geq 1}$. Then there exists a sequence $(Y_k)_{k\geq 1}$ of independent random variables such that Y_k has distribution G_k and

$$P(|X_k - Y_k| \geq \beta_k) \leq \beta_k \tag{4.13}$$

where $\beta_1 = 1$ and

$$\beta_k = 10 d_k T_k^{-1} \log T_k + 4\lambda_k^{\frac{1}{2}} T_k^{d_k} + \delta_k \tag{4.14}$$

for $k \geq 2$. In fact, Y_k is measurable with respect to the σ-field $\sigma(X_k, U_k)$ and Y_k is independent of $\sigma(X_1, \ldots, X_{k-1}, U_1, \ldots, U_{k-1})$, $k \geq 1$. Here $(U_k)_{k\geq 1}$ is an i.i.d. sequence of uniformly distributed random variables independent of $(X_k)_{k\geq 1}$.

We remark in passing that this result has been successfully applied to a large variety of dependence structures, such as lacunary trigonometric sequences, stationary Gaussian sequences with covariance tending to zero, Hilbert space valued martingale difference sequences and to sequences satisfying a strong mixing condition. In this last case, this is where the theorem, in conjunction with Corollary 3.10, will be applied in the proof of Theorem 4.3.

As noted above, we shall outline the proof of Theorem 4.3 only for the special case of mixing random variables uniformly distributed on $[0, 1]$. The empirical process $R(s, t)$ is a partial sum of indicator random variables, centered at expectations, i.e.,

$$R(s, t) = \sum_{n\leq t} \left[1_{[0,s]}(\xi_n) - s)\right], \qquad 0 \leq s \leq 1,\ t \geq 0. \tag{4.15}$$

Obviously, the summands are also mixing.

We now choose an increasing sequence $\{t_k, k \geq 1\}$ of positive integers such that $t_k = \exp(k^{1-\epsilon})$, for some $\epsilon > 0$. Next, for fixed k, we partition the unit interval $0 \leq s \leq 1$ into d_k subintervals $[s_j, s_{j+1})$, $1 \leq j \leq d_k$ each of length $1/d_k$ where $d_k \sim k^{\frac{1}{2}}$. Then

$$Z(j, k) = R(s_j, t_{k+1}) - R(s_j, t_k), \qquad 1 \leq j \leq d_k \tag{4.16}$$

can be considered as the components $Z(j, k)$ of random vectors $Z_k \in \mathbb{R}^{d_k}$. We call the sequence $\{Z_k, k \geq 1\}$ the skeleton process of the process $R(s, t)$.

At some point one needs to show that the variation of the empirical process $R(s, t)$ on the rectangles

$$\{(s, t) : s_j \leq s \leq s_{j+1}, t_k \leq t \leq t_{k+1}\} \tag{4.17}$$

is reasonably small. The following exponential bound together with several routine arguments will serve this purpose.

Theorem 4.7. *Let $\{\xi_n, n \geq 1\}$ be a strictly stationary sequence of real-valued random variables, centered at expectations and uniformly bounded by* 1. *Suppose that $\{\xi_n, n \geq 1\}$ satisfies a strong mixing condition with rate*

$$\alpha(n) \leq Cn^{-\rho}$$

for some $C > 0$ and $\rho > 2$. Then

$$\sigma^2 := E\xi_1^2 + 2\sum_{n\geq 2} E\xi_1\xi_n$$

is finite. Let $A \geq \sigma^2$ and $\kappa > 0$. Then for any $R > 0$

$$P\left(\left|\sum_{n\leq N} \xi_n\right| > 14\, R\, N^{1/2}\right) \ll \exp(-R^2/A) + N^{1+\kappa-\frac{1}{2}\rho}(A^{-2} + R^{-2}).$$

Remark. The constant implied by $\ll$ depends on C and ρ only. This result is from Philipp (1982). Using this theorem one can show that the variation of $R(s, t)$ over the rectangles (4.17) is $\ll t_k^{1/2}(\log t_k)^{-\epsilon_1}$, almost surely, uniformly for all j with $1 \leq j \leq d_k$. Here and for the rest of this section $\epsilon_i, i = 1, 2, \ldots$ are certain positive numbers whose values can be calculated explicitly. Hence for the purpose of proving the basic relation (4.6), the skeleton process contains all the needed information about the process R.

Next, we define a sequence $(p_k)_{k\geq 1}$ of positive integers tending to ∞ faster than any power of k but much slower than t_k. For the purpose of this outline we only remark that the p_k's are chosen such that $N_k = t_{k+1} - t_k - p_k$ tends to ∞ much faster than p_k. We set

$$X(j, k) = N_k^{-1/2}(R(s_j, t_{k+1}) - R(s_j, t_k + p_k)), \qquad 1 \leq j \leq d_k. \tag{4.18}$$

Again we consider $X(j, k)$ as the j-th component of the random vector X_k. $X(j, k)$ is the properly normalized sum of N_k mixing random variables, centered at their expectations and uniformly bounded by 1. By Theorem 3.23 the distribution of the vector X_k tends to a multivariate normal distribution G_k with covariance matrix $\Gamma_{d_k} = (\Gamma(s_i, s_j), 1 \leq i, j \leq d_k)$. More precisely, the characteristic function of X_k satisfies

$$\left|E\{\exp(i\langle u, X_k\rangle)\} - \exp(-\frac{1}{2}\langle u, \Gamma_{d_k} u\rangle)\right| \ll N_k^{-\epsilon_2} \tag{4.19}$$

for all $u \in \mathbb{R}^{d_k}$ with $|u| \leq N_k^{\epsilon_3}$.

The purpose for introducing the sequence $\{p_k, k \geq 1\}$ was to achieve a certain amount of separation between the random vectors X_k. Let $\mathcal{F}_k = \mathcal{M}_1^{t_{k+1}}$. Then X_k is $\mathcal{F}_k$-measurable and it follows from Corollary 3.10 that

$$E\,|E\{\exp(i\langle u, X_k\rangle)|\mathcal{F}_{k-1})\} - E\{\exp(i\langle u, X_k\rangle)\}| \ll \alpha(p_k)$$

for all $u \in \mathbb{R}^{d_k}$. This together with (4.19) shows that condition (4.11) is satisfied with $g_k(u) = \exp(-\frac{1}{2}\langle u, \Gamma_{d_k} u\rangle)$. It is easy to see that $G_k(u)$ satisfies (4.12). Thus by (4.14), $\beta_k \ll t_k^{-\epsilon_4}$ so that $\sum \beta_k < \infty$. Hence by Theorem 4.6 there exists a sequence $\{Y_k, k \geq 1\}$ of independent random vectors with distribution G_k such that (4.13) holds. Thus by the Borel–Cantelli lemma

$$|X_k - Y_k| \ll \beta_k, \qquad k = 1, 2, \ldots \quad \text{a.s.} \tag{4.20}$$

By (4.15), (4.16) and (4.18), $N_k^{\frac{1}{2}} X_k - Z_k$ is the sum of p_k random vectors with components uniformly bounded by 1. Hence

$$|N_k^{\frac{1}{2}} X_k - Z_k| \leq d_k p_k.$$

This, together with (4.20) yields, after a small calculation,

$$\left| Z_k - (t_{k+1} - t_k)^{\frac{1}{2}} Y_k \right| \ll t_k^{\frac{1}{2}} / \log t_k \quad \text{a.s.} \tag{4.21}$$

This means that the skeleton process $\{Z_k, k \geq 1\}$ can be approximated by a sequence $\{(t_{k+1} - t_k)^{\frac{1}{2}} Y_k, k \geq 1\}$ of independent normal random vectors with covariance matrix $(t_{k+1} - t_k)\Gamma_{d_k}$. By Lemma 4.14 below this sequence can be extended to a Kiefer process $K(s, t)$ with covariance function $\min(t, t')\Gamma(s, s')$. The sequence $\{(t_{k+1} - t_k)^{\frac{1}{2}} Y_k,\ k \geq 1\}$ in turn can be considered as the skeleton process of K. One can show by means of a generalized Fernique lemma, due to Lai (1974) that the variation of $K(s, t)$ on the rectangles (4.17) is bounded with probability 1. Theorem 4.3 follows from this last remark, from (4.21) and the statements about the variation of the empirical process and the Kiefer process.

4.4 Generalizations. Results more general than Theorem 4.3 are known. We shall point out a few of them. Let $(A, \mathcal{A}, P)$ be a probability space, and let $\mathcal{C} \subset \mathcal{A}$ be a class of sets. For each $\delta > 0$, let $N_I(\delta, \mathcal{C}, P)$ be the smallest number of sets $A_1, \ldots, A_d$ with the following property: for each $C \in \mathcal{C}$ there exist A_r, A_s, $1 \leq r, s, \leq d$, with $A_r \subset C \subset A_s$ and $P(A_s \setminus A_r) < \delta$. Recall that $\log N_I$ is called a metric entropy with inclusion.

The empirical measure P_n is defined as

$$P_n(B) := \frac{1}{n} \sum_{j \leq n} 1_{\{\xi_j \in B\}}, \qquad B \in \mathcal{A}$$

and the normalized empirical measure ν_n as

$$\nu_n := n^{1/2}(P_n - P).$$

If $N_I(\delta)$ grows at most polynomially in $1/\delta$ and if the mixing rate (4.4) is sufficiently small, then the empirical process can be approximated with probability 1 uniformly by a sum of independent Gaussian processes indexed by $C \in \mathcal{C}$ (see Philipp (1982)).

More generally, let $\mathcal{F}$ be a class of functions. Let $N_I(\delta, \mathcal{F}, P)$ denote the smallest number m such that for some $f_1, \ldots, f_m$ and all $f \in \mathcal{F}$ there are $i, j \leq m$ with $f_i \leq f \leq f_j$ and $\int (f_j - f_i) dP < \delta$. Then $\log N_I$ is called a metric entropy with bracketing. An idea of this sort was advanced by Philipp (1971, pp. 18–19). There a uniform LIL was proved for classes of random variables with metric entropy with bracketing not growing faster than linear polynomials in $1/\delta$ and where the random variables satisfy a ψ-mixing condition with exponential rate. For a stronger version in the independent case see, Philipp (1973).

Other important structures of classes of sets and classes of functions are the Vapnik–Chervonenkis classes. A collection $\mathcal{C}$ of measurable sets is called a Vapnik–Chervonenkis class if for some finite n no set F with n elements has all its subsets of the form $F \cap C$, $C \in \mathcal{C}$. The Vapnik–Chervonenkis number $V(\mathcal{C})$ denotes the smallest such n. It does not depend on the underlying probability measure P, in contrast to metric entropy classes.

4.5 Empirical process of functionals. In this subsection, we will investigate the empirical process of functionals of an absolutely regular process. For later applications, we will consider the generalized empirical process, defined by

$$W_n(t) = \sqrt{n}\left(\frac{1}{n}\sum_{i=1}^{n} u_t(X_i) - \int u_t(x)dF(x)\right), \tag{4.22}$$

where $\mathcal{U} = \{u_t,\ t \in [0, t_0]\}$ is some one-parameter class of functions. Note that by taking $u_t = 1_{(-\infty,t]}$ we regain the usual empirical process. We will later specifically be interested in the functions $u_t(x) = P(|Y - x| \leq t)$, but at this point we will consider a general class $\mathcal{U}$, satisfying the following two properties for all $x \in \mathbb{R}$:

$$0 \leq u_t(x) \leq 1, \text{ for all } t \in [0, t_0], \tag{4.23}$$

$$u_s(x) \leq u_t(x), \text{ for } 0 \leq s \leq t \leq t_0. \tag{4.24}$$

Without loss of generality we may assume that $t_0 = 1$ so that the process W_n is indexed by $[0, 1]$. We define

$$G(t) = Eu_t(X_0) \text{ and } g_t(x) = u_t(x) - G(t).$$

We assume that G is a Lipschitz function on $[0, 1]$, i.e., that there exists a constant C such that, for all $t, s \in [0, 1]$,

$$|G(t) - G(s)| \leq C\,|t - s|. \tag{4.25}$$

We view $\{W_n(t), 0 \leq t \leq 1\}$ as a random element of $D[0, 1]$, the space of right-continuous functions on $[0, 1]$ with left limits, equipped with the Skorohod topology. Recall the definition of an r-approximating functional (Definition 3.13) as well as of a p-continuous function (Definition 3.15). This section is based on Borovkova et al. (2001), where further details can be found.

Theorem 4.8. *Let $(X_n)_{n\in\mathbb{N}}$ be a 1-approximating functional with constants $(a_k)_{k\geq 0}$ of an absolutely regular process with mixing coefficients $(\beta_k)_{k\geq 0}$ satisfying*

$$\sum_{k=1}^{\infty} k^2 \beta_k^{\delta/(2+\delta)} < \infty, \tag{4.26}$$

for some $0 < \delta < 1$. Let $\mathcal{U}$ be a class of functions satisfying (4.23) and (4.24), and assume that $u_t, t \in [0, 1]$ are 1-continuous in the sense of Definition 3.15, with the same γ for all t. Assume, moreover, that

$$\sum_{k=1}^{\infty} k^2 a_k^{\delta/(2+2\delta)} < \infty \text{ and } \sum_{k=1}^{\infty} k^2 \left(\gamma(\sqrt{2\bar{a}_k})\right)^{\delta/(1+\delta)} < \infty. \tag{4.27}$$

Assume, moreover, that the function $G(t)$ satisfies the condition (4.25). *Then the empirical process $\{W_n(t),\ t \in [0,1]\}_{n\in\mathbb{N}}$, defined in* (4.22), *converges weakly in $D[0,1]$ to the mean-zero Gaussian process $\{W(t),\ t \in [0,1]\}$ with covariance structure*

$$\begin{aligned}\mathrm{Cov}(W(s), W(t)) &= \mathrm{Cov}(u_s(X_1), u_t(X_1)) \\ &+ \sum_{k=1}^{\infty} \mathrm{Cov}(u_s(X_1), u_t(X_{k+1})) + \sum_{k=1}^{\infty} \mathrm{Cov}(u_s(X_{k+1}), u_t(X_1)).\end{aligned} \tag{4.28}$$

Moreover, the series on the r.h.s. of (4.28) *converges absolutely, and the limiting process W has continuous sample paths with probability* 1.

The key tool in the proof of Theorem 4.8 is the following lemma bounding fourth moments of partial sums $S_n = \sum_{i=1}^{n} Y_i$ of functionals of an absolutely regular process.

Lemma 4.9. *Let $(Y_k)_{k\in\mathbb{Z}}$ be a* 1*-approximating functional with constants $(a_k)_{k\geq 0}$ of an absolutely regular process with mixing coefficients $(\beta_k)_{k\geq 0}$ satisfying*

$$\sum_{k=1}^{\infty} k^2 \left(a_k^{\delta/(1+\delta)} + \beta_k^{\delta/(2+\delta)} \right) < \infty.$$

Suppose, moreover, that $|Y_k| \leq 1$ and $EY_k = 0$. Then for the partial sums $S_n = Y_1 + \cdots + Y_n$, we have

$$ES_n^4 \leq C \left(n \, \|Y_0\|_{2+\delta}^{(2+\delta)/(1+\delta)} + n^2 \, \|Y_0\|_{2+\delta}^{(4+2\delta)/(1+\delta)} \right),$$

for some constant C depending only on $(a_k)_{k\geq 0}$ and $(\beta_k)_{k\geq 0}$.

Proof. Due to stationarity of the process $(Y_n)_{n\in\mathbb{Z}}$ we obtain

$$ES_n^4 \leq 4!\, n \sum_{i,j,k\geq 0,\, i+j+k\leq n-1} E(Y_0 Y_i Y_{i+j} Y_{i+j+k}).$$

By Lemma 3.14 and using $EY_0 = 0$, we get the following upper bounds on the expected values inside the sum on the r.h.s.:

$$\begin{aligned}|E\left\{Y_0(Y_i Y_{i+j} Y_{i+j+k})\right\}| &\leq 6\,\beta_{\lfloor \frac{i}{3} \rfloor}^{\delta/(2+\delta)} \|Y_0\|_{2+\delta}^2 + 8\, a_{\lfloor \frac{i}{3} \rfloor}^{\delta/(1+\delta)} \|Y_0\|_{2+\delta}^{(2+\delta)/(1+\delta)}, \\ |E\left\{(Y_0 Y_i Y_{i+j}) Y_{i+j+k}\right\}| &\leq 6\,\beta_{\lfloor \frac{k}{3} \rfloor}^{\delta/(2+\delta)} \|Y_0\|_{2+\delta}^2 + 8\, a_{\lfloor \frac{k}{3} \rfloor}^{\delta/(1+\delta)} \|Y_0\|_{2+\delta}^{(2+\delta)/(1+\delta)}.\end{aligned}$$

Alternatively, we can split $E(Y_0 Y_i Y_{i+j} Y_{i+j+k})$ in the middle as $E\{(Y_0 Y_i)(Y_{i+j} Y_{i+j+k})\}$ and apply (3.25) together with (3.23) to get the bound

$$\begin{aligned}\left|E\left\{Y_0 Y_i Y_{i+j} Y_{i+j+k}\right\}\right| &\leq E(Y_0 Y_i) E(Y_{i+j} Y_{i+j+k}) \\ &\quad + 4\,\beta_{\lfloor \frac{j}{3} \rfloor}^{\delta/(2+\delta)} \|Y_0\|_{2+\delta}^2 + 8\, a_{\lfloor \frac{j}{3} \rfloor}^{\delta/(1+\delta)} \|Y_0\|_{2+\delta}^{(2+\delta)/(1+\delta)} \\ &\leq \left(2\|Y_0\|_{2+\delta}^2 \beta_{\lfloor \frac{i}{3} \rfloor}^{\delta/(2+\delta)} + 4 a_{\lfloor \frac{i}{3} \rfloor}^{\delta/(1+\delta)} \|Y_0\|_{2+\delta}^{(2+\delta)/(1+\delta)} \right) \\ &\quad \times \left(2\|Y_0\|_{2+\delta}^2 \beta_{\lfloor \frac{k}{3} \rfloor}^{\delta/(2+\delta)} + 4\, a_{\lfloor \frac{k}{3} \rfloor}^{\delta/(1+\delta)} \|Y_0\|_{2+\delta}^{(2+\delta)/(1+\delta)} \right) \\ &\quad + 4\,\beta_{\lfloor \frac{j}{3} \rfloor}^{\delta/(2+\delta)} \|Y_0\|_{2+\delta}^2 + 8\, a_{\lfloor \frac{j}{3} \rfloor}^{\delta/(1+\delta)} \|Y_0\|_{2+\delta}^{(2+\delta)/(1+\delta)}.\end{aligned}$$

Much in the same way as in the proof of Lemma 3.22, we use these three bounds on $E(Y_0 Y_i Y_{i+j} Y_{i+j+k})$ to obtain an upper bound on the terms in the expansion of ES_n^4. We get

$$\begin{aligned} ES_n^4 \ &\leq \ Cn\Bigg\{ \sum_{0\leq j,k\leq i\leq n-1} \left(\beta_{\lfloor \frac{i}{3}\rfloor}^{\delta/(2+\delta)}\|Y_0\|_{2+\delta}^2 + a_{\lfloor \frac{i}{3}\rfloor}^{\delta/(1+\delta)}\|Y_0\|_{2+\delta}^{(2+\delta)/(1+\delta)}\right) \\ &\quad + \sum_{0\leq i,k\leq j\leq n-1} \left(\beta_{\lfloor \frac{i}{3}\rfloor}^{\delta/(2+\delta)}\|Y_0\|_{2+\delta}^2 + a_{\lfloor \frac{i}{3}\rfloor}^{\delta/(1+\delta)}\|Y_0\|_{2+\delta}^{(2+\delta)/(1+\delta)}\right) \\ &\quad \times \left(\beta_{\lfloor \frac{i}{3}\rfloor}^{\delta/(2+\delta)}\|Y_0\|_{2+\delta}^2 + a_{\lfloor \frac{i}{3}\rfloor}^{\delta/(1+\delta)}\|Y_0\|_{2+\delta}^{(2+\delta)/(1+\delta)}\right)\Bigg\}, \end{aligned} \tag{4.29}$$

where C is some universal constant. Now we note that

$$\sum_{0\leq j,k\leq i\leq n-1} \left(\beta_{\lfloor \frac{i}{3}\rfloor}^{\delta/(2+\delta)} + a_{\lfloor \frac{i}{3}\rfloor}^{\delta/(2+\delta)}\right) \leq \sum_{i=1}^{n} i^2 \left(\beta_{\lfloor \frac{i}{3}\rfloor}^{\delta/(2+\delta)} + a_{\lfloor \frac{i}{3}\rfloor}^{\delta/(2+\delta)}\right) < \infty$$

by the assumptions of the lemma. Hence

$$\begin{aligned} &\sum_{0\leq j,k\leq i\leq n-1} \left(\beta_{\lfloor \frac{i}{3}\rfloor}^{\delta/(2+\delta)}\|Y_0\|_{2+\delta}^2 + a_{\lfloor \frac{i}{3}\rfloor}^{\delta/(1+\delta)}\|Y_0\|_{2+\delta}^{(2+\delta)/(1+\delta)}\right) \\ &\qquad \leq \|Y_0\|_{2+\delta}^{(2+\delta)/(1+\delta)} \sum_{0\leq j,k\leq i\leq n-1} \left(\beta_{\lfloor \frac{i}{3}\rfloor}^{\delta/(2+\delta)} + a_{\lfloor \frac{i}{3}\rfloor}^{\delta/(1+\delta)}\right) \leq C\|Y_0\|_{2+\delta}^{(2+\delta)/(1+\delta)} \end{aligned} \tag{4.30}$$

as $(2+\delta)/(1+\delta) < 2$ and $\|Y_0\|_{2+\delta} \leq 1$. Concerning the second term on the r.h.s. of (4.29) we obtain the upper bound

$$\begin{aligned} &\sum_{0\leq i,k\leq j\leq n-1} \left(\beta_{\lfloor \frac{i}{3}\rfloor}^{\delta/(2+\delta)}\|Y_0\|_{2+\delta}^2 + a_{\lfloor \frac{i}{3}\rfloor}^{\delta/(1+\delta)}\|Y_0\|_{2+\delta}^{(2+\delta)/(1+\delta)}\right) \\ &\qquad \times \left(\beta_{\lfloor \frac{k}{3}\rfloor}^{\delta/(2+\delta)}\|Y_0\|_{2+\delta}^2 + a_{\lfloor \frac{k}{3}\rfloor}^{\delta/(1+\delta)}\|Y_0\|_{2+\delta}^{(2+\delta)/(1+\delta)}\right) \\ &\leq n\left\{\sum_{i=1}^{\infty} \left(\beta_{\lfloor \frac{i}{3}\rfloor}^{\delta/(2+\delta)}\|Y_0\|_{2+\delta}^2 + a_{\lfloor \frac{k}{3}\rfloor}^{\delta/(1+\delta)}\|Y_0\|_{2+\delta}^{(2+\delta)/(1+\delta)}\right)\right\}^2 \leq Cn\,\|Y_0\|_{2+\delta}^{(4+2\delta)/(1+\delta)}, \end{aligned}$$

where we have again used the fact that $(2+\delta)/(1+\delta) < 2$ and $\|Y_0\|_{2+\delta} \leq 1$. The last inequality, together with (4.29) and (4.30), proves the statement of the lemma. □

Proof of Theorem 4.8. First consider $W_n(t)$ for a fixed $t \in [0, 1]$. Observe that

$$W_n(t) = \frac{1}{\sqrt{n}} \sum_{i=1}^{n} g_t(X_i)$$

and that $(g_t(X_i))_{i\geq 1}$ are bounded random variables with mean zero and themselves functionals of an absolutely regular process $(Z_n)_{n\in\mathbb{Z}}$ satisfying the 1-approximation condition. Moreover, the summability condition (4.27) on the sequence $(\gamma(\bar{a}_l))_{l\geq 1}$ implies the requirements of Theorem 3.31. Thus, for fixed $t \in [0, 1]$, we get that $W_n(t)$ converges in distribution to a normal law with mean zero and variance

$$\sigma^2 = E\,(g_t(X_0))^2 + 2\sum_{k=1}^{\infty} \mathrm{Cov}\,(g_t(X_0), g_t(X_k))\,.$$

Using the Cramér–Wold device, we can show, for any $t_1, \ldots, t_k \in [0, 1]$, weak convergence of the vector $(W_n(t_1), \ldots, W_n(t_k))$ towards a multivariate Gaussian distribution with mean zero and covariance given by (4.29).

It remains to show that the sequence of processes $\{W_n(t),\ t \in [0, 1]\}_{n\in\mathbb{N}}$ is tight, i.e., that for any $\epsilon, \delta > 0$ we can find a grid $0 = t_0 < t_1 < \cdots < t_m = 1$ such that

$$\limsup_{n\to\infty} P\left\{ \max_{1\le i\le m} \sup_{t_{i-1}\le t<t_i} |W_n(t) - W_n(t_{i-1})| > \delta \right\} < \epsilon.$$

In order to show this, we proceed as in the proof of Proposition 2.11. Let $0 \le s \le t \le 1$ be given and apply Lemma 4.9 to the random variables $Y_i = g_t(X_i) - g_s(X_i)$, $i = 1, \ldots, n$. Observe that the conditions of Lemma 4.9 hold and that

$$\|g_t(X_i) - g_s(X_i)\|_{2+\delta} \le C|t - s|^{1/(2+\delta)}.$$

Hence we find

$$E\,|W_n(t) - W_n(s)|^4 \le C\left\{ |t - s|^{2/(1+\delta)} + \frac{|t - s|^{1/(1+\delta)}}{n} \right\}, \tag{4.31}$$

i.e., (2.3) with $\gamma = 4$, $\alpha = 2/(1+\delta) > 1$, $\beta = 1/(1+\delta)$ and $q = 1$. From here on we can almost literally follow the proof of Proposition 2.11, replacing at the appropriate places the identity $F(t) - F(s) = t - s$ by the inequality $0 \le G(t) - G(s) \le C(t - s)$, valid for $0 \le s \le t \le 1$. □

4.6 Approximation theorems. In this section we shall give a proof of Theorem 4.6 as well as of a similar result (Theorem 4.12 below). All of these results are based on a method developed by Berkes and Philipp (1979). The first result is a special case of a theorem of Monrad and Philipp (1991). Throughout this section, U will denote a random variable uniformly distributed over $[0, 1]$, and $\langle u, x\rangle$ will denote the inner product of the vectors $u, x \in \mathbb{R}^d$.

Theorem 4.10. *Let X be an $\mathbb{R}^d$-valued random variable defined on the probability space $(\Omega, \mathcal{S}, P)$ and let $\mathcal{F}$ be a countably generated sub-σ-field of $\mathcal{S}$. Assume that there exists a random variable U, uniformly distributed on $(0, 1)$ that is independent of the σ-field $\sigma(X)$ (this makes the σ-field rich enough). Let G be a distribution on $\mathbb{R}^d$ with characteristic function g. Suppose that for some nonnegative numbers λ, δ and $T \ge 10^8 d$,*

$$\int_{|u|\le T} E|E\{\exp(i\langle u, X\rangle)|\mathcal{F}\} - g(u)|\,du \le \lambda\,(2T)^d, \tag{4.32}$$

and that

$$G(\{x : \|x\| \ge T/4\}) < \delta. \tag{4.33}$$

Finally, let $\mathcal{H} \subset \mathcal{F}$ be a countably generated sub-σ-field. Then there exists an $\mathbb{R}^d$-valued random variable Y on $(\Omega, \mathcal{S}, P)$ with the following properties:

1. *Y is measurable with respect to $\sigma(X) \vee \sigma(U)$,*
2. *Y is independent of $\mathcal{H}$,*
3. *Y has distribution G,*

and

$$P(|X - Y| > \alpha) \leq \alpha$$

where

$$\alpha = 16\, d\, T^{-1} \log T + 2\, \lambda^{\frac{1}{2}}\, T^d + 2\delta. \tag{4.34}$$

An improvement of this and related results was recently obtained by Römersperger (1996, Remark 2.3).

Theorem 4.6 follows by induction. Let $(U_j)_{j\geq 1}$ be a sequence of independent random variables, independent of the sequence $(X_j)_{j\geq 1}$. Set $\mathcal{S}^* = \mathcal{S} \vee \sigma((U_j)_{j\geq 1})$ and $\mathcal{F}_k^* = \mathcal{F}_k \vee \sigma(U_1, \ldots, U_k)$. Redefine P accordingly. Then (4.11) remains valid with $\mathcal{F}_{k-1}$ replaced by $\mathcal{F}_{k-1}^*$.

Suppose that the random variables $Y_1, \ldots, Y_{k-1}$ have already been constructed, and set $\mathcal{H}_{k-1} = \sigma(X_1, \ldots, X_{k-1}, U_1, \ldots, U_{k-1})$. Then $\mathcal{H}_{k-1} \subset \mathcal{F}_{k-1}^*$ is countably generated, and condition (4.11) continues to hold with $\mathcal{F}_{k-1}^*$ replaced by $\mathcal{H}_{k-1}$. Hence by Theorem 4.10 there exists an $\mathbb{R}^{d_k}$-valued random variable Y_k defined on $(\Omega, \mathcal{S}^*, P)$ such that

1. Y_k is measurable with respect to $\sigma(X_k, U_k) \subset \mathcal{H}_k := \sigma(X_1, \ldots, X_k, U_1, \ldots, U_k)$,
2. Y_k is independent of $\mathcal{H}_{k-1}$,

and satisfies (4.13) and (4.14).

Applying Theorem 4.10 with G being the distribution of X we obtain, using Dvoretzky's lemma (see Corollary 3.10), the following result; see also Berkes and Philipp (1979).

Corollary 4.11. *Let X be an $\mathbb{R}^d$-valued random variable defined on $(\Omega, \mathcal{S}, P)$ and let $\mathcal{H} \subset \mathcal{S}$ be a countably generated sub-σ-field. Suppose there exists a random variable U uniformly distributed on $(0, 1)$ that is independent of $\sigma(X)$. Let $T \geq 10^8 d$. Then there exists an $\mathbb{R}^d$-valued random variable Y, independent of $\mathcal{H}$, measurable with respect to $\sigma(X) \vee \sigma(U)$, which has the same distribution as X and such that*

$$P(|X - Y| \geq \gamma) \leq \gamma$$

with

$$\gamma = 16\, d\, T^{-1} \log T + 2\, T^d\, \alpha(\mathcal{H}, \sigma(X))^{\frac{1}{2}} + P(|X| > T/4).$$

Given a Polish space (B, m), a set $A \subset B$ and $\rho \geq 0$, we write $A^\rho = \{x : \inf\{m(x, y) : y \in A\} \leq \rho\}$. As before, U is a random variable, defined on the underlying probability space that is uniformly distributed on $[0, 1]$. For random variables satisfying an absolute regularity condition of the form (3.3) the following theorem provides sharper bounds. Again we quote only a special case of a result of Monrad and Philipp (1991).

Theorem 4.12. *Let $\{B_k, m_k, k \geq 1\}$ be a sequence of Polish spaces, let $\mathcal{B}_k$ denote the Borel σ-field of B_k, and let $(X_k)_{k\geq 1}$ be a sequence of random variables, defined on $(\Omega, \mathcal{S}, P)$ and with X_k assuming values in B_k. Let $\{\mathcal{F}_k, k \geq 0\}$ be a nondecreasing sequence of sub-σ-fields of $\mathcal{S}$ such that X_k is $\mathcal{F}_k$-measurable for each $k \geq 1$. Assume there exists a random variable U, uniformly distributed on $(0, 1)$, that is independent of $\bigvee_{k\geq 0} \mathcal{F}_k$. Let $G_k(\cdot)$ be a distribution on $\mathcal{B}_k$, $k = 1, 2, \ldots$. Suppose there exist two sequences of nonnegative numbers $\{\alpha_k\}$ and $\{\beta_k\}$ such that for all $k \geq 1$*

$$E \sup_{A \in \mathcal{B}_k} \left\{ P\left(X_k \in A | \mathcal{F}_{k-1}\right) - G_k(A^{\alpha_k}) \right\} \leq \beta_k.$$

Then there exists a sequence $\{Y_k, k \geq 1\}$ *of independent random variables, defined on* $(\Omega, \mathcal{S}, P)$, *with* Y_k *assuming values in* B_k *and having distribution* G_k. *Moreover, for all* $k \geq 1$,

$$P(m_k(X_k, Y_k) > \alpha_k) \leq \beta_k.$$

Just as above, Theorem 4.12 follows by a repeated application of the following theorem.

Theorem 4.13. *Let* X *be a random variable, defined on the probability space* $(\Omega, \mathcal{S}, P)$, *taking values in the Polish space* (B, m) *with Borel-*σ*-field* $\mathcal{B}$. *Let* $\mathcal{F}$ *be a sub-*σ*-field of* $\mathcal{S}$ *and assume that there exists a random variable* U, *uniformly distributed on* $(0, 1)$, *that is independent of the* σ*-field* $\sigma(X)$. *Let* G *be a distribution on* $\mathcal{B}$ *and suppose that for some nonnegative numbers* α *and* β,

$$E \sup_{A \in \mathcal{B}} \left\{ P(X \in A \mid \mathcal{F}) - G(A^\alpha) \right\} \leq \beta. \tag{4.35}$$

Finally, let $\mathcal{H} \subset \mathcal{F}$ *be a countably generated sub-*σ*-field. Then there exists a random variable* Y *with values in* B, *defined on* $(\Omega, \mathcal{S}, P)$ *and measurable with respect to* $\sigma(X) \vee \sigma(U)$. *Moreover,* Y *is independent of* $\mathcal{H}$, *has distribution* G *and satisfies*

$$P(m(X, Y) > \alpha) \leq \beta.$$

The proof of Theorem 4.3 in Section 4.3 uses the following lemmas.

Lemma 4.14. *Let* S *and* T *be Polish spaces and* Q *a law on* $S \times T$, *with marginal* μ *on* S. *Let* $(\Omega, \mathcal{S}, P)$ *be a probability space and* X *be a random variable on* Ω *with values in* S *and distribution* μ. *Assume that there exists a random variable* U *on* Ω, *uniformly distributed on* $[0, 1]$ *and independent of* X. *Then there exists a random variable* Y *on* Ω *with values in* T *and* $\mathcal{L}(X, Y) = Q$.

This is Lemma 2.11 of Dudley and Philipp (1983). As observed in that paper it also follows from the proof of Theorem 1 of Skorohod (1976).

Lemma 4.15. *Let* F *and* G *be two probability distributions on* $\mathbb{R}^d$ *with characteristic functions* f *and* g, *respectively. Let* $T \geq 10^8 d$. *Then the Prohorov distance* $\pi(F, G)$ *satisfies*

$$\rho(F, G) < \left(\frac{T}{\pi}\right)^d \int_{|u| \leq T} |f(u) - g(u)|\, du + F(\{x : |x| \geq T/2\}) + 16\, d\, T^{-1} \log T.$$

Proof. We use a simple smoothing inequality of the form

$$\pi(F, G) \leq \pi(F * H, G * H) + 2 \max\left(r, H(\{x : |x| \geq r\})\right) \tag{4.36}$$

valid for any distributions F, G, H on $\mathbb{R}^d$ and for any $r > 0$. Relation (4.36) easily follows from the definition of the Prohorov distance. Let H be a distribution on $\mathbb{R}^d$ with density ν and characteristic function $h \in L^1$. We shall prove a general estimate for $\pi(F, G)$ in terms of H and specialize later to obtain the desired bound. Write $F_1 = F * H$ and $G_1 = G * H$. Then the characteristic functions of F_1 and G_1 are given by $f_1 = fh$ and $g_1 = gh$, respectively. Both belong to L^1. Hence we obtain for the densities ϕ and γ of F_1 and G_1, respectively,

$$\begin{aligned} |\phi(x) - \gamma(x)| &= (2\pi)^{-d} \left| \int_{\mathbb{R}^d} e^{-i(u,x)} (f_1(u) - g_1(u))\, du \right| \\ &\leq (2\pi)^{-d} \int_{\mathbb{R}^d} |f(u) - g(u)|\, |h(u)|\, du \\ &\leq (2\pi)^{-d} \left(I + 2 \int_{|u| \geq T} |h(u)|\, du \right), \end{aligned} \tag{4.37}$$

for all $x \in \mathbb{R}^d$. Here

$$I = \int_{|u| \le T} |f(u) - g(u)|\, du. \tag{4.38}$$

Hence for any Borel set B,

$$\begin{aligned} F_1(B) - G_1(B) &\le F_1(B \cap \{|x| \le T\}) - G_1(B \cap \{|x| \le T\}) + F_1(|x| \ge T) \\ &\le \int_{|x| \le T} |\phi(x) - \gamma(x)|\, dx + F(|x| \ge T/2) + H(|x| \ge T/2) \qquad (4.39) \\ &\le \left(\frac{T}{\pi}\right)^d \left(I + 2\int_{|u| \ge T} |h(u)| du\right) + F(|x| \ge T/2) \\ &\quad + H(|x| \ge T/2) \end{aligned}$$

by (4.37). We note that the r.h.s. of (4.39) is an upper bound for $\pi(F_1, G_1)$. Hence, if $r \le \frac{1}{2}T$ we obtain from (4.36)

$$\pi(F, G) \le \left(\frac{T}{\pi}\right)^d \left(I + 2\int_{|u| \ge T} |h(u)| du\right) + F(|x| \ge T/2) + 3\max(r, H(|x| \ge r)). \tag{4.40}$$

It remains to choose H and r suitably. Put $\sigma = 3d^{\frac{1}{2}} T^{-1} \log^{\frac{1}{2}} T$ and $r = 5dT^{-1} \log T$, and let H be a normal distribution with density

$$\nu(x) = (2\pi\sigma^2)^{-\frac{1}{2}d} \exp\left(-\frac{1}{2\sigma^2} \sum_{j \le d} x_j^2\right).$$

Then the characteristic function is given by $h(u) = \exp(-\frac{\sigma^2}{2} \sum_{j \le d} u_j^2)$. The lemma now follows from (4.38), (4.40) and the estimate

$$\begin{aligned} (2\pi)^{-\frac{1}{2}d} \int_{|u| \ge A} \exp\left(-\frac{1}{2} \sum_{j \le d} u_j^2\right) du &= P\left(\chi_d^2 \ge A\right) \\ &\le e^{-tA^2} E \exp(t\chi_d^2)\big|_{t=3/8} = e^{-3A^2/8}\, 2^d \end{aligned}$$

via elementary calculations. □

Combining Lemmas 4.14 and 4.15 and the Strassen–Dudley theorem (see, e.g., Dudley, 1989), we obtain the following.

Lemma 4.16. *Let X be a random variable with values in $\mathbb{R}^d$ and characteristic function f. Let G be a distribution on $\mathbb{R}^d$ with Fourier transform g. Assume there exists a random variable U, uniformly distributed on* $(0, 1)$, *independent of X. Then there exists a random variable Y with distribution G such that*

$$P(|X - Y| > \alpha) \le \alpha,$$

where

$$\alpha = \left(\frac{T}{\pi}\right)^d \int_{|u| \le T} |f(u) - g(u)|\, du + G\left(x : |x| \ge \frac{T}{2}\right) + 16\, d\, T^{-1} \log T$$

provided that $T \ge 10^8 d$.

We first prove Theorem 4.10 under the additional hypothesis that $\mathcal{H}$ is generated by a countable partition. We make heavy use of the ideas developed in Berkes and Philipp (1979).

Let $\epsilon > 0$ be chosen suitably later. By (4.32), Fubini's theorem and Markov's inequality,

$$\int_{|u|\le T} |E\{\exp(i\langle u, X\rangle)|\mathcal{H}\} - g(u)|du \le \epsilon\,(2T)^d \tag{4.41}$$

except on a set $A \in \mathcal{H}$ with $P(A) \le \lambda/\epsilon$.

Let D be any of the countably many atoms of $\mathcal{H}$ and keep it fixed. Let $\mathcal{F}^{(D)}$ denote the trace of $\mathcal{F}$ on D and define P_D by

$$P_D(E) = P(E|D), \quad E \in \mathcal{F}^{(D)}. \tag{4.42}$$

Note that $X1_D$ and $U1_D$ are still P_D-independent and that the P_D-distribution of $U1_D$ is still uniform. On D the conditional characteristic function

$$E\{\exp(i\langle u, X\rangle|\mathcal{H}\} = \frac{1}{P(D)}\int_D \exp(i\langle u, X\rangle)\,dP = f_D(u),$$

say, is a nonrandom function in u and can be interpreted as the Fourier transform of the P_D-distribution of X.

Thus, on the set D, either (4.41) holds in which case $D \subset A^c$, or else it fails in which case $D \subset A$. Assume first that $D \subset A^c$. Then by (4.41),

$$\int_{|u|\le T} |f_D(u) - g(u)|\,du \le \epsilon\,(2T)^d.$$

Hence by Lemma 4.16 there exists a random variable Y on $(D, \mathcal{F}^{(D)}, P_D)$ such that

$$P_D(Y \in B) = G(B), \quad B \in \mathcal{B}^d \tag{4.43}$$

and that

$$P_D(|X - Y| > \beta) \le \beta \tag{4.44}$$

if $D \subset A^c$, where

$$\beta = 16\,d\,T^{-1}\log T + \epsilon\,T^{2d} + \delta. \tag{4.45}$$

If on the other hand $D \subset A$, we choose Y with P_D distribution G but arbitrary otherwise. Then, obviously

$$P_D(|X - Y| > 0) \le 1 \quad \text{if } D \subset A.$$

As D runs through all the atoms of $\mathcal{H}$ we obtain a random variable Y defined on the whole space $(\Omega, \mathcal{F}, P)$ such that the conditional law $\mathcal{L}(Y|\mathcal{H}) = G$. Thus Y is independent of $\mathcal{H}$ and Y has distribution G. Moreover, Y is measurable $\sigma(X) \vee \sigma(U)$. Moreover, summing the relations (4.44) and (4.45) over all $D \in \mathcal{H}$, we obtain by (4.42)

$$P(|X - Y| > \beta) \le \beta + \lambda/\epsilon.$$

We choose $\epsilon = \lambda^{\frac{1}{2}}T^{-d}$ and obtain in view of (4.45) a result slightly stronger than claimed in (4.34).

For the proof of the case of general $\mathcal{H}$ we follow Monrad and Philipp (1991). Since $\mathcal{H}$ is countably generated, there exists a real-valued random variable W such that $\mathcal{H} = \sigma(W)$. For $n = 1, 2, \ldots$, let W_n denote the discrete random variable defined by

$$W_n := \sum_{-\infty<k<\infty} k\,2^{-n} 1_{\{k2^{-n} \leq W < (k+1)2^{-n}\}}$$

and let $\mathcal{H}_n = \sigma(W_n)$. Since $\mathcal{H}_n \subset \mathcal{H}$, Lemma 2.6 in Berkes and Philipp (1979) shows that condition (4.41) is satisfied with $\mathcal{H}_n$ in place of $\mathcal{H}$. By the result already proved above there exists an $\mathbb{R}^d$-valued random variable Y_n such that Y_n is independent of $\mathcal{H}_n$, Y_n has distribution G and Y_n is measurable $\sigma(X) \vee \sigma(U)$. Moreover,

$$P(|X - Y_n| > \alpha) \leq \alpha \tag{4.46}$$

where α is given in (4.34).

We now show that the sequence $\{\mathcal{L}(W, Y_n), n \geq 1\}$ of joint laws of W and Y_n converges weakly as $n \to \infty$. To see this, first note that for $j = 0, \pm 1, \pm 2, \ldots$ and $k = 1, 2, \ldots$, the events $\{W \in [j2^{-k}, (j+1)2^{-k})\} \in \mathcal{H}_k$. Thus for each dyadic interval I_k of rank k, for all $n \geq k$ and for all $B \in \mathcal{B}^d$ by independence

$$P(W \in I_k, Y_n \in B) = P(W \in I_k)\, G(B). \tag{4.47}$$

This proves the claim. It follows that the sequence $\{\mathcal{L}(X, W, Y_n),\ n \geq 1\}$ is a tight family of probability measures on $\mathcal{B}^{2d+1}$. Hence there exists a subsequence $\{n'\}$ such that

$$\mathcal{L}(X, W, Y_{n'}) \to Q$$

for some probability measure Q on $\mathcal{B}^{2d+1}$. Since $\mathcal{L}(X, W)$ is a marginal of Q it follows from Lemma 4.14 that there exists an $\mathbb{R}^d$-valued random variable Y such that $\mathcal{L}(X, W, Y) = Q$. Relation (4.47) implies that G is the conditional distribution of Y given W, and (4.46) implies

$$P(|X - Y| > \alpha) \leq \liminf_{n\to\infty} P(|X - Y_n| > \alpha) \leq \alpha.$$

□

Proof of Theorem 4.13. Again, we first prove Theorem 4.13 under the additional hypothesis that $\mathcal{H}$ is generated by a countable partition. The proof makes use of sketches of proofs of unconditional results in several earlier papers (see Philipp (1986) and its history given there).

Let D be any of the countably many atoms of $\mathcal{H}$ and note that on each D,

$$P(X \in A|\mathcal{H}) = P(X \in A|D)$$

is nonrandom. Hence we can rewrite (4.35) in the form

$$\sum_{D\in\mathcal{H}} \epsilon(D)\, P(D) \leq \beta, \tag{4.48}$$

where we set

$$\epsilon(D) = \sup_{A\in\mathcal{B}} P(X \in A|D) - G(A^\alpha). \tag{4.49}$$

In the context of Philipp (1986, Theorem 3.4), the usefulness of this observation for obtaining sharp constants was pointed out to us by Erich Berger. We thank him for this remark.

For the moment keep D fixed. We shall construct Y on each D separately. Define, for all $A \in \mathcal{B}$,

$$P_1(A) = P(X \in A|D) \text{ and } P_2(A) = G(A).$$

Then by (4.49) with $\epsilon = \epsilon(D)$,

$$P_1(A) \le P_2(A^\alpha) + \epsilon, \text{ for all } A \in \mathcal{B}.$$

Hence by the Strassen–Dudley theorem there exists a probability measure $Q = Q_D$ on $B \times B$ with marginals P_1 and P_2 such that

$$Q_D\{(x, y) : m(x, y) > \alpha\} \le \epsilon.$$

Hence by Lemma 4.14 there exists a random variable Y on $(D, \mathcal{F}^{(D)}, P_D)$ such that $\mathcal{L}(X, Y) = Q_D$ where $\mathcal{F}^{(D)}$ and P_D are defined as above. It follows that

$$P(m(X, Y) > \alpha, D) \le \epsilon(D)\, P(D). \tag{4.50}$$

As D runs through all atoms of $\mathcal{H}$ we obtain a random variable Y defined on the whole space $(\Omega, \mathcal{F}, P)$. We sum (4.50) over all sets D and obtain, in view of (4.48),

$$P(m(X, Y) > \alpha) \le \sum_{D \in \mathcal{H}} \epsilon(D) P(D) \le \beta.$$

We also note that Y is independent of $\mathcal{H}$ and has distribution G since P_2, the second marginal of Q_D, is the P_D-distribution of Y. This proves Theorem 4.13 in case that $\mathcal{H}$ is generated by a countable partition. The proof of the general case can be easily modeled after the proof of Theorem 4.10. □

5 Empirical U-processes

In empirical U-process theory, one investigates the asymptotic behavior of the process

$$\binom{n}{2}^{-1} \sum_{1 \le i < j \le n} f(X_i, X_j), \quad f \in \mathcal{F},$$

where $\mathcal{F}$ is a class of measurable, symmetric functions, $f : \mathbb{R}^{2k} \to \mathbb{R}$. Essentially, the same type of questions as in ordinary empirical process theory arises, for example about the validity of the Glivenko–Cantelli theorem and the weak convergence of the renormalized process

$$\sqrt{n} \left(\binom{n}{2}^{-1} \sum_{1 \le i < j \le n} f(X_i, X_j) - Ef(X_1, X_2) \right),$$

for independent and weakly dependent random variables with values in $\mathbb{R}^k$.

5.1 Motivation, dimension estimation. Let $h : \mathbb{R}^{2k} \to \mathbb{R}$ be a measurable function, symmetric in the sense that $h(x, y) = h(y, x)$, for all $x, y \in \mathbb{R}$. Then we define the bivariate U-statistic with kernel h by

$$U_n = \binom{n}{2}^{-1} \sum_{1 \le i < j \le n} h(X_i, X_j),$$

where $(X_n)_{n \ge 1}$ is a stationary sequence of $\mathbb{R}^k$-valued random variables with common distribution F. For independent observations, U-statistics have been introduced by Halmos (1946) and Hoeffding (1948) as unbiased estimators for functionals of the form $\theta(F) = \int_{\mathbb{R}^{2k}} h(x, y) dF(x) dF(y)$. In a nonparametric model, U_n is actually the minimum variance unbiased estimator $(UMVE)$ for $\theta(F)$.

Example 5.1. (i) The U-statistic associated with the kernel $h(x, y) = \frac{1}{2}(x - y)^2$, $x, y \in \mathbb{R}$, is the well-known sample variance

$$s_X^2 = \frac{1}{n-1} \sum_{i=1}^{n} (X_i - \overline{X})^2.$$

(ii) The kernel $h(x, y) = |x - y|$, $x, y \in \mathbb{R}$, gives rise to Gini's mean difference

$$U = \binom{n}{2}^{-1} \sum_{1 \le i < j \le n} |X_i - X_j|$$

which is a robust alternative to s_X^2 as estimator for the spread of the distribution (see, e.g., Denker (1985)).

(iii) Also nonsymmetric kernels can be of interest, as is shown by the kernel $h : \mathbb{R}^2 \times \mathbb{R}^2 \to \mathbb{R}$, defined by $h((x, y), (x', y')) = xy' - yx'$. The corresponding U-statistic, up to the norming constant $\binom{n}{2}$, is given by

$$\sum_{1 \le i < j \le n} h((X_i, Y_i), (X_j, Y_j)) = \sum_{1 \le i < j \le n} (X_i Y_j - X_j Y_i).$$

This process arises in the study of the random walk on the Heisenberg group, which is obtained by equipping $\mathbb{R}^3$ with the noncommutative multiplication

$$(x, y, z) \cdot (x', y', z') = (x + x', y + y', z + z' + (xy' - x'y)).$$

If $(X_i, Y_i, Z_i)_{i \ge 1}$ is an i.i.d. sequence of random variables, the random walk on the Heisenberg group is defined as the sequence of products

$$(X_1, Y_1, Z_1) \cdot \cdots \cdot (X_n, Y_n, Z_n) = \left(\sum_{i=1}^{n} X_i, \sum_{i=1}^{n} Y_i, \sum_{i=1}^{n} Z_i + \sum_{1 \le i < j \le n} (X_i Y_j - X_j Y_i) \right),$$

$n = 1, 2, \ldots$. The dominating term in the third coordinate is exactly the above mentioned U-statistics. For a detailed study of this random walk see Guivarc'h, Keane, and Roynette (1977).

U-statistics of dependent observations have attracted some attention in the past 20 years, the problem of dimension estimation serving as a main motivation. Estimation of the fractal dimension of a strange attractor from a chaotic time series has become one of the standard tools in the analysis of nonlinear deterministic time series; see for example Isham (1993).

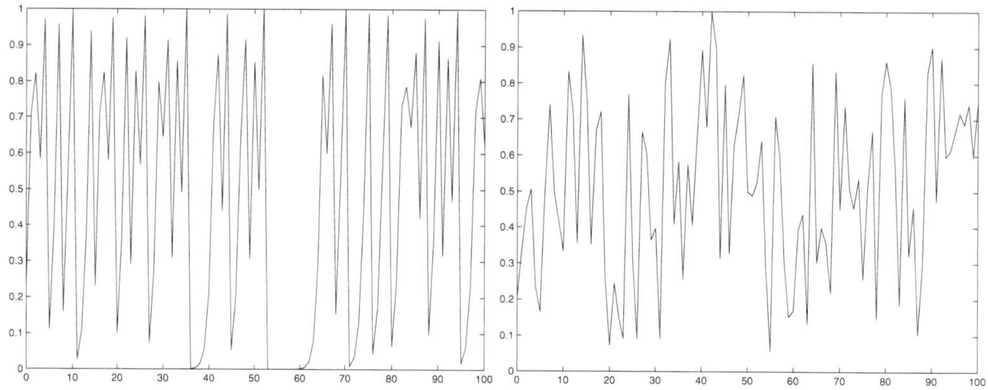

FIGURE 1. (i) Time series resulting from iteration of the logistic map. (ii) Autoregressive time series.

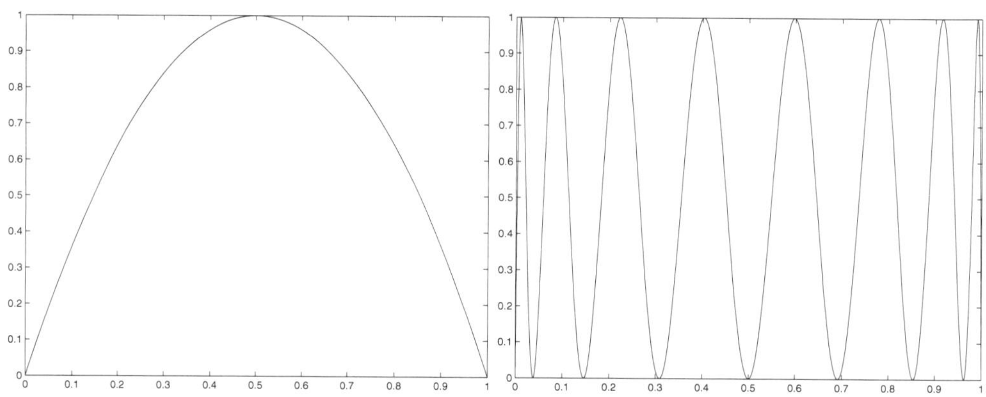

FIGURE 2. (i) Graph of the logistic map $f(x) = 4x(1-x)$. (ii) Graph of the fourth iterate of f.

The most popular model of a deterministic time series consists of a differentiable dynamical system $(\mathcal{X}, T)$ on a compact manifold $\mathcal{X}$, together with an observable function $y : \mathcal{X} \to \mathbb{R}$. An element $x \in \mathcal{X}$ describes the state of a system whose evolution is governed by the map $T : \mathcal{X} \to \mathcal{X}$. The state itself is hidden to the observer and only partially revealed via the observable $y(x)$. Starting from an initial state $x_0 \in \mathcal{X}$, the state at time n is given by $x_n = T^n x_0$, and the observed time series is

$$y_n = y(x_n) = y(T^n x_0).$$

Example 5.2. Define on [0, 1] the logistic map $Tx = 4x(1-x)$ and let $y(x) = x$. A plot of the resulting time series $(y_n)_{n\geq 1}$ is presented in Figure 1(i). At first sight this plot looks "random," and is not immediately distinguishable from a plot of a sequence of random numbers, as presented in Figure 1(ii). On the other hand, the sequence $(y_n)_{n\geq 0}$ is completely deterministic, i.e., a single value y_k determines the entire future. The explanation for the paradox that a completely deterministic process can have apparently random behavior, lies in the sensitive dependence on the initial value. A small uncertainty about the present value of the system translates into almost unpredictability of the value at a later point in time. This effect is seen very nicely from the graph of the iterate T^n, as shown in Figure 2

Given that observables of deterministic dynamical systems can have apparently random sample paths, it is an important problem to distinguish these from genuinely random systems. Note that any stationary process $(y_n)_{n\geq 0}$ can be represented in the above framework. Define $\mathcal{X} = \mathbb{R}^{\mathbb{N}_0}$, and $T : \mathcal{X} \to \mathcal{X}$ by $T((y_n)_{n\geq 0}) = (y_{n+1})_{n\geq 0}$, the one-sided shift map. If we let $y : \mathcal{X} \to \mathbb{R}$ be the projection onto the first coordinate, we obtain $y_n = y(T^n x_0)$. However, the state space is then infinite dimensional. Thus the problem of distinguishing deterministic from random processes is better formulated as distinguishing finite from infinite dimensional systems.

In the example of the process generated by the logistic map, the deterministic nature is easily detected by plotting the pairs (y_n, y_{n+1}) of consecutive random variables. These points all lie on the graph of the map $Tx = 4x(1-x)$, and thus on a one-dimensional subset of $\mathbb{R}^2$. For a genuinely random system, these points would fill a two-dimensional subset.

An extension of this technique works for systems of arbitrary dimension. We define the so-called k-dimensional reconstruction vectors,

$$X_n = (y_n, y_{n+1}, \ldots, y_{n+k-1}),$$

and note that these can be written as $X_n = Rec_k(x_n)$, where $Rec_k : \mathcal{X} \to \mathbb{R}^k$ denotes the reconstruction map

$$Rec_k(x) = (y(x), y(Tx), \ldots, y(T^{k-1}x)).$$

The reconstruction vectors lie in the image of $\mathcal{X}$ under the map Rec_k, and hence in a subspace whose dimension is bounded from above by the dimension of $\mathcal{X}$.

For many dynamical systems, the trajectories converge to a subset A of the state space, called the attractor. Often, the attractor is an object of fractal dimension so that topological dimension is not an appropriate notion. Of the many ways to define fractal dimensions, the correlation dimension is most popular, for a variety of reasons. The correlation dimension is defined for a probability distribution F on $\mathbb{R}^k$ via the so-called correlation integrals $C(r)$, given by

$$C(r) = F \times F\left(\left\{(x, y) \in \mathbb{R}^k \times \mathbb{R}^k : ||x - y|| \leq r\right\}\right).$$

Alternatively, the correlation integral can be written as $C(r) = \int F(B_r(x))dF(x)$, where $B_r(x)$ denotes the ball of radius r around x, or as $C(r) = P(||X - Y|| \leq r)$ where X and Y are independent F-distributed random variables.

If $C(r)$ behaves like a power function as $r \to 0$, i.e.,

$$C(r) \sim Cr^d, \tag{5.1}$$

we call d the correlation dimension of F. More precisely, we define

$$d = \lim_{r\to 0} \frac{\log C(r)}{\log r},$$

provided this limit exists.

Example 5.3. Let F denote Lebesgue measure on $[0, 1]$. Then $C(r) = 1 - (1-r)^2 = 2r - r^2$ and hence $d = 1$. In general, one can show that for measures with density, correlation dimension and topological dimension coincide.

In most situations, it is impossible to compute the correlation dimension of a given distribution, an added difficulty being that for most dynamical systems the invariant measure is not explicitly known. Thus one faces the statistical problem of estimating the correlation dimension from a finite sample $X_1, \ldots, X_n$ of random vectors with distribution F. In all practical situations, the observations will not be independent but will be a finite segment from a stationary ergodic process.

The natural estimator for the correlation integral is its empirical analogue, known as the sample correlation integral and defined by

$$\begin{aligned} C_n(r) &= \binom{n}{2}^{-1} \#\left\{1 \le i < j \le n : ||X_i - X_j|| \le r\right\} \\ &= \binom{n}{2}^{-1} \sum_{1 \le i < j \le n} 1_{\{||X_i - X_j|| \le r\}}. \end{aligned} \tag{5.2}$$

Note that $(C_n(r), 0 \le r \le r_0)$ is a U-process with a class of kernels $f(x, y) = 1_{\{||x-y|| \le r\}}$ indexed by $r \in [0, r_0]$.

An estimator of the correlation dimension is suggested by transforming the scaling property $C(r) \sim Cr^d$ into

$$\log C(r) \approx \log C + d \log r.$$

The latter relation can be used to estimate the correlation dimension by linear regression of $\log C_n(r)$ on $\log r$. The resulting estimator d_{GP} is known as the Grassberger–Procaccia estimator (Grassberger and Procaccia (1983)). It should be remarked here that computation of the Grassberger–Procaccia estimator still involves an element of choice, namely of the region of r-values in which to perform the linear regression. Here a delicate balance is to be made between the dangers of an r too large, resulting in bias because of deviation from exact scaling, and an r too small, resulting in increased variance.

Takens (1985) suggested an alternative estimator motivated by the maximum likelihood principle of parametric statistical inference. To derive the Takens estimator, assume for a moment that in (5.1) exact scaling holds, i.e., that

$$C(r) = Cr^d, \quad 0 \le r \le r_0.$$

If $R_1, \ldots, R_N$ denote independent copies of the distance $||X - Y||$, with X and Y themselves independent and F-distributed, we obtain

$$P\left(R_i \le r | R_i \le r_0\right) = \left(\frac{r}{r_0}\right)^d.$$

Hence conditionally given $R_i \le r_0$, the density of R_i is $f(r) = d\,(r/r_0)^{d-1}$. A small computation shows that the maximum likelihood estimator of d based on those distances R_i that do not exceed r is given by

$$\hat{d} = \left[-\frac{1}{\#\{1 \le i \le N : R_i \le r_0\}} \sum_{i=1}^{N} \log(R_i/r_0) I_{[0,r_0]}(R_i)\right]^{-1}.$$

Substituting actually observed interpoint distances $||X_i - X_j||$, $1 \le i < j \le n$, for the R_i's, we obtain Takens's estimator,

$$\hat{d}_T = \left[-\frac{1}{\sum_{1 \le i < j \le n} 1_{\{||X_i - X_j|| \le r_0\}}} \sum_{1 \le i < j \le n} \log(||X_i - X_j||/r_0) \cdot 1_{\{||X_i - X_j|| \le r_0\}}\right]^{-1}.$$

Though motivated by likelihood arguments, the Takens estimator is not the actual maximum likelihood estimator, as the observed distances $||X_i - X_j||$ are not independent at all. It should finally be mentioned that also the Takens estimator requires the choice of the range $[0, r_0]$ of distances to be taken into consideration.

Both the sample correlation integral and Takens's estimator involve U-processes, with classes of functions,

$$\begin{aligned}\mathcal{H}_1 &= \left\{f_r(x, y) = 1_{\{||x-y|| \le r\}},\ 0 \le r \le r_0\right\},\\ \mathcal{H}_2 &= \left\{f_r(x, y) = \log(||x - y||/r) 1_{\{||x-y|| \le r\}},\ 0 \le r \le r_0\right\}.\end{aligned}$$

For a single function $f \in \mathcal{F}$ the U-process is simply an ordinary U-statistic. Thus for the study of finite dimensional convergence of the process, we have to understand the asymptotics of U-statistics, both for dependent and for independent inputs. This will be the topic of the next two sections. We remark in passing that Dabrowski et al. (2002) have initiated a study of the tail of the U-processes as $r_0 = r_0(n) \to 0$. Under certain conditions, they obtained convergence towards a Poisson point process.

5.2 *U*-statistics law of large numbers. In this section we study the convergence of a U-statistic U_n to $\int\int h(x, y) dF(x) dF(y)$ where F denotes the marginal distribution of the (X_i)-process. In the i.i.d. case, the issue has been completely settled, and it turns out that the exact analogue of the strong law of large numbers for partial sums holds.

Theorem 5.4 (Hoeffding (1961), Berk (1966)). *If the input process $(X_i)_{i \ge 1}$ is i.i.d. and $h \in L_1(F \times F)$, we get*

$$U_n(h) \to \int\int h(x, y)\, dF(x)\, dF(y) \quad a.s.$$

The standard proof of this theorem makes use of the reverse martingale property of the U-statistics (U_n). In the case of dependent inputs, the law of large numbers for U-statistics does not hold, unless extra conditions are imposed. The following example shows what might go wrong.

Example 5.5. Let $T : [0, 1] \to [0, 1]$ be the period-doubling map, i.e., $Tx = 2x[\mathrm{mod} 1]$, and let F denote Lebesgue measure on $[0, 1]$. Let X_0 be a random variable, uniformly distributed on $[0, 1]$ and define $X_n = TX_{n-1}, n \ge 1$. Then $(X_n)_{n \ge 0}$ is a stationary ergodic process with marginal distribution F. Now, we define the set $G := \bigcup_{n \ge 0} \mathrm{graph}(T^n)$, and the kernel

$$h(x, y) = \frac{1}{2}(1_G(x, y) + 1_G(y, x)).$$

Note that G is an $F \times F$-nullset and that $(X_i, X_j) \in G$ for all indices $i \le j$. Thus $h(X_i, X_j) = 1$, and hence $U_n(h) \equiv 1$, whereas $\int\int h(x, y) dF(x) dF(y) = 0$, so that the law of large numbers is violated.

Analyzing the previous example, we see that the joint distribution of (X_i, X_k) converges to the product measure $F \times F$ in the topology of weak convergence, but not in the total variation norm. This makes plausible that we could find a bounded, but discontinuous kernel h for which $U_n(h)$ does not converge to $E_{F \times F} h(X, Y)$. If we want to exclude this possibility, we can either require h to be continuous or impose a mixing condition on the process (X_n) that will insure convergence of the pair distribution (X_i, X_k) to the product measure in total variation norm. As it turns out, absolute regularity is exactly the right condition for this.

Theorem 5.6 (Aaronson et al. (1996)). *Let $X_1, X_2, \ldots$ be stationary, ergodic with values in $\mathbb{R}^k$, having marginal distribution F, and let $h : \mathbb{R}^k \times \mathbb{R}^k \to \mathbb{R}$ be a bounded symmetric kernel function. Then*

$$U_n(h) \to \int_{\mathbb{R}^k \times \mathbb{R}^k} h(x, y)\, dF(x)\, dF(y) \quad a.s.$$

provided one of the following two conditions holds:

1. *The kernel h is $F \times F$-a.e. continuous.*

2. *The process $(X_i)_{i \in N}$ is absolutely regular.*

Proof. (1) First note that, because of boundedness of the kernel $h(x, y)$, it suffices to prove convergence of the so-called von Mises statistic

$$V_n(h) = \frac{1}{n^2} \sum_{1 \le i, j \le n} h(X_i, X_j).$$

Define the empirical pair measure μ_n^ω by

$$\mu_n^\omega = \frac{1}{n^2} \sum_{1 \le i, j \le n} \delta_{(X_i(\omega), X_j(\omega))}.$$

By the ergodic theorem, $\mu_n^\omega(A \times B) \to F(A)\, F(B)$, for any rectangle $A \times B$. As the class of rectangles is convergence determining (Billingsley, 1968), we get that $\mu_n^\omega \to F \times F$ weakly, and consequently

$$V_n(h) = \int\int h(x, y)\, d\mu_n^\omega(x, y) \to \int\int h(x, y)\, dF(x)\, dF(y)$$

for any bounded, $F \times F$-a.e. continuous kernel h.

(2) The key ingredient in the proof of the second part is an almost sure blockwise approximation of the original process by independent blocks. We define the sequence of (m, N) blocks of the process $(X_i)_{i \ge 0}$ by

$$\xi_s = (X_{(s-1)(m+N)+m+1}, \ldots, X_{s(m+N)}), \quad \text{for } s = 1, 2, 3, \ldots.$$

Lemma 5.7. *Let $(X_i)_{i \ge 0}$ be a stationary and absolutely regular process and let $(\xi_k)_{k \ge 0}$ be an (m, N)-blocking of $(X_i)_{i \ge 0}$. Then there exists a sequence of $\mathbb{R}^N$-valued i.i.d. vectors $(\xi'_k)_{k \ge 0}$ with the same marginal distribution as ξ_k, such that almost surely*

$$\limsup_{K \to \infty} \frac{1}{K} \#\{k \le K : \xi_k \ne \xi'_k\} \le \beta_m. \tag{5.3}$$

Proof. The proof of this lemma uses the Austro-Hungarian construction of Berkes and Philipp (1979) together with later refinements by Dehling and Philipp (1982) and Monrad and Philipp (1991); see Theorems 4.12 and 4.13 and their proofs in Section 4 above. The latter authors prove that an i.i.d. sequence $(\xi'_k)_{k \in N}$ exists with the same marginals as $(\xi_k)_{k \in \mathbb{N}}$ such that $P(\xi_k \ne \xi'_k) \le \beta_m$. Using their proof one can actually obtain a coupling with the property that the sequence of pairs (ξ_k, ξ'_k) is stationary and ergodic. Then (5.3) follows from the ergodic theorem. □

Given $\epsilon > 0$, we choose m, N such that $\beta_m \le \epsilon$ and $m/(m + N) \le \epsilon$. Subsequently, we introduce a new kernel $H : \mathbb{R}^N \times \mathbb{R}^N \to \mathbb{R}$ by

$$H(\xi, \eta) = \frac{1}{N^2} \sum_{1 \le i, j \le N} h(x_i, y_j),$$

where $\xi = (x_1, \ldots, x_N)$ and $\eta = (y_1, \ldots, y_N)$. By Lemma 5.7 we can define i.i.d. blocks $(B'_k)_{k\geq 1}$ with $\mathcal{L}(B_1) = \mathcal{L}(B'_1)$ satisfying (5.3). By the U-statistics law of large numbers for i.i.d. processes, we get

$$U_p(H) := \frac{1}{p(p-1)} \sum_{1\leq k\neq l\leq p} H(B'_k, B'_l) \to \int\int h(x, y)\, dF(x)\, dF(y).$$

Moreover one can show that $U_n(h) - U_{[n/(m+N)]}(H)$ is small—for details, see the original paper by Aaronson et al. (1996). □

5.3 U-statistics central limit theorems. The law of large numbers for U-statistics asserts that $U_n(h) - \theta \to 0$ a.s., where θ is defined by

$$\theta = \int\int h(x, y)\, dF(x)\, dF(y).$$

This leads to the question as to whether there exists a rescaled sequence $a_n(U_n(h)-\theta)$, converging to a nondegenerate limit. The standard tool for studying this question is the so-called Hoeffding decomposition. Given a kernel $h(x, y)$, we define

$$\begin{aligned} h_1(x) &= \int h(x, y)\, dF(y) - \theta, \\ h_2(x, y) &= h(x, y) - h_1(x) - h_1(y) - \theta. \end{aligned} \tag{5.4}$$

Observe that $Eh_1(X) = 0$ and

$$Eh_2(X, y) = Eh_2(x, Y) = 0 \tag{5.5}$$

for all y and all x. Kernels with the property (5.5) are called degenerate. We can rewrite (5.4) to obtain the following decomposition of $h(x, y)$,

$$h(x, y) = \theta + h_1(x) + h_1(y) + h_2(x, y).$$

With this decomposition of the kernel a decomposition of the U-statistic $U_n(h)$ is associated:

$$U_n(h) = \theta + \frac{2}{n}\sum_{i=1}^{n} h_1(X_i) + U_n(h_2).$$

Of the random terms on the right-hand side, $\frac{2}{n}\sum_{i=1}^{n} h_1(X_i)$, is simply the average of the random variables $h_1(X_i)$, $i = 1, \ldots, n$ that can be analyzed via central limit theorems for partial sums. At the same time, this linear term generally dominates in the Hoeffding decomposition, and one can show asymptotic normality of $\sqrt{n}(U_n(h) - \theta)$.

As motivation and guideline for the dependent case, we first want to explain the situation for i.i.d. input processes $(X_i)_{i\geq 0}$. In this case, the random variables $h_2(X_i, X_j)$ and $h_2(X_{i'}, X_{j'})$ are uncorrelated, unless $i = i'$ and $j = j'$, and hence

$$\mathrm{Var}(U_n(h_2)) = \binom{n}{2}^{-1} \mathrm{Var}\left(h_2(X_1, X_2)\right).$$

Thus $\sqrt{n}\, U_n(h_2) \to 0$ and we obtain the following theorem.

Theorem 5.8. *Let $(X_i)_{i\geq 0}$ be a process of i.i.d. random variables, and let h be a kernel satisfying $Eh^2(X_1, X_2) < \infty$. Then*

$$\sqrt{n}\,(U_n(h) - \theta) \to N(0, 4\sigma_{h_1}^2),$$

where $\sigma_{h_1}^2 = \mathrm{Var}(h_1(X))$.

A special situation arises when $\sigma_{h_1}^2 = 0$, i.e., when h itself is degenerate. Then the limit distribution of $\sqrt{n}\,(U_n(h) - \theta)$ is degenerate and one can actually show that $n\,(U_n(h) - \theta)$ converges in distribution. In this case, the limit distribution is a mixture of χ^2-distributions; see Serfling (1980). In the context of empirical U-processes, degenerate kernels play a less important role, and we will not pay further attention to them in this section.

Recall the notion of a 1-approximating functional of an absolutely regular process from Definition 3.13, the definition (3.21) of $\bar{a}_k$ and the notion of a p-continuous function from Definition 3.15.

Theorem 5.9. *Let $(X_n)_{n\in\mathbb{N}}$ be a 1-approximating functional of an absolutely regular process with mixing coefficients $(\beta_k)_{k\geq 0}$, and let h be a bounded 1-continuous kernel. Suppose that the sequences $(\beta_k)_{k\geq 1}$, $(\bar{a}_k)_{k\geq 1}$ and $(\gamma(\bar{a}_k))_{k\geq 1}$, satisfy the following summability condition:*

$$\sum_{k=1}^{\infty} k^2(\beta_k + \bar{a}_k + \gamma(\bar{a}_k)) < \infty. \tag{5.6}$$

Then the series

$$\sigma^2 = \mathrm{Var}\,(h_1(X_0))^2 + 2\sum_{k=1}^{\infty} \mathrm{Cov}\,(h_1(X_0), h_1(X_k))\,, \tag{5.7}$$

converges absolutely and, as $n \to \infty$,

$$\sqrt{n}(U_n - \theta) \xrightarrow{d} \mathcal{N}(0, 4\sigma^2).$$

Proof. We make use of the Hoeffding decomposition for U-statistics, and obtain

$$U_n = \theta + \frac{2}{n}\sum_{i=1}^{n} h_1(X_i) + \frac{2}{n(n-1)}\sum_{1\leq i<j\leq n} h_2(X_i, X_j)$$

where θ and h_1 are defined as above. The random variables $h_1(X_i)$ are bounded and have mean zero. As h is 1-continuous, $(h_1(X_n))_{n\geq 1}$ itself is a 1-approximating functional of $(Z_n)_{n\geq 1}$ (see Borovkova et al. (2001)). Moreover, the summability condition (5.6) on the corresponding sequence $(\gamma(\bar{a}_l))$ is that required by Theorem 3.31. Hence, from Theorem 3.31 it follows that

$$\frac{2}{\sqrt{n}}\sum_{i=1}^{n} h_1(X_i) \longrightarrow \mathcal{N}(0, 4\sigma^2),$$

where σ^2 is given by (5.7). Then the statement of the theorem will follow, once we can prove that $\sqrt{n}R_n \longrightarrow 0$ in probability, where R_n is defined by

$$R_n = \frac{2}{n(n-1)}\sum_{1\leq i<j\leq n} h_2(X_i, X_j).$$

This is indeed the case, as is established in the following lemma. □

Lemma 5.10. *Under the conditions of Theorem* 5.9, *we have*

$$\sup_n E\left\{\frac{1}{n}\sum_{1\le i<j\le n} h_2(X_i, X_j)\right\}^2 < \infty, \tag{5.8}$$

and hence $\sqrt{n}R_n \longrightarrow 0$ *in probability, as* $n \to \infty$.

Proof. By decomposing $\sum_{1\le i_1<j_1\le n}\sum_{1\le i_2<j_2\le n} h_2(X_{i_1}, X_{j_1})\, h_2(X_{i_2}, X_{j_2})$ into diagonal and off-diagonal terms, we obtain

$$\begin{aligned} E\left(\sum_{1\le i<j\le n} h_2(X_i, X_j)\right)^2 &= \sum_{1\le i<j\le n} E\left(h_2(X_i, X_j)\right)^2 \\ &+ \sum_{\substack{1\le i_1<j_1\le n\ 1\le i_2<j_2\le n \\ i_1\ne i_2 \text{ or } j_1\ne j_2}} E\left(h_2(X_{i_1}, X_{j_1})h_2(X_{i_2}, X_{j_2})\right). \end{aligned} \tag{5.9}$$

The first sum on the r.h.s. contains $\binom{n}{2} \le n^2$ terms, all of which are bounded and thus the sum is bounded by $C\,n^2$. Concerning the second sum, we note that $h_2(x, y)$ is a degenerate kernel, and thus

$$E_{X_{j_2}}\left(h_2(X_{i_1}, X_{j_1})h_2(X_{i_2}, X_{j_2})\right) = \int_{\mathbb{R}} h_2(X_{i_1}, X_{j_1})h_2(X_{i_2}, y)\, dF(y) = 0,$$

and similarly for $E_{X_{i_1}}, E_{X_{i_2}}, E_{X_{j_1}}$. As both h and h_1 are bounded and 1-continuous, this also holds for $g(x_1, x_2, x_3, x_4) = h_2(x_1, x_2)h_2(x_3, x_4)$, so that we may apply Lemma 3.18 with $r = \infty$ and $s = 1$. We now put the indices i_1, j_1, i_2, j_2 in increasing order and note that at least one of these indices is different from the remaining three. We denote by d_1 the largest difference between two consecutive indices and by d_2 the second largest. The location of the largest difference splits the set $\{i_1, j_1, i_2, j_2\}$ into two nonempty sets I_1 and I_2 in such a way that for $l_1 \in I_1$ and $l_2 \in I_2$ we have $l_2 - l_1 \ge d_1$. We can then apply Lemma 3.18 to get

$$\begin{aligned} &\left|E\left(h_2(X_{i_1}, X_{j_1})\, h_2(X_{i_2}, X_{j_2})\right)\right| \\ &\le \left|E_{X_{I_1}} E_{X_{I_2}}\left(h_2(X_{i_1}, X_{j_1})\, h_2(X_{i_2}, X_{j_2})\right)\right| + C\left(\beta_{\lfloor d_1/3\rfloor} + \bar{a}_{\lfloor d_1/3\rfloor} + \gamma(\bar{a}_{\lfloor d_1/3\rfloor})\right). \end{aligned}$$

At this point we have to distinguish two cases: either I_1 or I_2 is a singleton, or $|I_1| = |I_2| = 2$. In the first case, $E_{X_{I_1}} E_{X_{I_2}}\{h(X_{i_1}, X_{j_1})h(X_{i_2}, X_{j_2})\} = 0$, by degeneracy of h_2, as pointed out earlier. Let $\tilde{\sum}$ denote the sum over this set of indices and observe that there are at most $4!nk^2$ such indices with $d_1 = k$. Thus we find

$$\tilde{\sum}\left|E\left(h_2(X_{i_1}, X_{j_1})\, h_2(X_{i_2}, X_{j_2})\right)\right| \le C\,n\sum_{k=1}^{n} k^2\left(\beta_k + \bar{a}_k + \gamma(\bar{a}_k)\right).$$

In the second case, we apply Lemma 3.18 twice to get

$$\begin{aligned} &\left|E\left(h_2(X_{i_1}, X_{j_1})\, h_2(X_{i_2}, X_{j_2})\right)\right| \\ &\le C\left(\beta_{\lfloor d_1/3\rfloor} + \bar{a}_{\lfloor d_1/3\rfloor} + \gamma(\bar{a}_{\lfloor d_1/3\rfloor})\right) + C\left(\beta_{\lfloor d_2/3\rfloor} + \bar{a}_{\lfloor d_2/3\rfloor} + \gamma(\bar{a}_{\lfloor d_2/3\rfloor})\right). \end{aligned}$$

Let $\hat{\sum}$ denote the sum over the corresponding indices and observe that there are at most $4!\, n\, l^2$ such indices with $d_1 = k$ and $d_2 = l$. Thus we get

$$\hat{\sum} \left| E\left(h_2(X_{i_1}, X_{j_1})\, h_2(X_{i_2}, X_{j_2})\right)\right|$$
$$\leq C n \sum_{1\leq l\leq k\leq n} l^2 \left((\beta_k + \bar{a}_k + \gamma(\bar{a}_k)) + (\beta_l + \bar{a}_l + \gamma(\bar{a}_l))\right).$$

Combining all the estimates above with the summability conditions on $\beta_k, \bar{a}_k$ and $\gamma(\bar{a}_k)$, we get that

$$\sum_{1\leq i_1<j_1\leq n} \sum_{1\leq i_2<j_2\leq n} E[h_2(X_{i_1}, X_{j_1})\, h_2(X_{i_2}, X_{j_2})] \leq C\, n^2.$$

From here, we obtain directly (5.8), which implies weak convergence of $\sqrt{n} R_n$ to 0. □

5.4 Empirical U-process limit theorems. This section is devoted to the study of the empirical distribution function of data sets of the type $\{h(X_{i_1}, \ldots, X_{i_m}), 1 \leq i_1 < \cdots < i_m \leq n\}$, for some symmetric kernel $h : \mathbb{R}^m \to \mathbb{R}$. Our main result will be an invariance principle for the empirical process when the underlying observations are functionals of an absolutely regular process.

Definition 5.11. *Let $(X_n)_{n\geq 1}$ be a stationary stochastic process and let $h : \mathbb{R}^m \to \mathbb{R}$ be a measurable symmetric function. Then*

$$U_n(t) := \binom{n}{m}^{-1} \sum_{1\leq i_1<\cdots<i_m\leq n} 1_{\{h(X_{i_1},\ldots,X_{i_m})\leq t\}}, \quad t \in \mathbb{R},$$

is called a U-statistic empirical distribution function *or* empirical distribution function of U-statistic type.

Observe that $U_n : \mathbb{R} \to [0, 1]$ is the distribution function of the empirical U-statistic distribution $\binom{n}{m}^{-1} \sum \delta_{h(X_{i_1},\ldots,X_{i_m})}$, obtained by placing equal mass at each of the $\binom{n}{m}$ points $h(X_{i_1}, \ldots, X_{i_m})$, $1 \leq i_1 < \cdots < i_m \leq n$. Also note that for fixed $t \in \mathbb{R}$, $U_n(t)$ is simply a U-statistic with kernel $1_{\{h(x_1,\ldots,x_m)\leq t\}}$.

Example 5.12. (i) For $m = 1$ and $h(x) = x$, the corresponding U-statistic empirical distribution function is simply the empirical distribution function of the original observations $X_1, \ldots, X_n$, i.e.,

$$F_n(t) = \frac{1}{n} \sum_{i=1}^{n} 1_{\{X_i\leq t\}}.$$

(ii) The sample correlation integral, as defined in (5.2), is a U-statistic empirical distribution function with $m = 2$ and kernel $h(x, y) = \|x - y\|$.

The U-statistic empirical distribution function is the natural estimator for

$$U(t) = P(h(Y_1, \ldots, Y_m) \leq t)$$

where $Y_1, \ldots, Y_m$ are independent random variables, with the same one-dimensional marginal distributions as the process $(X_n)_{n\geq 1}$. In the case of independent observations, $U_n(t)$ has many nice properties. Among other things, $U_n(t)$ is an unbiased and strongly consistent estimator of

$U(t)$. The next theorem shows that consistency continues to hold in the stationary ergodic case. In the formulation of this and the following theorems, we will restrict attention to bivariate U-statistics, i.e., to the case $m = 2$. We remark that the results still hold for general m, with essentially unchanged proofs.

Theorem 5.13. *Let $(X_n)_{n\in\mathbb{Z}}$ be a stationary ergodic process, and let $h : \mathbb{R}^2 \to R$ be a symmetric kernel satisfying $\int\int 1_{\{h(x,y)=t\}} dF(x)dF(y) = 0$ for all t. Then, as $n \to \infty$,*

$$\sup_{t\in\mathbb{R}} |U_n(t) - U(t)| \to 0 \qquad \textit{almost surely.}$$

Proof. This result can be established in exactly the same way as the Glivenko–Cantelli lemma for the ordinary empirical distribution function. The only difference is that the U-statistic ergodic theorem of Aaronson et al. (1996), see Theorem 5.6 above, has to be employed to get pointwise convergence of $U_n(t)$ towards $U(t)$. □

In many statistical applications, approximations to the distribution of the centered process $\{U_n(t) - U(t); t \in \mathbb{R}\}$ are required. This is, e.g., the case if one wants to determine confidence bands for $\{U(t); t \in \mathbb{R}\}$, or if hypothesis tests are based on the U-statistic empirical distribution function. This motivates the study of the *empirical process of U-statistic structure*, or short *U-process*, $\{W_n(t); t \in \mathbb{R}\}$, defined by

$$W_n(t) = \sqrt{n}\,(U_n(t) - U(t)).$$

Weak convergence of empirical processes of U-statistic structure to a Gaussian process was established by Silverman (1983) and Serfling (1984) for the case of i.i.d. random variables. Further properties of U-processes were investigated by Dehling, Denker and Philipp (1987), Helmers, Janssen, and Serfling (1988), Nolan and Pollard (1988), and others. For weakly dependent observations, weak convergence of U-processes was established by Arcones and Yu (1994) and independently by Borovkova (1995). Their results hold for absolutely regular processes. In this chapter we will establish an invariance principle for functionals of an absolutely regular process.

As noted above, for fixed t, $U_n(t)$ is simply a U-statistic, with kernel given by

$$h(x, y; t) = 1_{\{h(x,y)\le t\}}. \tag{5.10}$$

Thus, a U-process can be viewed as a family of U-statistics, indexed by the parameter $t \in \mathbb{R}$. More generally, one can study U-processes indexed by any class $\mathcal{H}$ of kernels. In what follows we will consider 1-dimensional classes of kernels that can be parameterized by a real parameter $t \in [0, 1]$. We will denote the kernels by $h(x, y; t)$, but these are not necessarily of the form (5.10).

Let $\mathcal{H} = \{h(x, y; t) : t \in [0, 1]\}$ denote a class of kernel functions, and let $(X_n)_{n\in\mathbb{Z}}$ be a stationary stochastic process. We then define the processes $(\{U_n(t), t \in [0, 1]\})_{n\ge1}$, $\{U(t), t \in [0, 1]\}$ and $(\{W_n(t), t \in [0, 1]\})_{n\ge1}$ by

$$\begin{aligned}
U_n(t) &= \frac{2}{n(n-1)} \sum_{1\le i<j\le n} h(X_i, X_j, t),\\
U(t) &= \int h(x, y, t)\, dF(x)\, dF(y),\\
W_n(t) &= \sqrt{n}\,(U_n(t) - U(t)).
\end{aligned}$$

In our analysis, we need several assumptions on the class of kernels. We first assume that for all $x, y \in \mathbb{R}$ and $t \in [0, 1]$ we have

(A1) $0 \le h(x, y; t) \le 1$, $h(x, y, 0) = 0$.

(A2) $h(x, y; t)$ is increasing in t, for fixed $x, y \in \mathbb{R}$.

We will again apply the Hoeffding decomposition and define

$$h_1(x; t) = \int_{\mathbb{R}} h(x, y; t)dF(y)$$

(observe that, in contrast with earlier definitions of h_1, we do not center here at expectations). Note that, if the kernel function h satisfies the conditions (A1) and (A2), then the function h_1 satisfies analogous conditions, namely $0 \le h_1(x; t) \le 1$, $h_1(x; 0) = 0$ and h_1 is increasing in t for fixed x.

In addition, we have to impose uniform Lipschitz conditions on the integrated kernels $E_{X_0, X_k} h(X_0, X_k; t)$ and $E_{X_0} E_{X_k} h(X_0, X_k; t)$. We assume that there exists a constant $C > 0$ such that

$$|U(t) - U(s)| \le C|t - s|, \tag{5.11}$$

$$|Eh(X_0, X_k; t) - Eh(X_0, X_k; s)| \le C|t - s|, \tag{5.12}$$

hold for all $s, t \in [0, 1]$ and any $k \ge 1$.

In this section we want to study the behavior of the U-process $(\{W_n(t), 0 \le t \le 1\})_{n \ge 1}$ when the underlying sequence is a functional of an absolutely regular process. Under some regularity conditions, we prove weak convergence of the U-process to a Gaussian process with mean zero and covariance structure

$$E(W(s)W(t)) = 4\,\mathrm{Cov}\,(h_1(X_1; s), h_1(X_1; t)) \\ +4\sum_{k=1}^{\infty} \{\mathrm{Cov}\,(h_1(X_1; s), h_1(X_{k+1}; t)) + \mathrm{Cov}\,(h_1(X_1; t), h_1(X_{k+1}; s))\}. \tag{5.13}$$

The main result of this section is contained in the following theorem.

Theorem 5.14. *Let $(X_n)_{n \in \mathbb{Z}}$ be a 1-approximating functional with constants $(a_k)_{k \ge 0}$ of an absolutely regular process with mixing rate $(\beta_k)_{k \ge 0}$. Suppose that the class $\mathcal{H}$ of kernels satisfies* (A1) *and* (A2)*, that h_1 is uniformly 1-continuous and that* (5.11) *and* (5.12) *hold. Moreover, assume that the following summability conditions hold:*

$$\sum_{k=1}^{\infty} k^2(\bar{a}_k + \beta_k)^{1/2} < \infty, \tag{5.14}$$

$$\sum_{k=1}^{\infty} k^5 \gamma(\bar{a}_k) < \infty, \tag{5.15}$$

in addition to (4.26) *and* (4.27)*. Then the sequence of processes*

$$W_n = \sqrt{n}(U_n(t) - U(t)),\ t \in [0, 1],\ n \ge 1,$$

converges in distribution in $D[0, 1]$ to the mean-zero Gaussian process $(W(t), t \in [0, 1])$ with covariance structure given by (5.13)*. Moreover, the series* (5.13) *converges absolutely, and the limit process W has continuous sample paths on* $[0, 1]$ *with probability* 1.

Proof. For a fixed value of t, we apply Hoeffding's decomposition to the U-statistic $U_n(t)$, and get

$$U_n(t) = U(t) + \frac{2}{n}\sum_{i=1}^{n}(h_1(X_i, t) - U(t)) + R_n(t),$$

where

$$R_n(t) = \frac{2}{n(n-1)} \sum_{1\le i<j\le n} (h(X_i, X_j, t) - h_1(X_i, t) - h_1(X_j, t) + U(t)). \tag{5.16}$$

Hence the process $W_n(t)$ can be decomposed in the same way, as

$$W_n(t) = \frac{2}{\sqrt{n}}\sum_{i=1}^{n}(h_1(X_i, t) - U(t)) + \sqrt{n}R_n(t) = W_n'(t) + \sqrt{n}R_n(t),$$

where W_n' is defined by

$$W_n'(t) = \frac{2}{\sqrt{n}}\sum_{i=1}^{n}(h_1(X_i, t) - U(t)). \tag{5.17}$$

It will turn out that $W_n'(t)$ is the dominating linear term and that $\sqrt{n}R_n(t)$ is the remainder term. Note that the process $\{W_n'(t), 0 \le t \le 1\}$ has the form of an empirical process indexed by the class of functions $\{h_1(x, t),\ 0 \le t \le 1\}$. Thus, Theorem 4.8 implies that W_n' converges weakly in $D[0, 1]$ to the mean-zero Gaussian process W with covariance structure given by (5.13). By Slutsky's theorem, this establishes weak convergence of $(W_n)_{n\ge 1}$ to the same limit, provided we can show that the remainder terms $\{\sqrt{n}R_n(t); 0 \le t \le 1\}$ converge uniformly in probability to 0. This convergence of the remainder term to 0 is the essential part of the proof of weak convergence of U-processes. We present its proof in the rest of this section.

5.5 A useful chaining lemma. In this subsection, we present a lemma that establishes conditions under which an arbitrary sequence of processes indexed by [0, 1] uniformly converges to zero in probability. This lemma, established by Borovkova et al. (2001) does not require a specific form of the process or a specific dependence structure, and might thus be of interest in other contexts. A variant of the lemma was used earlier in the work of Dehling and Taqqu (1989) on empirical processes of long-range dependent observations.

Recall that a function f on $[0, 1]$ is called Lipschitz-continuous if for all $s, t \in [0, 1]$ there is a constant C such that

$$|f(t) - f(s)| \le C\,|t - s|.$$

In what follows we adopt the convention that C denotes a generic positive constant, not always taking the same value at each occurrence.

Lemma 5.15. *Let $\{X_n(t),\ t \in [0, 1]\}_{n\in\mathbb{N}}$ be a sequence of real-valued stochastic processes such that $X_n(0) = 0$ almost surely. Suppose that for some positive constants α, β, γ with $\alpha \le 1$, $\gamma < 1$ and $\alpha(1-\gamma) < \beta$, the following two conditions hold:*

(i) *There exists a Lipschitz-continuous function f on $[0, 1]$ such that*

$$E|X_n(t) - X_n(s)|^2 \le \frac{|f(t) - f(s)|^\gamma}{n^\beta} \quad \forall s, t \in [0, 1]. \tag{5.18}$$

(ii) *There exists a Lipschitz-continuous function g, a monotone Lipschitz-continuous function Λ and a sequence of stochastic processes $\{Y_n(t), t \in [0,1]\}_{n\geq 1}$ satisfying*

$$E|Y_n(t) - Y_n(s)|^r \leq C\left[(g(t)-g(s))^{1+h} + \frac{1}{n}(g(t)-g(s))^{\frac{1+h}{2}}\right] \quad \forall s,t \in [0,1] \tag{5.19}$$

for some $r > 0$, $0 < h < 1$, such that for all s, t, δ with $0 \leq s \leq t < s+\delta \leq 1$:

$$|X_n(t) - X_n(s)| \leq |X_n(s+\delta) - X_n(s)| + |Y_n(s+\delta) - Y_n(s)| + |\Lambda(s+\delta) - \Lambda(s)|\, n^\alpha. \tag{5.20}$$

Then, as $n \to \infty$,

$$\sup_{0\leq t\leq 1} |X_n(t)| \longrightarrow 0 \text{ in probability.}$$

Proof. For $k = 0, 1, \ldots, K$, where K will be chosen later, we introduce refining partitions of $[0,1]$ into 2^k subintervals, $0 = s_0^{(k)} < s_1^{(k)} < \cdots < s_{2^k}^{(k)} = 1$, in such a way that for any $i = 1, \ldots, 2^k$

$$|s_i^{(k)} - s_{i-1}^{(k)}| \leq 2^{-k}. \tag{5.21}$$

Due to Lipschitz-continuity of Λ we have

$$|\Lambda(s_i^{(k)}) - \Lambda(s_{i-1}^{(k)})| \leq C\, 2^{-k}. \tag{5.22}$$

For any $t \in [0,1]$ we denote by $i_k(t)$ the index of the largest point of the k-th partition that is to the left of t, i.e., such that $s_{i_k(t)}^{(k)} \leq t < s_{i_k(t)+1}^{(k)}$. Note that $s_{i_k(t)}^{(k)} \leq s_{i_{k+1}(t)}^{(k+1)}$, as the $(k+1)$-st partition is a refinement of the k-th partition, and thus contains more grid points. In this way we obtain a chain linking 0 and t:

$$0 = s_{i_0(t)}^{(0)} \leq s_{i_1(t)}^{(1)} \leq \cdots \leq s_{i_K(t)}^{(K)} \leq t < s_{i_K(t)+1}^{(K)},$$

and we can write

$$X_n(t) = \sum_{k=1}^{K} \left\{X_n(s_{i_k(t)}^{(k)}) - X_n(s_{i_{k-1}(t)}^{(k-1)})\right\} + X_n(t) - X_n(s_{i_K(t)}^{(K)}).$$

From this we get the following upper bound on $\sup_{0\leq t\leq 1} X_n(t)$:

$$\begin{aligned}\sup_{0\leq t\leq 1} |X_n(t)| \quad &\leq \quad \sum_{k=1}^{K} \max_{i_k(t), i_{k-1}(t)} \left|X_n(s_{i_k(t)}^{(k)}) - X_n(s_{i_{k-1}(t)}^{(k-1)})\right| \\ &\qquad + \sup_{s_{i_K(t)}^{(K)} \leq t < s_{i_K(t)+1}^{(K)}} \left|X_n(t) - X_n(s_{i_K(t)}^{(K)})\right|.\end{aligned}$$

For the second term on the r.h.s. we can find an upper bound by applying the inequalities (5.20) and (5.22),

$$\begin{aligned}\sup_{s_{i_K(t)}^{(K)} \leq t < s_{i_K(t)+1}^{(K)}} \left|X_n(t) - X_n(s_{i_K(t)}^{(K)})\right| \quad &\leq \quad \max_{i_K(t)} \left|X_n(s_{i_K(t)+1}^{(K)}) - X_n(s_{i_K(t)}^{(K)})\right| \\ &\qquad + \max_{i_K(t)} \left|Y_n(s_{i_K(t)+1}^{(K)}) - Y_n(s_{i_K(t)}^{(K)})\right| + C\frac{n^\alpha}{2K}.\end{aligned}$$

Now, let $\epsilon > 0$ be given. We choose $K = [\alpha \log_2 n + \log_2(2C/\epsilon)] + 1$, where $\log_2$ denotes the logarithm with base 2. Thus $C n^\alpha / 2^K < \epsilon/2$, and hence, as $\sum_{k=1}^{\infty} \frac{1}{(k+2)^2} < 1/2$,

$$\begin{aligned}
P&\left(\sup_{0\le t\le 1} |X_n(t)| > \epsilon\right) \\
&\le \sum_{k=1}^{K} P\left(\max_{i_{k-1}(t), i_k(t)} \left|X_n(s_{i_k(t)}^{(k)}) - X_n(s_{i_{k-1}(t)}^{(k-1)})\right| > \frac{\epsilon}{(k+2)^2}\right) \\
&+P\left(\max_{i_K(t)} \left|X_n(s_{i_K(t)+1}^{(K)}) - X_n(s_{i_K(t)}^{(K)})\right| > \frac{\epsilon}{(K+3)^2}\right) \\
&+P\left(\max_{i_K(t)} \left|Y_n(s_{i_K(t)+1}^{(K)}) - Y_n(s_{i_K(t)}^{(K)})\right| > \frac{\epsilon}{(K+4)^2}\right).
\end{aligned} \tag{5.23}$$

By the dyadic refining structure of our partitions, there are two possibilities for consecutive points in the chain: either $s_{i_{k-1}(t)}^{(k-1)} = s_{i_k(t)}^{(k)}$ or $s_{i_{k-1}(t)}^{(k-1)} = s_{i_k(t)+1}^{(k)}$. Hence using Markov's inequality, (5.18) and (5.21), we obtain

$$\begin{aligned}
P&\left(\max_{i_{k-1}(t), i_k(t)} \left|X_n(s_{i_k(t)}^{(k)}) - X_n(s_{i_{k-1}(t)}^{(k-1)})\right| > \frac{\epsilon}{(k+2)^2}\right) \\
&\le \sum_{i=1}^{2^k} P\left(\left|X_n(s_i^{(k)}) - X_n(s_{i-1}^{(k)})\right| > \frac{\epsilon}{(k+2)^2}\right) \\
&\le \sum_{i=1}^{2^k} \frac{\left|f(s_i^{(k)}) - f(s_{i-1}^{(k)})\right|^\gamma (k+2)^4}{n^\beta \epsilon^2} \\
&\le C\frac{(k+2)^4}{n^\beta\epsilon^2} \sum_{i=1}^{2^k} \left|s_i^{(k)} - s_{i-1}^{(k)}\right|^\gamma \le C\frac{(k+2)^4}{n^\beta \epsilon^2} 2^{k(1-\gamma)}.
\end{aligned} \tag{5.24}$$

In exactly the same way we get

$$P\left(\max_{i_K(t)} \left|X_n(s_{i_K(t)+1}^{(K)}) - X_n(s_{i_K(t)}^{(K)})\right| > \frac{\epsilon}{(K+3)^2}\right) \le C\frac{(K+3)^4}{n^\beta\epsilon^2} 2^{K(1-\gamma)}. \tag{5.25}$$

Finally, we can bound the third term on the r.h.s. of (5.23), by using Markov's inequality and (5.19),

$$\begin{aligned}
P&\left(\max_{i_K(t)} \left|Y_n(s_{i_K(t)+1}^{(K)}) - Y_n(s_{i_K(t)}^{(K)})\right| > \frac{\epsilon}{(K+4)^2}\right) \\
&\le \sum_{i=1}^{2^K} P\left(\left|Y_n(s_i^{(K)}) - Y_n(s_{i-1}^{(K)})\right| > \frac{\epsilon}{(K+4)^2}\right) \\
&\le \frac{(K+4)^{2r}}{\epsilon^r} \sum_{i=1}^{2^K} E\left|Y_n(s_i^{(K)}) - Y_n(s_{i-1}^{(K)})\right|^r \\
&\le C\frac{(K+4)^{2r}}{\epsilon^r} 2^K \left(\left(\frac{1}{2^K}\right)^{1+h} + \frac{1}{n}\left(\frac{1}{2^K}\right)^{\frac{1+h}{2}}\right) \\
&\le C\frac{(K+4)^{2r}}{\epsilon^r} 2^{-Kh},
\end{aligned} \tag{5.26}$$

as $1/n \leq C(1/2^K)^{(1+h)/2}$ by our choice of K and by the conditions imposed on the parameters h and α. In the final step, we combine the inequalities (5.23), (5.24), (5.25) and (5.26) with our conditions on α, β, γ to obtain

$$\begin{aligned} P\left(\sup_{0\leq t\leq 1}|X_n(t)|>\epsilon\right) &\leq C\,\frac{K(K+3)^4}{n^\beta\epsilon^2}2^{K(1-\gamma)}+C\,\frac{(K+4)^{2r}}{\epsilon^r}2^{-Kh} \\ &\leq C\,\frac{(K+3)^5}{\epsilon^2}n^{\alpha(1-\gamma)-\beta}+C\,\frac{(K+4)^{2r}}{\epsilon^r}n^{-\alpha h}\longrightarrow 0, \end{aligned}$$

because $(K+3)^p = (\alpha\log_2 n + \log_2(2C/\epsilon)+4)^p = o(n^\tau)$ for any $\tau, p > 0$. □

Chaining techniques as explored in the proof of Lemma 5.15 are well known and widely applied in the theory of empirical processes. In distinction to known results, Lemma 5.15 allows for an extra random term Y_n in the bound (5.20). This makes Lemma 5.15 applicable to U-processes where Y_n will be the first order term in the Hoeffding expansion. The second distinctive feature is the fact that only second moment bounds are required, thus avoiding technically involved higher moment calculations. This is made possible by the extra $n^{-\beta}$ term in (5.18), an idea exploited earlier by Dehling and Taqqu (1989) in their proof of the invariance principle for the empirical process of long-range dependent sequences.

5.6 Proof of Theorem 5.14. The remainder of this section is devoted to the proof of Theorem 5.14. What remains to be done is to show that the remainder term $R_n(t)$ in the Hoeffding decomposition, as defined in (5.16) converges uniformly to 0.

Lemma 5.16. *Under the conditions of Theorem 5.14,*

$$\sup_{t\in[0,1]}\sqrt{n}\,|R_n(t)|\longrightarrow 0 \text{ in probability, as } n\longrightarrow\infty.$$

In the proof of this lemma we will make use of Lemma 5.15, which requires an upper bound on the increments of R_n. We formulate this bound as a separate lemma.

Lemma 5.17. *Under the conditions of Theorem 5.14, there exists a nonnegative constant C such that*

$$E\left(R_n(t)-R_n(s)\right)^2 \leq C\,\frac{|t-s|^{1/2}}{n^2} \tag{5.27}$$

for all $s, t \in [0, 1]$.

Proof. We can write the increment $R_n(t) - R_n(s)$ of the remainder process (5.16) as

$$R_n(t)-R_n(s)=\frac{2}{n(n-1)}\sum_{1\leq i<j\leq n}h_2(X_i,X_j),$$

where the terms inside the sum are given by

$$\begin{aligned} h_2(x,y)=h_{2;s,t}(x,y) &= (h(x,y;t)-h(x,y;s))-(h_1(x;t)-h_1(x;s)) \\ &\quad -(h_1(y;t)-h_1(y;s))+(U(t)-U(s)) \end{aligned}$$

(in what follows we will drop the index s, t on $h_2(x, y)$ hoping that this will not give rise to confusion). Then the l.h.s. of (5.27) becomes

$$E(R_n(t)-R_n(s))^2=\frac{4}{n^2(n-1)^2}\sum_{1\leq i_1<j_1\leq n,\,1\leq i_2<j_2\leq n}E\left(h_2(X_{i_1},X_{j_1})h_2(X_{i_2},X_{j_2})\right).$$

Note that h_2 is a degenerate kernel and that hence $E_{X_{j_2}} h_2(X_{i_1}, X_{j_1}) h_2(X_{i_2}, X_{j_2}) = 0$ provided j_2 is different from the other three indices, and likewise for any other index occurring only once in $\{i_1, j_1, i_2, j_2\}$.

We put the indices i_1, j_1, i_2, j_2 into increasing order and note that at least one of these indices is different from the remaining three. We denote by d_1 the largest difference between two consecutive indices and by d_2 the second largest. The location of the largest difference splits the set $\{i_1, j_1, i_2, j_2\}$ into two nonempty sets I_1 and I_2 in such a way that for $l_1 \in I_1$ and $l_2 \in I_2$ we have $l_2 - l_1 \geq d_1$. At this point we have to distinguish two cases; either I_1 or I_2 is a singleton or $|I_1| = |I_2| = 2$. In the first case, $E_{X_{I_1}} E_{X_{I_2}} h_2(X_{i_1}, X_{j_1}) h_2(X_{i_2}, X_{j_2}) = 0$ and thus we may apply Lemma 3.18 with $r = s = \frac{1}{2}$ to obtain

$$|E\left\{h_2(X_{i_1}, X_{j_1}) h_2(X_{i_2}, X_{j_2})\right\} \leq 4\, M_{s,t} \left(\bar{a}_{\lfloor \frac{d_1}{3} \rfloor} + \beta_{\lfloor \frac{d_1}{3} \rfloor}\right)^{\frac{1}{2}} + 2\gamma(\bar{a}_{\lfloor \frac{d_1}{3} \rfloor}),$$

where

$$\begin{aligned} M_{s,t} &= \max\Big\{(E|h_2(X_{i_1}, X_{j_1}) h_2(X_{i_2}, X_{j_2})|^2)^{1/2}; \\ &\quad (E_{X_{j_2}} E_{X_{i_1} X_{j_1} X_{i_2}} |h_2(X_{i_1}, X_{j_1}) h_2(X_{i_2}, X_{j_2})|^2)^{1/2}\Big\}. \end{aligned}$$

Note that $h(x, y; t)$, $h_1(x; t)$ and $U(t)$ are increasing functions of t, taking values between 0 and 1, and that $Eh_1(X; t) = U(t)$. Hence $|J(x, y)| \leq 2$ and thus

$$E|h_2(X_{i_1}, X_{j_1}) h_2(X_{i_2}, X_{j_2})|^2 \leq 8\, E|h_2(X_{i_1}, X_{j_1})|,$$

and similarly

$$E_{X_{j_2}} E_{X_{i_1}, X_{i_2}, X_{j_2}} |h_2(X_{i_1}, X_{j_1}) h_2(X_{i_2}, X_{j_2})|^2 \leq 8\, E|h_2(X_{i_1}, X_{j_1})|.$$

Moreover, we get

$$\begin{aligned} E\left|h_2(X_{i_1}, X_{j_1})\right| &\leq \left\{E\left(h(X_{i_1}, X_{j_1}; t) - h(X_{i_1}, X_{j_1}; s)\right)\right. \\ &\qquad \left. +2\, E\left(h(X_{i_1}; t) - h(X_{i_1}; s)\right) + (U(t) - U(s))\right\} \\ &\leq \left\{E\left(h(X_{i_1}, X_{j_1}; t) - h(X_{i_1}, X_{j_1}; s)\right) + 3\,(U(t) - U(s))\right\}. \end{aligned}$$

This, together with the conditions imposed on h and U, implies the following upper bound on $M_{s,t}$:

$$M_{s,t} \leq C|t - s|^{1/2},$$

and thus

$$\left|E\left\{h_2(X_{i_1}, X_{j_1}) h_2(X_{i_2}, X_{j_2})\right\}\right| \leq C|t - s|^{1/2} \left(\bar{a}_{\lfloor \frac{d_1}{3} \rfloor} + \beta_{\lfloor \frac{d_1}{3} \rfloor}\right)^{\frac{1}{2}} + 2\gamma(\bar{a}_{\lfloor \frac{d_1}{3} \rfloor}). \tag{5.28}$$

We apply this inequality for $d_1 > d_0$, where $d_0 = d_0(n, s, t)$ denotes an integer whose specific value is to be chosen later. For $d_1 \leq d_0$, i.e., when the maximal separation between consecutive indices is small, we use the fact that $|h_2(x, y)| \leq 2$ to get the bound

$$E|h_2(X_{i_1}, X_{j_1}) h_2(X_{i_2}, X_{j_2})| \leq 2\, E|h_2(X_{i_1}, X_{j_1})| \leq C\,|t - s|. \tag{5.29}$$

Let now $\tilde{\sum}$ denote the sum over those indices where either I_1 or I_2 is a singleton so that the inequality (5.28) applies. We then find, observing that there are at most $4!\,n\,k$ such indices with $d_1 = k$,

$$\begin{aligned}
&\tilde{\sum} \left| E\left\{h_2(X_{i_1}, X_{j_1})h_2(X_{i_2}, X_{j_2})\right\}\right| \\
&\quad = \left(\tilde{\sum}_{d_1 \le d_0} + \tilde{\sum}_{d_1 > d_0}\right) \left| E\left\{h_2(X_{i_1}, X_{j_1})h_2(X_{i_2}, X_{j_2})\right\}\right| \\
&\quad \le C\left\{\tilde{\sum}_{d_1 \le d_0} |t-s| + \tilde{\sum}_{d_1 > d_0} \left[|t-s|^{1/2}\left(\bar{a}_{\lfloor \frac{d_1}{3}\rfloor} + \beta_{\lfloor \frac{d_1}{3}\rfloor}\right)^{\frac{1}{2}} + 2\,\gamma(\bar{a}_{\lfloor \frac{d_1}{3}\rfloor})\right]\right\} \\
&\quad \le C\,n\left\{ d_0^3|t-s| + \sum_{k=1}^{n} k^2\left(\bar{a}_{\lfloor \frac{k}{3}\rfloor} + \beta_{\lfloor \frac{k}{3}\rfloor}\right)^{1/2} |s-t|^{1/2} + \sum_{k=d_0}^{n} k^2 \gamma(\bar{a}_{\lfloor \frac{k}{3}\rfloor})\right\}.
\end{aligned}$$

Note that (5.15) implies that

$$\sum_{k=d_0}^{\infty} k^2 \gamma(\bar{a}_{\lfloor k/3\rfloor}) \le C \sum_{k=d_0}^{\infty} k^2\gamma(\bar{a}_k) \le \frac{C}{d_0^3}\sum_{k=d_0}^{\infty} k^5 \gamma(\bar{a}_k) \le C\, d_0^{-3}.$$

Thus, choosing $d_0 := |t-s|^{-1/6}$, we may conclude that

$$\tilde{\sum} \left| E h_2(X_{i_1}, X_{j_1})h_2(X_{i_2}, X_{j_2})\right| \le C\,n\,|t-s|^{1/2}. \tag{5.30}$$

We now turn to the second case, i.e., where $|I_1| = |I_2| = 2$, and we denote the corresponding sum by $\hat{\sum}$. Again, we let $d_0 = d_0(n, s, t)$ denote an integer whose value will be chosen later. For $d_1 > d_0$, we apply the correlation inequality of Lemma 3.18 to get

$$\begin{aligned}
|E h_2(X_{i_1}, X_{j_1})h_2(X_{i_2}, X_{j_2})| \;\le\; & |E_{X_{i_1}, X_{i_2}} E_{X_{j_1}, X_{j_2}} h_2(X_{i_1}, X_{j_1}) h_2(X_{i_2}, X_{j_2})| \\
& + 4\, M_{s,t}\left(\bar{a}_{\lfloor \frac{d_1}{3}\rfloor} + \beta_{\lfloor \frac{d_1}{3}\rfloor}\right)^{1/2} + 2\,\gamma\left(\bar{a}_{\lfloor \frac{d_1}{3}\rfloor}\right)
\end{aligned}$$

where again as above we find that $M_{s,t} \le C\,|t-s|^{1/2}$. For $d_1 \le d_0$, i.e., for small separation between consecutive indices, we use (5.29). In total we find

$$\begin{aligned}
&\hat{\sum} \left| E h_2(X_{i_1}, X_{j_1})h_2(X_{i_2}, X_{j_2})\right| \\
&\quad \le \hat{\sum} \left| E_{X_{I_1}} E_{X_{I_2}} h_2(X_{i_1}, X_{j_1})h_2(X_{i_2}, X_{j_2})\right| + C\,n\,|t-s|^{1/2}.
\end{aligned}$$

In treating the terms inside the sum on the r.h.s., we have to distinguish the cases $d_2 \le d_0$ and $d_2 > d_0$. In the first case, we apply (5.29). In the second case, we apply the correlation inequality of Lemma 3.18 to find

$$\left| E_{X_{I_1}} E_{X_{I_2}} h_2(X_{i_1}, X_{j_1})h_2(X_{i_2}, X_{j_2})\right| \le 4\,M_{s,t}\left(\bar{a}_{\lfloor \frac{d_2}{3}\rfloor} + \beta_{\lfloor \frac{d_2}{3}\rfloor}\right)^{1/2} + 2\,\gamma\left(\bar{a}_{\lfloor \frac{d_2}{3}\rfloor}\right).$$

Taken together, this yields the following bound for $\hat{\sum}$:

$$\begin{aligned}
&\hat{\sum} \left| E\left\{h_2(X_{i_1}, X_{j_1})h_2(X_{i_2}, X_{j_2})\right\}\right| \\
&\quad \le C \hat{\sum}_{d_2 > d_0} \left(M_{s,t}\left(\bar{a}_{\lfloor \frac{d_2}{3}\rfloor} + \beta_{\lfloor \frac{d_2}{3}\rfloor}\right)^{1/2} + \gamma(\bar{a}_{\lfloor \frac{d_2}{3}\rfloor})\right) \\
&\qquad + \hat{\sum}_{d_2 < d_0} C\,|t-s| + C\,n\,|t-s|^{1/2}
\end{aligned}$$

where again $M_{s,t} \le C\,|t-s|^{1/2}$. The first of the three sums on the r.h.s. is bounded by

$$C\sum_{k=1}^{n} n^2 k \left(\bar a_{\lfloor \frac{k}{3}\rfloor} + \beta_{\lfloor \frac{k}{3}\rfloor}\right)^{1/2} |t-s|^{1/2} \le C\,n^2\,|t-s|^{1/2}.$$

We now choose $d_0 = |t-s|^{-1/4}$, and get the following upper bound on the second term:

$$\hat{\sum}_{d_2\ge d_0} \gamma(\bar a_{\lfloor \frac{d_2}{3}\rfloor}) \le \sum_{k=d_0}^{n} n^2\,k\,\gamma(\bar a_{\lfloor \frac{k}{3}\rfloor}) = n^2 \sum_{k=d_0}^{n} k\,\gamma(\bar a_{\lfloor \frac{k}{3}\rfloor}) \le C\,\frac{n^2}{d_0^2} = C\,n^2\,|t-s|^{1/2}.$$

Finally, we obtain

$$\hat{\sum}_{d_2<d_0} C\,|t-s| \le n^2\,d_0^2\,|t-s| \le C\,n^2\,|t-s|^{1/2}.$$

All these estimates together prove the statement of the lemma. □

Proof of Lemma 5.16. In our proof, we apply Lemma 5.15 to the processes $X_n(t) = \sqrt{n}R_n(t)$. Observe that, according to (5.27), condition (5.18) of Lemma 5.15 holds with $\gamma = \frac{1}{2}$ and $\beta = 2$. Moreover, for any $s, t, \delta > 0$ with $0 \le s \le t \le s+\delta \le 1$ we have

$$\begin{aligned}
&R_n(t) - R_n(s) \qquad (5.31)\\
&= \frac{2}{n(n-1)} \sum_{1\le i\ne j\le n} [(h(X_i,X_j,t) - h(X_i,X_j,s))\\
&- (h_1(X_i,t) - h_1(X_i,s)) - (h_1(X_j,t) - h_1(X_j,s)) + (U(t) - U(s))]\\
&\le \frac{2}{n(n-1)} \sum_{1\le i\ne j\le n} [(h(X_i,X_j,s+\delta) - h(X_i,X_j,s))\\
&- (h_1(X_i,s+\delta) - h_1(X_i,s)) - (h_1(X_j,s+\delta) - h_1(X_j,s)) + (U(s+\delta) - U(s))]\\
&+ \frac{2}{n(n-1)} \sum_{1\le i\ne j\le n} [(h_1(X_i,s+\delta) - h_1(X_i,s)) + (h_1(X_j,s+\delta) - h_1(X_j,s))]\\
&\le |R_n(s+\delta) - R_n(s)| + \frac{2}{n}\Big|\sum_{i=1}^{n} [(h_1(X_i,s+\delta) - U(s+\delta))\\
&- (h_1(X_i,s) - U(s))]\Big| + 2|U(s+\delta) - U(s)|\\
&\le |R_n(s+\delta) - R_n(s)| + \frac{1}{\sqrt{n}}|W_n'(s+\delta) - W_n'(s)| + 2|U(s+\delta) - U(s)|
\end{aligned}$$

(recall the definition of $W_n'(t)$ from (5.17)). In the same way we get

$$\begin{aligned}
&R_n(s) - R_n(t) \qquad (5.32)\\
&\le |R_n(s+\delta) - R_n(s)| + \frac{2}{n(n-1)} \sum_{1\le i\ne j\le n} [(h_1(X_i,t) - h_1(X_i,s))\\
&+(h_1(X_j,t) - h_1(X_j,s)) + (h_1(X_j,s+\delta) - h_1(X_j,s))\\
&+(h_1(X_i,s+\delta) - h_1(X_i,s))]\\
&\le \frac{4}{n}\sum_{i=1}^{n} [(h_1(X_i,s+\delta) - U(s+\delta)) - (h_1(X_i,s) - U(s))] + 4\,|U(s+\delta) - U(s)|\\
&\le |R_n(s+\delta) - R_n(s)| + \frac{2}{\sqrt{n}}|W_n'(s+\delta) - W_n'(s)| + 4\,|U(s+\delta) - U(s)|.
\end{aligned}$$

We conclude that

$$\sqrt{n}\,|R_n(t) - R_n(s)| \leq \sqrt{n}\,|R_n(s+\delta) - R_n(s)| + 2\,|W_n'(s+\delta) - W_n'(s)| + 4\sqrt{n}\,|U(s+\delta) - U(s)|. \tag{5.33}$$

Note that (5.33) is the representation (5.20) of Lemma 5.15 for the process $\sqrt{n}R_n(t)$. Finally, condition (5.19) of Lemma 5.15 holds for $W_n'(t)$ with $r = 4$ by an application of the fourth moment bound (4.31). Thus, by applying Lemma 5.15 to the process $\sqrt{n}R_n(t)$, the statement of the lemma follows. □

6 Empirical processes of long-range dependent data

Long-range dependence of stochastic processes is a phenomenon often encoutered in real-life data. One of the first to observe this was Hurst (1951) who found in his analysis of Nile river flood data that the behavior of maximal flood heights over varying periods of time could not be explained by traditional weakly dependent time series models. Other phenomena that are characteristic of long-range dependence are slow polynomial decay of autocorrelations and a pole of the spectral density at zero. The whole area of long-range dependence has been immensily popularized by the publications of Mandelbrot (1982) who found evidence for this phenomenon in processes encountered in economics, finance, engineering and nature. In this section, we will introduce a Gaussian model of long-range dependence and investigate the behavior of the corresponding empirical process.

6.1 Introduction and models. So far, we have entirely focused our attention on concepts of weak dependence. More or less, we found that, under suitable mixing assumptions, the results from the i.i.d. case continued to hold. For example, the partial sums will still converge weakly to a normal law and the empirical process has a limit that resembles a Brownian bridge. In this section we want to look at long-range dependent (also called long memory) processes, where very different phenomena occur. It is, however, not so easy to give a general demarkation line between weak dependence and long-range dependence. Roughly speaking, long-range dependence persists in the limit. A good picture is that if you look at the process from a distance, of course normalized so that one still sees the process, then in the weakly dependent case one gets independent noise whereas in the long-range dependent case the limit process still exhibits dependence.

Formalizing the above idea somewhat further, let us consider a stationary mean-zero stochastic process $(X_j)_{j\geq 0}$. Given a positive integer N, we define the block renormalization process $(Z_k^N)_{k\geq 1}$ by

$$Z_k^N := \frac{1}{d_N} \sum_{(k-1)N<j\leq kN} X_j, \tag{6.1}$$

$k = 1, 2, \ldots$. The Z_k^N are normalized partial sums over a block of N consecutive integers and capture the idea that one looks at the process from a distance. Suppose that, as $N \to \infty$, the processes $(Z_k^N)_{k\geq 1}$ converge weakly to a process $(Z_k)_{k\geq 1}$ in the sense of weak convergence of the finite dimensional distributions. We could call the process weakly dependent if the renormalization limit is a process of independent identically distributed random variables and long-range dependent if the limit process is dependent. We should stress here that the weak spot in this definition is the fact that it is not at all clear whether the process $(X_j)_{j\geq 1}$ admits a block renormalization that converges in distribution. Still, this informal definition serves its purpose as intuitive guideline.

Often one can tell from the covariance function of a process if a possible limit of renormalized blocks can be independent or not. Suppose for the moment that the autocovariance function of $(X_j)_{j\geq 1}$ satisfies

$$r(k) = EX_0X_k = k^{-d}L(k), \tag{6.2}$$

where the exponent d satisfies $0 < d < 1$, and where $L : \mathbb{R}_+ \to \mathbb{R}_+$ is a slowly varying function, i.e., $L(cx)/L(x) \to 1$ as $x \to \infty$ for every $c > 0$. For simplicity we also assume that $r(0) = \text{Var}(X_j) = 1$. Then the variance of the partial sum $S_N := \sum_{j=1}^N X_j$ is given by

$$\begin{aligned}
\text{Var}(S_N) = \sum_{1\leq j,k\leq N} r(|j-k|) &= N\,r(0) + 2\sum_{k=1}^{N-1}(N-k)\,r(k) \\
&= N + 2N\sum_{k=1}^{N-1} k^{-d}L(k) - 2\sum_{k=1}^{N-1} k^{1-d}L(k) \\
&\sim N + 2N\frac{N^{1-d}}{1-d}L(N) - 2\frac{N^{2-d}}{2-d}L(N) \\
&\sim \frac{2}{(1-d)(2-d)}N^{2-d}L(N) \\
&=: C_d N^{2H} L(N),
\end{aligned}$$

where $H := 1 - \frac{d}{2}$ is called the Hurst coefficient and $C_d = \frac{2}{(1-d)(2-d)}$. Here we used Karamata's theorem in order to evaluate the asymptotic behavior of the above sums; see Bingham et al. (1987). Observe that by the assumption on d we have $\frac{1}{2} < H < 1$. Also note that the behavior of the variance of the partial sum S_N is very different if the power d in (6.2) is larger than 1. In that case the autocovariances are absolutely summable, i.e., $\sum_{k=1}^{\infty} |r(k)| < \infty$, and thus

$$\begin{aligned}
\text{Var}(S_N) &= N\left(r(0) + 2\sum_{k=1}^{N}(1 - \frac{k}{N})\,r(k)\right) \\
&\sim N\left(r(0) + 2\sum_{k=1}^{\infty} r(k)\right)
\end{aligned}$$

so that we have linear growth of the variance, indicative of weak dependence.

We return to the case of nonsummable autocovariances, i.e., $0 < d < 1$ or $\frac{1}{2} < H < 1$. Then we have seen above that $\text{Var}(S_N) \sim C_d N^{2H} L(N)$. Hence

$$d_N = N^H \sqrt{C_d L(N)}$$

is an appropriate renormalization for N-blocks. With this choice of d_N, we get

$$\text{Var}(Z_1^N + Z_2^N) = \frac{1}{d_N^2}\text{Var}(S_{2N}) = \frac{C_d(2N)^{2H}L(2N)}{C_d N^{2H} L(N)} \to 2^{2H}.$$

Thus, assuming that the 2nd moments of $(Z_k^N)_{k\geq 1}$ converge to those of $(Z_k)_{k\geq 1}$ as $N \to \infty$, we get $\text{Var}(Z_1 + Z_2) = 2^{2H} > 2 = \text{Var}(Z_1) + \text{Var}(Z_2)$. Hence the (Z_k)-process cannot be independent.

Instead of the block renormalized process, we can also study the renormalized partial sum process $(S_{[Nt]}/d_N, t \geq 0)$. Its variance satisfies

$$\begin{aligned}\operatorname{Var}\left(\frac{1}{d_N} S_{[Nt]}\right) &\sim \frac{C_d (Nt)^{2H} L(Nt)}{d_N^2} \\ &= \frac{(Nt)^{2H} L(Nt)}{N^{2H} L(N)} \sim t^{2H},\end{aligned}$$

as $N \to \infty$. If $(S_{[Nt]}/d_N, t \geq 0)$ converges weakly to a limit process $(Z(t), t \geq 0)$ and if the 2nd moments converge, too, then

$$\operatorname{Var}(Z(t)) = t^{2H}, \tag{6.3}$$

and this being nonlinear, $(Z(t), t \geq 0)$ does not have independent increments. This is again in sharp contrast with the case of summable autocovariances, i.e., $d > 1$, where $\operatorname{Var}(Z(t)) = t$ so that increments are at least uncorrelated. Summarizing these considerations, we see that in case the autocovariances satisfy (6.2), there is a clear demarkation line between weak and long-range dependence that lies at $d = 1$.

So far, the considerations have been vague, especially as none of the assumed convergences really has to hold. If one wants to become more specific, one has to study specific models. The simplest of these is a mean-zero Gaussian process $(X_j)_{j \geq 1}$ with autocovariance function satisfying (6.2). Without loss of generality we may assume that $r(0) = 1$, i.e., that the marginal distributions are standard normal. The following result, establishing weak convergence of the partial sum process in this model is due to Taqqu (1975).

Theorem 6.1. *Let $(X_j)_{j \geq 1}$ be a mean-zero Gaussian process with autocovariance function $r(k) = EX_0 X_k$ satisfying* (6.2), *where $0 < d < 1$, and where L is a slowly varying function. Assume moreover that $r(0) = 1$. Then the normalized partial sum processes $\{\frac{1}{d_N} \sum_{j \leq Nt} X_j, 0 \leq t \leq 1\}$ converge in distribution towards a mean-zero Gaussian process $(Z(s), s \geq 0)$ with covariance function*

$$E[Z(s)Z(t)] = \frac{1}{2}\left(|s|^{2H} + |t|^{2H} - |s-t|^{2H}\right). \tag{6.4}$$

The limit process $(Z(t); t \geq 0)$ is called fractional Brownian motion.

Note that convergence of the finite-dimensional distributions in Theorem 6.1 is a direct consequence of convergence of the variances, as all distributions are Gaussian. Taqqu's (1975) contribution was to establish tightness of the partial sum process. We finally remark that the autocovariance function (6.4) of the limit process $(Z(t))_{t \geq 0})$ can be computed from (6.3) via the representation

$$EZ(s)Z(t) = \frac{1}{2}\left\{EZ^2(s) + EZ^2(t) - E(Z(s) - Z(t))^2\right\}.$$

Fractional Brownian motion was first introduced by Kolmogorov in the 1940s and studied in detail by Mandelbrot and Van Ness (1968).

6.2 Noncentral limit theorems. The most popular and best studied model of long-range dependent phenomena is given as an instantaneous functional of a stationary Gaussian process $(X_j)_{j \geq 1}$. Such a process is defined by

$$Y_j = H(X_j),$$

where $H \in L_2(\mathbb{R}, \phi(x)dx)$, and where $\phi(x) = \frac{1}{\sqrt{2\pi}} e^{-x^2/2}$ denotes the standard normal density. We still assume that the autocovariance function $r(k)$ of (X_j) satisfies (6.2) and $r(0) = 1$, with $0 < d < 1$.

The main tool in the analysis of this process is the Hermite expansion of H. This is an expansion in terms of an orthonormal basis of polynomials in the space $L_2(\mathbb{R}, \phi(x)dx)$. The Hermite polynomials are defined by

$$H_q(x) = \frac{d^q}{dx^q} e^{-x^2/2} (-1)^q e^{x^2/2},$$

for all nonnegative integers q. The first Hermite polynomials are thus $H_0(x) = 1$, $H_1(x) = x$, $H_2(x) = x^2 - 1$ and $H_3(x) = x^3 - 3x$. Any $L_2(\mathbb{R}, \phi(x)dx)$-function H can be expanded into a series

$$H(x) = \sum_{q=0}^{\infty} \frac{J_q}{q!} H_q(x) \tag{6.5}$$

where the coefficients J_q are given by $J_q = EH(X_1)H_q(X_1)$.

From the Hermite expansion of H we obtain an expansion of the partial sums of the Y_j's as

$$\sum_{j=1}^{N} Y_j = \sum_{j=1}^{N} H(X_j) = \sum_{q=0}^{\infty} \sum_{j=1}^{N} \frac{J_q}{q!} H_q(X_j).$$

Some of the coefficients J_q in the expansion might be zero and the corresponding indices may be dropped from the expansion. We define the Hermite rank of H as the smallest integer m with $J_m \neq 0$, so that

$$\sum_{j=1}^{N} H(X_j) = \sum_{q=m}^{\infty} \frac{J_q}{q!} \sum_{j=1}^{N} H_q(X_j). \tag{6.6}$$

An indispensable tool for the further study of the Hermite expansion and of the terms in the r.h.s. of (6.6) is a formula for the covariances of the random variables $H_q(X_j)$, $q \in \mathbb{N}_0$, $j \geq 1$. This formula, which is a special case of the so-called diagram formula (see Major (1981)) states that

$$E\left[H_q(X_j) H_{q'}(X_{j'})\right] = \delta_{qq'}\, q!\, (EX_j X_{j'})^q, \tag{6.7}$$

where $\delta_{qq'}$ denotes the Dirac delta function. From (6.7) we may conclude two facts.

- First, the terms inside the outer sum on the r.h.s. of (6.6) are mutually orthogonal. In the analysis of the limit behavior of the partial sums $\sum_{j=1}^{n} Y_j$ we can thus study each of the terms $\sum_{j=1}^{N} H_q(X_j)$ separately.
- Second, the autocovariance function of the process $H_q(X_j)$ is given by

$$E\left[H_q(X_j) H_q(X_{j+k})\right] = q!\, (L(k))^q\, k^{-dq}. \tag{6.8}$$

Thus, as long as $0 < qd < 1$, we get for the variance of the summands on the r.h.s. of the Hermite expansion (6.6),

$$\mathrm{Var}\left(\sum_{j=1}^{N} H_q(X_j)\right) \sim C_{qd}\, q!\, (L(N))^q\, N^{2-qd}, \tag{6.9}$$

using calculations similar to the ones carried out in determining $\mathrm{Var}(\sum_{j=1}^{N} X_j)$.

The most important conclusion from (6.9) is that the terms $\sum_{j=1}^{N} H_q(X_j)$ in the Hermite expansion (6.6) are all of a different order of magnitude. The lowest order term, i.e., $\sum_{j=1}^{N} H_m(X_j)$, has the largest variance and is thus dominating the others. A rigorous statement of this observation is known as the weak reduction principle of Taqqu (1975). Here we state a functional version of this result.

Theorem 6.2. *Let $(X_j)_{j\geq 1}$ be a mean-zero Gaussian process with autocovariance function satisfying* (6.2), *where $0 < d < \frac{1}{m}$ and where L is a slowly varying function. Assume moreover that $r(0) = 1$. Let $H \in L_2(\mathbb{R}, \phi(x)dx)$ have Hermite rank m and define*

$$d_N^2(m) := \mathrm{Var}\Big(\sum_{j=1}^{N} H_m(X_j)\Big). \tag{6.10}$$

Then, as $N \to \infty$,

$$\sup_{0\leq t\leq 1} \frac{1}{d_N(m)} \left| \sum_{j=1}^{[Nt]} H(X_j) - \frac{J_m}{m!} \sum_{j=1}^{[Nt]} H_m(X_j) \right| \to 0$$

in probability.

The weak reduction principle thus reduces the investigation of the asymptotic behavior of the partial sum process $\sum_{j=1}^{[Nt]} X_j$ to that of the partial sum process $\sum_{j=1}^{[Nt]} H_m(X_j)$ in the Hermite polynomials. Dobrushin and Major (1979) and independently Taqqu (1979) finally established convergence of the process $\frac{1}{d_N(m)} \sum_{j=1}^{N} H_m(X_j)$.

Theorem 6.3. *Let $(X_j)_{j\geq 1}$ be a mean-zero Gaussian process with autocovariance function satisfying $r(0) = 1$ and* (6.2), *where $0 < d < \frac{1}{m}$ and where L is a slowly varying function. Let $d_N^2(m) := \mathrm{Var}(\sum_{j=1}^{N} H_m(X_j))$. Then, as $N \to \infty$,*

$$\left\{ \frac{1}{d_N(m)} \sum_{j=1}^{[Nt]} H_m(X_j), 0 \leq t \leq 1 \right\} \to \{Z_m(t), 0 \leq t \leq 1\}$$

in the sense of weak convergence in the function space $D[0,1]$ with the Skorohod topology. Here $\{Z_m(t), t \geq 0\}$ is an m-th order Hermite process which can be represented as a multiple Wiener–Itô stochastic integral

$$Z_m(t) := \int \cdots \int f_t(x_1, \ldots, x_m)\, dB(x_1) \cdots dB(x_m)$$

with respect to Brownian motion $B(t)$.

For the exact definition of the Hermite processes as well as the general theory of multiple stochastic integration see Major (1981). Note that the first order Hermite process is fractional Brownian motion, introduced in the investigation of the limit behavior of $\sum_{j=1}^{N} X_j$. As $H_1(x) = x$, Theorem 6.1 is in fact a special case of Theorem 6.3, for $m = 1$. Except for this case, the Hermite processes are non-Gaussian stationary processes.

6.3 Long-range dependent empirical process. In this section we examine the empirical distribution function $F_N(x) = \frac{1}{N}\sum_{j=1}^{N} 1_{\{Y_j \le x\}}$ of a process $(Y_j)_{j\ge 1}$ that is defined as instantaneous functional

$$Y_j = G(X_j), \quad j \ge 1,$$

of a stationary Gaussian process $(X_j)_{j\ge 1}$ whose autocovariance function satisfies (6.2) with $0 < d < 1$ and where L is a slowly varying function. The function $G : \mathbb{R} \to \mathbb{R}$ is measurable, but not necessarily an L_2-function. The sequence $(Y_j)_{j\ge 1}$ is stationary ergodic and thus we get by the ergodic theorem $F_N(x) \to F(x) = P(G(X_1) \le x)$ almost surely. Standard techniques can be used to show that this convergence even holds uniformly in x, i.e., that $\sup_x |F_N(x) - F(x)| \to 0$ almost surely. Thus we have established a Glivenko–Cantelli theorem for the empirical distribution function of $(Y_j)_{j\ge 1}$.

We now turn our attention to the corresponding empirical process, which, up to a normalization factor, is defined as

$$N(F_N(x) - F(x)) = \sum_{j=1}^{N} \left[1_{\{G(X_j)\le x\}} - P(G(X_j) \le x)\right], \quad x \in \mathbb{R}. \tag{6.11}$$

Observe that, for a fixed value x, the empirical process is a partial sum of another instantaneous functional of the original stationary Gaussian process $(X_j)_{j\ge 1}$. The corresponding functional is

$$H(s) = 1_{\{G(s)\le x\}} - P(G(X_j) \le x), \tag{6.12}$$

where one should keep in mind that H still depends on x.

For fixed x, convergence of $N(F_N(x) - F(x))$ can thus be studied by an application of the methods introduced in the previous section. We will consider the Hermite expansion

$$1_{\{G(X_j)\le x\}} - F(x) = \sum_{q=0}^{\infty} \frac{J_q(x)}{q!} H_q(X_j),$$

where

$$J_q(x) = E\left[1_{\{G(X_j)\le x\}} - F(x))H_q(X_j)\right] = \int_{\{s:G(s)\le x\}} \phi(s)H_q(s)\,ds, \tag{6.13}$$

for $q \ge 1$. The second identity follows from the fact that $EH_q(X_j) = 0$, again for $q \ge 1$. In addition $J_0(x) = E(1_{\{G(X_j)\le x\}} - F(x)) = 0$. The Hermite rank $m(x)$ of the functions H will, at least in general, depend on the value of x.

Using the weak reduction principle, we find thus that the asymptotic behavior of the empirical process, for a fixed value of x, is governed by the partial sum of the lowest order term in the Hermite expansion, i.e.,

$$[Nt](F_{[Nt]}(x) - F(x)) \approx \frac{J_m(x)}{m!} \sum_{j\le[Nt]} H_m(X_j).$$

Thus an application of Theorem 6.3, again for a fixed value of x, yields that the properly normalized empirical process $[Nt](F_{[Nt]}(x) - F(x))$ converges in distribution, i.e.,

$$d_N^{-1}(m)[Nt](F_{[Nt]}(x) - F(x)) \to \frac{J_m(x)}{m!} Z_m(t),$$

where Z_m is an m-th order Hermite process. A functional version of this result, i.e., convergence uniformly in x, was established by Dehling and Taqqu (1989).

Definition 6.4. *We define the Hermite rank of the class of functions*

$$\left\{1_{\{G(s)\leq x\}} - P(G(X_j) \leq x),\ \ x \in \mathbb{R}\right\}$$

as the smallest integer m such that $J_m(x) \neq 0$ *for some* $x \in \mathbb{R}$.

Observe that in terms of the local Hermite ranks $m(x)$, we have $m = \min_x m(x)$. Also note that $J_0(x) = 0$, for all $x \in \mathbb{R}$, and thus $m \geq 1$.

Example 6.5. (i) The function $G(x) = x$ gives rise to the class of functions $1_{\{s\leq x\}} - \Phi(x)$, $-\infty < x < \infty$, where Φ is the standard normal distribution function. This class has Hermite rank $m = 1$, since

$$J_1(x) = EH_1(X)(1_{\{X\leq x\}} - \Phi(x)) = EX1_{\{X\leq x\}} = -\phi(x)$$

is nonzero.

(ii) The function $G(x) = x^2$ gives rise to the class of functions $1_{\{s^2\leq x\}} - (\Phi(\sqrt{x}) - \Phi(-\sqrt{x}))$, $0 \leq x < +\infty$. Observe that for $x \geq 0$,

$$J_1(x) = \int_{-\infty}^{\infty} s\left(1_{\{-\sqrt{x}\leq s\leq\sqrt{x}\}} - (\Phi(\sqrt{x}) - \Phi(-\sqrt{x})\right)\phi(s)\,ds = 0,$$

because of symmetry, and that

$$\begin{aligned}
J_2(x) &= \int_{-\infty}^{\infty}\left(s^2-1\right)\left(1_{\{-\sqrt{x}\leq s\leq\sqrt{x}\}} - \left(\Phi(\sqrt{x}) - \Phi(-\sqrt{x})\right)\right)\phi(s)\,ds \\
&= \int_{-\sqrt{x}}^{\sqrt{x}} (s^2-1)\,\phi(s)\,ds \\
&= -s\phi(s)|_{-\sqrt{x}}^{\sqrt{x}} \\
&= -\sqrt{2\pi^{-1}x}e^{-x/2}.
\end{aligned}$$

For $x \leq 0$, both $J_1(x)$ and $J_2(x)$ are equal to zero. Thus, for the empirical process of the sequence $(X_j^2)_{j\geq 1}$ the Hermite rank of the corresponding class of functions is 2.

(iii) In general, Dehling and Taqqu (1992) could show that there are instantaneous functionals $(G(X_j))_{j\geq 1}$ such that the associated class of functions has arbitrary Hermite rank. A short proof of this fact uses Lyapunov's theorem on the convexity of the range of a vector-valued random measure. Dehling and Taqqu (1992) gave an alternative proof that shows that G can even be taken to be continuous.

Theorem 6.6. *Let* $(X_j)_{j\geq 1}$ *be a stationary, mean-zero Gaussian process with autocovariance function satisfying (6.2), let the class of functions* $1_{\{G(s)\leq x\}} - F(x)$, $-\infty < x < \infty$, *have Hermite rank m and let* $0 < d < \frac{1}{m}$. *Then*

$$\left(d_N^{-1}(m)[Nt]\left(F_{[Nt]}(x) - F(x)\right);\ -\infty \leq x \leq \infty,\ 0 \leq t \leq 1\right)$$

converges weakly in $D([-\infty,\infty]\times[0,1])$, *equipped with the sup-norm, to*

$$\left(\frac{J_m(x)}{m!}Z_m(t);\ -\infty \leq x \leq \infty,\ 0 \leq t \leq 1\right). \tag{6.14}$$

Here $d_N(m)$ *is defined as in (6.10) and* $(Z_m(t), 0 \leq t \leq 1)$ *is an m-th order Hermite process.*

Observe that the limit process occurring in the empirical process invariance principle in the long range dependent case, Theorem 6.6, is markedly different from what one gets in the independent or weakly dependent case. The process is basically deterministic with fixed shape $J_m(x)/m!$ and a random amplitude $Z_m(t)$. In this way there is a separation of the two variables x and t involved in the two-parameter limit process. This is completely different from the Kiefer process limit encountered in the i.i.d. or the weak dependent case. In addition, for $m \geq 2$, the limit process is non-Gaussian.

Example 6.7. We consider again $G(x) = x$, in which case $F_N(x)$ is simply the empirical distribution function of a Gaussian process. Here, as we have seen above, $m = 1$ and $J_1(x) = -\phi(x)$. According to the uniform reduction principle, the empirical process $[Nt](F_{[Nt]}(x) - F(x))$ is close to the partial sum process $-\phi(x)\sum_{j\leq Nt} X_j$. Thus, in the limit the variables x and t in the empirical process separate into $-\phi(x)$ and $\sum_{j\leq[Nt]} X_j$.

In light of the remarks about convergence of the empirical process at fixed values of x, the natural way to prove (6.6) might seem to establish tightness of the process $d_N^{-1}(m)[Nt](F_{[N]}(x)-F(x))$. Given the results of Section 2, this would necessitate bounds on the fourth moments of increments of the empirical process. These are, however, quite difficult to obtain. An alternative line of proof was suggested by Dehling and Taqqu (1989). They established first a uniform version of the reduction principle.

Proposition 6.8 (Uniform weak reduction principle). *Under the conditions of Theorem 6.6, there are constants $C, \kappa > 0$ such that*

$$P\left(\sup_{0\leq t\leq 1, x\in\mathbb{R}} d_N(m)^{-1}\left|[Nt]\left(F_{[Nt]}(x) - F(x)\right) - \frac{J_m(x)}{m!}\sum_{j\leq[Nt]} H_m(X_j)\right| \geq \epsilon\right)$$
$$\leq CN^{-\kappa}\left(1+\epsilon^{-3}\right) \tag{6.15}$$

holds for any $0 < \epsilon \leq 1$.

As a consequence of (6.15) we get

$$\sup_{0\leq t\leq 1, x\in\mathbb{R}} d_N^{-1}(m)\left|[Nt]\left(F_{[Nt]}(x) - F(x)\right) - \frac{J_m(x)}{m!}\sum_{j\leq[Nt]} H_m(X_j)\right| \to 0$$

in probability. This is the uniform analogue of Taqqu's (1975) weak reduction principle. Using this uniform reduction principle, the investigation of the asymptotic behavior of the empirical process can be reduced to that of $(J_m(x)/m!)\sum_{j\leq[Nt]} H_m(X_j)$. Thus, Theorem 6.6 becomes an immediate consequence of the noncentral limit theorem, Theorem 6.3.

Proof of Proposition 6.8. For the proof of the uniform weak reduction principle we first define the doubly indexed processes

$$S_N(n,x) = d_N^{-1}(m)\sum_{j\leq n}\left[1_{\{Y_j\leq x\}} - F(x) - \frac{J_m(x)}{m!}H_m(X_j)\right].$$

In order to show uniform convergence of this process to 0, we will apply a chaining technique of the kind outlined in Section 2. This requires bounds on the moments of increments of S_N. At this point, a crucial observation is that bounds on the second moments suffice, essentially as a result of the fact that S_N itself converges to 0 like a power of N. The necessary moment inequality is contained in the following lemma.

Lemma 6.9. *Under the conditions of Theorem 6.6 there exist constants $\gamma > 0$ and C such that for $n \leq N$,*

$$E\,|S_N(n,y) - S_N(n,x)|^2 \leq C\frac{n}{N}\frac{F(y)-F(x)}{N^\gamma}.$$

Details of the proofs can be found in Dehling and Taqqu (1989).

6.4 Applications. The use of empirical process theory in statistics relies largely on the fact that many sample statistics can be written as functionals of the empirical distribution function. If the associated functional is sufficiently smooth, the asymptotic distribution of the sample statistics can be derived from empirical process theory. This program can be carried out in general, i.e., both for weakly dependent as well as for long-range dependent processes. As a consequence of the peculiar character of the limit process in Theorem 6.6, limit theorems for sample statistics of long-memory data are often quite different from those encountered with short memory data. In this section, we want to explore this effect for U-statistics and von-Mises statistics. More examples can be found, e.g., in the survey paper by Beran (1992).

Given a kernel $h : \mathbb{R}^2 \to \mathbb{R}$ and a stationary process $(Y_j)_{j\geq 1}$, we have earlier defined the bivariate U-statistics with kernel h as

$$U_N(h) = \binom{N}{2}^{-1} \sum_{1\leq i<j\leq N} h(Y_i, Y_j).$$

In a similar way, we define the bivariate von-Mises statistics $V_N(h)$ by

$$V_N(h) = \frac{1}{N^2} \sum_{1\leq i,j\leq N} h(Y_i, Y_j).$$

Apart from the normalization factor, the difference between the U-statistics and the von-Mises statistics lies in the inclusion of the diagonal terms in $V_N(h)$. Thus limit results for one of these statistics can easily be translated into results for the other. It turns out that in some situations U-statistics have more convenient mathematical properties, whereas in others the von-Mises statistics have. It is for this reason that we consider both of them here.

In what follows, we will always assume that $h(x, y)$ is a symmetric kernel, i.e., that $h(x, y) = h(y, x)$. A very special role in U-statistics theory is played by so-called degenerate kernels that were implicitly already introduced in Section 5.

Definition 6.10. *A kernel $h(x, y)$ is called degenerate if*

$$\int h(x,y)\,dF(x) = \int h(x,y)\,dF(y) = 0$$

for all $x, y \in \mathbb{R}$, where F is the marginal distribution function of the underlying process $(Y_j)_{j\geq 1}$.

A prototype of degenerate kernels is a product kernel $h(x, y) = f(x)f(y)$ where f satisfies $\int f(x)dF(x) = 0$. Actually, any degenerate L_2-kernel can be expanded into a series $\sum_{i=1}^{\infty} \lambda_i f_i(x) f_i(y)$, where the f_i's are orthogonal in $L_2(\mathbb{R}, dF)$ and satisfy $\int f_i(x)dF(x) = 0$ and where $\sum_{i=1}^{\infty} \lambda_i^2 < \infty$ (see Denker (1985)). Degenerate kernels occur also naturally in statistics, e.g., in connection with χ^2-testing of goodness-of-fit.

Example 6.11. Let $a_1, \ldots, a_M$ be real numbers and let $\mathbb{R} := \bigcup_{i=1}^{M} A_i$ be a disjoint partition of $\mathbb{R}$ into M sets. Define $G : \mathbb{R} \to \mathbb{R}$ by

$$G(x) = \sum_{i=1}^{M} a_i 1_{A_i}(x).$$

Then $Y_j = G(X_j)$ has M different outcomes $a_1, \ldots, a_M$ with corresponding probabilities $p_i = P(G(X_1) = a_i) = P(X_1 \in A_i)$, for $i = 1, \ldots, M$. We consider the χ^2-test for goodness-of-fit based on the observations $Y_1, \ldots, Y_N$. Let $p_i^{(N)} := \#\{j \leq N : Y_j = a_i\}$ be the observed frequencies and define the χ^2-test statistic

$$R_N = \sum_{i=1}^{M} \frac{(p_i^{(N)} - Np_i)^2}{Np_i}.$$

Observe that R_N/N is a von-Mises statistic with kernel

$$h(x, y) = \sum_{i=1}^{M} \frac{1}{p_i} \left(1_{\{x=a_i\}} - p_i\right) \left(1_{\{y=a_i\}} - p_i\right),$$

and that h is degenerate. Let m be the Hermite rank of the class of functions $\Delta_x(s) = 1_{\{G(s) \leq x\}}$, $x \in \mathbb{R}$. Actually, m can be computed in terms of the sets $A_1, \ldots, A_M$, as follows:

$$m = \min \left\{ q : \int_{A_i} H_q(s)\phi(s) ds \neq 0 \text{ for some } i = 1, \ldots, M \right\}.$$

As a result of Theorem 6.12 (see below), we obtain the following limit theorem for the von-Mises partial sum process $[Nt]R_{[Nt]}$:

$$d_N^{-2}(m)\, [Nt]\, R_{[Nt]} \to C(Z_m(t))^2,$$

where the constant $C = C(h, m)$ can be computed as follows:

$$\begin{aligned}
(m!)^2 C &= \int\int h(x, y)\, dJ_m(x)\, dJ_m(y) \\
&= \int\int h(G(u), G(v))\, H_m(u)\, H_m(v)\, \phi(u)\, \phi(v)\, du\, dv \\
&= \sum_{i=1}^{M} \frac{1}{p_i} \int\int (1_{\{u \in A_i\}} - p_i)\, (1_{\{v \in A_i\}} - p_i)\, H_m(u)\, H_m(v)\, \phi(u)\, \phi(v)\, du\, dv \\
&= \sum_{i=1}^{M} \frac{1}{p_i} \left(\int \left(1_{\{u \in A_i\}} - p_i\right) H_m(u)\, \phi(u)\, du \right)^2 \\
&= \sum_{i=1}^{M} \frac{1}{p_i} \left(\int_{A_i} H_m(u)\phi(u) du \right)^2.
\end{aligned}$$

It is interesting to note two important differences with the usual theory of χ^2-tests. For one thing, the limit distribution of $d_N^{-2} N R_N$ may not be a χ^2-distribution. In addition, even if we get a limiting χ^2-distribution, which is the case when $m = 1$, this distribution has only one degree of freedom.

The second order term $U_N(h_2)$ in the Hoeffding decomposition of a U-statistic has a degenerate kernel, as was observed in Section 5. Thus, limit theorems for a general U-statistic $U_N(h)$ can be obtained by studying separately the linear first order term and a U-statistic with degenerate kernel, explaining the prominent role of degenerate U-statistics in limit theorems.

In the context of von-Mises statistics, degenerate kernels allow for a representation of $V_N(h)$ as integral with respect to the empirical process. We get namely, for a degenerate kernel h,

$$\begin{aligned} V_N(h) &= \int\int h(x,y)dF_N(x)dF_N(y) \\ &= \int\int h(x,y)d(F_N(x)-F(x))d(F_N(y)-F(y)). \end{aligned} \tag{6.16}$$

This representation suggests deriving weak convergence of $V_n(h)$ from weak convergence of the empirical process $F_N(x) - F(x)$, after a proper normalization. This program will be carried out in this section for the case of long-range dependent data. We remark already here that some sophistication is required in this, as the map $G \mapsto \int\int h(x,y)dG(x)dG(y)$ is generally not continuous so that we cannot simply apply the continuous mapping principle.

For i.i.d. observations, weak convergence of $NV_N(h)$ was established by Filippova (1961) who showed that the limit can indeed be represented as a doubly stochastic integral

$$\int\int h(x,y)dW^{(0)}(x)dW^{(0)}(y),$$

where $W^{(0)}$ is a Brownian bridge process. A functional version of this, i.e., weak convergence of the von-Mises partial sum process $\frac{1}{N}[Nt]^2 V_{[Nt]}(h)$ and of the corresponding U-statistic partial sum process, was obtained by Denker, Grillenberger, and Keller (1985) and independently by Mandelbaum and Taqqu (1984). Strong approximation results as well as the law of the iterated logarithm were obtained in a series of papers by Dehling, Denker, and Philipp (1984, 1986) and by Dehling (1989). The long-range dependent case was studied by Dehling and Taqqu (1989) who obtained the following result.

Theorem 6.12. *Let $(X_j)_{j\geq 1}$ be a stationary, mean-zero Gaussian process with autocovariance function satisfying (6.2), let $Y_j = G(X_j)$ for some measurable function G, let the class of functions $\{1_{\{G(s)\leq x\}} - F(x),\ -\infty < x < \infty\}$, have Hermite rank m and $0 < d < \frac{1}{m}$. In addition, let h be a degenerate symmetric kernel with bounded total variation and assume that $h(x,y)$ and $F(x)F(y)$ have no joint discontinuities. Then the $D[0,1]$-valued random elements*

$$\left(d_N^{-2}(m) \sum_{1\leq i,j\leq [Nt]} h(Y_i,Y_j),\ 0\leq t\leq 1 \right)$$

converge weakly to the process

$$\left\{ \left(\frac{1}{m!}\right)^2 \int_{-\infty}^{\infty}\int_{-\infty}^{\infty} h(x,y)\,dJ_m(x)\,dJ_m(y) \right\} (Z_m(t))^2,$$

where $J_m(x)$ is defined as in (6.13) and $Z_m(t)$ denotes the m-th order Hermite process. The same limit result holds for the corresponding U-statistic partial sum process

$$\left(d_N^{-2}(m) \sum_{1\leq i\neq j\leq [Nt]} h(Y_i,Y_j),\ 0\leq t\leq 1 \right).$$

Sketch of the proof. The proof of weak convergence of the von-Mises statistics uses the representation (6.16) and integration by parts to get

$$
\begin{aligned}
&d_N^{-2}(m)[Nt]^2 V_{[Nt]}(h)\\
&= \int\int h(x,y)d\left(d_N^{-1}(m)[Nt](F_{[Nt]}(x)-F(x))\right)d\left(d_N^{-1}(m)[Nt](F_{[Nt]}(y)-F(y))\right)\\
&= \int\int \left(d_N^{-1}(m)[Nt](F_{[Nt]}(x)-F(x))\right)\left(d_N^{-1}(m)[Nt](F_{[Nt]}(y)-F(y))\right)dh(x,y).
\end{aligned}
$$

Under the assumptions on h, the map $G \mapsto \int\int G(x)G(y)dh(x,y)$ is continuous. Thus we can apply the continuous mapping theorem and empirical process weak convergence to get convergence of

$$
\int\int d_N^{-1}(m)[Nt](F_{[Nt]}(x)-F(x))d_N^{-1}(m)[Nt](F_{[Nt]}(y)-F(y))dh(x,y)
$$

to $\left(\frac{1}{m!}\right)^2 \int\int h(x,y)dJ_m(x)dJ_m(y)(Z_m(t))^2$. □

The condition that h should have bounded total variation is quite restrictive and excludes many common kernels used in statistics. Using truncation techniques, Dehling and Taqqu (1991) were able to relax the assumptions on h to locally bounded variation together with a moment condition.

Observe that, in contrast with the case of independent or weakly dependent observations, the limit process does not involve a doubly stochastic integral. This is a result of the deterministic character of the limit process in Theorem 6.6. The fact that the limit distribution of the χ^2-test statistic, in the case $m = 1$, has only one degree of freedom is a direct consequence of this observation.

The second special feature arising in the context of long-range dependent observations is the fact that U-statistics and von-Mises statistics have the same limit. In the independent and weakly dependent case the limits differ by $Eh(X_1, X_1)$ as a result of the interplay of the law of large numbers for the diagonal terms and the central limit theorem for the off-diagonal terms that both have the same $1/N$ normalization.

A third distinctive feature of long-range dependent U-statistics is the fact that both the linear term as well as the second order term in the Hoeffding decomposition can contribute to the limit distribution. This is in sharp contrast with the situation when observations are independent or weakly dependent, where the linear term is dominating and the second order term only contributes in the degenerate case when the linear term vanishes.

Example 6.13. An example where both the linear and the second order term contribute to the limit distribution is given by the variance estimator. Suppose that we have an affine linear function G, i.e., $G(x) = \sigma x + \mu$ and that the process under consideration is $Y_j = G(X_j)$. Then, the sample variance of $Y_1, \ldots, Y_N$ is given by

$$
s_N^2 = \frac{1}{N-1}\sum_{i=1}^{N}(Y_i - \bar{Y})^2 = \frac{1}{N(N-1)}\sum_{1\le i\ne j\le N}\frac{1}{2}(Y_i - Y_j)^2.
$$

The latter representation shows that s_N^2 is indeed a U-statistic with kernel $h(x,y) = (x-y)^2/2$. In order to determine the asymptotic distribution of $s_N^2 - \sigma^2$, we introduce

$$
U_N = N(N-1)(s_N^2 - \sigma^2) = \sum_{1\le i\ne j\le N}\left(\frac{1}{2}(Y_i - Y_j)^2 - \sigma^2\right),
$$

which is again a U-statistic, now with kernel $h(x,y) = \frac{1}{2}(x-y)^2 - \sigma^2$. The Hoeffding decomposition of h is given by

$$h(x,y) = \frac{1}{2}((x-\mu)^2 - \sigma^2) + \frac{1}{2}((y-\mu)^2 - \sigma^2) - (x-\mu)(y-\mu),$$

which, after some rewriting yields the following Hoeffding decomposition of U_N:

$$\begin{aligned} U_N &= \frac{1}{2} \sum_{1 \le i \ne j \le N} ((Y_i - \mu)^2 - \sigma^2) + \frac{1}{2} \sum_{1 \le i \ne j \le N} ((Y_j - \mu)^2 - \sigma^2) \\ &\quad - \sum_{1 \le i \ne j \le N} (Y_i - \mu)(Y_j - \mu) \\ &= (N-1) \sum_{i=1}^{N} ((Y_i - \mu)^2 - \sigma^2) - \sum_{1 \le i \ne j \le N} (Y_i - \mu)(Y_j - \mu). \end{aligned}$$

The first term in this decomposition is simply a partial sum of nonlinear functionals of the underlying Gaussian process. The Hermite decomposition of the functional is particularly simple in this case, as $(Y_j - \mu)^2 - \sigma^2 = \sigma^2 H_2(X_i)$, and thus by the noncentral limit theorem

$$N^{-1} d_N^{-1}(2)(N-1) \sum_{i=1}^{N} \left((Y_i - \mu)^2 - \sigma^2 \right) \to \sigma^2 Z_2(1),$$

in distribution, as $N \to \infty$. Regarding the second order term, we obtain, by Theorem 6.12,

$$d_N^{-2}(1) \sum_{1 \le i \ne j \le N} (Y_i - \mu)(Y_j - \mu) \to C(Z_1(1))^2,$$

where the constant C is given by $C = \sigma^2 \int \int (xy)^2 \phi(x)\phi(y)dxdy = \sigma^2$, and where $Z_1(t)$ is fractional Brownian motion. Observe that

$$d_N^{-2}(1) \approx N^{D-2} L^{-1}(N) = N^{-1} N^{D-1} L^{-1}(N) \approx N^{-1} d_N^{-1}(2)$$

so that, in fact, both terms in the Hoeffding decomposition of U_N contribute to the limit. Thus, the limit distribution of $s_N^2 - \sigma^2$, suitably normalized, is a linear combination of $Z_2(1)$ and $(Z_1(1))^2$.

6.5 Linear processes with long memory. In the previous sections, we have exclusively studied long-memory processes that were defined as instantaneous functionals of a stationary Gaussian process. An alternative class of models for long-range dependence is given by linear processes with slowly decaying coefficients. Let $(\epsilon_i)_{i \in \mathbb{Z}}$ be a sequence of independent random variables with common marginal distribution function $H(x)$, satisfying $E\epsilon_i = 0$ and $E\epsilon_i^2 < \infty$. Moreover, let $(a_i)_{i \ge 0}$ be a sequence of constants satisfying $a_i = i^{-\beta} L(i)$ for some $\beta \in (\frac{1}{2}, 1)$ and for some slowly varying function L. We define the one-sided moving average process $(X_n)_{n \ge 1}$ by

$$X_n = \sum_{i=0}^{\infty} a_i \epsilon_{n-i}, \quad n \ge 1.$$

Observe that X_n is well defined and has finite variance, as $\sum_{i=0}^{\infty} a_i^2 < \infty$, and that $(X_n)_{n=1}^{\infty}$ is a stationary stochastic process. However, as $\sum_{i=0}^{\infty} |a_i| = \infty$, the spectral density of $(X_n)_{n=1}^{\infty}$ has a singularity at $\lambda = 0$, indicating the presence of long-range dependence in this model.

Partial sums of nonlinear functionals of such moving-average processes have been studied by Surgailis (1983) as well as by Avram and Taqqu (1987), who established noncentral limit theorems. In these results, the role of the Hermite polynomials is taken over by so-called Appell polynomials.

The empirical process of long-range dependent moving average processes was studied by Ho and Hsing (1996) as well as by Giraitis and Surgailis (1999). The following is a simplified version of a result of Giraitis and Surgailis.

Theorem 6.14. *Suppose that* $E\exp(iu\epsilon) \le C/((1+|u|)^{\delta})$, $u \in \mathbb{R}$ *and that* $E|\epsilon|^r < \infty$ *for some* $r > 2$. *Then*

$$d_N^{-1} N\,(F_N(x) - F(x)) \to c\,f(x)\,Z \tag{6.17}$$

weakly, as elements of the function space $D(\bar{\mathbb{R}})$, *where* $f(x) = F'(x)$ *and* Z *is standard normally distributed.*

Note that for Gaussian innovations ϵ_i, the limit in (6.17) coincides with what we found in Example 6.7.

Ho and Hsing (1996) even have an asymptotic expansion of $d_N^{-1}N(F_N(x) - F(x))$ of which $-f(x)Z$ is the first order term. An overview on the theory of linear processes with long memory is provided by Koul and Surgailis (2002).

7 Pair correlations

Pair correlations have been studied over the last 70 years in various forms and disguises in mathematics, statistics, fluid mechanics, electrical engineering and various branches of physics. Let $\{X_j : 1 \le j \le N\}$ be real numbers or vectors, or random variables. In some form or another, pair correlations measure the average closeness of the $N(N-1)$ pairs (X_i, X_j), $1 \le i \ne j \le N$. Here "closeness" is to be understood in a very general sense, not just with respect to Euclidean distance. For random sequences $\{X_j, j \ge 1\}$, i.e., for sequences of random variables or random vectors the concept of pair correlations as well as the closely related correlation integrals all fall under the umbrella of empirical processes of U-statistics structure or U-processes.

Let $\{X_j, j \ge 1\}$ be a sequence of identically distributed p-dimensional random vectors with common distribution function F. Let $\mathcal{H}$ be a class of kernel functions $h : \mathbb{R}^p \times \mathbb{R}^p \times \mathbb{R} \to \mathbb{R}$. Usually it is assumed that h is symmetric in the first two variables, i.e., $h(x, y, t) = h(y, x, t)$ for all $x, y \in \mathbb{R}^p$ and $t \in \mathbb{R}$. However, in some of the results mentioned below we even need nonsymmetric kernels. More generally, the X_j's can be Banach space valued, the parameter t can be replaced by classes of functions, such as Vapnik–Chervonenkis classes, and the multivariate situation is a straightforward extension of the present case (m=2). The U-process indexed by the class $\mathcal{H}$ can be defined as

$$U_N(t) = U_N(h; t) := \sum_{1 \le i < j \le N} h(X_i, X_j, t), \qquad N \ge 2, h \in \mathcal{H}. \tag{7.1}$$

In the independent case $\frac{2}{N(N-1)} U_N(t)$ is an unbiased estimator for

$$U(t) = \int_{\mathbb{R}^{2p}} h(x, y, t)\,dF(x)\,dF(y).$$

We shall make two major assumptions. First, we restrict ourselves to the case of random variables X_j all of which have uniform distribution on [0, 1], i.e.,

$$P(X_j \leq x) = x, \quad 0 \leq x \leq 1,\ j \geq 1. \tag{7.2}$$

Stationarity of the sequence $\{X_j, j \geq 1\}$ is not assumed in the case of dependent random variables. Second, we shall restrict ourselves to the following two kernels:

$$h(x, y, t) = 1(|x - y| \leq t) \tag{7.3}$$

and

$$\begin{aligned} h(x, y, t) &= 1(0 \leq \{x\} - \{y\} \leq t \mod 1) \\ &= 1(\{x - y\} \leq t) = 1_{[0,t]}(x - y). \end{aligned} \tag{7.4}$$

Here $1(A)$ denotes the indicator of the set A, and $1_A(\cdot)$ denotes the indicator of the set A, extended with period 1. The quantity $\{x\} = x - [x]$ denotes the fractional part of x. In (7.3), $|\cdot|$ is the Euclidean distance in $\mathbb{R}$, whereas in (7.4), $\{x\}$ and $\{y\}$ are considered points on the one-dimensional torus group. Notice that the kernel defined in (7.4) is not symmetric.

For the kernel h defined in (7.3) (and for a general distribution function F) the corresponding U-process

$$U_N(t) := \sum_{1 \leq i < j \leq N} 1(|X_j - X_i| \leq t) \tag{7.5}$$

divided by $\frac{N(N-1)}{2}$ is called the empirical or sample correlation integral; see relation (5.2). If, in addition, $t = t_N \to 0$ the corresponding limit theorems often appear under the heading "limit theorems for short distances," as well as under the heading "pair correlations." (See, e.g., Eastwood and Horváth (1999), Horváth (1991), Rudnick and Sarnak (1998), Rudnick and Zaharescu (1999) as well as Dabrowski, Dehling, Mikosch and Sharipov (2002).) For stationary ergodic sequences $\{X_j, j \geq 1\}$, with a general common distribution function F, and for a very general class of kernels h, Borovkova, Burton, and Dehling (1999) established a Glivenko–Cantelli-type theorem proving almost sure uniform convergence of $\binom{N}{2}^{-1} U_N(t)$ to $U(t)$. For strictly stationary absolutely regular sequences $\{X_j, j \geq 1\}$ they proved a weak invariance principle with the approximating process being a multiple stochastic integral with integrators being specified Kiefer processes. This extends an earlier result of Dehling, Denker, and Philipp (1987) who obtained this result for sequences of independent, identically distributed random variables.

In this section, which is based on a recent paper by Berkes, Philipp, and Tichy (2001), we prove almost sure uniform estimates on $U_N(t)$, given in (7.5), over short intervals. We shall provide error terms which, except for powers of logarithm, are sharp.

In case that the kernel h is given by (7.3) the corresponding U-process is given by (7.5). We set, for $0 \leq \alpha \leq 2$,

$$\begin{aligned} \Gamma_N(\alpha) &:= \sup\left(\left|U_N(t) - \binom{N}{2}(2t - t^2)\right| : t \leq N^{-\alpha}\right) \\ &= \sup\left(\left|\sum_{1 \leq i < j \leq N}\left[1(|X_j - X_i| \leq t) - (2t - t^2)\right]\right| : t \leq N^{-\alpha}\right). \end{aligned} \tag{7.6}$$

Observe that for independent random variables X, Y with uniform distribution over [0, 1],

$$E1(|X-Y| \leq t) = P(|X-Y| \leq t) = 2t - t^2. \tag{7.7}$$

Since we will estimate $\Gamma_N(\alpha)$ using second moments only and since $E(Z-b)^2$ is minimal for $b = EZ$, the term in (7.7) is the appropriate amount to be subtracted from $U_N(t)$ in (7.6). The kernel h defined in (7.3) is symmetric in x and y. Thus if in (7.5) the range of summation is replaced by $1 \leq i \neq j \leq N$, then the resulting U-process is exactly twice the U-process given in (7.5). As a consequence, after this change, Theorems 7.1, 7.10 and 7.11 below will remain valid as they stand.

The situation is somewhat different in case that the kernel h is given by (7.4). We denote the corresponding U-process by

$$V_N = V_N(L) := \sum_{1 \leq i < j \leq N} 1_L \left(X_j - X_i\right), \tag{7.8}$$

for some interval $L \subseteq [0, 1]$. Here V_N does not denote the von-Mises statistic introduced in Section 6. There the summation is extended over all pairs (i, j) with $1 \leq i, j \leq N$. Here we denote by $V_N^* = V_N^*(L)$ the process defined by (7.8) but with the range of summation extended over all pairs (i, j) with $1 \leq i \neq j \leq N$. For $0 \leq \alpha \leq 2$ we set

$$\Delta_N(\alpha) := \sup \left\{ \left| V_N(L) - \binom{N}{2} |L| \right| : L \subseteq [0, N^{-\alpha}] \right\}, \tag{7.9}$$

and we denote by $\Delta_N^*(\alpha)$ the analogous quantity, but with V_N and $\binom{N}{2}$ replaced by V_N^* and $N(N-1)$, respectively. Here $|L|$ denotes the length of the interval L.

7.1 Pair correlations for weakly dependent random variables. We first consider the random variables X_j, defined by (1.1) in Section 1. Let $\{n_j, j \geq 1\}$ be a sequence of integers. As in (1.1) we set for $0 \leq \omega < 1$,

$$X_j = n_j \omega \quad \mod 1. \tag{7.10}$$

Now $\{X_j, j \geq 1\}$ is a sequence of weakly dependent random variables, defined on $(\Omega, \mathcal{F}, P) = ([0, 1), \mathcal{B}, P)$ where $\mathcal{B}$ are the Lebesgue sets and P is Lebesgue measure. As is easy to see, each X_j has uniform distribution over [0, 1).

Except for Proposition 7.8, we assume that $\{n_j, j \geq 1\}$ satisfies an Erdős gap condition, i.e.,

$$n_{j+1}/n_j \geq 1 + cj^{-\rho}, \; j \geq 1, \tag{7.11}$$

for some $c > 0$ and $0 \leq \rho < 1$. Notice that if we set $\rho = 0$, then condition (7.11) reduces to the Hadamard gap condition. We shall prove the following results which at least in the case of independent random variables are sharp save for a few powers of $\log N$, according to Theorem 7.11 below.

Theorem 7.1. *Let $\{n_j, j \geq 1\}$ be a sequence of integers satisfying the Erdős gap condition (7.11). Let X_j be defined by (7.10) and set $\Gamma_N(\alpha) = \Gamma_N(\alpha, \rho)$ as in (7.6). Then with probability* 1,

$$\Gamma_N(\alpha) \ll \left(N^{1-\frac{1}{2}\alpha+\frac{3}{2}\rho} + N^{2-2\alpha}\right) (\log N)^{11/2}.$$

Notice that in the formulation of Theorem 7.1 the term t^2 could be dropped in the corresponding definition of $\Gamma_N(\alpha)$ as it obviously could be absorbed in the error term.

Proposition 7.2. *Let $\{n_j, j \geq 1\}$ be a sequence of integers satisfying the Erdős gap condition* (7.11). *Let X_j be defined by* (7.10) *and set $\Delta_N(\alpha) = \Delta_N(\alpha, \rho)$ as in* (7.9). *Then with probability* 1,

$$\Delta_N(\alpha) \ll N^{1-\frac{1}{2}\alpha+\frac{3}{2}\rho}(\log N)^{11/2}.$$

Moreover, the same estimate holds for $\Delta_N^*(\alpha)$.

We remark that in case that $\alpha \leq 3\rho$, the estimate for $\Delta_N^*(\alpha)$ is superseded by Proposition 7.8 below.

Proposition 7.2 and similar results for different sequences $\{n_j, j \geq 1\}$ follow from the following proposition which is similar in spirit to Theorem 1 of Rudnick and Zaharescu (1999). First, we introduce some notation. Let $\{n_j, j \geq 1\}$ be an increasing sequence of integers. Let $1 \leq Q \leq H$ be integers and let $d > 0$ and $\sigma \geq 0$. Consider the system of diophantine equations

$$\begin{cases} a(n_j - n_i) = b(n_l - n_k) & H \leq j < H+Q,\ 1 \leq i < j, \\ & H \leq l < H+Q,\ 1 \leq k < l, \\ 1 - dH^{-\sigma} \leq \left|\frac{a}{b}\right| \leq 1 + dH^{-\sigma} & a, b \in \mathbb{Z}. \end{cases} \tag{7.12}$$

We call a solution $(a, b, n_i, n_j, n_k, n_l)$ trivial if both $j = l$ and $i = k$. For fixed $a \in \mathbb{Z}$ there are at most $2HQ$ trivial solutions, even if $d = \infty$.

Proposition 7.3. *Let $\{n_j, j \geq 1\}$ be a sequence of integers. Suppose there exist constants $c > 0, C > 0, \sigma > 0, \gamma \geq 0, \tau \geq 1$ with the following property. For each pair (Q, H) with $1 \leq Q \leq H$ and for each fixed $a \in \mathbb{Z}$ with*

$$0 < |a| \leq H^{20} \tag{7.13}$$

the system (7.12) *of diophantine equations has at most $CH^\tau Q(\log H)^\gamma$ solutions. Then with probability* 1

$$\Delta_N(\alpha) \ll N^{\frac{1}{2}(1+\sigma+\tau-\alpha)}(\log N)^{\frac{1}{2}(\gamma+9)}.$$

Moreover, the same estimate holds for $\Delta_N^*(\alpha)$.

To prepare for the proof of the estimate for $\Delta_N^*(\alpha)$, we first define V_N' by (7.8), but with the summation extended over all pairs (i, j) with $1 \leq j < i \leq N$. We define $\Delta_N'(\alpha)$ by (7.9), but with V_N replaced by V_N'. We have

$$\Delta_N^*(\alpha) \leq \Delta_N(\alpha) + \Delta_N'(\alpha)$$

as $V_N^* = V_N + V_N'$. The system of diophantine equations needed for the treatment of $\Delta_N'(\alpha)$ is analogous to (7.12) but with the roles of i and j and of k and l interchanged. However, this amounts to keeping the system (7.12) as it stands, except that the equation in (7.12) needs to be multiplied by -1. Hence the proof of the bound for $\Delta_N'(\alpha)$ will be the same as the forthcoming proof of the bound for $\Delta_N(\alpha)$.

The proof of Proposition 7.3 requires two lemmas. First we introduce some more notation. Recall that 1_L has period 1. We write

$$S = S(H, Q; L; \omega) = \sum_{H \leq j < H+Q} \sum_{1 \leq i < j} \left(1_L((n_j - n_i)\omega) - |L|\right).$$

Lemma 7.4. *Under the hypotheses of Proposition* 7.3, *we have*

$$ES^2 \ll |L|H^{\sigma+\tau}Q(\log H)^{\gamma+1} + H^{-16},$$

where the constant implied by $\ll$ only depends on C.

Proof. We set

$$f(\omega) = 1_L(\omega) - |L| = \sum_{a\neq 0} c_a e^{2\pi i a\omega},$$
$$f_H^*(\omega) = \sum_{0<|a|\leq H^{20}} c_a e^{2\pi i a\omega}$$

and

$$S_H^*(\omega) := \sum_{H\leq j<Q+H} \sum_{1\leq k<j} f_H^*((n_j - n_k)\omega).$$

Clearly $|c_a| \leq |a|^{-1}$ and thus by Minkowski's inequality

$$||S - S_H^*||_2 \leq Q(Q+H)H^{-10} \ll H^{-8}. \tag{7.14}$$

To estimate $||S_H^*||_2$, we note that for some constant θ,

$$||S_H^*||_2 \leq \sum_{0\leq u\leq \theta H^\sigma \log H} \left\| \sum_a c_a \sum_{H\leq j<H+Q} \sum_{1\leq k<j} e^{2\pi i a(n_j - n_k)\omega} \right\|_2,$$

where $\sum_a$ is extended over all a with $(1 + dH^{-\sigma})^u \leq |a| \leq (1 + dH^{-\sigma})^{u+1}$. Thus and since $|xy| \leq \frac{1}{2}(x^2 + y^2)$ we obtain

$$||S_H^*||_2$$
$$\leq \sum_{0\leq u\leq \theta H^\sigma \log H} \left(\sum_{a,b} c_a \bar{c}_b \sum_{H\leq j,l<H+Q} \sum_{1\leq i<j,\, 1\leq k<l} 1\{a(n_j - n_i) = b(n_l - n_k)\} \right)^{\frac{1}{2}}$$
$$\leq \sum_{u=0}^{\theta H^\sigma \log H} \left(\sum_a |c_a|^2 \sum_b \sum_{H\leq j,l<H+Q} \sum_{1\leq i<j,\, 1\leq k<l} 1\{a(n_j - n_i) = b(n_l - n_k)\} \right)^{\frac{1}{2}}.$$

The innermost triple sum does not exceed the number of solutions of the system (7.12). Thus using the main hypothesis of Proposition 7.3, we obtain, by Cauchy's inequality and Parseval's identity,

$$||S_H^*||_2 \leq (CH^\tau Q(\log H)^\gamma)^{\frac{1}{2}} \sum_{0\leq u\leq \theta H^\sigma \log H} \left(\sum_a |c_a|^2 \right)^{\frac{1}{2}}$$
$$\ll (H^\tau Q(\log H)^\gamma)^{\frac{1}{2}} (H^\sigma \log H)^{\frac{1}{2}} \cdot |L|^{\frac{1}{2}}.$$

The lemma follows now from (7.14). □

Lemma 7.5. *Fix* $0 \leq \alpha \leq 2$, *and set*

$$\Delta = \Delta(H, Q; \alpha; \omega) = \sup\left\{ |S(H, Q; L; \omega| : L \subset [0, H^{-\alpha}] \right\}.$$

Then for each $R \geq 1$,

$$P(\Delta \geq R) \ll R^{-2} Q H^{\sigma+\tau-\alpha} (\log H)^{\gamma+4}.$$

Proof. Denote a typical interval L by $L = H^{-\alpha}[0, s], 0 \leq s \leq 1$. We write s in its binary expansion

$$s = \sum_{m \geq 1} \varepsilon_m 2^{-m} \qquad \varepsilon_m \in \{0, 1\}, \tag{7.15}$$

and we set

$$M = [4 \log H]. \tag{7.16}$$

Then

$$s = \sum_{m=1}^{M} \varepsilon_m 2^{-m} + \theta 2^{-M}, \tag{7.17}$$

where $0 \leq \theta \leq 1$. For $0 \leq u < v \leq 1$ we set

$$Z(u, v) = Z(u, v; H, Q) = \sum_{H \leq j < H+Q} \sum_{1 \leq i < j} \left(1_{H^{-\alpha}[u,v)}((n_j - n_i)\omega) - (v - u)H^{-\alpha} \right).$$

Then for each $0 \leq s \leq 1$,

$$|Z(0, s)| \leq \sum_{m=1}^{M} |Z(d_m 2^{-m}, (d_m + 1)2^{-m})| + 2Q(H + Q)2^{-M} H^{-\alpha}, \tag{7.18}$$

where the d_m 's are integers with $0 \leq d_m < 2^m$ for $1 \leq m \leq M$. This is a special case of the familiar chaining argument in empirical process theory, dating back at least 50 years; see, e.g., Cassels (1951) or Chung (1949) and also Section 2 above. Hence

$$\begin{aligned} P\left(\sup_{0 \leq s \leq 1} |Z(0, s)| \geq 2R \right) &\leq \sum_{m=1}^{M} P\left(\max_{0 \leq d < 2^m} |Z(d2^{-m}, (d + 1)2^{-m})| \geq R/M \right) \\ &\leq \sum_{m=1}^{M} \frac{M^2}{R^2} 2^m (2^{-m} H^{-\alpha} Q H^{\sigma+\tau} (\log H)^{\gamma+1} + H^{-16}) \\ &\ll R^{-2} Q H^{\sigma+\tau-\alpha} (\log H)^{\gamma+4}, \end{aligned} \tag{7.19}$$

thus establishing the claim of the lemma. □

We now can finish the proof of Proposition 7.3. Fix $r \geq 0$. We shall estimate

$$\max_{0 \leq Q < 2^r} \Delta(2^r, Q; \alpha).$$

Write Q in binary expansion. Then for each $0 \leq Q < 2^r$,

$$\Delta(2^r, Q; \alpha) \leq \sum_{l=1}^{r} \Delta(2^r + m_l 2^{l-1}, 2^{l-1}; \alpha),$$

where $0 \leq m_l \leq 2^{r-l}$. This argument goes back to at least Gál and Koksma (1950). Hence setting $\beta = \frac{1}{2}(1 + \sigma + \tau - \alpha)$

$$\begin{aligned} &P\left(\max_{0 \leq Q < 2^r} \Delta(2^r, Q; \alpha) \geq 2^{r\beta} r^{\frac{\gamma+9}{2}} \right) \\ &\leq \sum_{l=1}^{r} P\left(\max_{0 \leq m \leq 2^{r-l}} \Delta(2^r + m2^{l-1}, 2^{l-1} \alpha) \geq 2^{r\beta} r^{\frac{\gamma+7}{2}} \right). \end{aligned}$$

We estimate these probabilities using Lemma 7.5 with $R = 2^{r\beta} r^{\frac{1}{2}(\gamma+7)}$, $H = 2^r + m2^{l-1} \le 2^{r+1}$ and $Q = 2^{l-1}$. Then the probability in question is

$$\ll \sum_{l \le r} 2^{r-l} 2^{-2r\beta} r^{-\gamma-7} 2^l 2^{r(\sigma+\tau-\alpha)} r^{\gamma+4} \ll r^{-2}.$$

Replacing the last exponent -2 by $-1-\varepsilon$ will give a smaller power of $\log N$. Thus by the Borel–Cantelli lemma with probability 1,

$$\max_{0<Q<2^r} \Delta(2^r, Q; \alpha) \ll 2^{r\beta} r^{\frac{1}{2}(\gamma+9)}.$$

If N is given, define n by $2^{n-1} \le N < 2^n$. Then with probability 1,

$$\Delta_N(\alpha) \ll \sum_{r<n} \max_{0 \le Q \le 2^r} \Delta(2^r, Q; \alpha) \ll \sum_{r<n} 2^{r\beta} r^{\frac{1}{2}(\gamma+9)} \ll N^\beta (\log N)^{\frac{1}{2}(\gamma+9)},$$

which yields Proposition 7.3. □

Notice that in the last estimate we used the fact that as a function of r the suprema over intervals contained in $[0, 2^{-r\alpha}]$ are nonincreasing.

The proof of Proposition 7.2 can be completed using the following two lemmas. The first lemma is well known and the second one is proved via elementary arguments; for proofs, see, e.g., Berkes, Philipp, and Tichy (2001).

Lemma 7.6. *Suppose that* $\{n_j, j \ge 1\}$ *satisfies* (7.11). *Let* $A < B$. *Then the number of* n_j*'s with* $j \le N$ *and* $A \le n_j \le B$ *is* $\ll N^\rho \log(B/A)$, *where the constant implied by* $\ll$ *only depends on* c *and* ρ.

Lemma 7.7. *Suppose that* $\{n_j, j \ge 1\}$ *satisfies* (7.11). *Let* $1 \le Q \le H$ *be given and fix* $a \in \mathbb{Z}$ *with*

$$0 < |a| \le H^{20}.$$

Set $\sigma = \rho$ *and* $d = \frac{c}{4}$. *Then the system* (7.12) *of diophantine equations has at most* $CH^{1+2\rho} Q (\log H)^2$ *solutions. Here* $C > 0$ *depends only on* c *and* ρ.

We apply Proposition 7.3 with $\sigma = \rho$, $\tau = 1 + 2\rho$ and $\gamma = 2$ and obtain Proposition 7.2. It is interesting to note that the case $\alpha = 0$ requires no "fast" growth condition.

Proposition 7.8. *Let* $\{n_j, j \ge 1\}$ *be any increasing sequence of integers. Let* X_j *be defined by* (7.10). *Then with probability* 1,

$$\Delta_N^*(0) \ll N (\log N)^{2+\varepsilon}, \quad \varepsilon > 0.$$

Remark. In Berkes, Philipp and Tichy (2001) this was erroneously stated for $\Delta_N(0)$ instead of $\Delta_N^*(0)$.

Proof. By the Erdős–Turán inequality (see, e.g., Drmota and Tichy (1997, p. 15) or Kuipers and Niederreiter (1974, p. 112), we have, for each $M > 0$,

$$\begin{aligned} \Delta_N^*(0) &\le 6N^2 M^{-1} + 6 \sum_{h=1}^{M} \frac{1}{h} \left| \sum_{1 \le j \ne k \le N} e^{2\pi i (n_j - n_k) h\omega} \right| \\ &\le 6N^2 M^{-1} + 6 \sum_{h=1}^{M} \frac{1}{h} \left(\left| \sum_{1 \le j \le N} e^{2\pi i n_j h\omega} \right|^2 + N \right). \end{aligned} \tag{7.20}$$

A straightforward adaption of Baker's (1980, pp. 37–38) argument yields the result. The details are as follows. By Hunt's (1968) maximal inequality (see also Baker (1980, relation (10)),

$$E\left(\max_{\frac{1}{2}n\le r\le n}\left|\sum_{j\le r} e^{2\pi i n_j h\omega}\right|^2\right)\le Cn,$$

where C is a numerical constant. Thus by (7.20) with $M = N(= m)$, we have, for $k \ge 1$,

$$E\left(\max_{2^{k-1}\le m<2^k}\Delta_m^*(0)\right)\le 6\,2^k+6\sum_{h\le 2^k}\frac{1}{h}\left(E\max_{2^{k-1}\le r<2^k}\left|\sum_{j\le r} e^{2\pi i n_j h\omega}\right|^2+2^k\right)\ll 2^k\,k.$$

Thus by Markov's inequality and the Borel–Cantelli lemma, we have, with probability 1,

$$\max_{2^{k-1}\le m<2^k}\Delta_m^*(0)\ll 2^k k^{2+\varepsilon},$$

which finally yields Proposition 7.8. □

We finally complete the proof of Theorem 7.1. The only difference between U_N and V_N is that in V_N the indicator 1_L has period 1, whereas the indicator in the definition of U_N does not. We shall estimate the relevant difference, as follows:

$$\begin{aligned}0 &\le \sum_{1\le i<j\le N}\left(1_{[0,t]}(X_j-X_i)-1(0\le X_j-X_i\le t)\right)\\ &= \sum_{1\le i<j\le N}\left(1(0\le X_j-X_i\le t \mod 1)-1(0\le X_j-X_i\le t)\right)\\ &= \sum_{1\le i<j\le N}1(0\le X_j\le X_i-1+t).\end{aligned}\tag{7.21}$$

The last equality follows from the fact that

$$0\le X_j-X_i\le t \mod 1$$

if and only if either

$$0\le X_j-X_i\le t,\quad \text{i.e.,}\quad X_i\le X_j\le X_i+t$$

or

$$X_i-1\le X_j\le X_i-1+t,\quad \text{i.e.,}\quad 0\le X_j\le X_i-1+t.$$

In view of (7.21) the following rather crude estimate will yield Theorem 7.1.

Lemma 7.9. *With probability* 1, *we have*

$$\sum_{1\le i<j\le N}1\left(0\le X_j\le X_i-1+N^{-\alpha}\right)\ll\left(N^{2-2\alpha}+N^{1+\rho-\alpha}\right)(\log N)^2.$$

Proof. By monotonicity it is enough to prove the lemma for $N = 2^r$, replacing $N^{-\alpha}$ by $2\,2^{-r\alpha}$. To apply Markov's inequality, we need to estimate

$$P(0 \leq \{n_j\omega\} \leq \{n_i\omega\} - 1 + 2 \cdot 2^{-r\alpha}).$$

Suppose that $i < j$. By a well-known elementary estimate we have for integrable f and $0 \leq a < b \leq 1$ and some $|\theta| \leq 4$,

$$\int_a^b f(\{n_j\omega\})\, d\omega = (b-a) \int_0^1 f(\omega)\, d\omega + \theta n_j^{-1} \int_0^1 |f(\omega)|\, d\omega. \tag{7.22}$$

The above probability is bounded by

$$\int_0^1 1(1 - 2^{-r\alpha+1} \leq \{n_j\omega\})\, 1(\{n_i\omega\} \leq 2^{-r\alpha+1})\, d\omega. \tag{7.23}$$

Now $1(0 \leq \{n_i\omega\} \leq 2^{-r\alpha+1})$ is the indicator of a union of n_i disjoint intervals of length $2^{-r\alpha+1}n_i^{-1}$. Thus using (7.22) n_i times with $f(\omega) = 1(1 - 2^{-r\alpha+1} \leq \{n_j\omega\})$ and (a, b) the corresponding n_i intervals we obtain for (7.23) the bound

$$P(1 - 2^{-r\alpha+1} \leq \{n_j\omega\})\, P(\{n_i\omega\} \leq 2^{-r\alpha+1}) + 4n_i n_j^{-1} P(1 - 2^{-r\alpha+1} \leq \{n_j\omega\})$$

$$\leq 2^{-2r\alpha+2} + 4\,2^{-r\alpha+1} n_i/n_j. \tag{7.24}$$

Now

$$\begin{aligned} \sum_{1 \leq i < j < 2^r} n_i/n_j &\leq \sum_{1 \leq i < j < 2^r} (1 + cj^{-\rho})^{i-j} \\ &\leq \sum_{j \leq 2^r} (1 + cj^{-\rho})(1 - 1/(1 + cj^{-\rho}))^{-1} \\ &\ll \sum_{j \leq 2^r} j^{-\rho} \ll 2^{r(1+\rho)}. \end{aligned}$$

Summing (7.24) over $1 \leq i < j \leq 2^r$ we obtain

$$\sum_{1 \leq i < j \leq 2^r} P\left(0 \leq \{n_j\omega\} \leq \{n_i\omega\} - 1 + 2\,2^{-r\alpha}\right) \ll 2^{2r(1-\alpha)} + 2^{r(1+\rho-\alpha)}.$$

By Markov's inequality this implies that with probability 1,

$$\sum_{1 \leq i < j \leq 2^r} 1\left(\{n_j\omega\} \leq \{n_i\omega\} - 1 + 2^{-r\alpha+1}\right) \ll \left(2^{2r(1-\alpha)} + 2^{r(1+\rho-\alpha)}\right) r^2.$$

This proves the lemma for $N = 2^r$, and thus as noted above for N in general. □

Hence by Proposition 7.2 and (7.21), we have, with probability 1,

$$\Gamma_N(\alpha) \ll \left(N^{1-\frac{1}{2}\alpha+\frac{3}{2}\rho} + N^{2-2\alpha}\right) (\log N)^{11/2},$$

which proves Theorem 7.1. □

7.2 Pair correlations for independent random variables. We define U_N, V_N, Γ_N and Δ_N as in the previous section. With this notation we have the following results.

Theorem 7.10. *If* $\{X_j, j \geq 1\}$ *is a sequence of independent random variables with uniform distribution* (7.2), *then*

$$\Gamma_N(\alpha) \ll \left(N^{\frac{2-\alpha}{2}} + N^{\frac{3-3\alpha}{2}}\right)(\log N)^{9/2}$$

with probability 1.

Save for a few powers of $\log N$ this estimate is sharp.

Theorem 7.11. *Let* $0 \leq \alpha \leq 2$. *Then with probability* 1,

$$\limsup_{N\to\infty} (N^{\frac{2-\alpha}{2}} + N^{\frac{3-3\alpha}{2}})^{-1}\Gamma_N(\alpha) \geq c > 0,$$

for some constant c.

Proposition 7.12. *If* $\{X_j, j \geq 1\}$ *is a sequence of independent random variables with uniform distribution* (7.2), *then*

$$\Delta_N(\alpha) \ll N^{1-\frac{1}{2}\alpha}(\log N)^{9/2}$$

with probability 1.

Remark. In fact, both Proposition 7.12 and Theorem 7.10 remain valid for $\alpha > 2$. Again, except for powers of logarithm the estimate is sharp.

Proposition 7.13. *Let* $0 \leq \alpha \leq 2$. *Then with probability* 1,

$$\limsup_{N\to\infty} N^{-(1-\frac{1}{2}\alpha)}\Delta_N(\alpha) \geq c > 0$$

for some constant c.

Remark. It is perhaps interesting to compare the estimates for Δ_N and Γ_N. The additional term $N^{\frac{3-3\alpha}{2}}$ stems from the fact that the kernel corresponding to Δ_N is periodic with period 1, whereas the other kernel is not.

The proof of Proposition 7.12 is an easy modification of the proof of Proposition 7.3. For $1 \leq Q \leq H$ set

$$S = S(H, Q; L; \omega) = \sum_{H \leq j < H+Q} \sum_{1 \leq i < j} (1_L(X_j - X_i) - |L|).$$

We first show that the conclusion of Lemma 7.4 remains valid.

Lemma 7.14. *We have*

$$ES^2 \ll |L| H Q \log H + H^{-16}.$$

Proof. We define f, f_H^* and, with the obvious modification, S_H^* as in the proof of Lemma 7.4. Then (7.14) remains valid. To estimate $\|S_H^*\|_2$ we proceed as in the proof of Lemma 7.4. We observe that because of independence the expectation of each term in the relevant sum

$$\sum_b \sum_{H \leq j,l < H+Q} \sum_{1 \leq h < j,\, 1 \leq k < l} e^{2\pi i(a(X_j - X_h) - b(X_l - X_k))}$$

vanishes unless $j = l$, $h = k$ and $b = a$. Thus for fixed a this sum reduces to $Q(H+Q)$. This yields the conclusion of the lemma. □

The remainder of the proof of Proposition 7.12 is identical with that of Proposition 7.3, in the special case $\sigma = 0$, $\tau = 1$ and $\gamma = 0$.

For the proof of Theorem 7.11 it is enough to show that there is a positive constant c such that for all $N \geq 1$,

$$P\left(\left|\sum_{1\leq i<j\leq N} (1(|X_i - X_j| \leq N^{-\alpha}) - (2N^{-\alpha} - N^{-2\alpha}))\right| \geq c\,(N^{1-\frac{1}{2}\alpha} + N^{(3-3\alpha)/2})\right) \geq c \tag{7.25}$$

since then

$$P\left(\Gamma_N(\alpha) \geq c\,(N^{1-\frac{1}{2}\alpha} + N^{(3-3\alpha)/2})\right) \geq c, \tag{7.26}$$

and thus by the Kolmogorov zero-one law the right-hand side of (7.26) can be replaced by 1. But relation (7.25) follows from Horváth's (1991, Theorem 2) with $c(N) = N^{-\alpha}$.

For the proof of Proposition 7.13 again it is enough to show the existence of a positive constant c such that for all $N \geq 1$,

$$P\left(\left|\sum_{1\leq i<j\leq N} \left(1_{[0,N^{-\alpha}]}(X_j - X_i) - N^{-\alpha}\right)\right| \geq c\,N^{1-\frac{1}{2}\alpha}\right) \geq c. \tag{7.27}$$

We need two well-known facts from analysis and martingale theory. The first is the Paley–Zygmund theorem.

Theorem A. *Let X be a nonnegative random variable with finite second moment. Suppose that*

$$||X||_2 \leq \frac{1}{a}\,EX$$

for some $0 \leq a \leq 1$. Let $0 \leq b \leq a$. Then

$$P(X \geq b\,||X||_2) \geq (b-a)^2.$$

The next result is a special case of Burkholder's (1973, Theorem 3.1) square function inequality.

Theorem B. *Let $\{Y_n, \mathcal{F}_n, n \geq 1\}$ be a real-valued martingale. Then for every $1 < p < \infty$, there are constants $c_p > 0$ and C_p such that*

$$c_p\,E\left\{\sup_{n\geq 1}|Y_n|^p\right\} \leq E\left\{\left(Y_1^2 + \sum_{n\geq 1}(Y_{n+1} - Y_n)^2\right)^{p/2}\right\} \leq C_p\,E\left\{\sup_{n\geq 1}|Y_n|^p\right\}.$$

Let $L = [0, N^{-\alpha}]$ and write $g(z) = 1_L(z) - |L|$.

Lemma 7.15. *Let $\mathcal{F}_k := \sigma(X_1, \ldots, X_k)$, $k \geq 1$, be the natural filtration of the sequence $\{X_j, j \geq 1\}$. Let*

$$v_k := \sum_{1\leq j<k} g(X_k - X_j).$$

Then $\{v_k, \mathcal{F}_k, k \geq 1\}$ is a martingale difference sequence and $\{V_k^*, \mathcal{F}_k, k \geq 1\}$ is a martingale. Here $V_k^* = V_N - EV_N$, where V_N is defined in (7.8).

Proof. Since 1_L has period 1, the conditional expectation

$$\begin{aligned} E(v_k|\mathcal{F}_{k-1}) &= \sum_{1\le j<k} E\{g(X_k - X_j)|\mathcal{F}_{k-1}\} \\ &= \sum_{1\le j<k} \int_0^1 g(u - X_j)\, du \\ &= \sum_{1\le j<k} \int_0^1 (1_L(u) - |L|)\, du = 0, \end{aligned}$$

which completes the proof of the lemma. □

Lemma 7.16. *We have, for $N \ge N_0(\alpha)$,*

$$EV_N^{*2} \ge \frac{1}{4} N^{2-\alpha}.$$

Proof. Since the sequence $\{v_k, \mathcal{F}_k, k \ge 1\}$ is a martingale difference sequence, the random variables v_k are orthogonal. Thus

$$EV_N^{*2} = \sum_{k\le N} E v_k^2.$$

Now

$$\begin{aligned} E(v_k^2|\mathcal{F}_{k-1}) &= E\left\{\left(\sum_{1\le j<k} g(X_k - X_j)\right)^2 |\mathcal{F}_{k-1}\right\} \\ &= \int_0^1 \left(\sum_{1\le j<k} g(u - X_j)\right)^2 du \\ &= \sum_{1\le i,j<k} \int_0^1 g(u - X_j)\, g(u - X_i)\, du. \end{aligned}$$

For $i \ne j$ the expectation of the integrals vanish by Fubini's theorem and independence. Thus by periodicity

$$\begin{aligned} E v_k^2 &= E(E(v_k^2|\mathcal{F}_{k-1})) = \sum_{1\le j<k} E \int_0^1 g^2(u - X_j)\, du \\ &= (k-1)E\left(\int_0^1 1_L(u)du - |L|^2\right) \\ &= (k-1)(N^{-\alpha} - N^{-2\alpha}), \end{aligned}$$

which yields the lemma. □

The next step is to estimate EV_N^{*4} and to apply Theorem A. For this we need the following result.

Lemma 7.17. *We have, for* $1 \le m < n \le N$,

$$E v_n^4 \ll n^2 N^{-2\alpha} + n\, N^{-\alpha} \quad \text{and} \quad E v_m^2 v_n^2 \ll m\, n\, N^{-2\alpha},$$

where the constants implied by $\ll$ *are absolute.*

Before we prove Lemma 7.17 we finish the proof of Proposition 7.13. We first apply Theorem B to the martingale $\{V_n^*, \mathcal{F}_n, 1 \le n \le N\}$ with $p = 4$. We obtain

$$\begin{aligned} E V_N^{*4} &\le c_4^{-1} E\left\{\left(\sum_{n \le N} v_n^2\right)^2\right\} \\ &\ll \sum_{n \le N} E v_n^4 + \sum_{m < n \le N} E v_m^2 v_n^2 \\ &\ll \sum_{n \le N} n^2 N^{-2\alpha} + \sum_{n \le N} n N^{-\alpha} + \sum_{m < n \le N} m\, n\, N^{-2\alpha} \ll N^{4-2\alpha}. \end{aligned}$$

Hence by Lemma 7.16 and by Theorem A applied to V_N^{*2} we obtain the conclusion of Proposition 7.13.

Thus it remains to prove Lemma 7.17. We have

$$\begin{aligned} E(v_n^4 | \mathcal{F}_{n-1}) &= E\left\{\left(\sum_{1 \le k < n} g(X_n - X_k)\right)^4 | \mathcal{F}_{n-1}\right\} \\ &= \int_0^1 \left(\sum_{1 \le k < n} g(u - X_k)\right)^4 du \\ &= \sum_{1 \le h, i, j, k < n} \int_0^1 g(u - X_h)\, g(u - X_i)\, g(u - X_j)\, g(u - X_k)\, du. \end{aligned}$$

By independence and Fubini's theorem the expectation of the integral vanishes unless (h, i, j, k) come in pairs (not necessarily different). By the proof of Lemma 7.16,

$$E \int_0^1 g^2(u - X_k) du = N^{-\alpha} - N^{-2\alpha}$$

and

$$E \int_0^1 g^4(u - X_k)\, du \le E \int_0^1 g^2(u - X_k)\, du \le N^{-\alpha}.$$

Thus

$$E v_n^4 = E(E(v_n^4 | \mathcal{F}_{n-1})) \ll n^2 N^{-2\alpha} + n N^{-\alpha},$$

since there are at most $2n^2$ different pairs and n quadruples.

The proof of the estimate for the mixed moments is similar, but more intricate. We have, for $m < n$,

$$\begin{aligned} E(v_m^2 v_n^2) &= E\left\{\left(\sum_{1\le i<m} g(X_m - X_i)\right)^2 \left(\sum_{1\le k<n} g(X_n - X_k)\right)^2\right\} \\ &= \sum_{(*)} \int_0^1 E\left(g(X_m - X_i)\, g(X_m - X_j)\, g(u - X_k)\, g(u - X_l)\right) du, \end{aligned}$$

where the last summation $(*)$ is extended over $1 \le i, j < m$ and $1 \le k, l < n$. We break up this sum according to the following ranges of summation. $\Sigma^{(1)}$ contains all terms with $k, l > m$, $\Sigma^{(2)}$ contains all terms with $k = m, l > m$, $\Sigma^{(3)}$ contains all terms with $k = l = m$, $\Sigma^{(4)}$ contains all terms with $k = m, l < m$ and finally $\Sigma^{(5)}$ contains all terms with $k, l < m$. A typical term in $\Sigma^{(1)}$ equals

$$\int_0^1 E\left(g(v - X_i)\, g(v - X_j)\right) dv \int_0^1 E\left(g(u - X_k)\, g(u - X_l)\right) du.$$

Such a term will vanish unless both $i = j$ and $k = l$. Thus

$$\Sigma^{(1)} \ll mN^{-\alpha}\, nN^{-\alpha} \ll m\, n\, N^{-2\alpha}.$$

Next, each term in $\Sigma^{(2)}$ vanishes since l is isolated and $E(g(u - X_l)) = 0$ for fixed u. Similarly, a typical term in $\Sigma^{(3)}$ equals

$$\int_0^1 \int_0^1 E\left(g(v - X_i)\, g(v - X_j)\right) g^2(u - v)\, du\, dv.$$

Such a term will vanish unless $i = j$. In this case the integral equals

$$\int_0^1 \int_0^1 \int_0^1 g^2(v - w)\, g^2(u - v)\, du\, dv\, dw \ll N^{-2\alpha}.$$

Thus

$$\Sigma^{(3)} \ll mN^{-2\alpha}.$$

Now a typical term in $\Sigma^{(4)}$ equals

$$\int_0^1 \int_0^1 E\left(g(v - X_i)\, g(v - X_j)\, g(u - v)\, g(u - X_l)\right) du\, dv.$$

This expectation vanishes unless $i = j = l$. In this case the integral reduces to

$$\int_0^1 \int_0^1 \int_0^1 g^2(v - w)\, g(u - v)\, g(u - w)\, du\, dv\, dw \ll N^{-2\alpha}$$

and so

$$\Sigma^{(4)} \ll mN^{-2\alpha}.$$

Finally, in $\Sigma^{(5)}$ a typical term equals

$$\int_0^1 \int_0^1 E\left(g(v - X_i)\, g(v - X_j)\, g(u - X_k)\, g(u - X_l)\right) du\, dv.$$

If one index is different from any of the other three the expectation vanishes. Thus for the expectation not to vanish the indices must come in pairs or as a quadruple. In either case the integral does not exceed $N^{-2\alpha}$. Consequently

$$\Sigma^{(5)} \ll m^2 N^{-2\alpha}.$$

This completes the proof of Lemma 7.17, and thus of Proposition 7.13. □

In view of relation (7.21) above, the following estimate will yield Theorem 7.10.

Lemma 7.18. *With probability* 1,

$$\sup_{0 \le t \le N^{-\alpha}} \left| \sum_{1 \le i < j \le N} \left(1(0 \le X_j \le X_i - 1 + t) - \frac{1}{2} t^2 \right) \right| \ll N^{\frac{3}{2}(1-\alpha)} (\log N)^4.$$

The proof uses the arguments of Lemma 7.5 and Proposition 7.3. Let $0 \le s \le 1$ and let the binary expansion be given by (7.15). Let $1 \le Q \le H$ and define M by (7.16). Then (7.17) holds. For $0 \le u < v \le 1$ we define

$$Y(u, v) = Y(u, v; H, Q) = \sum_{H \le j < H+Q} \sum_{1 \le i < j} \eta_{ik},$$

where we set

$$\eta_{ij} := 1\left(X_i - 1 + uH^{-\alpha} \le X_j \le X_i - 1 + vH^{-\alpha}\right) - \frac{1}{2}(v^2 - u^2) H^{-2\alpha}.$$

Then for each $0 \le s \le 1$, relation (7.18) holds with Z replaced by Y:

$$|Y(0, s)| \le \sum_{m=1}^{M} |Y(d_m 2^{-m}, (d_m + 1) 2^{-m})| + 2Q(H + Q) 2^{-M} H^{-2\alpha}. \tag{7.28}$$

Here the d_m's are integers with $0 \le d_m < 2^m$, $1 \le m \le M$. Similarly, relation (7.19) holds with Z replaced by Y.

To apply Chebyshev's inequality we need to estimate $E\eta_{ij}\eta_{kl}$. By independence and since $E\eta_{ij} = 0$, we have $E\eta_{ij}\eta_{kl} = 0$, if all four indices are different. In the other cases, we get

$$\left| E\eta_{ij}\eta_{kl} \right| < \begin{cases} \frac{1}{2}(v^2 - u^2) H^{-2\alpha} & \text{if } i = k,\ j = l, \\ (v - u)^2 H^{-4\alpha} & \text{if } i = k \text{ and } j \ne l \text{ or if } i \ne k \text{ and } j = l. \end{cases}$$

The first estimate is clear and for the second estimate we note that if $i = k$ and $j \ne l$ the expectation of the product of the indicators figuring in the definition of η_{ij} and η_{kl} equals, because of independence,

$$\begin{aligned} &P(X_i - 1 + uH^{-\alpha} \le X_j, X_l \le X_i - 1 + vH^{-\alpha}) \\ &= \int_{1 - vH^{-\alpha}}^{1} P(s - 1 + uH^{-\alpha} \le X_j, X_l \le s - 1 + vH^{-\alpha}) ds \\ &= \int_{1 - vH^{-\alpha}}^{1} (P(s - 1 + uH^{-\alpha} \le X_j \le s - 1 + vH^{-\alpha}))^2 ds \\ &< (v - u)^2 H^{-4\alpha}. \end{aligned}$$

Thus for fixed $0 \le d < 2^m$ and $m \le M$ we obtain

$$\begin{aligned} E(Y(d2^{-m}, (d+1)2^{-m}))^2 &\le H(H+Q)H^{-2\alpha}((d+1)^2 - d^2)\,2^{-2m} \\ &\quad +(H+Q)Q^2\,2^{-2m}H^{-3\alpha} + (H+Q)^2Q2^{-2m}H^{-3\alpha} \\ &\ll H^{1-2\alpha}Q\,2^{-m} + H^{2-3\alpha}Q\,2^{-2m}. \end{aligned}$$

Thus by (7.28) and (7.19)

$$\begin{aligned} P(\sup_{0\le s\le 1} |Y(0,s)| \ge 2R) &\ll \sum_{m=1}^{M} \frac{M^2}{R^2} 2^m (H^{1-2\alpha}Q\cdot 2^{-m} + H^{2-3\alpha}Q\cdot 2^{-2m}) \\ &\ll R^{-2}H^{1-2\alpha}Q(1+H^{1-\alpha})(\log H)^3. \end{aligned}$$

We now can complete the proof of the lemma in the same way as the proof of Proposition 7.3. □

Combining Lemma 7.18 with Proposition 7.12 we get Theorem 7.10.

Acknowledgments. The authors would like to thank the Centre for Mathematical Physics and Stochastics (MaPhySto) and its Director, as well as the organizers of the Instructional Workshop on *Empirical Process Techniques for Dependent Data* for the opportunity to present the material here. They also thank Thomas Mikosch and Olimjon Sharipov for carefully reading the manuscript and for many comments that led to the improvement of the presentation.

References

[1] J. Aaronson, R.M. Burton, H.G. Dehling, D. Gilat, T. Hill and B. Weiss: Strong laws for L- and U-statistics. *Transactions of the American Mathematical Society* **348** (1996), 2845–2865.

[2] M.A. Arcones and B. Yu: Central limit theorems for empirical processes and U-processes of stationary mixing sequences. *Journal of Theoretical Probability* **7** (1994), 47–71.

[3] F. Avram and M.S. Taqqu: Noncentral limit theorems and Appell polynomials. *Annals of Probability* **15** (1987), 767–775.

[4] R.C. Baker: Metric number theory and the large sieve. *Journal of the London Mathematical Society* **24** (1981), 34–40.

[5] J. Beran: Statistical methods for data with long-range dependence (with discussions). *Statistical Science* **7** (1992), 404–427.

[6] H.C.P. Berbee: *Random Walks with Stationary Increments and Renewal Theory*. Mathematical Centre Tracts **112**, Mathematisch Centrum, Amsterdam, 1979.

[7] R.H. Berk: Limit behavior of posterior distributions when the model is incorrect. *Annals of Mathematical Statistics* **37** (1966), 51–58.

[8] I. Berkes and W. Philipp: An almost sure invariance principle for the empirical distribution function of mixing random variables. *Zeitschrift für Wahrscheinlichkeitstheorie und Verwandte Gebiete* **41** (1977), 115–137.

[9] I. Berkes and W. Philipp: Approximation theorems for independent and weakly dependent random vectors. *Annals of Probability* **7** (1979), 29–54.

[10] I. Berkes and W. Philipp: The size of trigonometric and Walsh series and uniform distribution mod 1. *Journal of the London Mathematical Society* **50** (1994), 454–464.

[11] I. Berkes, W. Philipp and R. Tichy: The pair correlation for independent and weakly depedent random variables. *Illinois Journal of Mathematics* **45** (2001), 559–580.

[12] P. Billingsley: *Convergence of Probability Measures.* John Wiley & Sons, New York, 1968 (2nd edition: J. Wiley, New York, 1999).

[13] P. Billingsley: *Probability and Measure.* 3rd edition, John Wiley & Sons, New York, 1995.

[14] N.H. Bingham, C.M. Goldie and J.L. Teugels: *Regular Variation.* Encyclopedia of Mathematics and Its Applications **27**. Cambridge University Press, Cambridge, 1987.

[15] J.R. Blum, D.L. Hanson and L.H. Koopmans: On the strong law of large numbers for a class of stochastic processes. *Zeitschrift für Wahrscheinlichkeitstheorie und verwandte Gebiete* **2** (1963), 1–11.

[16] S.A. Borovkova: Weak convergence of the empirical process of U-statistics structure for dependent observations. *Theory of Stochastic Processes* **18** (1995), 115–124.

[17] S.A. Borovkova, R.M. Burton and H.G. Dehling: Consistency of the Takens estimator for the correlation dimension. *Annals of Applied Probability* **9** (1999), 376–390.

[18] S.A. Borovkova, R.M. Burton and H.G. Dehling: Limit theorems for functionals of mixing processes with applications to U-statistics and dimension estimation. *Transactions of the American Mathematical Society* **353** (2001), 4261–4318.

[19] S.A. Borovkova, R.M. Burton and H.G. Dehling: From Dimension Estimation to Asymptotics of Dependent U-Statistics. *Bolyai Society Mathematical Studies X, Limit Theorems, Balatonlelle*, Budapest 2001, 1–34.

[20] R.C. Bradley: *Introduction to Strong Mixing Conditions*, vol. I. Technical Report, Department of Mathematics, Indiana University, Bloomington, 2001.

[21] L. Breiman: *Probability.* Addison Wesley, Reading, MA, 1968.

[22] D. L. Burkholder: Distribution function inequalities for martingales. *Annals of Probability* **1** (1973), 19–42.

[23] L. Carleson: On convergence and growth of partial sums of Fourier series. *Acta Mathematica* **116** (1966), 135–157.

[24] J.W.S. Cassels: Some metrical theorems of Diophantine approximation III. *Proceedings of the Cambridge Philosophical Society* **46** (1950), 219–225.

[25] J. W. S. Cassels: An extension of the law of the iterated logarithm. *Proceedings of the Cambridge Philosophical Society* **47** (1951), 55–64.

[26] K. L. Chung: An estimate concerning the Kolmogorov limit distribution. *Transactions of the American Mathematical Society* **67** (1949), 36–50.

[27] M. Csörgő and P. Révész: A new method to prove Strassen type laws of invariance principle II. *Zeitschrift für Wahrscheinlichkeitstheorie und verwandte Gebiete* **31** (1975), 261–269.

[28] A.R. Dabrowski and H.G. Dehling: Estimating conditional occupation time distributions for dependent sequences. *The Canadian Journal of Statistics* **24** (1996), 55–65.

[29] A.R. Dabrowski, H.G. Dehling, T. Mikosch and O. Sharipov: Poisson limits for U-statistics. *Stochastic Processes and Their Applications* **99** (2002), 137–157.

[30] Yu.A. Davydov: The invariance principle for stationary processes. *Theory of Probability and Its Applications* **15** (1970), 487–498.

[31] H.G. Dehling: *Grenzwertsätze für Summen schwach abhängiger vektor-wertiger Zufallsvariablen.* Dissertation, Georg-August-Universität Göttingen, 1981.

[32] H.G. Dehling: Limit theorems for sums of weakly dependent Banach space valued random variables. *Zeitschrift für Wahrscheinlichkeitstheorie und verwandte Gebiete* **63** (1983), 393–432.

[33] H.G. Dehling: The functional law of the iterated logarithm for von-Mises functionals and multiple Wiener integrals. *Journal of Multivariate Analysis* **28** (1989), 177–189.

[34] H.G. Dehling: Complete convergence of triangular arrays and the law of the iterated logarithm for U-statistics. *Statistics and Probability Letters* **7** (1989), 319–321.

[35] H.G. Dehling and W. Philipp: Almost sure invariance principles for weakly dependent vector-valued random variables. *Annals of Probability* **10** (1982), 689–701.

[36] H.G. Dehling, M. Denker and W. Philipp: Invariance principles for von Mises and U-statistics. *Zeitschrift für Wahrscheinlichkeitstheorie und verwandte Gebiete* **67** (1984), 139–167.

[37] H.G. Dehling, M. Denker and W. Philipp: A bounded law of the iterated logarithm for Hilbert space valued martingales and its application to U-statistics. *Probability Theory and Related Fields* **72** (1986), 111–131.

[38] H.G. Dehling, M. Denker and W. Philipp: The almost sure invariance principle for the empirical process of U-statistic structure. *Annales de l'Institut Henri Poincaré* **23** (1987), 121–134.

[39] H.G. Dehling and M.S. Taqqu: The limit behavior of empirical processes and symmetric statistics. *Bulletin of the International Statistical Institute* **52** (1987), volume 4, 217–234.

[40] H.G. Dehling and M.S. Taqqu: The empirical process of some long-range dependent sequences with an application to U-statistics. *Annals of Statistics* **17** (1989), 1767–1783.

[41] H.G. Dehling and M.S. Taqqu: Bivariate symmetric statistics of long-range dependent observations. *Journal of Statistical Planning and Inference* **28** (1991), 153–165.

[42] H.G. Dehling and M.S. Taqqu: Continuous functions whose level sets are orthogonal to all polynomials of a given degree. *Acta Mathematica Hungarica* **60** (1992), 217–224.

[43] M. DENKER: *Asymptotic Distribution Theory in Nonparametric Statistics.* Vieweg Verlag, Braunschweig, 1985.

[44] M. DENKER and G. KELLER: On U-statistics and von Mises statistics for weakly dependent processes. *Zeitschrift für Wahrscheinlichkeitstheorie und verwandte Gebiete* **64** (1983), 505–522.

[45] M. DENKER and G. KELLER: Rigorous statistical procedures for data from dynamical systems. *Journal of Statistical Physics* **44** (1986), 67–93.

[46] M. DENKER, CHR. GRILLENBERGER and K. SIEGMUND: *Ergodic Theory on Compact spaces.* Lecture Notes in Mathematics **527**, Springer Verlag, Berlin, 1976.

[47] M. DENKER, C. GRILLENBERGER and G. KELLER: A note on invariance principles for von Mises statistics. *Metrika* **32** (1985), 197–214.

[48] C.M. DEO: A note on empirical processes of strong-mixing sequences. *Annals of Probability* **1** (1973), 870–875.

[49] R.L. DOBRUSHIN and P. MAJOR: Non-central limit theorems for non-linear functionals of Gaussian fields. *Zeitschrift für Wahrscheinlichkeitstheorie und verwandte Gebiete* **50** (1979), 27–52.

[50] M.D. DONSKER: An invariance principle for certain probability limit theorems. *Memoirs of the American Mathematical Society* **6** (1951).

[51] M.D. DONSKER: Justification and extension of Doob's heuristic approach to the Kolmogorov–Smirnov theorems. *Annals of Mathematical Statistics* **23** (1952), 277–281.

[52] J.L. DOOB: Heuristic approach to the Kolmogorov–Smirnov theorems. *Annals of Mathematical Statistics* **20** (1949), 393–403.

[53] J.L. DOOB: *Stochastic Processes.* J. Wiley & Sons, New York, 1953.

[54] P. DOUKHAN: *Mixing: Properties and Examples.* Lecture Notes in Statistics **85**, Springer Verlag, 1994.

[55] P. DOUKHAN and F. PORTAL: Principe d'invariance faible avec vitesse pour un processus empirique dans un cadre multidimensionnel et fortement mélangeant. *Comptes Rendus Academie des Sciences Paris, Série I* **297** (1983), 505–508.

[56] P. DOUKHAN and F. PORTAL: Principe d'invariance faible pour la fonction de répartition empirique dans un cadre multidimensionnel et mélangeant. *Probability and Mathematical Statistics* **8** (1987), 117–132.

[57] M. DRMOTA and R.F. TICHY: *Sequences, Discrepancies and Applications.* Springer Lecture Notes in Mathematics **1651**, 1997.

[58] R.M. DUDLEY *Real Analysis and Probability.* Wadsworth, Belmont, California, 1989.

[59] R.M. DUDLEY and W. PHILIPP: Invariance principles for sums of Banach space valued random elements and empirical processes. *Zeitschrift für Wahrscheinlichkeitstheorie verwandte Gebiete* **62** (1983), 509–552.

[60] A. DVORETZKY: Asymptotic normality for sums of dependent random variables. In: *Proceedings of the Sixth Berkeley Symposium on Mathematical Statistics and Probability Theory* (1970), vol. II, 513–535.

[61] V. EASTWOOD and L. HORVÁTH: Limit theorems for short distances in $\mathbb{R}^m$. *Statistics and Probability Letters* **45** (1999), 261–268.

[62] P. ERDŐS: Problems and results in diophantine approximations. *Compositio Mathematica* **16** (1964), 52–65.

[63] P. ERDŐS and I.S. GÁL: On the law of the iterated logarithm, I-II. *Indagationes Mathematicae* **17** (1955), 65–76, 77–84.

[64] P. ERDŐS and M. KAC: On certain limit theorems in the theory of probability. *Bulletin of the American Mathematical Society* **52** (1946), 292–302.

[65] P. ERDŐS and J.F. KOKSMA: On the uniform distribution modulo 1 of sequences $(f(n, \theta))$. *Indagationes Mathematicae* **11** (1949), 299–302.

[66] P. ERDŐS and P. TURÁN: On a problem in the theory of uniform distribution I. *Indagationes Mathematicae* **10** (1948), 370–378.

[67] J. ESARY, F. PROSCHAN and D. WALKUP: Association of random variables with applications. *Annals of Mathematical Statistics* **38** (1967), 1466–1474.

[68] H. FIEDLER, W. JURKAT and O. KÖRNER: Asymptotic expansions of certain theta series. *Acta Arithmetica* **32** (1977), 129–146.

[69] A.A. FILIPPOVA: Mises's theorem on the asymptotic behavior of functionals of empirical distribution functions and its statistical applications. *Theory of Probability and Its Applications* **7** (1961), 24–57.

[70] H. FINKELSTEIN: The law of the iterated logarithm for empirical distributions. *Annals of Mathematical Statistics* **42** (1971), 607–615.

[71] I. S. GÁL and J.F. KOKSMA: Sur l'ordre de grandeur des fonctions sommables. *Proceedings Koninklijke Nederlandse Akademie van Wetenschappen* **53** (1950), 638–653.

[72] L. GIRAITIS and D. SURGAILIS: Central limit theorem for the empirical process of a linear sequence with long memory. *Journal of Statistical Planning and Inference* **80** (1999), 81–93.

[73] P. GRASSBERGER and I. PROCACCIA: Characterization of strange attractors. *Physical Review Letters* **50** (1983), 346–349.

[74] Y. GUIVARC'H, M. KEANE and B. ROYNETTE: *Marches aléatoires sur les groupes de Lie.* Lecture Notes in Mathematics **624**, Springer Verlag, Berlin, 1977.

[75] P. HALMOS: The theory of unbiased estimation. *Annals of Mathematical Statistics* **17** (1946), 34–43.

[76] R. HELMERS, P. JANSSEN and R. SERFLING: Glivenko–Cantelli properties of some generalized empirical DFs and strong convergence of generalized L-statistics. *Probability Theory and Related Fields* **79** (1988), 75–93.

[77] H.-C. HO and T. HSING: On the asymptotic expansion of the empirical process of long-memory moving averages. *Annals of Statistics* **24** (1996), 992–1024.

[78] W. HOEFFDING: A class of statistics with asymptotically normal distribution. *Annals of Mathematical Statistics* **19** (1948), 293–325.

[79] W. HOEFFDING: The strong law of large numbers for U-statistics. *University of North Carolina Mimeo Report* No. 302, 1961.

[80] J. HOFFMANN-JØRGENSEN: *Stochastic Processes on Polish Spaces. Various Publication Series* **39**, Aarhus Universitet, Aarhus, Denmark, 1991.

[81] L. HORVÁTH: Short distances on the line. *Stochastic Processes and Their Applications* **39** (1991), 65–80.

[82] R.A. HUNT: *On the convergence of Fourier series, orthogonal expansions and their continuous analogues.* Southern Illinois University Press (1968), 235–255.

[83] H.E. HURST: Long-term storage capacity of reservoirs. *Transactions of the American Society of Civil Engineers* **116** (1951), 770–808.

[84] I.A. IBRAGIMOV: Some limit theorems for stochastic processes stationary in the strict sense. *Doklady Akademii Nauk SSSR* **125** (1959), 711–714.

[85] I.A. IBRAGIMOV: Some limit theorems for stationary processes. *Theory of Probability and Its Applications* **7** (1962), 349–382.

[86] I.A. IBRAGIMOV and YU.V. LINNIK: *Independent and Stationary Sequences of Random Variables.* Wolters-Noordhoff Publishing, Groningen, 1971.

[87] M. IOSIFESCU: A very simple proof of a generalization of the Gauss–Kuzmin–Lévy theorem on continued fractions and related questions. *Revue Roumaine Mathematiques Pures et Appliques* **37** (1992), 901–914.

[88] V. ISHAM: Statistical aspects of chaos: a review. In: *Networks and Chaos - Statistical and Probabilistic Aspects* (O.E. Barndorff-Nielsen, J.L. Jensen and W.S. Kendall, eds.). Chapman & Hall, London, 1993.

[89] J. KIEFER: Skorohod embedding of multivariate rv's and the sample df. *Zeitschrift für Wahrscheinlichkeitstheorie und verwandte Gebiete* **24** (1972), 1–35.

[90] A. KOLMOGOROFF and G. SELIVERSTOFF: Sur la convergence de series de Fourier. *Atti Accad. Naz. Lincei Ser. 6* **3** (1926), 307–310.

[91] A.N. KOLMOGOROV and YU.A. ROZANOV: On strong mixing conditions for stationary Gaussian processes. *Theory of Probability and Its Applications* **5** (1960), 204–208.

[92] J. KOMLÓS, P. MAJOR and G. TUSNADY: An approximation of partial sums of independent RV's and the sample DF. I. *Zeitschrift für Wahrscheinlichkeitstheorie und verwandte Gebiete* **32** (1975), 111–131.

[93] J. KOMLÓS, P. MAJOR and G. TUSNADY: An approximation of partial sums of independent RV's and the sample DF. II. *Zeitschrift für Wahrscheinlichkeitstheorie und verwandte Gebiete* **34** (1975), 33–58.

[94] H. Koul and D. Surgailis: Asymptotic expansion of the empirical process of long memory moving averages. In: *Empirical Processes for Dependent Data*, (H.G. Dehling, T. Mikosch and M. Sørensen, eds.). Birkhäuser, Boston (2002).

[95] J. Kuelbs: Kolmogorov's law of the iterated logarithm for Banach space valued random variables. *Illinois Journal of Mathematics* **21** (1977), 784–800.

[96] J. Kuelbs and W. Philipp: Almost sure invariance principles for partial sums of mixing B-valued random variables. *The Annals of Probability* **8** (1980), 1003-1036.

[97] L. Kuipers and H. Niederreiter: *Uniform Distribution of Sequences*. John Wiley, New York, 1974.

[98] T.L. Lai: Reproducing kernel Hilbert spaces and the law of the iterated logarithm for Gaussian processes. *Zeitschrift für Wahrscheinlichkeitstheorie und verwandte Gebiete* **29** (1974), 7–19.

[99] P. Major: *Multiple Wiener–Itô Integrals*. Lecture Notes in Mathematics **849**, Springer Verlag, Berlin, 1981.

[100] A. Mandelbaum and M.S. Taqqu: Invariance principle for symmetric statistics. *Annals of Statistics* **12** (1984), 483–496.

[101] B.B. Mandelbrot: *The Fractal Geometry of Nature*. W.H.Freeman and Company, New York, 1982.

[102] B.B. Mandelbrot and J.W. van Ness: Fractional Brownian motion, fractional noises and applications. *SIAM Review* **10** (1968), 909–918.

[103] D. Monrad and W. Philipp: Nearby variables with nearby conditional laws and a strong approximation theorem for Hilbert space valued martingales. *Probability Theory and Related Fields* **88** (1991), 381–404.

[104] D. Nolan and D. Pollard: Functional limit theorems for U-processes. *Annals of Probability* **16** (1988), 1291–1298.

[105] W. Philipp: The central limit problem for mixing sequences of random variables. *Zeitschrift für Wahrscheinlichkeitstheorie und verwandte Gebiete* **12** (1969), 155–171.

[106] W. Philipp: Mixing sequences of random variables and probabilistic number theory. *Memoirs of the American Mathematical Society* **114** (1971).

[107] W. Philipp: Empirical distribution functions and uniform distribution mod 1. In: *Diophantine Approximation and Its Applications*. (C. Osgood, ed.), Academic Press, New York (1973), 211–234.

[108] W. Philipp: Limit theorems for lacunary series and uniform distribution mod 1. *Acta Arithmetica* **26** (1974), 241–251.

[109] W. Philipp: A functional law of the iterated logarithm for empirical distribution functions of weakly dependent random variables. *Annals of Probability* **5** (1977), 319–350.

[110] W. Philipp: Invariance principles for sums of mixing random elements and the multivariate empirical process. In: *Colloquium Mathematicae Societatis Janos Bolyái: Limit Theorems in Probability and Statistics*, Veszprem (1982).

[111] W. PHILIPP: Invariance principles for independent and weakly dependent random variables. In: *Dependence in Probability and Statistics.* (Eberlein, E., Taqqu, M.S., eds.) Progress in Probability and Statistics **11** (1986), Birkhäuser, Boston, 225–269.

[112] W. PHILIPP: Empirical distribution functions and strong approximation theorems for dependent random variables. A problem of Baker in probabilistic number theory. *Transactions of the American Mathematical Society* **345** (1994), 705–727.

[113] W. PHILIPP and W.F. STOUT: Almost sure invariance principles for partial sums of weakly dependent random variables. *Memoirs of the American Mathematical Society* **161** (1975).

[114] A. PLESSNER: Über Konvergenz von trigonometrischen Reihen. *Journal für reine und angewandte Mathematik* **155** (1926), 15–25.

[115] D. POLLARD: *Convergence of Stochastic Processes.* Springer Verlag, New York, 1984.

[116] P. REVÉSZ: *The Laws of Large Numbers.* Academic Press, New York, London, 1968.

[117] E. RIO: Covariance inequalities for strongly mixing processes. *Annales de l'Institut Henri Poincaré Probability and Statistics* **29** (1993), 587–597.

[118] E. RIO: About the Lindeberg method for strongly mixing sequences. *ESAIM: Probabability and Statistics* **1** (1995), 35–61.

[119] E. RIO: *Théorie asymptotique des processus aléatoires faiblement dépendants.* Mathématiques et applications **31**, Springer Verlag, 2000.

[120] M. RÖMERSPERGER: A note on nearby variables with nearby conditional laws. *Probability Theory and Related Fields* **106** (1996), 371–377.

[121] M. ROSENBLATT: A central limit theorem and a strong mixing condition. *Proceedings of the National Academy of Sciences USA* **42** (1956), 43–47.

[122] W. RUDIN: *Principles of Mathematical Analysis.* McGraw–Hill, 1976.

[123] Z. RUDNICK and P. SARNAK: The pair correlation function of fractional parts of polynomials. *Communications in Mathematical Physics* **194** (1998), 61–70.

[124] Z. RUDNICK and A. ZAHARESCU: A metric result on the pair correlation of fractional parts of sequences. *Acta Arithmetica* **89** (1999), 283–293.

[125] R. SALEM and A. ZYGMUND: On lacunary trigonometric series. *Proceedings of the National Academy of Sciences USA* **33** (1947), 333–338.

[126] R. SALEM and A. ZYGMUND: La loi du logarithme itéré pour les séries trigonométriques lacunaires. *Bulletin des Sciences Mathématiques (2)* **74** (1950), 220–224.

[127] P.K. SEN: A note on weak convergence of empirical processes for sequences of ϕ-mixing random variables. *Annals of Mathematical Statistics* **42** (1971), 2131–2133.

[128] P.K. SEN: Almost sure behavior of U-statistics and von Mises' differentiable statistical functionals. *Annals of Statistics* **2** (1974), 387–395.

[129] R.J. SERFLING: *Approximation Theorems of Mathematical Statistics.* John Wiley & Sons, New York, 1980.

[130] R. SERFLING: Generalized L-, M- and R-statistics. *Annals of Statistics* **12** (1984), 76–86.

[131] B.W. SILVERMAN: Limit theorems for dissociated random variables. *Advances in Applied Probability* **8** (1976), 806–819.

[132] B.W. SILVERMAN: Convergence of a class of empirical distribution functions of dependent random variables. *Annals of Probability* **11** (1983), 745–751.

[133] B.W. SILVERMAN and T. BROWN: Short distances, flat triangles and Poisson limits. *Journal of Applied Probability* **15** (1978), 815–825.

[134] A.V. SKOROHOD: On a representation of random variables. *Theory of Probability and Its Applications* **21** (1976), 628–632.

[135] D.A. SOTRES and M. GHOSH: Strong convergence of linear rank statistics for mixing processes. *Sankya, Series B* **39** (1977), 1–11.

[136] W.F. STOUT: *Almost Sure Convergence.* Academic Press, New York-London, 1974.

[137] V. STRASSEN and R.M. DUDLEY: The central limit theorem and ε-entropy. In: *Lecture Notes in Mathematics* **89**, Springer-Verlag, 1969, 224–231.

[138] D. SURGAILIS: Zones of attraction of self-similar multiple integrals. *Lithuanian Mathematical Journal* **22** (1983), 327–340.

[139] F. TAKENS: Detecting strange attractors in turbulence. In: *Dynamical Systems and Turbulence.* Lecture Notes in Mathematics **898**, Springer-Verlag, 1981, 336–381.

[140] F. TAKENS: On the numerical determination of the dimension of the attractor. In: *Dynamical Systems and Bifurcations.* Lecture Notes in Mathematics **1125**, Springer-Verlag, 1985, 99–106.

[141] M.S. TAQQU: Weak convergence to fractional Brownian motion and to the Rosenblatt process. *Zeitschrift für Wahrscheinlichkeitstheorie und verwandte Gebiete* **31** (1975), 287–302.

[142] M.S. TAQQU: Convergence of integrated processes of arbitrary Hermite rank. *Zeitschrift für Wahrscheinlichkeitstheorie und verwandte Gebiete* **50** (1979), 53–83.

[143] V.A. VOLKONSKII and YU.A. ROZANOV: Some limit theorems for random functions I. *Theory of Probability and Its Applications* **4** (1959), 178–197.

[144] A. WALFISZ: Ein metrischer Satz über Diophantische Approximationen. *Fundamenta Mathematicae* **16** (1930), 361–385.

[145] H. WEYL: Über die Gleichverteilung von Zahlen mod eins. *Mathematische Annalen* **77** (1916), 313–352.

[146] C.S. WITHERS: Convergence of empirical processes of mixing random variables. *Annals of Statistics* **3** (1975), 1101–1108.

[147] K. YOSHIHARA: Limiting behaviour of U-statistics for stationary, absolutely regular processes. *Zeitschrift für Wahrscheinlichkeitstheorie und verwandte Gebiete* **35** (1976), 237–252.

[148] H. Yu: A Glivenko–Cantelli lemma and weak convergence for empirical processes of associated sequences. *Probability Theory and Related Fields* **95** (1993), 357–370.

[149] V.V. Yurinskii: On the error of the Gaussian approximation for convolutions. *Theory of Probability and Its Applications* **22** (1977), 236–247.

Herold Dehling
Fakultät für Mathematik
Ruhr-Universität Bochum
Universitätsstrasse 150
44780 Bochum, Germany
herold.dehling@ruhr-uni-bochum.de

Walter Philipp
Department of Statistics
University of Illinois at Urbana-Champaign
725 S. Wright Street
Champaign, IL 61820, U.S.A.
philipp@stat.uiuc.edu

Part II

Techniques for the Empirical Process of Stationary Sequences

Weak Dependence: Models and Applications

PATRICK ANGO NZE AND PAUL DOUKHAN

ABSTRACT. This paper aims at a systematic introduction to a new weak dependence condition. We show that some standard models satisfy this property, including stationary Markov models, bilinear models, and more generally, Bernoulli shifts. In some cases no mixing properties can be expected without additional regularity assumption on the distribution of the innovations distribution for which a weak dependence condition can be easily derived. We apply the theory to derive a weak Donsker invariance principle and the empirical CLT and we present an application to kernel estimates for density and regression functions.

1 Introduction

In this paper we aim at introducing an alternative to standard mixing conditions (see, e.g., [24]) for stationary sequences commonly used in statistics.

Consider a stationary sequence of random variables. A naive but complicated way of defining weak dependence would be to require that certain asymptotic results which are valid for an independent sequence remain correct for the stationary sequence.

We intend to define a weak convergence concept in a simpler way. The following two requirements should be a guideline.

- The weak convergence condition must be verifiable on some standard models used in practice. Our experience (see [11] and [2]) shows that it is in general very difficult to apply strong mixing or absolute regularity conditions to concrete models.
- The weak convergence condition must be adapted to the applications in statistics one has in mind.

Our main goal is to derive a weak convergence concept which allows one to establish two probabilistic limit theorems: the Donsker invariance principle and the empirical CLT. They are major tools in statistical applications, for example for the estimation of change points in the mean of a stationary sequence, for Kolmogorov–Smirnov goodness-of-fit tests and for functional kernel estimation problems.

This paper is organized as follows. The first section is devoted to the definitions of our new version of a weak dependence. The main sources are the seminal works [13] and [4].

In the second section the weak dependence property, in the newly defined sense, is verified for some standard stationary sequences used in statistics: Markovian models, associated sequences, and Bernoulli shifts. In addition, the new dependence condition gives access to some classes of models which could not be studied by standard tools such as mixing.

In a third section, we prove some moment inequalities which are applied in the weak dependence framework. A moment inequality for sums can be proved by some combinatorial arguments. This condition also yields some subexponential inequalities described in the application to functional estimation.

Next, we then turn to the Donsker invariance principle and the empirical CLT. We also present a general method leading to CLT results for triangular arrays. The technical trick is the Lindeberg–Rio (see [26]) method, which can easily be handled in this weak dependence context as well. The latter method is compared with Bernstein's blocking technique (see [17]). Each method provides a CLT. The dependence rates for CLT convergence are compared later on in the functional estimation frame. Surprisingly, there is no uniformly best method.

Finally, we consider functional estimation. Besides the convergence results which are interesting for their own sake, the previous approach allows one to compare

- the weak dependence conditions with standard mixing,
- the different methods for proving CLTs,
- the almost sure convergence rates of kernel estimators, in contrast to the optimal rates in the the i.i.d. case.

The number of applications is relatively limited due to the fact that the conditions have been developed quite recently. We hope that the scope of statistical applications can be enlarged through future research.

2 Independence and dependence

2.1 Independence and correlation. Recall that the random variables (r.v.'s) X, Y with values in $\mathbb{R}^d$ are *independent* if

$$\forall f, g \text{ bounded measurable: } \mathrm{Cov}\,(f(X), g(Y)) = 0. \tag{2.1}$$

It actually suffices to consider classes of continuous or more regular functions f, g. The mixing conditions were introduced by weakening (2.1): instead of zero, the right-hand expression is supposed to be "small."

Also notice that in some cases, orthogonality yields independence:

(a) Bernoulli r.v.'s.

(b) Gaussian vectors: $(X, Y) \in \mathbb{R}^{a+b}$.

(c) Associated random vectors.

The first case only yields pairwise independence but not mutual independence in the case of more than two r.v.'s; it follows from orthogonality of the four pairs (X, Y), $(X, 1-Y)$, $(1-X, Y)$ and $(1-X, 1-Y)$ if the pair (X, Y) is orthogonal and the variables (X, Y) are both supported in $\{0, 1\}$. But with the exception of the Bernoulli case, properties inherited from covariances are not easily translated into properties of sigma-fields. In the two latter cases (b) and (c), additional inequalities may be proved for Lipschitz functions:

$$|\,\mathrm{Cov}\,(f(X), g(Y))\,| \le c(f, g) \sum_{i,j} |\,\mathrm{Cov}\,\left(X_i, Y_j\right)\,|.$$

2.2 The weak dependence conditions. Covariances of r.v.'s are much easier to compute than mixing coefficients. Therefore we describe dependence properties for a process $(X_n)_{n\in\mathbb{Z}}$ in terms of covariances instead of sigma-fields. Let $\mathbb{L}^\infty$ be the set of real-valued bounded measurable functions on some space $\mathbb{R}^u$ and $\|\cdot\|_\infty$ the corresponding norm. We define the Lipschitz modulus of a function $h : \mathbb{R}^u \to \mathbb{R}$:

$$\mathrm{Lip}(h) = \sup_{x\neq y} \frac{|h(x) - h(y)|}{\|x - y\|_1}, \qquad \text{where } \|(z_1, \dots, z_u)\|_1 = |z_1| + \cdots + |z_u|.$$

Define

$$\mathcal{L} = \left\{h \in \mathbb{L}^\infty;\ \|h\|_\infty \leq 1, \mathrm{Lip}(h) < \infty\right\}. \tag{2.2}$$

Definition 2.1 (Doukhan and Louhichi [13]). *The sequence $(X_n)_{n\in\mathbb{Z}}$ is said to be $(\theta, \mathcal{L}, \psi)$-*weakly dependent *if there exists a sequence $\theta = (\theta_r)_{r\in\mathbb{N}}$ decreasing to zero at infinity, and a function ψ with arguments $(h, k, u, v) \in \mathcal{L}^2 \times \mathbb{N}^2$ such that for any u-tuple $(i_1, \dots, i_u)$ and any v-tuple $(j_1, \dots, j_v)$ with $i_1 \leq \cdots \leq i_u < i_u + r \leq j_1 \leq \cdots \leq j_v$, one has*

$$|\mathrm{Cov}(h(X_{i_1}, \dots, X_{i_u}), k(X_{j_1}, \dots, X_{j_v}))| \leq \psi(h, k, u, v)\theta_r, \tag{2.3}$$

where the functions h and k are defined on $\mathbb{R}^u$ and $\mathbb{R}^v$, respectively.

Examples of interest involve the function $\psi_1'(h, k, u, v) = v\,\mathrm{Lip}(k)$ (e.g., in causal linear processes), $\psi_1(h, k, u, v) = u\,\mathrm{Lip}(h) + v\,\mathrm{Lip}(k)$, (e.g., in noncausal linear processes), $\psi_2(h, k, u, v) = uv\,\mathrm{Lip}(h)\,\mathrm{Lip}(k)$ (e.g., in associated processes), and $\psi_2'(h, k, u, v) = v\,\mathrm{Lip}(h)\,\mathrm{Lip}(k)$. The prime exponents are conventional notation.

Notice that the sequence θ depends on both the class $\mathcal{L}$ and the function ψ. The function ψ may really depend on all its arguments, in contrast to the case of bounded mixing sequences. This definition is hereditary through images by convenient functions, as was noticed in [36] in order to check a weak dependence property more easily.

The following example, due to Rosenblatt [30], describes the case where mixing fails to hold. This is the (Markov) $AR(1)$-process with Bernoulli innovation, satisfying $\mathbb{P}(\xi_0 = 0) = \mathbb{P}(\xi_0 = 1) = \frac{1}{2}$,

$$X_n = \frac{1}{2}(X_{n-1} + \xi_n).$$

It also determines the Bernoulli shift (see definition below) $X_n = H(\xi_n)$ with

$$H(x) = \sum_{k=0}^{\infty} 2^{-(k+1)} x_k.$$

This model has the stationary uniform distribution on the interval (0, 1) but it satisfies no mixing condition (the past may entirely be recovered from the present). For a rigorous proof see [17], p. 375.

Define also the class

$$\mathcal{I} = \left\{\bigotimes_{i=1}^{u} g_{x_i};\ x_i \in \mathbb{R}_+^*, u \in \mathbb{N}^*\right\}, \quad \text{where } g_x(y) = \mathbf{1}_{x\leq y} - \mathbf{1}_{y\leq -x}, \forall x \in \mathbb{R}_+^*. \tag{2.4}$$

The following lemma links $\mathcal{I}$-weak dependence with $\mathcal{L}$-weak dependence. Indeed, examples are proved to satisfy a weak dependence condition w.r.t. the class $\mathcal{L}$. Consider the weaker $\mathcal{L}_0 \cap \mathcal{C}_b^1$-weak dependence condition defined with

$$\mathcal{L}_0 = \left\{\bigotimes_{i=1}^{u} f_i;\ f_i \in \mathcal{L},\ f_i : \mathbb{R} \to \mathbb{R}, i = 1, \dots, u,\ u \in \mathbb{N}^*\right\}.$$

Here $\mathcal{C}_b^1$ stands for the set of partially differentiable functions with bounded partial derivatives and the function ψ is either ψ_1 or ψ_2. This latter condition will thus imply $\mathcal{I}$-weak dependence under concentration assumptions.

Lemma 2.2. *Let (X_n) be a sequence of r.v.'s. Suppose that, for some real $\alpha > 0$,*

$$C(\lambda) = \sup_{x \in \mathbb{R}} \sup_i \mathbb{P}(x \le X_i \le x + \lambda) \le C\lambda^\alpha. \tag{2.5}$$

Suppose that the sequence (X_n) is

- *$(\theta, \mathcal{L}_0 \cap \mathcal{C}_b^1, \psi_1)$-weakly dependent, then it is $(\theta_\mathcal{I}, \mathcal{I}, \psi)$-weakly dependent with*

$$\theta_{\mathcal{I},r} = \theta_r^{\frac{\alpha}{1+\alpha}} \text{ and } \psi(h,k,u,v) = 2(8C)^{\frac{1}{1+\alpha}}(u+v).$$

- *$(\theta, \mathcal{L}_0 \cap \mathcal{C}_b^1, \psi_2)$-weakly dependent, then it is $(\theta_\mathcal{I}, \mathcal{I}, \psi)$-weakly dependent with*

$$\theta_{\mathcal{I},r} = \theta_r^{\frac{\alpha}{2+\alpha}} \text{ and } \psi(h,k,u,v) = (8C)^{\frac{2}{2+\alpha}}(u+v)^{\frac{2(1+\alpha)}{2+\alpha}}.$$

In order to better understand the mechanism of proof, we recall the proof of this lemma.

Proof of Lemma 2.2. First recall the following useful inequality, valid for real numbers $0 \le x_i, y_i \le 1$,

$$|x_1 \cdots x_m - y_1 \cdots y_m| \le \sum_{i=1}^m |x_i - y_i|.$$

Let $g, f \in \mathcal{I}$. Then

$$g(y_1, \ldots, y_u) = g_{x_1}(y_1) \cdots g_{x_u}(y_u) \quad \text{and} \quad f(y_1, \ldots, y_v) = g_{x'_1}(y_1) \cdots g_{x'_v}(y_v)$$

for some $u, v \in \mathbb{N}^*$ and $x_i, x'_j \ge 0$. For fixed $x > 0$ and $a > 0$, let

$$f_x(y) = \mathbf{1}_{y>x} - \mathbf{1}_{y \le -x} + \left(\frac{y}{a} - \frac{x}{a} + 1\right)\mathbf{1}_{x-a<y<x} + \left(\frac{y}{a} + \frac{x}{a} - 1\right)\mathbf{1}_{-x<y<-x+a}.$$

Therefore $\mathrm{Lip}(f_x) = a^{-1}$ and $\|f_x\|_\infty = 1$ and $\mathrm{Lip}(h) \le a^{-1}$, $\mathrm{Lip}(k) \le a^{-1}$ if we set

$$h(y_1, \ldots, y_u) = f_{x_1}(y_1) \cdots f_{x_u}(y_u), \quad k(y_1, \ldots, y_v) = f_{x'_1}(y_1) \cdots f_{x'_v}(y_v).$$

Consider $i_1 \le \cdots \le i_u \le i_u + r \le j_1 \le \cdots \le j_v$ and set

$$\mathrm{Cov}(h,k) := \mathrm{Cov}(h(X_{i_1}, \ldots, X_{i_u}), k(X_{j_1}, \ldots, X_{j_v})).$$

- $(\theta, \mathcal{L}_0 \cap \mathcal{C}_b^1, \psi_1)$ (resp., $(\theta, \mathcal{L}_0 \cap \mathcal{C}_b^1, \psi_2)$)-weak dependence implies, with either $c(u,v) = (u+v)$ or $c(u,v) = (u+v)^2$,

$$|\mathrm{Cov}(h,k)| \le \frac{1}{a} c(u,v)\theta_r \qquad \left(\text{resp., } |\mathrm{Cov}(h,k)| \le \frac{1}{a^2} c(u,v)\theta_r\right).$$

- Inequality (2.5) yields $|\mathrm{Cov}(g,f) - \mathrm{Cov}(h,k)| \le 8Ca^\alpha(u+v)$. Hence $|\mathrm{Cov}(g,f)| \le 8Ca^\alpha(u+v) + \frac{1}{a}c(u,v)\theta_r$ (resp., $|\mathrm{Cov}(g,f)| \le 8Ca^\alpha(u+v) + \frac{1}{a^2}c(u,v)\theta_r$).

The lemma follows by setting

$$a = \left[\frac{c(u,v)\theta_r}{8C(u+v)}\right]^{1/(1+\alpha)} \qquad \left(\text{resp., } a = \left[\frac{c(u,v)\theta_r}{8C(u+v)}\right]^{1/(2+\alpha)}\right).$$

2.3 Mixing. For the sake of completeness, we recall some of the main mixing properties. For more details, the reader is referred to Doukhan [11].

Let $(\Omega, \mathcal{A}, \mathbb{P})$ be a probability space and let $\mathcal{U}, \mathcal{V}$ be two sub σ-algebras of $\mathcal{A}$. Various measures of dependence between $\mathcal{U}$ and $\mathcal{V}$ have been introduced:

$$\begin{aligned}
\alpha(\mathcal{U}, \mathcal{V}) &= \sup\{|\mathbb{P}(U \cap V) - \mathbb{P}(U)\mathbb{P}(V)|;\ U \in \mathcal{U}, V \in \mathcal{V}\},\\
\beta(\mathcal{U}, \mathcal{V}) &= \mathbb{E}\sup\{|\mathbb{P}(V|\mathcal{U}) - \mathbb{P}(V)|;\ V \in \mathcal{V}\},\\
\rho(\mathcal{U}, \mathcal{V}) &= \sup\{|\mathrm{Corr}(u, v)|;\ u \in \mathbb{L}^2(\mathcal{U}), v \in \mathbb{L}^2(\mathcal{V})\},\\
\phi(\mathcal{U}, \mathcal{V}) &= \sup\{|\mathbb{P}(V|U) - \mathbb{P}(V)|;\ U \in \mathcal{U}, V \in \mathcal{V}\}.
\end{aligned}$$

Those coefficients are, respectively, Rosenblatt's [29] strong mixing coefficient, $\alpha(\mathcal{U}, \mathcal{V})$, Wolkonski and Rozanov's absolute regularity coefficient in [37], $\beta(\mathcal{U}, \mathcal{V})$, Kolmogorov's and Rozanov's maximal correlation coefficient $\rho(\mathcal{U}, \mathcal{V})$ [18], and $\phi(\mathcal{U}, \mathcal{V})$ the uniform mixing coefficient from Ibragimov [17].

Let $\mathbf{X} = (X_n)_{n\in\mathbb{Z}}$ be a discrete time stationary process. We set $X_A = \{X_t; t \in A\}$ if $A \subset \mathbb{Z}$ for the A-marginal of X and $\sigma(Z)$ denotes the sigma-algebra generated by a random variable Z.

For any coefficient defined above, say $c(\cdot, \cdot)$, we shall call the process $\mathbf{X}$ a c-mixing process if $\lim_{k\to\infty} c_{\mathbf{X},k} = 0$ where $c_{\mathbf{X},k} = c(\sigma(X_{]-\infty,0]}), \sigma(X_{[k,+\infty[})$. The following implications are well known:

$$\phi\text{-mixing} \Rightarrow \rho\text{-mixing} \Rightarrow \alpha\text{-mixing} \quad \text{and} \quad \phi\text{-mixing} \Rightarrow \beta\text{-mixing} \Rightarrow \alpha\text{-mixing}$$

The converse implications are not valid.

3 Models

3.1 Markovian models. Let $(\xi_n)_{n\in\mathbb{Z}}$ be an i.i.d. sequence and M a measurable function. We turn our attention to models given by the equation

$$X_n = M(X_{n-1}, \ldots, X_{n-p}, \xi_n). \tag{3.1}$$

In order to justify the title of this section, notice that the vector-valued sequence $(X_n^{(p)})_{n\in\mathbb{Z}}$, where $X_n^{(p)} = (X_{n-1}, \ldots, X_{n-p})$, is a Markov process.

Such models are associated with ergodic Markov chains (see [21] for a review of the main models). Thus the stationarity assumption is feasible. An interesting subclass of examples is the following:

$$X_n = R(X_{n-1}, \ldots, X_{n-p}, \eta_n) + S(X_{n-1}, \ldots, X_{n-p})\xi_n,$$

where the function S satisfies $S(x_1, \ldots, x_p) \geq s > 0$ for some $s \in \mathbb{R}$, and any real numbers $x_1, \ldots, x_p$, both functions R and S essentially satisfy some contraction assumptions (see [11], [14], [2], or [35] for extensions) and (ξ_n) and (η_n) are two mutually independent i.i.d. sequences.

For instance, $ARMA(p, q)$ processes

$$Y_n = \sum_{i=1}^{p} a_i Y_{n-i} + \xi_n + \sum_{j=1}^{q} b_j \xi_{n-j}$$

have such a Markov representation. Indeed $X_n = (Y_n, Y_{n,n-1}, \ldots, Y_{n,n-\ell})$, where $Y_{n,j} = \mathbb{E}[Y_n \mid Y_{n+i} : i \leq n]$, and $\ell = \max\{p, q \mid 1\}$, is a Markov process. The mixing properties of these models were studied in [22].

Lipschitz models can be shown to be exponentially $\mathcal{L}$-weakly dependent.

Proposition 3.1 ([14]). *Let a vector-valued Markov model be defined through the recurrence relation* (3.1). *Assume that*

$$\mathbb{E}\|M(x,\xi_n)-M(y,\xi_n)\|^\varsigma \le a\|x-y\|^\varsigma \quad \text{and} \quad \mathbb{E}\|M(0,\xi_n)\|^\varsigma < \infty,$$

for some $0 \le a < 1$ *and* $\varsigma \ge 1$. *Then the sequence* $(X_n)_{n\in\mathbb{Z}}$ *is* $(\theta, \mathcal{L}, \psi)$*-weakly dependent. Here,* $\theta_r = \mathcal{O}(a^r)$ *and* $\psi(h,k,u,v) = v \operatorname{Lip} k$.

We note that no stationarity assumption was required.

More general $AR(p)$ models, $X_n = M(X_{n-1}, \dots, X_{n-p}; \xi_n)$, have the same properties: if, for example, $\mathbb{E}|M(0;\xi_n)| < \infty$ and, for some constants $a_j \ge 0$, $1 \le j \le p$ with $\sum_{j=1}^p a_j < 1$,

$$\mathbb{E}|M(x_1,\dots,x_p;\xi_n) - M(y_1,\dots,y_p;\xi_n)| \le \sum_{j=1}^p a_j|x_j - y_j|.$$

This model is geometrically $(\theta, \mathcal{L}, \psi)$-weakly dependent with function $\psi(h,k,u,v) = \min\{u \operatorname{Lip} h, v \operatorname{Lip} k\}$ as proved in [13].

Remark. Under the assumption that the distribution of ξ_0 has an integrable density f which does not vanish almost everywhere, the model with additive noise

$$M(X_{n-1},\dots,X_{n-p},\xi_n) = R(X_{n-1},\dots,X_{n-p}) + \xi_n$$

can be shown to be ergodic and mixing. More precisely, if the function R is continuous, and $\left|R(x_1,\dots,x_p)\right| \le A + a_1|x_1| + \dots + a_p|x_p|$ with $a_1 + \dots + a_p < 1$, under the invariant initial distribution, the sequence is absolutely regular with $\beta_n = \mathcal{O}(e^{-bn})$ (see [8]).

3.2 Bernoulli shifts.

Definition 3.2. *Let* $(\xi_i)_{i\in\mathbb{Z}}$ *be a sequence of i.i.d. real-valued r.v.'s and the function* $H : \mathbb{R}^{\mathbb{Z}} \to \mathbb{R}$ *be measurable. The sequence* $(X_n)_{n\in\mathbb{Z}}$ *is called a Bernoulli shift if it is defined by* $X_n = H(\xi_{n-j}, j \in \mathbb{Z})$.

Causal shifts are given as $X_n = H(\xi_n, \xi_{n-1}, \xi_{n-2}, \dots, \xi_0, \xi_{-1}, \xi_{-2}, \dots)$, i.e., $H : \mathbb{R}^{\mathbb{N}} \to \mathbb{R}$. Such shifts provide natural examples of weakly dependent, but nonmixing sequences (see [27]).

Notice that the stationary Markov setting provides such causal sequences. In fact, consider a Markov process driven by the recurrence equation $X_t = M(X_{t-1}, \xi_t)$, for some i.i.d. sequence $(\xi_t)_{t\in\mathbb{Z}}$. Then the function H is defined, *if it exists*, implicitly by the relation $H(x) = M(H(x'), x_0)$, where $x = (x_0, x_1, x_2, \dots)$, $x' = (x_1, x_2, x_3, \dots)$. Control theory yields tools to provide explicit Bernoulli shift representations (see [22] and [21]).

Chaotic representations. We now focus on chaotic expansions associated with the discrete chaos generated by the sequence $(\xi_i)_{i\in\mathbb{Z}}$; in a condensed formulation we write $H(x) = \sum_{k=0}^\infty H^{(k)}(x)$, where $H^{(k)}(x)$ denotes the k-th order chaos contribution, $H^{(0)}(x) = a_0^{(0)}$ is only a centering constant, and

$$H^{(k)}(x) = \sum_{j_1=-\infty}^{\infty} \sum_{j_2=-\infty}^{\infty} \cdots \sum_{j_k=-\infty}^{\infty} a^{(k)}_{j_1,\dots,j_k} x_{j_1} \times x_{j_2} \times \cdots \times x_{j_k}$$

or, in short, in vector notation, $H^{(k)}(x) = \sum_{j\in\mathbb{Z}^k} a_j^{(k)} x_j$.

In contrast with mixing conditions, it may be proved that even noncausal sequences may admit a $\mathcal{L}$-weakly dependent behaviour. This is a by-product of Proposition 3.4 to follow.

Definition 3.3. *For any integer $k > 0$, we denote by δ_k any number such that*

$$\sup_{i\in\mathbb{Z}} \mathbb{E}|H(\xi_{i-j}, j \in \mathbb{Z}) - H(\xi_{i-j}\mathbf{1}_{|j|<k}, j \in \mathbb{Z})| \le \delta_k.$$

Such sequences $(\delta_k)_{k\in\mathbb{Z}^+}$ are related to the modulus of uniform continuity of H. If, for instance, for positive constants $(a_i)_{i\in\mathbb{Z}}$, $0 < b \le 1$, the inequality

$$|H(u_i, i \in \mathbb{Z}) - H(v_i, i \in \mathbb{Z})| \le \sum_{i\in\mathbb{Z}} a_i |u_i - v_i|^b$$

holds for any sequences $(u_i), (v_i) \in \mathbb{R}^{\mathbb{Z}}$ and if the sequence $(\xi_i)_{i\in\mathbb{Z}}$ has a finite b-th order moment, then one can choose $\delta_k = \sum_{|i|\ge k} a_i \mathbb{E}|\xi_i|^b$.

Proposition 3.4 ([13]). *Bernoulli shifts are $(\theta, \mathcal{L}, \psi)$-weakly dependent with $\theta_r = 2\delta_{r/2}$ and with the function $\psi(h, k, u, v) = 4\,(u\|k\|_\infty \mathrm{Lip}(h) + v\|h\|_\infty \mathrm{Lip}(k))$. Under causality, this holds with $\theta_r = \delta_r$ and $\psi(h, k, u, v) = 2v\,\mathrm{Lip}\,k\|h\|_\infty$.*

Processes associated with a finite number of chaotic terms (i.e., $H^{(k)} = 0$ if $k > k_0$) are also called *Volterra processes*. A suitable bound for δ_r corresponds here to the stationarity condition

$$\delta_r = \sum_{k=0}^{\infty} \left\{ \sum_{j\in\mathbb{Z}^k;\, \|j\|_\infty > r} \left| a_j^{(k)} \right| \right\} \mathbb{E}|\xi_0|^k < \infty.$$

A prime example of a Volterra process is a linear process $X_t = \sum_{-\infty}^{\infty} a_k \xi_{t-k}$. A suitable sequence is $\theta_r = 2\mathbb{E}|\xi_0| \sum_{|k|>r/2} |a_k|$ with the previous function ψ_1.

The simple bilinear process, $X_t = (a + b\xi_{t-1})X_{t-1} + \xi_t$, is stationary if $c = \mathbb{E}|a + b\xi_0| < 1$ (see [34]). It is a Bernoulli shift with $H(x) = x_0 + \sum_{j=1}^{\infty} x_j \prod_{s=1}^{j} (a + bx_s)$, for $x = (x_i)_{i\in\mathbb{N}}$. We truncate the previous series up to rank r in order to obtain $\delta_r = \theta_r = c^r(r + 1)/(1 - c)$.

Giraitis, Koul, and Surgailis (see [28], [15]) consider $ARCH(\infty)$-models. We are given a nonnegative sequence $(b_j)_{j\ge1}$ and an i.i.d. sequence of nonnegative random variables $(\xi_j)_{j\ge0}$. The process is defined through the recurrence relation

$$X_t = \left[a + \sum_{j=1}^{\infty} b_j X_{t-j} \right] \xi_t.$$

Such models are proved to have a stationary representation with the chaotic expansion

$$X_t = a \sum_{\ell=1}^{\infty} \sum_{j_1=1}^{\infty} \cdots \sum_{j_\ell=1}^{\infty} b_{j_1} \cdots b_{j_\ell} \xi_{t-j_1} \cdots \xi_{t-[j_1+\cdots+j_\ell]}$$

under the simple assumption $\mathbb{E}\xi_0 \sum_{j=1}^{\infty} b_j < 1$. They extend the standard ARCH(p) model ($b_j = 0$ if $j > p$). Here we have $\theta_r = \left(\mathbb{E}\xi_0 \sum_{j=1}^{\infty} b_j \right)^r$.

Association.

Definition 3.5. *The sequence $(Z_t)_{t\in\mathbb{Z}}$ is associated, if for all coordinatewise increasing real-valued functions h and k,*

$$\mathrm{Cov}(h(Z_t, t \in A), k(Z_t, t \in B)) \ge 0$$

for all finite subsets A and B of $\mathbb{Z}$.

Associated sequences are $(\theta, \psi_2, \mathcal{L})$-weakly dependent with $\theta_r = \sup_{k\geq r} \operatorname{Cov}(X_0, X_r)$ (see [13]). Note that broad classes of examples of associated processes result from the fact that any independent sequence is associated and that monotonicity preserves association (for this, see [23]).

The case of Gaussian sequences is analogous by setting $\theta_r = \sup_{k\geq r} |\operatorname{Cov}(X_0, X_k)|$. Also, one may consider combinations of sums of Gaussian and associated sequences, or Bernoulli shifts driven by stationary, associated, instead of i.i.d. sequences.

Note that for associated or Gaussian sequences, ψ_2' replaces ψ_2 if the quantity $\theta_r = \sup_{k\geq r} |\operatorname{Cov}(X_0, X_k)|$ is replaced by $\theta_r = \sum_{k\geq r} |\operatorname{Cov}(X_0, X_k)|$.

Remark. The causal linear process $X_n = \sum_{t=0}^{\infty} a_t \xi_{n-t}$, satisfies a β-mixing condition if ξ_0 has density and (for some $\delta > 0$) the $(1+\delta)$-th order moment exists, together with the condition $\sum_{t=-\infty}^{\infty} |a_t|^{\delta} < \infty$. Then [25] proves that $\beta_n \leq C \sum_{l=n}^{\infty} \left(\sum_{k=l}^{\infty} |a_k|\right)^{\delta/1+\delta}$, for some $C > 0$.

Assume for instance that $a_j = \mathcal{O}(j^{-a})$. Then the previous regularity and moment conditions hold true if, for some $a > 2 + 1/\delta$, we have $a_j = \mathcal{O}(j^{-a})$. Indeed, $\beta_n \sim n^{-b}$ with $b = (a-2)\delta/(1+\delta)$. For example, $\sum_{n=0}^{\infty} \beta_n < \infty$ if $a > 3 + 2/\delta$. For $\delta = 1$ the latter condition turns into $a > 5$. It reads as $a > 3$ if $\delta = \infty$.

4 Algebraic moments of sums (see [13])

Let $(X_n)_{n\in\mathbb{N}}$ be a sequence of centered r.v.'s and let $S_n = \sum_{i=1}^{n} X_i$. In this section, we give moment bounds for $|\mathbb{E}S_n^q|$, when $q \in \mathbb{N}$ and $q \geq 2$. Let (X_n) be a sequence of centered r.v.'s and define for $r \geq 1$ the coefficients of weak dependence as nondecreasing sequences $(C_{r,q})_{q\geq 2}$ such that

$$\left|\operatorname{Cov}\left(X_{t_1} \times \cdots \times X_{t_m}, X_{t_{m+1}} \times \cdots \times X_{t_q}\right)\right| \leq C_{r,q}, \tag{4.1}$$

for all the successions $\{t_1, \cdots, t_q\}$ such that $1 \leq t_1 \leq \cdots \leq t_q \leq n$ and, for some integer $1 \leq m < q$, $t_{m+1} - t_m = r$. Explicit bounds $C_{r,q}$ are provided by [13] in order to build inequalities for the partial sums S_n. Two kinds of bounds are considered, either

$$\left|\operatorname{Cov}\left(X_{t_1} \times \cdots \times X_{t_m}, X_{t_{m+1}} \times \cdots \times X_{t_q}\right)\right| \leq c q^{\gamma} M^{q-2} \theta_r \tag{4.2}$$

or

$$\left|\operatorname{Cov}\left(X_{t_1} \times \cdots \times X_{t_m}, X_{t_{m+1}} \times \cdots \times X_{t_q}\right)\right| \leq c \int_0^{\theta_r} Q_{X_{t_1}}(x) \times \cdots \times Q_{X_{t_q}}(x) dx, \tag{4.3}$$

where Q_X denotes the quantile function of the distribution of X and $c, \gamma \geq 0$ are real numbers.

In the examples, the previous bound (4.2) holds for bounded sequences such that $\|X_n\|_\infty \leq M$. Under $(\theta, \mathcal{L}, \psi)$-weak dependence, this yields the bound

$$C_{r,q} = \max_{1\leq m<q} \psi(j^{\otimes m}, j^{\otimes(q-m)}, m, q-m) M^q \theta_r$$

where $j(x) = x\mathbf{1}_{|x|\leq 1} + \mathbf{1}_{x>1} - \mathbf{1}_{x<-1}$.

As in Lemma 2.2, we see that under $(\theta, \mathcal{L}, \psi)$-weak dependence with $\psi(h, k, u, v) = c(u, v) \operatorname{Lip}(h) \operatorname{Lip}(k)$ a bound is given by

$$C_{r,q} = \max_{1\leq m<q} c(m, q-m) M^{q-2} \theta_r.$$

For unbounded random variables,

$$C_{r,q} = c \int_0^{\theta_r} Q_{X_{t_1}}(x) \times \cdots \times Q_{X_{t_q}}(x)\mathrm{d}x$$

under $(\theta, \mathcal{I}, \psi)$-weak dependence.

An analogous bound was obtained by Rio for strongly mixing sequences (see [26]). The bound (4.3) holds for more general r.v.'s, using moment or tail assumptions.

A first consequence of inequality (4.2) is the following Marcinkiewicz–Zygmund inequality: Let $(X_n)_{n\in\mathbb{N}}$ be a sequence of centered r.v.'s satisfying the condition

$$C_{r,q} = \mathcal{O}(r^{-q/2}); \tag{4.4}$$

then there is a constant $B > 0$ not depending on n for which

$$|\mathbb{E}S_n^q| \le Bn^{q/2}. \tag{4.5}$$

The following lemma gives moment inequalities for $q \in \{2, 4\}$; it was essentially proved in Billingsley ([5], Lemmas 3 and 4, p. 172).

Lemma 4.1 ([13]). *If $(X_n)_{n\in\mathbb{N}}$ is a sequence of centered r.v.'s, then*

$$\mathbb{E}S_n^2 \le 2n\sum_{r=0}^{n-1} C_{r,2} \text{ and } \mathbb{E}S_n^4 \le 4!\left\{\left(n\sum_{r=0}^{n-1} C_{r,2}\right)^2 + n\sum_{r=0}^{n-1}(r+1)^2 C_{r,4}\right\}. \tag{4.6}$$

The following theorems deal with higher order moments.

Theorem 4.2 ([13]). *Let $q \ge 2$ be some integer. Suppose that the dependence coefficients $C_{r,p}$ associated with the sequence (X_n) satisfy for all integers $0 < p \le q$ and for some positive constants M, γ, C,*

$$(\mathcal{H}_1) \qquad C_{r,p} = Ce^{\gamma p}M^{p-2}\theta_r.$$

Then, for any integer $n \ge 2$,

$$|\mathbb{E}S_n^q| \le \frac{(2q-2)!}{(q-1)!}e^{q\gamma}\left\{\left(Cn\sum_{r=0}^{n-1}\theta_r\right)^{q/2} \vee \left(CM^{q-2}n\sum_{r=0}^{n-1}(r+1)^{q-2}\theta_r\right)\right\}. \tag{4.7}$$

Theorem 4.2 fits to bounded sequences. Now we consider the unbounded case. Let us define the class $\mathcal{I}$,

$$\mathcal{I} = \left\{\bigotimes_{i=1}^{u} g_{x_i};\, x_i \in \mathbb{R}_+^*, u \in \mathbb{N}^*\right\}, \text{ where } g_x(y) = \mathbf{1}_{x\le y} - \mathbf{1}_{y\le -x}, \forall x \in \mathbb{R}_+^*. \tag{4.8}$$

We shall consider $(\theta, \mathcal{I}, \psi)$-weak dependence where $\psi(h, k, u, v) = c(u, v)$.

Theorem 4.3 ([13]). *Let $q \ge 2$ be some integer. If $(X_n)_{n\in\mathbb{N}}$ is a centered and $(\theta, \mathcal{I}, \psi)$-weakly dependent sequence, set $C_q = (\max_{u+v\le q} c(u, v)) \vee 2$. Then*

$$|\mathbb{E}S_n^q| \le \frac{(2q-2)!}{(q-1)!}$$
$$\times\left(C_q\sum_{i=1}^{n}\int_0^1(\theta^{-1}(u)\wedge n)^{q-1}Q_i^q(u)du \vee \left(C_2\sum_{i=1}^{n}\int_0^1(\theta^{-1}(u)\wedge n)Q_i^2(u)du\right)^{q/2}\right).$$

In the special case of strongly mixing and stationary sequences, this is Theorem 1 in [26].

Remark. Exponential inequalities can be derived from Theorem 4.2. Define

$$M_{q,n} = n \sum_{r=0}^{n-1} (r+1)^{q-2} C_{r,q}.$$

We first suppose that for all integers $q \geq 2$ and n:

$$C_{r,p} = Ce^{\gamma p} M^{p-2} \theta_r, \quad M_{q,n} \leq A_n \frac{q!}{\beta^q},$$

where β is some constant and (A_n) is a sequence independent of q. Then the exponential inequality

$$\forall x \in \mathbb{R}_+^*, \qquad \mathbb{P}\left(|S_n| \geq x\sqrt{A_n}\right) \leq A \exp\left(-B\sqrt{\beta x}\right), \qquad (DP)$$

holds true with $A = e^{4+1/12}\sqrt{8\pi}$, and $B = e^{5/2-\gamma/2}$. Let (X_n) be a sequence of centered r.v.'s. Inequality (DP) is fulfilled if, for constants $C, \sigma, \gamma, b > 0$, we have $C_{r,q} = C\sigma^2 M^{q-2} e^{\gamma q} e^{-br}$ together with the bounds $\|X_n\|_\infty \leq M$ and $\|X_n\|_2 \leq \sigma$, for any integer $n \geq 0$. In this case $A_n = n\sigma^2$ in (DP). For example, it is enough to assume $(\theta, \mathcal{L}, \psi)$-weak dependence with $\theta_r = \mathcal{O}(e^{-br})$ and $\psi(h, k, u, v) \leq e^{\delta(u+v)} \operatorname{Lip}(h) \operatorname{Lip}(k))$ for some $\delta \geq 0$ in order to satisfy the previous requirements.

The use of combinatorics in the inequalities (DP) makes them relatively weak. As for a comparison, Bernstein's inequality, which is valid for independent sequences, allows one to replace the term $\sqrt{x}$ in the previous inequality by x^2 under the same assumption $n\sigma^2 \geq 1$. In the mixing cases, analogous exponential inequalities can also be obtained by using coupling arguments which are no longer available in a weak dependence framework.

5 Limit theorems

5.1 The Donsker process. Consider a stationary sequence $(X_n)_{n\in\mathbb{Z}}$ with mean zero. Denote by $[x]$ the integer part of a real number x. The *Donsker process* $(D_n(t))_{t\in[0,1]}$ is defined for any sample with size n as the following continuous time process:

$$D_n(t) = \begin{cases} \sum_{i=1}^k X_i & \text{for } t = k/n, k = 0, 1, \ldots, n, \\ \text{linearly interpolated} & \text{elsewhere.} \end{cases}$$

Let $(W_t)_{t\in[0,1]}$ be Brownian motion, that is W denotes the centered Gaussian real-valued process with covariance

$$\mathbb{E} W_s W_t = \min\{s, t\}.$$

We consider the following convergence result in the space $\mathcal{C}([0, 1])$ of continuous functions on the unit interval when the sample size n grows to infinity:

Theorem 5.1. *Suppose that the stationary sequence $(X_n)_{n\in\mathbb{Z}}$ satisfies $\mathbb{E}|X_0|^{4+\delta} < \infty$ for some $\delta > 0$. Assume a $(\theta, \mathcal{I}, \psi)$-weak dependence condition with the function $\psi_1(h, k, u, v)$ (respectively, ψ_2) and $\theta_r = \mathcal{O}(r^{-a})$ with $a \geq 2 + 4/\delta$ (respectively, $a > 2$).*

Then the following functional convergence holds in the space $\mathcal{C}([0, 1])$:

$$\frac{1}{\sqrt{n}} D_n(t) \to \sigma W_t.$$

The following series is assumed to be convergent:

$$\sigma^2 = \sum_{k=-\infty}^{\infty} \mathrm{Cov}(X_0, X_k).$$

The case $\sigma^2 = 0$ was considered in [24]. We shall assume here that $\sigma^2 \neq 0$.

Remark. Without any regularity condition on the innovations, Theorem 5.1 holds for a bounded Lipschitz function of a linear process $X_n = \sum_{k=-\infty}^{\infty} a_k \xi_{n-k}$ if $a_k = \mathcal{O}(k^{-D})$ when $D > 3$. The latter does not need to be causal while Ho and Hsing need this causality assumption in [16].

Proof of Theorem 5.1. Lemma 4.2 and a maximal inequality due to Moricz et al. (cf. [11], p. 40) yield

$$\mathbb{E}|S_n|^{2+\delta} = \mathcal{O}\left(n^{1+\delta/2}\right) \tag{5.1}$$

if for any increasing sequence of integers $0 \le i < j < k \le l$,

$$\sum_{m=0}^{\infty} m\, \mathbb{E}|X_0 X_m| < \infty \text{ and } \mathrm{Cov}\left(X_i X_j, X_k X_l\right) = \mathcal{O}\left((k-j)^{-2}\right). \tag{5.2}$$

Moreover, this entails that $\sigma^2 = n^{-1}\mathrm{Var}(S_n) > 0$ and the finite dimensional (fidi) convergence follows. The tightness of the process is a consequence of (5.2). The first part of (5.2) follows from the covariance bound $|\,\mathrm{Cov}\,(X_0, X_r)\,| = \mathcal{O}\left(\theta_r^{(2+\delta)/(4+\delta)}\right)$. The latter bound follows from $\mathrm{Cov}\left(X_i X_j, X_k X_l\right) = \mathcal{O}\left(\theta_{k-j}^{\delta/(4+\delta)}\right)$.

Remark. Let $\delta > 0$. Assume the existence of a moment of order $(2+\delta)$ for X_0. The Donsker functional CLT holds in the strong mixing case if $\sum_{n=0}^{\infty} n^{2/\delta}\alpha_n < \infty$ (see [26]). The condition $\sum_{n=0}^{\infty} \rho(2^n) < \infty$ with $\mathbb{E}X_0^2 < \infty$ implies the functional convergence, as Shao proved [31].

Remark. Here we consider conditions in terms of conditional expectations with respect to an adapted filtration. We first recall that Theorem 5.1 holds for martingales with stationary square integrable increments $\mathbb{E}X_0^2 < \infty$ (see [5]). Let $(\mathcal{M}_n)$ be a filtration adapted to the process $(X_n)_{n\in\mathbb{Z}}$: X_n is $\mathcal{M}_n$-measurable for any $n \in \mathbb{Z}$. Dedecker and Rio [9] prove that for a centered function, square integrable process, such that the sum $\sum_{n=0}^{\infty} X_0 \mathbb{E}\,(X_n|\mathcal{M}_0)$ is convergent in $\mathbb{L}^1$, the sequence $\mathbb{E}(X_0^2 + 2X_0 S_n|\mathcal{I})$ converges in $\mathbb{L}^1$ to some random variable η. The σ-field $\mathcal{I}$ is the tail σ-field. Moreover, conditionally to $\mathcal{I}$, $D_n(t)/\sqrt{n}$ converges to a Brownian motion ηW_t.

- This result provides a functional CLT to a limit process which is not Gaussian (see [6] for results related to the latter case).
- Note that ergodicity implies the triviality of the tail σ-field and a standard Donsker theorem follows.
- Standard results prove this theorem under a more restrictive $\mathbb{L}^2$ assumption: both series

$$\sum_{n=0}^{\infty} \mathbb{E}(X_n|\mathcal{M}_0) \text{ and } \sum_{n=0}^{\infty} (X_n - \mathbb{E}(X_n|\mathcal{M}_0))$$

converge.

- This theorem yields new results. Consider a stationary Markov sequence $(\xi_n)_{n\in\mathbb{Z}}$ with stationary distribution μ and transition operator P. Let g be a function such that $X_n = g(\xi_n)$ is a centered at its expectation, functional of (ξ_n). Then the assumption turns into the convergence of the series $\sum_{n=0}^{\infty} gP^n g$ in $\mathbb{L}^1(\mu)$.

- The previous result concerning strongly mixing sequences also appears as a consequence of this theorem.

5.2 Empirical process. Let us consider a stationary sequence $(X_n)_{n\in\mathbb{Z}}$.

We assume without loss of generality that the marginal distribution of this sequence is the uniform law on $[0, 1]$. The empirical process of the sequence (X_n) at time n is defined as $\frac{1}{\sqrt{n}} E_n(x)$, where

$$E_n(x) = \sum_{k=1}^{n} \left(\mathbf{1}_{(X_k \le x)} - \mathbb{P}(X_k \le x)\right).$$

Note that $E_n = n\,(F_n - F)$ if F_n and F denote the empirical d.f. and the marginal d.f., respectively. We consider the following convergence result in the Skohorod space $\mathcal{D}([0, 1])$ when the sample size n converges to infinity:

$$\frac{1}{\sqrt{n}} E_n(x) \to \bar{B}(x).$$

Here $(\bar{B}(x))_{x\in[0,1]}$ is the dependent analogue of a Brownian bridge, that is $\bar{B}$ denotes the centered Gaussian process with covariance given by

$$\mathbb{E}\bar{B}(x)\bar{B}(y) = \sum_{k=-\infty}^{\infty} \left(\mathbb{P}(X_0 \le x,\ X_k \le y) - \mathbb{P}(X_0 \le x)\mathbb{P}(X_k \le y)\right). \tag{5.3}$$

Note that for independent sequences with a marginal distribution function F, this turns into $\bar{B}(x) = B(x)$ for some standard Brownian bridge B; this justifies the name of generalized Brownian bridge.

Let (X_n) be a stationary sequence assumed to satisfy the following weak dependence condition:

$$\sup_{f\in\mathcal{F}} \left|\operatorname{Cov}\left(\prod_{i=1}^{2} f(X_{t_i}), \prod_{i=3}^{4} f(X_{t_i})\right)\right| \le \theta_r, \tag{5.4}$$

where $\mathcal{F} = \{x \to \mathbf{1}_{s<x\le t}$ for $s, t \in [0, 1]\}$, $0 \le t_1 \le t_2 \le t_3 \le t_4$ and $r = t_3 - t_2$ (in this case a weak dependence condition holds for a class of functions $\mathbb{R}^u \to \mathbb{R}$ working only with the values $u = 1$ or 2).

Proposition 5.2. *Let (X_n) be a stationary sequence such that (5.4) holds. Assume that there exists $\nu > 0$ such that*

$$\theta_r = \mathcal{O}(r^{-5/2-\nu}). \tag{5.5}$$

Then the sequence of processes $\left(\left\{\frac{1}{\sqrt{n}} E_n(t); t \in [0, 1]\right\}\right)_{n>0}$ is tight in the Skohorod space $\mathcal{D}([0, 1])$.

Theorem 5.3. *Suppose that the stationary sequence* (X_n) *is* $(\theta, \mathcal{L}_1, \psi_j)$*-weakly dependent, with either* $j = 1$ *and* $\theta_r = \mathcal{O}(r^{-15/2-\nu})$*, or* $j = 2$ *and* $\theta_r = \mathcal{O}(r^{-5-\nu})$*. Then the following empirical functional convergence holds true in the Skohorod space of real-valued càdlàg functions on the unit interval,* $\mathcal{D}([0, 1])$*:*

$$\frac{1}{\sqrt{n}} E_n(x) \to \bar{B}(x).$$

If the sequence (X_n) is i.i.d., then the series (5.3) reduces to a single term. This covariance has essentially $1/2$-order Hölder regularity. If the second order marginal distributions (X_0, X_k) have a joint, continuous density, then the k-th order term is a C^2-function. Hence the regularity of the covariance is determined by the central term. The trajectories of the process $\bar{B}$ cannot thus be expected to be more than $1/2$-Hölder regular.

Proof of Theorem 5.3. Thanks to the Rosenthal inequality in Theorem 4.2,

$$\begin{aligned} \|E_n(t) - E_n(s)\|_4 &\le C\sqrt{n} \sum_{r=0}^{n-1} \min\left\{r^{-5/2-\nu}, |t-s|\right\} + \left(n \sum_{r=0}^{n-1} (r+1)^2 \theta_r\right)^{1/4} \\ &\le \sqrt{n}\left(|t-s|^{(a-1)/2a} + n^{(2-a)/4}\right). \end{aligned}$$

The conclusion follows from Shao and Yu's tightness lemma [33]. The fidi convergence follows from a CLT whose proof uses blocking techniques due to Ibragimov [17].

Remark. If the sequence is strongly mixing, the summability condition $\sum_{n=0}^{\infty} \alpha_n < \infty$ implies fidi convergence. The empirical functional convergence holds if, in addition, for some $a > 1$, $\alpha_n = \mathcal{O}(n^{-a})$ (see [26]). In an absolutely regular framework, Doukhan, Massart and Rio (see [26]) obtain the empirical functional convergence when, for some $a > 2$, $\beta_n = \mathcal{O}(n^{-1}(\log n)^{-a})$. Shao and Yu [33] obtain the empirical functional convergence theorem when the maximal correlation coefficient satisfies the condition $\sum_{n=0}^{\infty} \rho(2^n) < \infty$.

5.3 A triangular CLT: The Lindeberg–Rio method. Let $X_{n,j}, 1 \le j \le n$ be a triangular array. We intend to describe a method for proving CLT convergence results for such arrays. We call it the Lindeberg–Rio method. We shall omit the first index n when possible without confusion. Consider a sequence $(W_j)_{j \in \mathbb{N}}$ of i.i.d. r.v.'s with standard normal law, and independent of $(X_n)_{t \in \mathbb{Z}}$. Define

$$\begin{aligned} \sigma_n^2 &= \operatorname{Var}(S_n), \\ S_k &= \sum_{j=1}^{k} X_j, \ 1 \le k \le n, \text{ and } S_0 = 0, \\ \tau_k &= \sum_{j=k}^{n} \sqrt{v_{n,j}} W_j = \sum_{j=k}^{n} V_j, \ 1 \le k \le n, \text{ and } \tau_{n+1} = 0, \end{aligned}$$

where we assume that $v_{n,j} = \operatorname{Var}(S_j) - \operatorname{Var}(S_{j-1}) > 0$.

Once one has proved that $\sigma_n^2 \to \sigma^2$, it remains to prove that for any three times differentiable function with bounded derivatives up to order 3, φ, say,

$$\Delta_n(\varphi) = \mathbb{E}\varphi(S_n) - \mathbb{E}\varphi(W_0) \to 0. \tag{5.6}$$

Consider the following:

$$U_j = S_{j-1} + \tau_{j+1}, \ R_j(x) = \varphi(U_j + x) - \varphi(U_j) - \frac{v_j}{2}\varphi''(U_j) \ (1 \le j \le n).$$

Clearly, we have

$$\begin{aligned}\Delta_n(\varphi) &= \sum_{k=1}^{n} \Delta_{n,k}(\varphi),\\ \Delta_{n,k}(\varphi) &= \mathbb{E}R_k(X_k) - \mathbb{E}R_k(V_k) \;=\; \Delta_k^1(\varphi) - \Delta_k^2(\varphi).\end{aligned}$$

A Taylor expansion yields

$$\begin{aligned}\Delta_k^2(\varphi) &= \mathbb{E}\left(\varphi\left(U_k+V_k\right)-\varphi\left(U_k\right)-V_k\varphi'\left(U_k\right)-\frac{V_k^2}{2}\varphi''\left(U_k\right)\right)\\ &= \frac{1}{6}\mathbb{E}\left(V_k^3\varphi^{(3)}\left(U_k+\vartheta_k V_k\right)\right),\ \text{with } 0<\vartheta_k<1,\end{aligned}$$

and $\left|\Delta_k^2(\varphi)\right| \le C\left(v_k\right)^{3/2}$. Moreover, with $0<\vartheta_k<1$,

$$\begin{aligned}\Delta_k^1(\varphi) &= \mathbb{E}\left(\varphi\left(U_k+X_k\right)-\varphi\left(U_k\right)-\frac{v_k}{2}\varphi''\left(U_k\right)\right)\\ &= \mathbb{E}\left(X_k\varphi'\left(U_k\right)+\frac{1}{2}\left(X_k^2-\frac{v_k}{2}\right)\varphi''\left(U_k\right)+\frac{1}{6}X_k^3\varphi^{(3)}\left(U_k+\vartheta_k X_k\right)\right).\end{aligned}$$

It then follows that

$$\begin{aligned}\sum_{k=1}^{n}\Delta_k^1(\varphi) &= \sum_{k=1}^{n}\sum_{j=1}^{k-1}\mathrm{Cov}\left(\varphi''\left(S_{k-1-j}+\tau_{j+1}\right)X_{k-j},X_k\right)\\ &+ \frac{1}{2}\sum_{k=1}^{n}\sum_{j=1}^{k-1}\mathrm{Cov}\left(\varphi^{(3)}\left(S_{k-1-j}+\tau_{j+1}+\vartheta_{k-j}X_{k-j}\right)X_{k-j}^2,X_k\right)\\ &+ \sum_{k=1}^{n}\mathrm{Cov}\left(X_k,\varphi'\left(\tau_{k+1}\right)\right)+\frac{1}{2}\sum_{k=1}^{n}\mathbb{E}\left(\varphi''\left(U_k\right)\left(X_k^2-\mathbb{E}X_k^2\right)\right)\\ &+ \mathbb{E}\left(\sum_{k=1}^{n-1}\mathrm{Cov}\left(X_0,X_k\right)\sum_{j=k+1}^{n}\varphi''\left(U_j\right)\right)+\frac{1}{6}\sum_{k=1}^{n}\mathbb{E}\left(\varphi^{(3)}\left(U_k+\iota_k X_k\right)X_k^3\right)\\ &= E_1+E_2+E_3+E_4+E_5+E_6. \qquad (5.7)\end{aligned}$$

Here $0<\iota_k<1$. The term E_3 vanishes. The other ones are controlled in two ways: either by uniform bounds or by dependence covariance bounds. See [7] for further details.

We can apply this method to density estimation on $\mathbb{R}$. Consider a stationary sequence $(X_t)_{t\in\mathbb{Z}}$ with marginal density f. Assume that the densities of the pairs (X_0,X_k), $k\in\mathbb{Z}^+$, exist, and are uniformly bounded: $\sup_{k>0}\|f_{(k)}\|_\infty<\infty$.

Let K be some kernel function with integral 1, Lipschitz and compactly supported. The kernel density estimator is defined by

$$\hat{f}(x)=\hat{f}_{n,h}(x)=\frac{1}{nh}\sum_{t=1}^{n}K\left(\frac{x-X_t}{h}\right).$$

Theorem 5.4. *Suppose that the stationary sequence $(X_t)_{t\in\mathbb{Z}}$ is $(\theta,\mathcal{L}_1,\psi_j')$-weakly dependent, with either $j=1,2$, $\theta_r=\mathcal{O}(r^{-a})$ and $a>2+j$, or $(\theta,\mathcal{L}_1,\psi_j)$-weakly dependent $j=1,2$,*

$\theta_r = \mathcal{O}(r^{-a})$ *and* $a > 2 + j + 1/\delta$, *where* $n^\delta h \to \infty$ *for some* $\delta \in]0, 1[$. *If* $f(x) > 0$, *then*

$$\sqrt{nh}\,\left(\hat{f}(x) - \mathbb{E}\hat{f}(x)\right) \xrightarrow{\mathcal{D}} \mathcal{N}\left(0, f(x)\int K^2(u)du\right).$$

Remark. In order to obtain a CLT for time series, the first paper using the Bernstein blocking technique seems to be due to Dehling [10]. This method is also described in [11] and it is used in [13]. The technique consists of dividing a sample $\{Z_1, \ldots, Z_n\}$ into blocks $U_m = \{Z_k\}_{k\in K_m}$ which are almost independent and such that Card $\left(\{1, \ldots, n\} \setminus \bigcup_m K_m\right) = o(n)$ as $n \to \infty$. This condition essentially allows one to work with the variables $T_m = \sum_{k\in K_m} Z_k$ (built from those blocks) exactly as in the classical Lindeberg method for the independent case.

Proof of Theorem 5.4. The details of the proof follow [3].

6 Functional estimation

We consider a stationary processes $(Z_t)_{t\in\mathbb{Z}}$ with $Z_t = (X_t, Y_t)$ where $X_t,\ Y_t \in \mathbb{R}$. The quantity of interest is the regression function $r(x) = \mathbb{E}(Y_0|X_0 = x)$. Let K be some kernel function with integral 1. Assume that K is a Lipschitz function with a compact support. The kernel estimator is defined by

$$\hat{f}(x) = \hat{f}_{n,h}(x) = \frac{1}{nh}\sum_{t=1}^{n} K\left(\frac{x - X_t}{h}\right),\ \hat{g}(x) = \hat{g}_{n,h}(x) = \frac{1}{nh}\sum_{t=1}^{n} Y_t K\left(\frac{x - X_t}{h}\right),$$

$$\hat{r}(x) = \hat{r}_{n,h}(x) = \frac{\hat{g}_{n,h}(x)}{\hat{f}_{n,h}(x)} \quad \text{if} \quad \hat{f}_{n,h}(x) \neq 0; \quad \hat{r}(x) = 0 \text{ otherwise.}$$

Here $h = (h_n)_{n\in\mathbb{N}}$ is a sequence of positive real numbers. We always assume that $h_n \to 0,\ nh_n \to \infty$ as $n \to \infty$.

Definition 6.1. *Let* $\rho = a + b$ *with* $(a, b) \in \mathbb{N}\times]0; 1]$. *Let* $\mathcal{C}^a$ *be the set of a times differentiable functions. Define the set of* ρ*-regular functions* $\mathcal{C}_\rho$ *by*

$$\mathcal{C}_\rho = \Big\{u : \mathbb{R} \to \mathbb{R};\ u \in \mathcal{C}^a \text{ and } \exists A \geq 0\ |u^{(a)}(x) - u^{(a)}(y)| \leq A|x - y|^b \text{ for all } x, y \text{ in any compact subset}\Big\}.$$

If $g \in \mathcal{C}_\rho$, one can choose a kernel function K of order ρ such that

$$b_h(x) = \mathbb{E}(\hat{g}(x)) - g(x) = \mathcal{O}(h^\rho),$$

uniformly on any compact subset of $\mathbb{R}$. We need some more assumptions.

- The marginal density f of X_0 exists and is continuous with $f(x) > 0$ for the points x where f is estimated.
- The regression function $r(\cdot) = \mathbb{E}(Y_0|X_0 = \cdot)$ exists and is continuous.
- For some $p \geq 1$, $x \to g_p(x) = f(x)\,\mathbb{E}(|Y_0|^p|X_0 = x)$ exists and is continuous.
- One of the following moment assumptions holds:

$$\mathbb{E}|Y_0|^S < \infty \quad \text{for some } S \geq p, \tag{6.1}$$

$$\mathbb{E}\exp(|Y_0|) < \infty. \tag{6.2}$$

6.1 Second order properties. We study the properties of $\hat{g}(x)$. Assume that the function g_2 exists and is continuous. Consider the following conditionally centered equivalent of g_2 which appears in the asymptotic variance of the estimator $\hat{r}$,

$$G_2(x) = f(x)\,\mathrm{Var}\,(Y_0|X_0 = x) = g_2(x) - f(x)r^2(x).$$

Assume that the densities of the pairs (X_0, X_k), $k \in \mathbb{Z}^+$, exist and are uniformly bounded: $\sup_{k>0} \|f_{(k)}\|_\infty < \infty$. Moreover, suppose that, uniformly over all $k \in \mathbb{Z}^+$, the functions

$$r_{(k)}(x, x') = \mathbb{E}\left(|Y_0 Y_k| \big| X_0 = x, X_k = x'\right) \tag{6.3}$$

are continuous. Under these assumptions, the functions $g_{(k)} = f_{(k)} r_{(k)}$ are locally bounded.

Theorem 6.2. *Suppose that the stationary sequence $(Z_t)_{t\in\mathbb{Z}}$ is $(\theta, \psi_j, \mathcal{L})$-weakly dependent with $\theta_r = \mathcal{O}(r^{-a})$ and $a > 2+j$, for $j = 1$ or $j = 2$. Assume that the sequence $(Z_t)_{t\in\mathbb{Z}}$ satisfies the conditions (6.2) and (6.3). Suppose that $n^\delta h \to \infty$ for some $\delta \in]0, 1[$. Then uniformly in x belonging to any compact subset of $\mathbb{R}$,*

$$\mathrm{Var}(\hat{g}(x)) = \frac{1}{nh}\, g_2(x) \int K^2(u)du \; + o\left(\frac{1}{nh}\right)$$

and

$$\mathrm{Var}\left(\hat{g}(x) - r(x)\hat{f}(x)\right) = \frac{1}{nh} G_2(x) \int K^2(u)du + o\left(\frac{1}{nh}\right).$$

6.2 Central limit theorems. We first consider the density estimator.

Theorem 6.3. *Suppose that the stationary sequence $(X_t)_{t\in\mathbb{Z}}$ is either*

- *$(\theta, \mathcal{L}_1, \psi'_j)$-weakly dependent, with $j = 1, 2$, $\theta_r = \mathcal{O}(r^{-a})$ and $a > 2 + j$,*
- *or $(\theta, \mathcal{L}_1, \psi_1)$-weakly dependent with $\theta_r = \mathcal{O}(r^{-a})$ and $a > 3(1 + \sqrt{5})/2$,*
- *or $(\theta, \mathcal{L}_1, \psi_2)$-weakly dependent with $\theta_r = \mathcal{O}(r^{-a})$ and $a > 6$.*

Assume, moreover, that $f(x) > 0$. Then as $nh \to \infty$,

$$\sqrt{nh}\,\left(\hat{f}(x) - \mathbb{E}\hat{f}(x)\right) \xrightarrow{\mathcal{D}} \mathcal{N}\left(0, f(x) \int K^2(u)du\right).$$

Proof. The CLT is obtained by the Lindeberg–Rio method described above and by Bernstein's blocking technique described in [17]. The theorem follows by comparing the rates obtained by both methods. See [3] for more details.

Theorem 6.4. *Assume that the stationary sequence $(Z_t)_{t\in\mathbb{Z}}$ is $(\theta, \psi_j, \mathcal{L})$-weakly dependent with $\theta_r = \mathcal{O}(r^{-a})$ with*

$$a > \min\left(\max\left(2 + j, 3(2 + j)\delta\right), \max\left(2 + j + \frac{1}{\delta}, \frac{2 + 2(2 + j)\delta}{1 + \delta}\right)\right),$$

for $j = 1$ or $j = 2$. Assume, moreover, that the sequence $(Z_t)_{t\in\mathbb{Z}}$ satisfies the conditions (6.2) and (6.3). Consider a positive kernel K. Let $f, g \in \mathcal{C}_\rho$ for some $\rho \in]0, 2]$, and $nh^{1+2\rho} \to 0$. Then, for all x belonging to any compact subset of $\mathbb{R}$,

$$\sqrt{nh}\,\left(\hat{r}(x) - r(x)\right) \xrightarrow{\mathcal{D}} \mathcal{N}\left(0, \frac{G_2(x)}{f^2(x)} \int K^2(u)du\right).$$

Proof. See [3]. They also prove the following.

Proposition 6.5. *Assume that the stationary sequence $(Z_t)_{t\in\mathbb{Z}}$ is $(\theta, \psi_j, \mathcal{L})$-weakly dependent with $\theta_r = \mathcal{O}(r^{-a})$ with $a > 9$ for $j = 1$ and $a > 12$ for $j = 2$. Suppose that $(Z_t)_{t\in\mathbb{Z}}$ satisfies the conditions (6.2) and (6.3). Consider a positive kernel K. Let $f, g \in \mathcal{C}_\rho$ for some $\rho \in]0, 2]$, and $nh \to \infty$. Then, for all x belonging to any compact subset of $\mathbb{R}$,*

$$\sqrt{nh}\left(\hat{r}(x) - \mathbb{E}\hat{r}(x)\right) \xrightarrow{\mathcal{D}} \mathcal{N}\left(0, \frac{G_2(x)}{f^2(x)} \int K^2(u)du\right).$$

Remark. The central limit Theorem 6.4 for the regression function holds true for strongly mixing sequences, under the moment assumption (6.1) if $h_n \to 0$, $nh_n/\log(n) \to \infty$ and $\alpha_n = \mathcal{O}(n^{-a})$ with $\alpha > 2S/(S-2)$ ([27]). No positivity assumption on the kernel K is required.

6.3 Almost sure convergence. For the sake of simplicity, we only consider the geometrically dependent case.

Theorem 6.6. *Let $(Z_t)_{t\in\mathbb{Z}}$ be a stationary sequence satisfying the conditions (6.2), (6.3), and that $(Z_t)_{t\in\mathbb{Z}}$ is either $(\theta, \psi_1, \mathcal{L})$- or $(\theta, \psi_2, \mathcal{L})$-weakly dependent with $\theta_r \le a^r$ for some $0 < a < 1$.*

(i) *If $nh/\log^4(n) \to \infty$, then for any $M > 0$, almost surely,*

$$\sup_{|x|\le M} |\hat{g}(x) - \mathbb{E}\hat{g}(x)| = O\left(\frac{\log^2(n)}{\sqrt{nh}}\right).$$

(ii) *For any $M > 0$, if $f, g \in \mathcal{C}_\rho$ for some $\rho \in]0, \infty[$, $h \sim \left(\frac{\log^4(n)}{n}\right)^{1/(1+2\rho)}$ and $\inf_{|x|\le M} f(x) > 0$, then, almost surely,*

$$\sup_{|x|\le M} |\hat{r}(x) - r(x)| = O\left\{\left(\frac{\log^4(n)}{n}\right)^{\rho/(1+2\rho)}\right\}.$$

Remark. Liebscher [19] proves the uniform almost sure convergence in a strongly mixing framework, at the optimal rate $\mathcal{O}\left(\left(\frac{\log(n)}{n}\right)^{\rho/(1+2\rho)}\right)$, if $\alpha_r = \mathcal{O}(r^{-a})$, with $a > 4 + 3/\rho$.

References

[1] P. Ango Nze: Critères d'ergodicité de modèles markoviens. Estimation non paramétrique sous des hypothèses de dépendance. *Ph.D. thesis: Université Dauphine, Paris,* 1995.

[2] P. Ango Nze: Critères d'ergodicité géométrique ou arithmétique de modèles linéaires perturbés à représentation markovienne. *Comptes Rendus de l'Académie des Sciences. Paris* **t. 326, Série 1** (1998), 371–376.

[3] P. Ango Nze, P. Bühlmann, and P. Doukhan: Weak dependence beyond mixing and asymptotics for nonparametric regression. Technical Report, Department of Statistics, ETH Zürich, available under http://www.stat.math.eth.ch/~buhlmann/bibliog.html; to appear in *Annals of Statistics*.

[4] P. Bickel and P. Bühlmann: Mixing property and functional central limit theorems for a sieve bootstrap in time series. Technical Report 440, Department of Statistics, University of California Berkeley, 1995.

[5] P. Billingsley: *Convergence of Probability Measures.* Wiley, New York, 1968.

[6] P. Breuer and P. Major: Central limit theorems for non-linear functionals of gaussian random fields. *Journal of Multivariate Analysis* **13** (1983), 425–441.

[7] C. Coulon-Prieur and P. Doukhan: A triangular CLT for weakly dependent sequences. *Statistics and Probability Letters* **47** (2000), 61–68.

[8] Y. Davydov: Mixing conditions for Markov chains. *Theory of Probability and Its Applications* **18** (1973), 313–328.

[9] J. Dedecker and E. Rio: On the functional central limit theorem for stationary processes. *Annales de l'Institut Henri Poincaré* **série B, 36** (2000), 1–34.

[10] H. Dehling: *Grenzwertsätze für Summen schwach abhängiger vektor-wertiger Zufallvariablen.* Doctoral Dissertation Göttingen, 1981.

[11] P. Doukhan: *Mixing: Properties and Examples.* Lecture Notes in Statistics 85, Springer Verlag, 1994.

[12] P. Doukhan: Limit theorems for stationary sequences. Preprint: University of Cergy Pontoise, 1998.

[13] P. Doukhan and S. Louhichi: A new weak dependence condition and application to moment inequalities. *Stochastic Processes and Their Applications* **84** (1999), 313–343.

[14] M. Duflo: *Méthodes récursives aléatoires.* Masson, Paris (English Edition, Springer, 1996).

[15] L. Giraitis and D. Surgailis: Multivariate Appell polynomials and the central limit theorem. In: E. Eberlein and M. Taqqu (Eds.) *Dependence in Probability and Statistics.* pp. 27–51, Birkhäuser, Boston, 1986.

[16] H.-C. Ho and T. Hsing: Limit theorems for functionals of moving averages. *Annals of Probability* **25** (1997), 1636–1669.

[17] I. A. Ibragimov: Some limit theorems for stationary processes. *Theory of Probability and Its Applications* **7** (1962), 349–382.

[18] A. N. Kolmogorov and Y. A. Rozanov: On the strong mixing conditions for stationary gaussian sequences. *Theory of Probability and Its Applications* **5** (1960), 204–207.

[19] E. Liebscher: Strong convergence of sums of α- mixing random variables with applications to density estimation. *Stochastic Processes and Their Applications* **65** (1996), 69–98.

[20] S. Louhichi: Weak convergence for empirical processes of associated sequences. Preprint 98-36, University of Paris Sud, Orsay, 1998.

[21] S. P. Meyn and R. L. Tweedie: *Markov Chains and Stochastic Stability.* Communications in Control and Engineering, Springer Verlag, 1993.

[22] A. Mokkadem: Propriétés de mélange des modèles autorégressifs polynômiaux. *Annales de l'Institut Henri Poincaré* **série B, 26-2** (1990), 219–260.

[23] C. M. Newman: Asymptotic independence and limit theorems for positively and negatively dependent random variables. In: Y. L. Tong (Ed.), *Inequalities in Statistics and Probability*, IMS Lecture Notes-Monograph Series 5, 127–140, 1984.

[24] M. Peligrad: Recent advances in the central limit theorem and its weak invarianace principle for mixing sequences of random variables. In E. Eberlein and M. S. Taqqu (Eds.), *Dependence in Probability and Statistics, a survey of recent results*. Oberwolfach, Birkhäuser, Boston, 1986.

[25] T. D. Pham and L. T. Tran: Some mixing properties of time series models. *Stochastic Processes and Their Applications* **19** (1985), 297–303.

[26] E. Rio: *Théorie asymptotique des processus aléatoires faiblement dépendants*. Mathématiques et applications 31, Springer Verlag, 2000.

[27] P. M. Robinson: Nonparametric estimators for time series. *Journal of Time Series Analysis* **4** (1983), 185–207.

[28] P. M. Robinson: Testing for strong serial correlation and dynamic conditional heteroskedasticity in multiple regression. *Journal of Econometrics* **47** (1991), 67–84.

[29] M. Rosenblatt: A central limit theorem and a strong mixing condition. *Proceedings of the National Academy of Sciences, U.S.A.*, **42** (1956), 43–47.

[30] M. Rosenblatt: *Stationary Sequences and Random Fields*. Birkhäuser, Boston, 1985.

[31] Q. M. Shao: Remark on the invariance principle for ρ-mixing sequences of random variables. *Chinese Annals of Mathematics* **9A** (1988), 409–412.

[32] Q. M. Shao: Maximal inequalities for partial sums of ρ-mixing sequences. *Annals of Probability* **23** (1995), 948–965.

[33] Q. M. Shao and H. Yu: Weak convergence for weighted empirical processes of dependent sequences. *Annals of Probability* **24** (1996), 2052–2078.

[34] H. Tong: A note on a Markov bilinear stochastic process in discrete time. *Journal of Time Series Analysis* **2** (1981), 279–284.

[35] A. Yu. Veretennikov: On polynomial mixing and convergence rate for stochastic difference and differential equation. *Theory of Probability and Its Applications*, **44**, (2000), 361–374.

[36] C. S. Withers: Central limit theorem for dependent variables. *Zeitschrift für Wahrscheinlichkeitstheorie und verwandte Gebiete* **57** (1981), 509–534.

[37] V. A. Wolkonski and Y. A. Rozanov: Some limit theorems for random functions, *Part I. Theory of Probability and Its Applications* **4** (1959), 178–197; *Part II Theory of Probability and Its Applications* **6** (1961), 186–198.

[38] H. Yu: A Glivenko–Cantelli Lemma and weak convergence for empirical processes of associated sequences. *Probability Theory and Related Fields* **95** (1993), 357–370.

PATRICK ANGO NZE
UNIVERSITÉ LILLE 3, UFR AES, BP 149
F-59653 VILLENEUVE D'ASCQ CEDEX, FRANCE
ANGONZE@UNIV-LILLE3.FR

PAUL DOUKHAN
UPRESA 8088 CNRS-MATHÉMATIQUES
UNIVERSITÉ DE CERGY PONTOISE
BÂTIMENT LES CHÊNES
33 BOULEVARD DU PORT
F-95011 CERGY-PONTOISE CEDEX, FRANCE
AND
CREST-LS, TIMBRE J340
3 AVENUE PIERRE LAROUSSE
F-92245 MALAKOFF, FRANCE
PAUL.DOUKHAN@MATH.U-CERGY.FR

Maximal Inequalities and Empirical Central Limit Theorems

Jérôme Dedecker and Sana Louhichi

Abstract. This work presents some recent developments concerning empirical central limit theorems for dependent sequences. We try to give general conditions under which an adaptive truncation in the chaining procedure works. We show that these conditions are satisfied for a large class of mixing processes satisfying suitable moment inequalities.

1 Introduction

Due to the wide range of its applications in statistics, from parametric to nonparametric estimation, the theory of empirical processes has been widely investigated since the early work of Donsker [9]. For independent observations, Donsker proved in 1952 the weak convergence of the empirical distribution function to a Brownian bridge, which provides as a straightforward consequence the asymptotic behavior of the Kolmogorov–Smirnov statistic.

Dudley [12], [13] extended this result to empirical processes indexed by a class of functions $\mathcal{F}$ and gave a precise definition of weak convergence in the nonseparable space $\ell^\infty(\mathcal{F})$ of bounded functionals from $\mathcal{F}$ to $\mathbb{R}$. When $\mathcal{F} = \{\mathbf{1}_C, C \in \mathcal{C}\}$ reduces to indicators of sets, he also provided two different kinds of conditions on $\mathcal{F}$ for the empirical process to converge to an $\ell^\infty(\mathcal{F})$-valued Gaussian random variable. The first one involves combinatorial arguments ($\mathcal{C}$ is a VC class) and is strongly related to the notion of universal entropy. The second one is an integrability condition on the entropy with inclusion of the class $\mathcal{C}$.

In 1982, Pollard [21] proved an empirical central limit theorem under an integrability condition on the universal entropy of $\mathcal{F}$. This condition is optimal in a sense: when replacing universal entropy by the classical metric entropy, it is the minimal condition (in terms of the metric entropy) ensuring the limiting Gaussian process to have continuous sample paths.

Independently, in 1984 Bass [4] obtained the convergence of set-indexed partial sum processes under a minimal condition in terms of entropy with inclusion. His proof is based on a delicate adaptive truncation in the chaining procedure. The main technique is a systematic use of Bernstein's inequality (or, equivalently, of Rosenthal's inequality) which provides the needed truncation level. The same tools and techniques were used in 1987 by Ossiander [18] in the context of empirical processes to obtain optimal conditions on $\mathcal{F}$ in terms of entropy with bracketing (the analogue of entropy with inclusion for classes of functions).

The theory of empirical processes was soon extended to different types of dependent variables. An early result of Billingsley [6] extended Donsker's theorem to (uniformly) ϕ-mixing sequences. For more general classes of functions, see for instance Phillip [19], Doukhan, Léon, and Portal [10], Massart [16], or Andrews and Pollard [2].

In terms of entropy with bracketing, the most complete work seems to be that of Doukhan, Massart and Rio [11] in a β-mixing framework. They showed, in particular, that the right norm to consider in order to measure the size of the brackets is related to the dependence structure of the variables (see also Rio [25] Chapter 8, for more details on this subject). After a careful reading of this paper, we infer that suitable maximal inequalities are the only essential tools to carry out the chaining procedure.

The paper is organized as follows: Section 2 is devoted to the mathematical background. In Section 3 we state our main result: Theorem 3.3 provides conditions in terms of maximal inequalities for the empirical process indexed by finite sets of functions which are sufficient to derive the asymptotic equicontinuity of the empirical process. In Section 4 we show how to derive such inequalities from Rosenthal-type inequalities. We obtain the tightness of the process under Ossiander's condition in the i.i.d. case and Doukhan, Massart and Rio's in the β-mixing case. The condition we propose for nonuniform ϕ-mixing sequences are new, to our knowledge. In Theorem 5.1 of Section 5, we recall some recent conditions yielding the finite-dimensional convergence of the empirical process, which together with Section 4 imply empirical central limit theorems under various dependence conditions (cf. Theorem 5.2). The proof of Theorem 3.3 (an adaptation of [11]) is postponed until Section 6. Some technical lemmas concerning the β-mixing case are presented in the appendix.

2 Empirical process, weak convergence and tightness in $\ell^\infty(\mathcal{F})$

Let $X = (X_i)_{i\in\mathbb{Z}}$ be a strictly stationary sequence of random elements with values in a Polish space $\mathcal{X}$ with common marginal distribution P. Denote by P_n the empirical probability measure and by Z_n the centered and normalized empirical measure:

$$P_n = \frac{1}{n}\sum_{i=1}^{n}\delta_{X_i}, \qquad Z_n = \sqrt{n}(P_n - P).$$

Let $\mathcal{F}$ be a class of measurable functions from $\mathcal{X}$ to $\mathbb{R}$. The space $\ell^\infty(\mathcal{F})$ is the set of all functions z from $\mathcal{F}$ to $\mathbb{R}$ such that $\|z\|_{\mathcal{F}} = \sup_{f\in\mathcal{F}}|z(f)|$ is finite. A random variable X with values in $\ell^\infty(\mathcal{F})$ is tight if for any positive ϵ there exists a compact set K_ϵ of $(\ell^\infty(\mathcal{F}), \|.\|_{\mathcal{F}})$ such that $\mathbb{P}(X \in K_\epsilon) \geq 1-\epsilon$.

Assume that for every x in $\mathcal{X}$, $\sup_{f\in\mathcal{F}}|f(x) - P(f)|$ is finite. Under this minimal condition, the empirical process $\{Z_n(f), f \in \mathcal{F}\}$ can be viewed as a variable with values in $\ell^\infty(\mathcal{F})$, although it may not be measurable with respect to the Borel σ-algebra generated by $\|.\|_{\mathcal{F}}$. Nevertheless, we say that Z_n converges weakly to an $\ell^\infty(\mathcal{F})$-valued random variable Z (i.e., Borel measurable) if, for every continuous bounded function h from $(\ell^\infty(\mathcal{F}), \|.\|_{\mathcal{F}})$ to $\mathbb{R}$, the outer expectation $\mathbb{E}^*(h(Z_n))$ converges to $\mathbb{E}(h(Z))$ (see for instance [27] p. 4 for the definition of outer expectations and measures, and more details about weak convergence for nonmeasurable maps).

The variable Z_n with values in $\ell^\infty(\mathcal{F})$ is asymptotically ρ-equicontinuous if there exists a semimetric ρ on $\mathcal{F}$ such that, for every $\epsilon > 0$,

$$\lim_{\delta\to 0}\limsup_{n\to\infty}\mathbb{P}^*\Big(\sup_{\rho(f,g)\leq\delta,\ f,g\in\mathcal{F}}|Z_n(f) - Z_n(g)| > \epsilon\Big) = 0,$$

where $\mathbb{P}^*$ stands for the outer probability. In the sequel, for short, we shall omit the stars. The $\ell^\infty(\mathcal{F})$-valued variable Z_n converges weakly to a tight limit in $\ell^\infty(\mathcal{F})$ if and only if it is asymptotically ρ-equicontinuous for a semimetric ρ such that $(\mathcal{F}, \rho)$ is totally bounded, and every one of its finite-dimensional marginals $(Z_n(f_1), \dots Z_n(f_k))$ converges weakly (see for instance [22], Theorem 10.2).

A random variable G with value in $\ell^\infty(\mathcal{F})$ is a Gaussian process if every one of its finite-dimensional marginals $(G(f_1), \ldots, G(f_k))$ is normally distributed. Denote by Γ its covariance function $\Gamma(f, g) = \mathrm{Cov}(G(f), G(g))$ and write $\Gamma(f) = \Gamma(f, f)$ for the associated quadratic form. Let K be the set of functions of $\ell^\infty(\mathcal{F})$ which are continuous with respect to $\Gamma(.)$. A zero-mean Gaussian process G is tight if and only if $\ell^\infty(\mathcal{F})$ is relatively compact with respect to $\Gamma(.)$ and G has almost sure continuous sample paths: $\mathbb{P}(G \in K) = 1$. If G is a tight Gaussian process, then it is also Gaussian as an $\ell^\infty(\mathcal{F})$-valued random variable: for every element d of the dual of $\ell^\infty(\mathcal{F})$, the real-valued random variable $d(G)$ is normally distributed.

3 From maximal inequalities to tightness

Before stating the main theorem of this paper, we need some more definitions.

Definition 3.1. *For any $p \geq 1$, let $\mathbb{L}^p(P)$ be the class of real-valued functions on $(\mathcal{X}, P)$ such that $\|f\|_p^p = P(|f|^p)$ is finite. We say that a normed space $(L, \|.\|)$ is an $\mathcal{L}$-space if*

1. *There exist two positive numbers a_1 and a_2 such that $\|f\|_1 \leq a_1\|f\|$ for any f in L and $\|f\| \leq a_2\|f\|_\infty$ for any f in $\mathbb{L}^\infty(P)$. In particular, this implies that $\|f - P(f)\| \leq (1 + a_1a_2)\|f\|$.*
2. *$\|.\|$ is nondecreasing: if f and g are two elements of L such that $|f| \leq |g|$, then $\|f\| \leq \|g\|$.*

Note that $\mathbb{L}^p(P)$ is an $\mathcal{L}$-space as soon as $p \geq 1$.

Definition 3.2. *Entropy with bracketing: let $\mathcal{F}$ be a subset of an $\mathcal{L}$-space $(L, \|.\|)$. Given (f, g) in $L \times L$ with $f \leq g$, the bracket $[f, g]$ is the set of all functions h with $f \leq h \leq g$. An ε-bracket is a bracket $[f, g]$ with $\|f - g\| \leq \varepsilon$. The entropy with bracketing $\mathbb{H}(\mathcal{F}, \varepsilon, \|.\|)$ is the logarithm of the smallest number of ε-brackets needed to cover $\mathcal{F}$.*

Theorem 3.3 below gives sufficient conditions for the process Z_n to be asymptotically equicontinuous. They are in fact general conditions under which Ossiander's method (cf. [18]) works. The main point is to find a truncating function for which the maximum of $Z_n(g)$ over a finite set of m-bounded functions is well controlled.

A typical application of Theorem 3.3 is the following: let $\mathcal{G}$ be any finite set of centered functions of $\mathbb{L}^\infty(P)$, and m, δ two positive numbers such that $\|g\|_\infty \leq m$ and $\|g\|_{2b/(b-1)} \leq \delta$ for any g in $\mathcal{G}$. Set $H = \max(1, \log(|\mathcal{G}|))$ and assume that there exists b in $[1, \infty]$ and a constant C depending only on X such that

$$\mathbb{E}\left(\max_{g\in\mathcal{G}} |Z_n(g)|\right) \leq C\left(\delta\sqrt{H} + \frac{m^{\frac{b+1}{b-1}}\delta^{\frac{2}{1-b}}H}{\sqrt{n}}\right). \tag{3.1}$$

Then the $\ell^\infty(\mathcal{F})$-valued variable Z_n is asymptotically equicontinuous as soon as

$$\int_0^1 \sqrt{\mathbb{H}(\mathcal{F}, u, \|.\|_{2b/(b-1)})}\, du < \infty.$$

We shall see in Section 4 that inequality (3.1) holds for i.i.d. sequences with $b = \infty$, and for some nonuniform ϕ-mixing sequences when b equals 1 or 2 (similar bounds may be obtained for b in $]1, 2[$ via interpolation arguments). In the latter case, the value of b is entirely determined by the decay of the ϕ-mixing coefficients.

The fact that the right norm to consider in order to measure the size of the brackets is related to the dependence structure of the variables has been clearly pointed out by Doukhan, Massart and Rio [11] for absolutely regular sequences (see Section 4.3). In this framework, one obtains a bound similar to (3.1) involving the exact rate of mixing of the sequence (cf. equation (4.12)). This inequality leads to the norm $\|.\|_{2,\beta}$ indexed by the whole sequence $\beta = (\beta_i)_{i\geq 0}$ of β-mixing coefficients of X (see equation (4.8) for the definition of $\|.\|_{2,\beta}$). From the inequality $\|g\|_2 \leq \|g\|_{2,\beta}$, we infer that the $\mathcal{L}$-space $\mathcal{L}_{2,\beta}(P) = \{g : \|g\|_{2,\beta} < \infty\}$ is continuously embedded in $\mathbb{L}^2(P)$. Conversely, we shall see that for b in $[1, \infty]$, the space $\mathbb{L}^{2b/(b-1)}(P)$ is continuously embedded in $\mathcal{L}_{2,\beta}(P)$ as soon as the β-mixing coefficients of X satisfy $\sum k^{b-1}\beta(k) < \infty$. This coincides with the i.i.d. case when $b = \infty$, and yields the same rate as for nonuniform ϕ-mixing sequences when b equals 1 or 2.

In each case, the asymptotic equicontinuity of Z_n may be obtained via the following general theorem.

Theorem 3.3. *Let X be a stationary sequence with marginal distribution P. Let $(L, \|.\|)$ be an $\mathcal{L}$-space and $\mathcal{F}$ a class of functions of L with envelope function F (i.e., $|f| \leq F$ for any f in $\mathcal{F}$).*

Assume that there exist two functions, m_1 from $\mathbb{N}^* \times \mathbb{R}_+^* \times \mathbb{N}^*$ to $\mathbb{R}^+ \cup \{+\infty\}$, nondecreasing in the second variable and nonincreasing in the third, and m_2 from $\mathbb{N}^* \times \mathbb{N}^*$ to $\mathbb{R}^+$ nonincreasing in the second variable, such that both Conditions 1 and 2 hold:

1. *For any (n, δ, k, l) in $\mathbb{N}^* \times \mathbb{R}_+^* \times \mathbb{N}^* \times \mathbb{N}^*$ and any finite subset $\mathcal{G}$ of centered functions of L with cardinality $|\mathcal{G}| \leq k$, whose elements satisfy $\|g\| \leq \delta$ and $\|g\|_\infty \leq \frac{m_1(n,\delta,l)}{m_2(n,k)}$, we have*
$$\mathbb{E}\Big(\max_{g\in\mathcal{G}} |Z_n(g)|\Big) \leq C_1\delta\left(\sqrt{\max(1, \log(k))} + \frac{\max(1, \log(k))}{\sqrt{\max(1, \log(l))}}\right)$$
for some constant C_1 depending only on X.

2. *For any (n, δ, k) in $\mathbb{N}^* \times \mathbb{R}_+^* \times \mathbb{N}^*$ and any g of L such that $\|g\| \leq \delta$, we have, setting $m(n, \delta, k) = \frac{m_1(n,\delta,k)}{m_2(n,k)}$,*
$$\sqrt{n}\|g\mathbf{1}_{4g>m(n,\delta,k)}\|_1 \leq C_2\delta\sqrt{\max(1, \log(k))},$$
for some constant C_2 depending only on X.

Assume furthermore that there exists a function M from $\mathbb{N}^* \times \mathbb{R}_+^* \times \mathbb{N}^*$ to $\mathbb{R}^+ \cup \{+\infty\}$ such that both Conditions 3 and 4 hold:

3. *For any finite subset $\mathcal{G}$ of centered functions of L with cardinality $|\mathcal{G}| \leq k$, whose elements satisfy $\|g\| \leq \delta$ and $\|g\|_\infty \leq M(n, \delta, k)$, we have*
$$\mathbb{E}\Big(\max_{g\in\mathcal{G}} |Z_n(g)|\Big) \leq C_3\delta\sqrt{\max(1, \log(k))} + R(n, \delta, k),$$
for some constant C_3 depending only on X, and $R(n, \delta, k)$ tending to zero as n tends to infinity.

4. *The envelope function F of the class $\mathcal{F}$ satisfies*
$$\text{for any } (\delta, k), \quad \lim_{n\to\infty} \sqrt{n}\|F\mathbf{1}_{4F>M(n,\delta,k)}\|_1 = 0.$$

Then the $\ell^\infty(\mathcal{F})$-valued variable Z_n is asymptotically $\|.\|$-equicontinuous as soon as

$$\int_0^1 \sqrt{\mathbb{H}(\mathcal{F}, u, \|.\|)}\, du < \infty. \tag{3.2}$$

Remark 3.4. It is well known that (3.2) implies the existence of an envelope function F belonging to L. In many cases the choice $M \equiv m$ ensures that Condition 4 is satisfied for any F in L (in that case Condition 3 is realized with $C_3 = 2C_1$ and $R \equiv 0$). However, it may happen (see Section 4.3) that the good truncating function M depends on the envelope F.

Remark 3.5. If the function $m_1 \equiv \infty$ satisfies the assumptions of Theorem 3.3 (note that it always satisfies Condition 2), then one can take $M \equiv \infty$ as well, so that Conditions 3 and 4 are automatically realized (see the preceding remark). In such a case, the result can be improved by replacing entropy with bracketing by metric entropy (i.e., the logarithm of the smallest number of balls of radius u with respect to $\|.\|$ necessary to cover $\mathcal{F}$).

Remark 3.6. We have written the theorem with $\sqrt{.}$ because this function appears when considering classical Rosenthal-type inequalities. In fact the result remains true when replacing $\sqrt{.}$ by any increasing subadditive function $\varphi(.)$ in Conditions 1, 2, and 3. The entropy condition (3.2) then becomes

$$\int_0^1 \varphi(\mathbb{H}(\mathcal{F}, u, \|.\|))\, du < \infty.$$

Remark 3.7. Let m_1, m_2, M and R be four functions satisfying Conditions 1, 2 and 3 of Theorem 3.3. It follows from Proposition 6.2 that there exists a positive constant K not depending on n and δ such that, for any class $\mathcal{F}$ whose elements satisfy $\|f\| \leq \delta$,

$$\mathbb{E}\Big(\sup_{f\in\mathcal{F}} |Z_n(f)|\Big) \leq K \int_0^\delta \sqrt{1 \vee \mathbb{H}(\mathcal{F}, u, \|.\|)}du + 2\sqrt{n}\|F\mathbf{1}_{2F>M(n,\delta)}\|_1 + 2R(n, \delta),$$

where F is the envelope function of $\mathcal{F}$ and we set $R(n, \delta) = R(n, (1+a_1a_2)\delta, \mathbb{N}(\delta))$, $M(n, \delta) = M(n, (1+a_1a_2)\delta, \mathbb{N}(\delta))$ and $\mathbb{N}(\delta) = \exp(\mathbb{H}(\mathcal{F}, \delta, \|.\|))$. In particular, if (3.1) holds, then $\|.\| = \|.\|_{2b/(b-1)}$, $a_1 = a_2 = 1$, $R_n \equiv 0$ and

$$M(n, \delta) = 2\delta \left(\frac{n}{1 \vee \mathbb{H}(\mathcal{F}, \delta, \|.\|_{2b/(b-1)})}\right)^{\frac{b-1}{2b+2}}.$$

In the i.i.d. case (i.e., $b = \infty$), this is the same bound as in Theorem 2.14.2 of [27].

4 From Rosenthal's inequality to maximal inequalities

4.1 The i.i.d. case. In this section, we assume that the sequence $(X_i)_{i\in\mathbb{Z}}$ is i.i.d. If f is a centered function of $\mathbb{L}^p(P)$ for $p \geq 2$, Rosenthal's inequality yields (see for instance Pinelis [20])

$$\|Z_n(f)\|_p \leq K\Big(\sqrt{p}\|f\|_2 + \frac{p}{\sqrt{n}}\Big\|\max_{1\leq i\leq n} |f(X_i)|\Big\|_p\Big), \tag{4.1}$$

for some absolute constant K. Now let $\mathcal{G}$ be any finite set of centered functions of $\mathbb{L}^\infty(P)$, and m, δ two positive numbers such that $\|g\|_\infty \leq m$ and $\|g\|_2 \leq \delta$ for any g in $\mathcal{G}$. From (4.1) we get that

$$\max_{g\in\mathcal{G}} \|Z_n(g)\|_p \leq K\Big(\sqrt{p}\delta + \frac{pm}{\sqrt{n}}\Big). \tag{4.2}$$

Since, for $p \geq 2$,

$$\mathbb{E}\Big(\max_{g\in\mathcal{G}}|Z_n(g)|\Big) \leq \Big\|\max_{g\in\mathcal{G}}|Z_n(g)|\Big\|_p \leq \Big(\sum_{g\in\mathcal{G}}\mathbb{E}|Z_n(g)|^p\Big)^{1/p} \leq |\mathcal{G}|^{1/p}\max_{g\in\mathcal{G}}\|Z_n(g)\|_p,$$

we obtain, applying (4.2) with $p = 2H = 2\max(1, \log|\mathcal{G}|)$,

$$\mathbb{E}\left(\max_{g\in\mathcal{G}}|Z_n(g)|\right) \leq C\left(\delta\sqrt{H} + \frac{mH}{\sqrt{n}}\right) \tag{4.3}$$

for some absolute constant C.

Set $H(k) = \max(1, \log(k))$. From (4.3) it is clear that the functions defined by

$$m_2 \equiv 1, \quad m(n,\delta,k) = m_1(n,\delta,k) = \delta\sqrt{\frac{n}{H(k)}}$$

satisfy Conditions 1 and 2 of Theorem 3.3 for the space $L = \mathbb{L}^2(P)$ equipped with the norm $\|.\|_2$ (note that Condition 2 is satisfied with $C_2 = 4$). Clearly, the choice $M \equiv m$ ensures that both Conditions 3 and 4 are satisfied, so that the $\ell^\infty(\mathcal{F})$-valued variable Z_n is asymptotically $\|.\|_2$-equicontinuous as soon as

$$\int_0^1 \sqrt{\mathbb{H}(\mathcal{F}, u, \|.\|_2)}du < \infty.$$

This result was first obtained by Ossiander [18]. Andersen, Giné, Ossiander, and Zinn [1] weakened the bracketing assumption and used majorizing measure instead of metric entropy.

4.2 An extension to martingales. Let $(X_i)_{i\in\mathbb{Z}}$ be a stationary sequence and define the σ field $\mathcal{M}_i = \sigma(X_k, k \leq i)$. For any f in $\mathbb{L}^1(P)$, define the variable

$$W_n(f) = \frac{1}{\sqrt{n}}\sum_{i=1}^n\Big(f(X_i) - \mathbb{E}(f(X_i)|\mathcal{M}_{i-1})\Big)$$

and the two numbers $d_1(f)$ and $d_2(f)$,

$$d_1(f) = \left\|\mathbb{E}\left(|f(X_1)||\mathcal{M}_0\right)\right\|_\infty \text{ and } d_2(f) = \left\|\mathbb{E}\left((f(X_1))^2|\mathcal{M}_0\right)\right\|_\infty^{1/2}.$$

From [20] again, we have the bounds

$$\|W_n(f)\|_p \leq K\left(\sqrt{p}d_2(f) + \frac{p}{\sqrt{n}}\Big\|\max_{1\leq i\leq n}|f(X_i) - \mathbb{E}(f(X_i)|\mathcal{M}_{i-1})|\Big\|_p\right). \tag{4.4}$$

Arguing as in Section 4.1 we infer that, for any finite set $\mathcal{G}$ of functions of $\mathbb{L}^\infty(P)$, and m, δ two positive numbers such that $\|g\|_\infty \leq m$ and $d_2(g) \leq \delta$ for any g in $\mathcal{G}$,

$$\mathbb{E}\left(\max_{g\in\mathcal{G}}|W_n(g)|\right) \leq C\left(\delta\sqrt{H} + \frac{mH}{\sqrt{n}}\right), \tag{4.5}$$

where $H = \max(1, \log|\mathcal{G}|)$.

Looking at the proof of Theorem 3.3, we see that it may be adapted to the process W_n by replacing the $\mathbb{L}^1$ norm in Conditions 2 and 4 by the norm d_1. Now, since $d_1(f\mathbf{1}_{4f>m}) \leq$

$4m^{-1}(d_2(f))^2$, the function m of the preceding section works. We infer that the $\ell^\infty(\mathcal{F})$-valued variable W_n is asymptotically d_2-equicontinuous as soon as

$$\int_0^1 \sqrt{\mathbb{H}(\mathcal{F}, u, d_2)}du < \infty. \tag{4.6}$$

If furthermore any function f of $\mathcal{F}$ satisfies $\mathbb{E}(f(X_1)|\mathcal{M}_0) = P(f)$, then the $\ell^\infty(\mathcal{F})$-valued variable W_n coincides with Z_n, and the latter is asymptotically equicontinuous under the entropy condition (4.6).

Let us give two applications of this result.

1. Assume that $X_i = (Y_i, Y_{i-1})$ where $(Y_i)_{i\in\mathbb{Z}}$ is a stationary Markov chain with values in an arbitrary space $(E, \mathcal{E})$ and with transition kernel K. Let $\mathcal{F}$ be a class of $\mathbb{L}^2(P)$ such that $\mathbb{E}(f(X_1)|\mathcal{M}_0) = P(f)$ for any f in $\mathcal{F}$. In that case, the $\ell^\infty(\mathcal{F})$-valued variable Z_n is asymptotically d_2-equicontinuous as soon as (4.6) holds, where the norm d_2 is given by

$$(d_2(f))^2 = \sup_{x\in E} \int |f(x, y)|^2 K(x, dy).$$

2. Assume that $X_i = (\varepsilon_i, Y_i)$, where $\varepsilon = (\varepsilon_i)_{i\in\mathbb{Z}}$ is stationary with marginal distribution P_ε, $Y = (Y_i)_{i\in\mathbb{Z}}$ is i.i.d. with marginal distribution P_Y, and ε is independent of Y. Set $\|g\|_{\varepsilon,p}^p = P_\varepsilon(|g|^p)$ and $\|h\|_{Y,p}^p = P_Y(|h|^p)$. Let $\mathcal{F}_\varepsilon$ be a class of M-bounded functions of $\mathbb{L}^\infty(P_\varepsilon)$, $\mathcal{F}_Y$ be a class of centered functions of $\mathbb{L}^2(P_Y)$ with envelope F_Y and

$$\mathcal{F} = \mathcal{F}_\varepsilon \times \mathcal{F}_Y = \{g \times h, g \in \mathcal{F}_\varepsilon, h \in \mathcal{F}_Y\}.$$

Clearly, $\mathbb{E}(f(X_1)|\mathcal{M}_0) = 0$ for any f in $\mathcal{F}$. Now, if $g_l \le g \le g_u$ and $h_l \le h \le h_u$, then we have $(g \times h)_l \le g \times h \le (g \times h)_u$, where

$$\begin{aligned}(g_{l+} \times h_{l+}) + (g_{u-} \times h_{u-}) + (g_{l-} \times h_{u+}) + (g_{u+} \times h_{l-}) &= (g \times h)_l,\\ (g_{l-} \times h_{l-}) + (g_{u+} \times h_{u+}) + (g_{l+} \times h_{u-}) + (g_{u-} \times h_{l+}) &= (g \times h)_u,\end{aligned}$$

$f_+ = f \vee 0$, and $f_- = f \wedge 0$. If furthermore $|h_u| \vee |h_l| \le F_Y$ and $|g_u| \vee |g_l| \le M$, we easily obtain that

$$d_2\left((g \times h)_u - (g \times h)_l\right) \le 4(\|F_Y\|_{Y,2}\|g_l - g_u\|_{\epsilon,\infty} + M\|h_l - h_u\|_{Y,2}).$$

From this inequality, we infer that

$$\mathbb{H}(\mathcal{F}, 4u(M + \|F_Y\|_{2,Y}), d_2) \le \mathbb{H}(\mathcal{F}_\varepsilon, u, \|.\|_{\varepsilon,\infty}) + \mathbb{H}(\mathcal{F}_Y, u, \|.\|_{Y,2}).$$

We conclude that the $\ell^\infty(\mathcal{F})$-valued variable Z_n is asymptotically d_2-equicontinuous as soon as

$$\int_0^1 \sqrt{\mathbb{H}(\mathcal{F}_\varepsilon, u, \|.\|_{\varepsilon,\infty})}du < \infty \quad \text{and} \quad \int_0^1 \sqrt{\mathbb{H}(\mathcal{F}_Y, u, \|.\|_{Y,2})}du < \infty.$$

Setting $\mathcal{M}_{\varepsilon,0} = \sigma(\varepsilon_i, i \le 0)$, the conclusion remains true if we use the norm $(d_{\varepsilon,2}(g))^2 = \|\mathbb{E}(|g(\varepsilon_1)|^2|\mathcal{M}_{\varepsilon,0})\|_\infty$ instead of $\|.\|_{\varepsilon,\infty}$ in the first entropy condition.

The result of this section is a particular case of a remarkable work of Nishiyama [17] (see Theorem 4.2 and Corollary 4.3 therein). In this paper no stationarity assumption is made, the natural random distances

$$d_{2,n}(f) = \Big(\frac{1}{n}\sum_{i=1}^n \mathbb{E}(|f(X_i)|^2|\mathcal{M}_{i-1})\Big)^{1/2}$$

are used instead of d_2, and the condition on $\mathcal{F}$ is expressed in terms of partitioning entropy (in the spirit of Theorem 2.11.9 of [27]). The general context of $\ell^\infty(\mathcal{F})$-valued martingale difference arrays is also considered.

4.3 Absolutely regular sequences. We first recall the definition of the β-mixing coefficient between two σ-algebras (cf. Rozanov and Volkonskii [26]). Let $(\Omega, \mathcal{A}, \mathbb{P})$ be a probability space. Given two σ-algebras $\mathcal{U}$ and $\mathcal{V}$ of $\mathcal{A}$, the β-mixing coefficient is defined by

$$2\beta(\mathcal{U}, \mathcal{V}) = \sup \sum_{(i,j)\in I\times J} |\mathbb{P}(U_i \cap V_j) - \mathbb{P}(U_i)\mathbb{P}(V_j)|,$$

where the supremum is taken over all the finite partitions $(A_i)_{i\in I}$ and $(B_j)_{j\in J}$, respectively, $\mathcal{U}$ and $\mathcal{V}$-measurable.

Now, let $X = (X_i)_{i\in\mathbb{Z}}$ be a stationary sequence with marginal distribution P and $\mathcal{M}_0$ be the σ-algebra defined by $\mathcal{M}_0 = \sigma(X_i, i \le 0)$. Define the coefficients $\beta_{\infty,k}$ of X by

$$\beta_{\infty,k}(n) = \sup\{\beta\left(\mathcal{M}_0, \sigma(X_{i_1}, \dots, X_{i_k})\right), 0 < n \le i_1 \le \dots \le i_k\}. \tag{4.7}$$

In this section we shall only consider $\beta_{\infty,\infty}$ coefficients. For the sake of simplicity we omit the indexes and put

$$\beta(0) = 0 \quad \text{and for } n > 0, \quad \beta(n) = \beta_{\infty,\infty}(n).$$

The sequence X is said to be absolutely regular as soon as $\beta(n)$ tends to zero as n tends to infinity. Let β^{-1} be the inverse cadlag of the decreasing function $t \to \beta([t])$, and Q_g the inverse cadlag of the tail function $t \to P(|g| > t)$.

From now, assume that the series $\sum \beta(k)$ is finite. This implies that β^{-1} is an integrable function with respect to Lebesgue measure on $[0, 1]$. Following Doukhan, Massart and Rio [11], define the set $\mathcal{L}_{2,\beta}(P)$ as the set of functions g for which

$$\|g\|_{2,\beta}^2 = \int_0^1 \beta^{-1}(u) Q_g^2(u)\, du < \infty. \tag{4.8}$$

It is proved in Lemma 1 of [11] that $(\mathcal{L}_{2,\beta}(P), \|.\|_{2,\beta})$ is a normed subspace of $\mathbb{L}^2(P)$. Since β^{-1} is greater than 1, we get that $\|g\|_2 \le \|g\|_{2,\beta}$. On the other hand, for any function g in $\mathbb{L}^\infty(P)$, Q_g is bounded by $\|g\|_\infty$, and (4.8) implies that $\|g\|_{2,\beta}^2 \le \|g\|_\infty^2 \sum \beta(k)$. Collecting the above facts and noting that $|f| \le |g|$ implies $Q_f \le Q_g$, we infer that $(\mathcal{L}_{2,\beta}(P), \|.\|_{2,\beta})$ is an $\mathcal{L}$-space.

We now try to find a function m which satisfies the assumptions of Theorem 3.3. To this end, we approximate the original sequence by a sequence of independent random variables (see again [11]). The main tool to perform this approximation is Berbee's coupling lemma (cf. [5]), which we recall hereafter:

Lemma 4.1. *Let X and Y be two random variables taking their values in the Borel spaces S_1 and S_2, respectively, and let U be a random variable with uniform distribution on* $[0, 1]$, *independent of (X, Y). There exists a random variable $Y^* = f(X, Y, U)$, where f is a measurable function from $S_1 \times S_2 \times [0, 1]$ into S_2, such that*

1. *Y^* is independent of X and has the same distribution as Y.*
2. $\mathbb{P}(Y \ne Y^*) = \beta(\sigma(X), \sigma(Y))$.

Let q be any positive integer. Starting from Lemma 4.1 we construct by induction a sequence of random variables $(X_i^0)_{i>0}$ such that

(a) For any $i \geq 0$, the random variable $U_i^0 = (X_{iq+1}^0, \ldots, X_{iq+q}^0)$ has the same distribution as $U_i = (X_{iq+1}, \ldots, X_{iq+q})$.

(b) The sequence $(U_{2i}^0)_{i\geq 0}$ is i.i.d. and so is $(U_{2i+1}^0)_{i\geq 0}$.

(c) For any $i \geq 0$, $\mathbb{P}(U_i \neq U_i^0) \leq \beta(q)$.

Let $\mathcal{G}$ be any finite set of centered functions of $\mathbb{L}^\infty(P)$ and m, δ two positive numbers such that $\|g\|_\infty \leq m$ and $\|g\|_{2,\beta} \leq \delta$ for any g in $\mathcal{G}$. Setting $Z_n^0 = n^{-1/2}\sum_{i=1}^n (\delta_{X_i^0} - P)$, we have

$$\mathbb{E}\left(\max_{g\in\mathcal{G}} |Z_n(g)|\right) \leq \mathbb{E}\left(\max_{g\in\mathcal{G}} |Z_n^0(g)|\right) + \mathbb{E}\left(\max_{g\in\mathcal{G}} |Z_n(g) - Z_n^0(g)|\right). \tag{4.9}$$

Clearly,

$$|Z_n(g) - Z_n^0(g)| \leq 2\frac{\|g\|_\infty}{\sqrt{n}} \sum_{i=1}^n \mathbf{1}_{X_i \neq X_i^0},$$

and, since $\mathbb{P}(X_i \neq X_i^0) \leq \beta(q)$, we obtain

$$\mathbb{E}\left(\max_{g\in\mathcal{G}} |Z_n(g) - Z_n^0(g)|\right) \leq 2m\beta(q)\sqrt{n}. \tag{4.10}$$

It remains to control the first term on the right-hand side of (4.9). Define the random variables $Y_k^0(g)$ by

$$Y_k^0(g) = \sum_{i=kq+1}^{(k+1)q\wedge n} g(X_i^0).$$

For $Y_k^0(g)$, we have the upper bounds

$$\|Y_k^0(g)\|_\infty \leq q\|g\|_\infty \leq qm \quad \text{and} \quad \|Y_k^0(g)\|_2 \leq 4\|g\|_{2,\beta} \leq 4\delta,$$

the second inequality following from Rio's covariance inequality [23]. Since the random variables $(Y_{2k}^0)_{k\geq 0}$ are independent (as well as $(Y_{2k+1}^0)_{k\geq 0}$), we obtain, applying twice (4.5),

$$\mathbb{E}\left(\max_{g\in\mathcal{G}} |Z_n^0(g)|\right) \leq D\Big(\delta\sqrt{H} + \frac{mqH}{\sqrt{n}}\Big),$$

for some absolute constant D. Combining this inequality with (4.10) yields

$$\mathbb{E}\left(\max_{g\in\mathcal{G}} |Z_n(g)|\right) \leq D\Big(\delta\sqrt{H} + \frac{mqH}{\sqrt{n}} + \sqrt{n}m\beta(q)\Big). \tag{4.11}$$

We now choose the integer q: define $q(x) = \min\{p \in \mathbb{N}^* : \beta(p) \leq px\}$. Taking $q = q(H/n)$ in (4.11), we obtain

$$\mathbb{E}\left(\max_{g\in\mathcal{G}} |Z_n(g)|\right) \leq 2D\Big(\delta\sqrt{H} + \frac{mHq(H/n)}{\sqrt{n}}\Big). \tag{4.12}$$

Set $H(k) = \max(1, \log(k))$. From (4.12) it is clear that the functions defined by

$$m_1(n, \delta, k) = 8\delta\sqrt{\frac{n}{H(k)}}, \quad m_2(n, k) = q(H(k)/n), \quad m(n, \delta, k) = \frac{8\delta}{q(H(k)/n)}\sqrt{\frac{n}{H(k)}}$$

satisfy Condition 1 of Theorem 3.3 for the space $(L, \|.\|) = (\mathcal{L}_{2,\beta}(P), \|.\|_{2,\beta})$. In fact it also satisfies Condition 2 (cf. Lemma 4 in [11] and Lemma 7.2 in the appendix). There seems to be no reason why the function m should be a good truncating function for any envelope F in $\mathcal{L}_{2,\beta}(P)$. We shall see in the appendix (Lemma 7.3) how to find the second function M from inequality (4.11).

Combining the above facts with the results of the appendix, we infer that the assumptions of Theorem 3.3 are satisfied for the $\mathcal{L}$-space $(L, \|.\|) = (\mathcal{L}_{2,\beta}(P), \|.\|_{2,\beta})$. This implies that the $\ell^\infty(\mathcal{F})$-valued variable Z_n is asymptotically $\|.\|_{2,\beta}$-equicontinuous as soon as

$$\int_0^1 \sqrt{\mathbb{H}(\mathcal{F}, u, \|.\|_{2,\beta})}du < \infty. \tag{4.13}$$

This result is due to [11] and extends Ossiander's [18] for i.i.d. sequences (in which case $\|.\|_{2,\beta} = \|.\|_2$). Let us see how it applies to standard mixing rates, (cf. [11] for a much deeper discussion):

1. If $\sum k^{b-1}\beta(k)$ is finite for some $b \geq 1$, then (4.13) holds as soon as

$$\int_0^1 \sqrt{\mathbb{H}(\mathcal{F}, u, \|.\|_{2b/(b-1)})}du < \infty.$$

2. Assume that $\beta(k) = O(b^k)$ for some b in $]0, 1[$, and denote by φ the convex function $\varphi(x) = x^2\log(1 + |x|)$. The space $\mathbb{L}(\varphi, P)$ of real-valued functions f such that

$$\|f\|_\varphi = \inf\Big\{c > 0,\ \mathbb{E}\Big(\varphi\Big(\frac{|f(X_0)|}{c}\Big)\Big) \leq 1\Big\} < \infty$$

is continuously embedded in $\mathcal{L}_{2,\beta}(P)$. In particular, (4.13) holds as soon as

$$\int_0^1 \sqrt{\mathbb{H}(\mathcal{F}, u, \|.\|_\varphi)}du < \infty.$$

We conclude this section with some bibliographic notes: Rio [25], Chapter 8, considers the minimal case $\sum \beta(k) < \infty$. If $\mathcal{F}$ is a class of bounded functions, he gives entropy conditions on $\mathcal{F}$ which imply asymptotic tightness of Z_n. The proof uses Goldstein's coupling Theorem (see [14]), an analogue of coupling for Harris chains. Comparing to (4.13), the conditions that he obtains are less good for classes of smooth functions but strictly better if $\mathcal{F}$ consists of indicators of sets. As an application, he proves a central limit theorem for the empirical distribution function under the minimal condition $\sum \beta(k) < \infty$, which cannot be obtained via (4.13).

The same author (Rio [24]) uses again Goldstein's theorem together with symmetrization arguments to obtain conditions in terms of uniform entropy of the class $\mathcal{F}$. In particular, Theorem 3 of this paper is an extension to the absolutely regular case of Pollard's empirical central limit theorem (cf. [21]). Results of this type are of a special interest for VC classes of sets or VC-Major classes (see for instance [27], Chapter 2.6). For these particular classes in a β-mixing context, see also Arcones and Yu [3].

4.4 Nonuniform ϕ-mixing sequences. We first recall the definition of the ϕ-mixing coefficient between two σ-algebras (cf. Ibragimov [15]). Let $(\Omega, \mathcal{A}, \mathbb{P})$ be a probability space. Given two σ-algebras $\mathcal{U}$ and $\mathcal{V}$ of $\mathcal{A}$, the ϕ-mixing coefficient is defined by

$$\phi(\mathcal{U}, \mathcal{V}) = \sup\{\|\mathbb{P}(V|\mathcal{U}) - \mathbb{P}(V)\|_\infty, \ V \in \mathcal{V}\}.$$

Now, let $X = (X_i)_{i\in\mathbb{Z}}$ be a stationary sequence with marginal distribution P and $\mathcal{M}_0$ be the σ-algebra defined by $\mathcal{M}_0 = \sigma(X_i, i \leq 0)$. As in (4.7), define the coefficients $\phi_{\infty,k}$ of X by

$$\phi_{\infty,k}(n) = \sup\{\phi\left(\mathcal{M}_0, \sigma(X_{i_1}, \dots, X_{i_k})\right), 0 < n < i_1 < \cdots < i_k\}. \tag{4.14}$$

Recall that $\phi_{\infty,k}$-mixing is more restrictive than $\beta_{\infty,k}$-mixing in the sense that $\beta_{\infty,k}(n) \leq \phi_{\infty,k}(n)$. When $k < \infty$, we call these coefficients nonuniform, because they only control the dependence between the past σ-algebra $\mathcal{M}_0$ and any k-tuple $(X_{i_1}, \dots, X_{i_k})$ of the future, while the classical uniform $\phi_{\infty,\infty}$ or $\beta_{\infty,\infty}$ coefficients control the dependence between the past and the whole future $\sigma(X_i, i \geq n)$. Nonuniform coefficients are of a special interest for stationary random fields, for which classical uniform coefficients are much too restrictive. All the results we shall give in this section may be extended to random fields indexed by $\mathbb{Z}^d$ with the appropriate extension of (4.14). Note also that dealing with $\phi_{\infty,k}$ allows us to consider nonergodic sequences (or fields), while $\phi_{\infty,\infty}$-mixing sequences as well as absolutely regular sequences are necessarily ergodic (cf. Section 5).

Now from Proposition 1 and Corollary 4(a) in Dedecker [8] (see also Lemma 3 therein) we have, for any centered function f of $\mathbb{L}^\infty(P)$ and $p \geq 3$:

1. Suppose that $\sum k\phi_{\infty,2}(k) < \infty$. Then, for any centered function f of $\mathbb{L}^\infty(P)$ and any two numbers m, δ such that $\|f\|_\infty \leq m$ and $\|f\|_4 \leq \delta$,

$$\|Z_n(f)\|_p \leq K_1(\phi)\left(\delta\sqrt{p} + \frac{pm^3}{\delta^2\sqrt{n}}\right).$$

Let $\mathcal{G}$ be any finite set of centered functions of $\mathbb{L}^\infty(P)$, and m, δ two positive numbers such that $\|g\|_\infty \leq m$ and $\|g\|_4 \leq \delta$ for any g in $\mathcal{G}$. Arguing as in Section (4.1) and setting $H = \max(1, \log(|\mathcal{G}|))$, we obtain

$$\mathbb{E}\left(\max_{g\in\mathcal{G}} |Z_n(g)|\right) \leq C_1(\phi)\left(\delta\sqrt{H} + \frac{m^3 H}{\delta^2\sqrt{n}}\right). \tag{4.15}$$

Set $H(k) = \max(1, \log(k))$. From (4.15) it is clear that the functions defined by

$$m_2 \equiv 1, \quad (m(n, \delta, k))^3 = (m_1(n, \delta, k))^3 = \delta^3\sqrt{\frac{n}{H(k)}}$$

satisfy Conditions 1 and 2 of Theorem 3.3 for the space $L = \mathbb{L}^4(P)$ equipped with the norm $\|.\|_4$ (note that Condition 2 is satisfied with $C_2 = 4^3$). Clearly, the choice $M \equiv m$ ensures that both Conditions 3 and 4 are satisfied, so that the $\ell^\infty(\mathcal{F})$-valued variable Z_n is asymptotically $\|.\|_4$-equicontinuous as soon as

$$\int_0^1 \sqrt{\mathbb{H}(\mathcal{F}, u, \|.\|_4)}du < \infty.$$

2. Assume that $\sum k^{b-1}\phi_{\infty,2}(k) < \infty$ for some b in $]1,2[$. Then, for any centered function f of $\mathbb{L}^\infty(P)$ and any two numbers m, δ such that $\|f\|_\infty \leq m$ and $\|f\|_{2b/(b-1)} \leq \delta$,

$$\|Z_n(f)\|_p \leq K_2(\phi)\left(\delta\sqrt{p} + \frac{pm^3\delta^{-2} \vee pm^{\frac{b+1}{b-1}}\delta^{\frac{2}{1-b}}}{\sqrt{n}}\right).$$

Let $\mathcal{G}$ be any finite set of centered functions of $\mathbb{L}^\infty(P)$, and m, δ two positive numbers such that $\|g\|_\infty \leq m$ and $\|g\|_{2b/(b-1)} \leq \delta$ for any g in $\mathcal{G}$. Arguing as above, we obtain

$$\mathbb{E}\left(\max_{g\in\mathcal{G}}|Z_n(g)|\right) \leq C_2(\phi)\left(\delta\sqrt{H} + \frac{Hm^3\delta^{-2} \vee Hm^{\frac{b+1}{b-1}}\delta^{\frac{2}{1-b}}}{\sqrt{n}}\right). \tag{4.16}$$

Define $G_\delta(m) = m^3\delta^{-2} \vee m^{\frac{b+1}{b-1}}\delta^{\frac{2}{1-b}}$ and take

$$m_2 \equiv 1, \quad m(n,\delta,k) = m_1(n,\delta,k) = G_\delta^{-1}\left(\delta\sqrt{\frac{n}{H(k)}}\right),$$

so that Condition 1 of Theorem 3.3 is satisfied. Next, write

$$\begin{aligned}\sqrt{n}\|g\mathbf{1}_{4g>m(n,\delta,k)}\|_1 &\leq \frac{\sqrt{H(k)}}{\delta}\|gG_\delta(4g)\|_1\\ &\leq \sqrt{H(k)}\Big((4/\delta)^3P(|g|^4) + (4/\delta)^{\frac{b+1}{b-1}}P\big(|g|^{\frac{2b}{b-1}}\big)\Big).\end{aligned}$$

Since $\|g\|_4 \leq \|g\|_{2b/(b-1)}$, the last inequality implies that Condition 2 is satisfied with $C_2 = 4^3+4^{(b+1)/(b-1)}$ for the space $L = \mathbb{L}^{2b/(b-1)}(P)$ equipped with the norm $\|.\|_{2b/(b-1)}$. Taking $M \equiv m$, the same arguments apply to Conditions 3 and 4. We conclude that the $\ell^\infty(\mathcal{F})$-valued variable Z_n is asymptotically $\|.\|_{2b/(b-1)}$-equicontinuous as soon as

$$\int_0^1 \sqrt{\mathbb{H}(\mathcal{F},u,\|.\|_{2b/(b-1)})}du < \infty.$$

3. Assume that $\sum \phi_{\infty,1}(k)$ is finite. Then, for any function f in $\mathbb{L}^\infty(P)$,

$$\|Z_n(f)\|_p \leq K_3(\phi)\sqrt{p}\|f\|_\infty.$$

Arguing as above, we infer that the function $m = \infty$ satisfies the assumptions of Theorem 3.3 for the space $L = \mathbb{L}^\infty(P)$ equipped with the norm $\|.\|_\infty$. Consequently, the $\ell^\infty(\mathcal{F})$-valued variable Z_n is asymptotically $\|.\|_\infty$-equicontinuous as soon as

$$\int_0^1 \sqrt{\mathbb{H}(\mathcal{F},u,\|.\|_\infty)}du < \infty.$$

These results are new, to our knowledge. The comparison between uniform β-mixing (see Section 4.3) and nonuniform ϕ-mixing allows us to make the following conjecture: if, for a positive integer n, the series $\sum_{k>0} k^{n-1}\phi_{\infty,n}(k)$ is finite and the class $\mathcal{F}$ satisfies

$$\int_0^1 \sqrt{\mathbb{H}(\mathcal{F},u,\|.\|_{2n/(n-1)})}du < \infty,$$

then the $\ell^\infty(\mathcal{F})$-valued variable Z_n is asymptotically $\|.\|_{2n/(n-1)}$-equicontinuous.

5 Empirical central limit theorems

We first recall a central limit theorem for nonuniform mixing sequences (cf. Dedecker [7], Corollary 3):

Theorem 5.1. *Let $X = (X_i)_{i\in\mathbb{Z}}$ be a strictly stationary sequence of random variables with value in a Polish space $\mathcal{X}$ with common marginal distribution P. Let $T : \mathcal{X}^{\mathbb{Z}} \mapsto \mathcal{X}^{\mathbb{Z}}$ be the shift operator: $(T(x))_i = x_{i+1}$. Denote by $\mathcal{I}$ the σ-algebra of T-invariant Borel sets of $\mathcal{X}^{\mathbb{Z}}$. Let f be an element of $\mathbb{L}^2(P)$. If the mixing coefficients of X satisfy*

$$\sum_{k>0} \int_0^{\beta_{\infty,1}(k)} Q_f^2(u)du < \infty,$$

then $Z_n(f)$ converges weakly to $\varepsilon\sqrt{\Gamma_{\mathcal{I}}(f)}$, where ε is a standard Gaussian random variable independent of $X^{-1}(\mathcal{I})$ and

$$\Gamma_{\mathcal{I}}(f) = \sum_{k\in\mathbb{Z}} \mathrm{Cov}(f(X_0), f(X_k)|X^{-1}(\mathcal{I})). \tag{5.1}$$

Applying the nonergodic version of Ibragimov–Billingsley's central limit theorem for martingales with stationary differences, we infer that the conclusion of Theorem 5.1 remains true if $\mathbb{E}(f(X_1)|\mathcal{M}_0) = P(f)$ (in that case, the sum in (5.1) reduces to the conditional variance term). Note also that, if $\beta_{\infty,2}(n)$ tends to zero as n tends to infinity, then $\sigma(X_0, X_k)$ is independent of $X^{-1}(\mathcal{I})$ and consequently $\Gamma_{\mathcal{I}}(f)$ is constant. The above facts together with the results of Section 4 yield the following.

Theorem 5.2. *Let $X = (X_i)_{i\in\mathbb{Z}}$ be a stationary sequence with marginal distribution P and $\mathcal{F}$ be a class of functions of $\mathbb{L}^2(P)$. Consider the following assumptions:*

1. $\mathbb{E}(f(X_1)|\mathcal{M}_0) = P(f)$ *for any f in $\mathcal{F}$, and* $\int_0^1 \sqrt{\mathbb{H}(\mathcal{F}, u, d_2)}du < \infty$.
2. $\sum_{k>0} \beta(k) < \infty$ *and* $\int_0^1 \sqrt{\mathbb{H}(\mathcal{F}, u, \|.\|_{2,\beta})}du < \infty$.
3. $\sum k^{b-1}\phi_{\infty,2}(k) < \infty$ *for some b in* $]1, 2]$, *and* $\int_0^1 \sqrt{\mathbb{H}(\mathcal{F}, u, \|.\|_{2b/(b-1)})}du < \infty$.
4. $\sum_{k>0} \phi_{\infty,1}(k) < \infty$ *and* $\int_0^1 \sqrt{\mathbb{H}(\mathcal{F}, u, \|.\|_{\infty})}du < \infty$.

If either 2 or 3 holds, then the $\ell^\infty(\mathcal{F})$-valued variable Z_n converges weakly to a zero-mean tight Gaussian process with covariance function

$$\Gamma(f, g) = \sum_{k\in\mathbb{Z}} \mathrm{Cov}(f(X_0), g(X_k)).$$

If either 1 or 4 holds, then the $\ell^\infty(\mathcal{F})$-valued variable Z_n converges weakly to a tight random variable whose conditional distribution with respect to $X^{-1}(\mathcal{I})$ is that of a zero-mean tight Gaussian process with covariance function

$$\Gamma_{\mathcal{I}}(f, g) = \sum_{k\in\mathbb{Z}} \mathrm{Cov}(f(X_0), g(X_k)|X^{-1}(\mathcal{I})).$$

6 Chaining

The chaining procedure we borrow here is an extension of Ossiander's method due to Doukhan, Massart, and Rio [11].

Our purpose is to prove the stochastic equicontinuity of the empirical process over a class of functions $\mathcal{F}$ equipped with the norm $\|.\|$ and fulfilling the entropy condition (3.2). More precisely, we shall prove that

$$\lim_{\sigma\to 0}\limsup_{n\to+\infty}\mathbb{E}\left(\sup_{\|f-g\|\leq\sigma}|Z_n(f)-Z_n(g)|\right)=0. \tag{6.1}$$

Set $\mathcal{F}^\sigma := \{f-g,\ \ f,g\in\mathcal{F} \text{ and } \|f-g\|\leq\sigma\}$, and consider the family of bounded functions $\mathcal{F}^\sigma(M) := \{(f-g)\mathbf{1}_{4F\leq M},\ \ f,g\in\mathcal{F} \text{ and } \|f-g\|\leq\sigma\}$. Clearly, for each positive M,

$$\mathbb{E}\left(\sup_{f\in\mathcal{F}^\sigma}|Z_n(f)|\right)\leq\mathbb{E}\left(\sup_{f\in\mathcal{F}^\sigma(M)}|Z_n(f)|\right)+4\sqrt{n}\|F\mathbf{1}_{4F>M}\|_1. \tag{6.2}$$

Let $\mathbb{N}(\sigma)=\exp(\mathbb{H}(\mathcal{F}^\sigma,\sigma,\|.\|))$ and take

$$M=M_n=M(n,(1+a_1a_2)\sigma,\mathbb{N}(\sigma)). \tag{6.3}$$

From (6.2) and Condition 4 of Theorem 3.3 we infer that

$$\limsup_{n\to\infty}\mathbb{E}\left(\sup_{f\in\mathcal{F}^\sigma}|Z_n(f)|\right)\leq\limsup_{n\to\infty}\mathbb{E}\left(\sup_{f\in\mathcal{F}^\sigma(M_n)}|Z_n(f)|\right),$$

so that (6.1) holds as soon as

$$\lim_{\sigma\to 0}\limsup_{n\to\infty}\mathbb{E}\left(\sup_{f\in\mathcal{F}^\sigma(M_n)}|Z_n(f)|\right)=0. \tag{6.4}$$

Note that if f belongs to $\mathcal{F}^\sigma(M_n)$, then $\|f\|\leq\sigma$ and $2\|f\|_\infty\leq M_n$. Note also that we have $\mathbb{H}(\mathcal{F}^\sigma(M_n),x)\leq\mathbb{H}(\mathcal{F}^\sigma,x)\leq 2\mathbb{H}(\mathcal{F},x/2)$. From these two elementary remarks, we infer that (6.4) will follow from Proposition 6.2 below.

Before stating this proposition, we introduce as in [11] an appropriate upper bound H for the entropy $\mathbb{H}$.

Lemma 6.1. *For any class of functions $\mathcal{F}$, there exists a nonincreasing upper bound H of $\mathbb{H}(\mathcal{F},.,\|.\|)$ such that $x\to x^4H(x)$ is nondecreasing and, for any δ in* $[0,1]$,

$$\psi(\delta)=\int_0^\delta\sqrt{1\vee H(u)}du\leq 4\int_0^\delta\sqrt{1\vee\mathbb{H}(\mathcal{F},u,\|.\|)}du.$$

See [11] for a proof.

Proposition 6.2. *Let m_1, m_2, M and R be four functions satisfying Conditions* 1, 2, *and* 3 *of Theorem* 3.3. *Let $\mathcal{F}$ be any class of functions of L satisfying* (3.2). *Define* $\mathbb{N}(\sigma)=\exp(\mathbb{H}(\mathcal{F},\sigma,\|.\|))$ *and $M_n=M(n,(1+a_1a_2)\sigma,\mathbb{N}(\sigma))$. Assume that any function f of $\mathcal{F}$ satisfy the inequalities $2\|f\|_\infty\leq M_n$ and $\|f\|\leq\sigma$. Then there exists a constant K, not depending on (n,σ), such that*

$$\mathbb{E}\left(\sup_{f\in\mathcal{F}}|Z_n(f)|\right)\leq K\psi(\sigma)+2R(n,(1+a_1a_2)\sigma,\mathbb{N}(\sigma)).$$

Proof. We start with some preliminary notations.

Definitions 6.3. *Let* $\delta_0 = \sigma$ *and* $\delta_k = 2^{-k}\delta_0$. *For each* k, *we choose a covering of* $\mathcal{F}$ *by brackets* $B_{j,k} := [g_{j,k}, h_{j,k}]$ *for* $1 \le j \le J_k$, *with* $\|h_{j,k} - g_{j,k}\| \le \delta_k$ *and* $J_k \le \mathbb{N}(\delta_k)$. *Since* $2|f| \le M_n$ *for all* $f \in \mathcal{F}$, *we may assume that* $|h_{j,k} - g_{j,k}| \le M_n$. *In each bracket* $B_{j,k}$, *fix a point* $f_{j,k}$ *belonging to* $\mathcal{F}$. *Let* $\pi_k f := f_{\psi_k f,k}$ *where* ψ_k *is a mapping from* $\mathcal{F}$ *to* $[1, J_k]$ *defined by* $\psi_k f := \min\{j \in [1, J_k] : f \in B_{j,k}\}$. *Set* $\Delta_k f := h_{\psi_k f,k} - g_{\psi_k f,k}$. *Clearly,*

$$|f - \pi_k f| \le \Delta_k f \quad \textit{and} \quad \|\Delta_k f\| \le \delta_k. \tag{6.5}$$

Finally define $\mathbf{H}(\delta) = \sum_{\delta_k \ge \delta} H(\delta_k)$, $\mathbf{N}(\delta) := [\exp(\mathbf{H}(\delta))]$, *and* $N(\delta_k) := [\exp(H(\delta_k))]$.

Definitions 6.4. *Set* $2\alpha = 3(1 + a_1 a_2)$. *Define the nonincreasing sequences* (b_j) *and* (q_j) *and the sequence* (m_j) *by*

$$b_k = m_1(n, \alpha\delta_k, \mathbf{N}(\delta_{k+1})), \quad q_k = 4m_2(n, \mathbf{N}(\delta_{k+1})), \quad m_k = \frac{b_k}{q_k}. \tag{6.6}$$

For any $f \in \mathcal{F}$, *define* $\nu := \nu(f) := N \wedge \min\{j \ge 0, \ \ \Delta_j f > m_j\}$, *where* N *is defined by*

$$N := \min\left\{k \ge 0 : \delta_k \le \frac{\psi(\sigma)}{\sqrt{n}}\right\} \vee 1.$$

Without loss of generality, suppose that $\mathbb{H}(\mathcal{F}, \sigma, \|.\|) \ge 1$ (note that the same is necessarily true for $H(\sigma)$ and $\mathbf{H}(\sigma)$). We now list all the inequalities which are useful to prove Proposition 6.2.

Facts 6.5.

$$\text{If} \quad |g| \le h, \quad \text{then} \quad |Z_n(g)| \le Z_n(h) + 2\sqrt{n}\|h\|_1. \tag{6.7}$$

$$\text{For all } k \in \mathbb{N}, \quad \mathbf{H}(\delta_{k+1}) \le 17\mathbf{H}(\delta_k). \tag{6.8}$$

There exist positive constants $K_1, \dots, K_5$ such that, for any integer $1 \le k \le N-1$, we have

$$\mathbb{E}\left(\sup_{f\in\mathcal{F}}\left|Z_n\left(\Delta_{k-1} f \mathbf{1}_{\Delta_k f > \frac{b_{k-1}}{q_k}} \prod_{i=0}^{k-1} \mathbf{1}_{\Delta_i f \le m_i}\right)\right|\right) \le K_1 \delta_k \sqrt{\mathbf{H}(\delta_k)}, \tag{6.9}$$

$$\mathbb{E}\left(\sup_{f\in\mathcal{F}}\left|Z_n\left(\Delta_k f \mathbf{1}_{m_k < \Delta_k f \le \frac{b_{k-1}}{q_k}} \prod_{i=0}^{k-1} \mathbf{1}_{\Delta_i f \le m_i}\right)\right|\right) \le K_2 \delta_k \sqrt{\mathbf{H}(\delta_k)}, \tag{6.10}$$

$$\mathbb{E}\left(\sup_{f\in\mathcal{F}}\left|Z_n\left((\pi_k(f) - \pi_{k-1}(f))\mathbf{1}_{\Delta_k f \le \frac{b_{k-1}}{q_k}} \prod_{i=0}^{k-1} \mathbf{1}_{\Delta_i f \le m_i}\right)\right|\right) \le K_3 \delta_k \sqrt{\mathbf{H}(\delta_k)}, \tag{6.11}$$

$$\mathbb{E}\left(\sup_{f\in\mathcal{F}}\left|Z_n\left(\Delta_{N-1} f \prod_{i=0}^{N-1} \mathbf{1}_{\Delta_i f \le m_i}\right)\right|\right) \le K_4 \delta_N \sqrt{\mathbf{H}(\delta_N)}, \tag{6.12}$$

$$\text{for } 0 \le k \le N-1, \quad \sqrt{n}\left\|\Delta_k f \mathbf{1}_{\Delta_k f > m_k}\right\|_1 \le K_5 \delta_k \sqrt{\mathbf{H}(\delta_k)}. \tag{6.13}$$

Proof of Facts 6.5. Inequality (6.7) is straightforward, and (6.8) follows directly from Lemma 6.1.

We first prove (6.9). Clearly the supremum in (6.9) reduces to a maximum over a finite set of functions with cardinality less than $\mathbf{N}(\delta_k)$. Moreover if g is any function belonging to this set, we have $\|g - P(g)\| \le (1 + a_1 a_2)\delta_{k-1} \le \alpha\delta_{k-1}$ and

$$\|g - P(g)\|_\infty \le m_{k-1} \le \frac{m_1(n, \alpha\delta_{k-1}, \mathbf{N}(\delta_k))}{m_2(n, \mathbf{N}(\delta_k))},$$

and (6.9) follows from Condition 1 of Theorem 3.3.

We now prove (6.10). As for the proof of (6.9), the supremum in (6.10) reduces to a maximum over a finite set of functions with cardinality less than $\mathbf{N}(\delta_k) \le \mathbf{N}(\delta_{k+1})$. If g is any function belonging to this set, we have $\|g - P(g)\| \le (1 + a_1 a_2)\delta_k \le \alpha\delta_{k-1}$ and

$$\|g - P(g)\|_\infty \le \frac{b_{k-1}}{q_k} \le \frac{m_1(n, \alpha\delta_{k-1}, \mathbf{N}(\delta_k))}{m_2(n, \mathbf{N}(\delta_{k+1}))}.$$

Condition 1 of Theorem 3.3 yields then

$$\mathbb{E}\left(\sup_{f\in\mathcal{F}}\left|Z_n\left(\Delta_k f \mathbf{1}_{m_k < \Delta_k f \le \frac{b_{k-1}}{q_k}} \prod_{i=0}^{k-1} \mathbf{1}_{\Delta_i f \le m_i}\right)\right|\right) \le \alpha C_1 \delta_{k-1}\left(\sqrt{\mathbf{H}(\delta_{k+1})} + \frac{\mathbf{H}(\delta_{k+1})}{\sqrt{\mathbf{H}(\delta_k)}}\right).$$

The last inequality, together with (6.8) yields (6.10).

To prove (6.11), note that the supremum in (6.11) reduces to a maximum over a finite set of functions with cardinality less than $\mathbf{N}(\delta_k) \le \mathbf{N}(\delta_{k+1})$. If g is any function belonging to this set, we have $\|g - P(g)\| \le (1 + a_1 a_2)(\delta_k + \delta_{k-1}) = \alpha\delta_{k-1}$ and

$$\|g - P(g)\|_\infty \le \frac{4b_{k-1}}{q_k} = \frac{m_1(n, \alpha\delta_{k-1}, \mathbf{N}(\delta_k))}{m_2(n, \mathbf{N}(\delta_{k+1}))},$$

and we conclude as for (6.10).

To prove (6.12), note that the supremum in (6.12) reduces to a maximum over a finite set of functions with cardinality less than $\mathbf{N}(\delta_{N-1}) \le \mathbf{N}(\delta_N)$. If g is any function belonging to this set, we have $\|g - P(g)\| \le (1 + a_1 a_2)\delta_{N-1} \le \alpha\delta_{N-1}$ and

$$\|g - P(g)\|_\infty \le m_{N-1} \le \frac{m_1(n, \alpha\delta_{N-1}, \mathbf{N}(\delta_N))}{m_2(n, \mathbf{N}(\delta_N))},$$

and we conclude as in (6.9).

To prove (6.13) note that $\|\Delta_k f\| \le \delta_k \le \alpha\delta_k$. Since

$$m_k = \frac{m_1(n, \alpha\delta_k, \mathbf{N}(\delta_{k+1}))}{4m_2(n, \mathbf{N}(\delta_{k+1}))},$$

(6.13) follows from (6.8) and Condition 2 of Theorem 3.3. This completes the proof of Facts 6.5.

Chaining. Arguing as in Doukhan, Massart and Rio [11], we have the decomposition

$$\mathbb{E}\left(\sup_{f\in\mathcal{F}}|Z_n(f)|\right) \le \mathbb{E}_1 + \mathbb{E}_2 + \mathbb{E}_3 + \mathbb{E}_4 + \mathbb{E}_5 + \mathbb{E}_6, \tag{6.14}$$

where

$$\mathbb{E}_1 = \mathbb{E}\left(\sup_{f\in\mathcal{F}} |Z_n(\pi_0 f)|\right),$$

$$\mathbb{E}_2 = \mathbb{E}\left(\sup_{f\in\mathcal{F}} \left|Z_n(f-\pi_0 f)\mathbf{1}_{\nu(f)=0}\right|\right),$$

$$\mathbb{E}_3 = \sum_{k=1}^{N-1}\mathbb{E}\left(\sup_{f\in\mathcal{F}} \left|Z_n(f-\pi_{k-1} f)\mathbf{1}_{\nu(f)=k}\mathbf{1}_{\Delta_k f > \frac{b_{k-1}}{q_k}}\right|\right),$$

$$\mathbb{E}_4 = \sum_{k=1}^{N-1}\mathbb{E}\left(\sup_{f\in\mathcal{F}} \left|Z_n(f-\pi_{k} f)\mathbf{1}_{\nu(f)=k}\mathbf{1}_{\Delta_k f \le \frac{b_{k-1}}{q_k}}\right|\right),$$

$$\mathbb{E}_5 = \mathbb{E}\left(\sup_{f\in\mathcal{F}} \left|Z_n(f-\pi_{N-1} f)\mathbf{1}_{\nu(f)=N}\right|\right),$$

$$\mathbb{E}_6 = \sum_{k=1}^{N-1}\mathbb{E}\left(\sup_{f\in\mathcal{F}} \left|Z_n\left((\pi_k(f)-\pi_{k-1}(f))\mathbf{1}_{\Delta_k f \le \frac{b_{k-1}}{q_k}}\mathbf{1}_{\nu(f)\ge k}\right)\right|\right).$$

Control of $\mathbb{E}_1$. When f runs over the set $\mathcal{F}$, $\pi_0 f$ runs over a finite class of functions with cardinality less than $\mathbb{N}(\sigma)$. Moreover, we have the bounds $\|\pi_0 f - P(\pi_0 f)\|_\infty \le M_n$ and $\|\pi_0 f - P(\pi_0 f)\| \le (1+a_1a_2)\sigma$. Hence according to Condition 3 of Theorem 3.3, we have

$$\mathbb{E}_1 \le C_3(1+a_1a_2)\sigma\sqrt{H(\sigma)} + R(n,(1+a_1a_2)\sigma,\mathbb{N}(\sigma)). \tag{6.15}$$

Control of $\mathbb{E}_2$. We deduce from $\{\nu(f)=0\}=\{\Delta_0 > m_0\}$, (6.5), and (6.7) that

$$\mathbb{E}_2 \le \mathbb{E}\left(\sup_{f\in\mathcal{F}} \left|Z_n\left(\Delta_0 f\mathbf{1}_{\Delta_0 f>m_0}\right)\right|\right) + 2\sqrt{n}\sup_{f\in\mathcal{F}}\left\|\Delta_0 f\mathbf{1}_{\Delta_0 f>m_0}\right\|_1. \tag{6.16}$$

Using inequality (6.13), the second term in the right-hand side is bounded by

$$\sqrt{n}\sup_{f\in\mathcal{F}}\left\|\Delta_0 f\mathbf{1}_{\Delta_0 f>m_0}\right\|_1 \le K_5\delta_0\sqrt{\mathbf{H}(\delta_0)}. \tag{6.17}$$

Note that

$$\left\|\Delta_0 f\mathbf{1}_{\Delta_0 f>m_0} - \mathbb{E}\left(\Delta_0 f\mathbf{1}_{\Delta_0 f>m_0}\right)\right\| \le (1+a_1a_2)\sigma \quad \text{using (6.5), and}$$
$$\left\|\Delta_0 f\mathbf{1}_{\Delta_0 f>m_0} - \mathbb{E}\left(\Delta_0 f\mathbf{1}_{\Delta_0 f>m_0}\right)\right\|_\infty \le M_n.$$

To control the first term of inequality (6.16), note that the supremum reduces to a maximum over a finite set of functions with cardinality less than $\mathbb{N}(\sigma)$. Hence according to Condition 3 of Theorem 3.3, we get

$$\mathbb{E}\left(\sup_{f\in\mathcal{F}} \left|Z_n\left(\Delta_0 f\mathbf{1}_{\Delta_0 f>m_0}\right)\right|\right) \le C_3(1+a_1a_2)\sigma\sqrt{H(\sigma)} + R(n,(1+a_1a_2)\sigma,\mathbb{N}(\sigma)). \tag{6.18}$$

Collecting inequalities (6.16), (6.17), (6.18), we deduce

$$\mathbb{E}_2 \le C\delta_0\sqrt{\mathbf{H}(\delta_0)} + R(n,(1+a_1a_2)\sigma,\mathbb{N}(\sigma)). \tag{6.19}$$

Control of $\mathbb{E}_3$. We deduce from (6.5), (6.7) that

$$\mathbb{E}_3 \le \sum_{k=1}^{N-1} \mathbb{E}\Big(\sup_{f\in\mathcal{F}} \Big|Z_n(\Delta_{k-1}(f)\mathbf{1}_{\nu(f)=k}\mathbf{1}_{\Delta_k f > \frac{b_{k-1}}{q_k}}\Big|\Big) + \sum_{k=1}^{N-1} 2\sqrt{n} \sup_{f\in\mathcal{F}} \Big\|\Delta_{k-1}(f)\mathbf{1}_{\nu(f)=k}\mathbf{1}_{\Delta_k f > \frac{b_{k-1}}{q_k}}\Big\|_1. \tag{6.20}$$

In order to control the first term on the right-hand side of (6.20), note that

$$\{\nu(f)=k\} = \{m_k < \Delta_k f,\ \Delta_i f \le m_i,\ \ i = 0\dots, k-1\}.$$

Hence inequality (6.9) yields

$$\sum_{k=1}^{N-1} \mathbb{E}\Big(\sup_{f\in\mathcal{F}} \Big|Z_n(\Delta_{k-1}(f)\mathbf{1}_{\nu(f)=k}\mathbf{1}_{\Delta_k f > \frac{b_{k-1}}{q_k}}\Big|\Big) \le K_1 \sum_{k=1}^{N-1} \delta_k \sqrt{\mathbf{H}(\delta_k)}. \tag{6.21}$$

On the other hand, since

$$\mathbf{1}_{\nu(f)=k}\mathbf{1}_{\Delta_k f > \frac{b_{k-1}}{q_k}} \le \mathbf{1}_{\Delta_k f > \frac{b_{k-1}}{q_k}}\mathbf{1}_{\Delta_{k-1} f \le \frac{b_{k-1}}{q_{k-1}}} \le \mathbf{1}_{\Delta_k f > \Delta_{k-1} f}\mathbf{1}_{\Delta_k f > \frac{b_k}{q_k}},$$

we deduce, using inequality (6.13),

$$\sqrt{n}\sup_{f\in\mathcal{F}} \Big\|\Delta_{k-1}(f)\mathbf{1}_{\nu(f)=k}\mathbf{1}_{\Delta_k f > \frac{b_{k-1}}{q_k}}\Big\|_1 \le \sqrt{n}\sup_{f\in\mathcal{F}} \Big\|\Delta_k(f)\mathbf{1}_{\Delta_k f > \frac{b_k}{q_k}}\Big\|_1 \le K_5\delta_k\sqrt{\mathbf{H}(\delta_k)}.$$

This last bound together with (6.21) and (6.20) yields

$$\mathbb{E}_3 \le (K_1 + 2K_5)\sum_{k=1}^{N-1} \delta_k \sqrt{\mathbf{H}(\delta_k)}. \tag{6.22}$$

Control of $\mathbb{E}_4$. We deduce from (6.5) and (6.7) that

$$\mathbb{E}_4 \le \sum_{k=1}^{N-1} \mathbb{E}\Big(\sup_{f\in\mathcal{F}} \Big|Z_n(\Delta_k(f)\mathbf{1}_{\nu(f)=k}\mathbf{1}_{\Delta_k f \le \frac{b_{k-1}}{q_k}}\Big|\Big) + \sum_{k=1}^{N-1} 2\sqrt{n} \sup_{f\in\mathcal{F}} \Big\|\Delta_k(f)\mathbf{1}_{\nu(f)=k}\mathbf{1}_{\Delta_k f \le \frac{b_{k-1}}{q_k}}\Big\|_1. \tag{6.23}$$

In order to control the first term on right-hand side of (6.23), note that

$$\{\nu(f)=k\} = \{m_k < \Delta_k f,\ \Delta_i f \le m_i,\ \ i = 0\dots, k-1\}.$$

Hence inequality (6.10) yields

$$\sum_{k=1}^{N-1} \mathbb{E}\Big(\sup_{f\in\mathcal{F}} \Big|Z_n(\Delta_k(f)\mathbf{1}_{\nu(f)=k}\mathbf{1}_{\Delta_k f \le \frac{b_{k-1}}{q_k}}\Big|\Big) \le K_2 \sum_{k=1}^{N-1} \delta_k \sqrt{\mathbf{H}(\delta_k)}. \tag{6.24}$$

On the other hand, since

$$\mathbf{1}_{\nu(f)=k}\mathbf{1}_{\Delta_k f\le \frac{b_{k-1}}{q_k}} \le \mathbf{1}_{\Delta_k f>\frac{b_k}{q_k}},$$

we deduce, using inequality (6.13),

$$\sqrt{n}\sup_{f\in\mathcal{F}}\left\|\Delta_k(f)\mathbf{1}_{\nu(f)=k}\mathbf{1}_{\Delta_k f\le \frac{b_{k-1}}{q_k}}\right\|_1 \le \sqrt{n}\sup_{f\in\mathcal{F}}\left\|\Delta_k(f)\mathbf{1}_{\Delta_k f>\frac{b_k}{q_k}}\right\|_1 \le K_5\delta_k\sqrt{\mathbf{H}(\delta_k)}.$$

This last bound together with (6.24) and (6.23) yields

$$\mathbb{E}_4 \le (K_2+2K_5)\sum_{k=1}^{N-1}\delta_k\sqrt{\mathbf{H}(\delta_k)}. \tag{6.25}$$

Control of $\mathbb{E}_5$. We deduce from (6.5), (6.7) that

$$\mathbb{E}_5 \le \mathbb{E}\left(\sup_{f\in\mathcal{F}}|Z_n(\Delta_{N-1}(f)\mathbf{1}_{\nu(f)=N}|\right) + 2\sqrt{n}\sup_{f\in\mathcal{F}}\|\Delta_{N-1}(f)\|_1\,. \tag{6.26}$$

Since $\prod_{i=0}^{N-1}\mathbf{1}_{\Delta_i f\le m_i} = \mathbf{1}_{\nu(f)=N}$, we infer from inequality (6.12) that

$$\mathbb{E}\left(\sup_{f\in\mathcal{F}}|Z_n(\Delta_{N-1}(f)\mathbf{1}_{\nu(f)=N}|\right) \le K_4\delta_N\sqrt{\mathbf{H}(\delta_N)}. \tag{6.27}$$

Now

$$\sqrt{n}\sup_{f\in\mathcal{F}}\|\Delta_{N-1}(f)\|_1 \le a_1\sqrt{n}\sup_{f\in\mathcal{F}}\|\Delta_{N-1}(f)\| \le a_1\sqrt{n}\delta_{N-1} \le 2a_1\psi(\sigma),$$

the last inequality following from the definition of N (cf. Definitions 6.4). This last bound together with (6.27) and (6.26) yields

$$\mathbb{E}_5 \le K_4\delta_N\sqrt{\mathbf{H}(\delta_N)} + 4a_1\psi(\sigma). \tag{6.28}$$

Control of $\mathbb{E}_6$. We deduce from

$$\mathbf{1}_{\Delta_k f\le \frac{b_{k-1}}{q_k}}\mathbf{1}_{\nu(f)\ge k} = \mathbf{1}_{\Delta_k f\le \frac{b_{k-1}}{q_k}}\prod_{i=0}^{k-1}\mathbf{1}_{\Delta_i f\le m_i}$$

and from inequality (6.11) that

$$\mathbb{E}_6 \le K_3\sum_{k=1}^{N-1}\delta_k\sqrt{\mathbf{H}(\delta_k)}. \tag{6.29}$$

End of the proof of Proposition 6.2. Collecting inequalities (6.14), (6.15), (6.19), (6.22), (6.25), (6.28), and (6.29), we obtain

$$\mathbb{E}\left(\sup_{f\in\mathcal{F}}|Z_n(f)|\right) \le K\left\{\psi(\sigma)+\sum_{k=1}^{N}\delta_k\sqrt{\mathbf{H}(\delta_k)}\right\} + 2R(n,(1+a_1a_2)\sigma,\mathbb{N}(\sigma)), \tag{6.30}$$

for a positive constant K. From the definition of $\mathbf{H}$, we get

$$\begin{aligned}\sum_{k=1}^{N} \delta_k \sqrt{\mathbf{H}(\delta_k)} &\leq \sum_{k\geq 0} \delta_k \Big(\sum_{j\leq k} \sqrt{H(\delta_j)}\Big) \\ &\leq 2\sum_{j\geq 0} \delta_j \sqrt{H(\delta_j)} \leq 4\int_0^{\sigma} \sqrt{H(u)}du = 4\psi(\sigma),\end{aligned}$$

which together with inequality (6.30) completes the proof of Proposition 6.2.

7 Appendix

As in [11] define the functions

$$B(t) = \int_0^t \beta^{-1}(u)du \quad \text{and} \quad \delta_h(\epsilon) = \sup_{t\leq\epsilon} Q_h(t)\sqrt{B(t)}.$$

The following lemmas and their proofs are due to Doukhan, Massart, and Rio [11].

Lemma 7.1. *Let h be any nonnegative function in $\mathbb{L}^1(P)$. Then, for any ϵ in $]0, 1]$, the following inequality holds:*

$$\|h\mathbf{1}_{h>Q_h(\epsilon)}\|_1 \leq \frac{2\epsilon\delta_h(\epsilon)}{\sqrt{B(\epsilon)}}. \tag{7.1}$$

In particular, if $\delta_h(1) \leq \delta$ and $a\sqrt{\epsilon\beta^{-1}(\epsilon)} \geq \delta$, then

$$\|h\mathbf{1}_{h>a}\|_1 \leq 2\delta\sqrt{\frac{\epsilon}{\beta^{-1}(\epsilon)}}. \tag{7.2}$$

Proof. Recall that for any function f in $\mathbb{L}^1(P)$, $\|f\|_1 = \int_0^1 Q_f(u)du$. The quantile function of $h\mathbf{1}_{h>a}$ being equal to $Q_h(u)\mathbf{1}_{u\leq Q_h^{-1}(a)}$, we obtain

$$\|h\mathbf{1}_{h>Q_h(\epsilon)}\|_1 = \int_0^{Q_h^{-1}(Q_h(\epsilon))} Q_h(t)dt \leq \int_0^{\epsilon} Q_h(t)dt.$$

Bearing in mind the definition of $\delta_h(\epsilon)$ and using the concavity of B, we infer that

$$\|h\mathbf{1}_{h>Q_h(\epsilon)}\|_1 \leq \delta_h(\epsilon)\int_0^{\epsilon} \frac{dt}{\sqrt{B(t)}} \leq \delta_h(\epsilon)\sqrt{\frac{\epsilon}{B(\epsilon)}}\int_0^{\epsilon} \frac{dt}{\sqrt{t}},$$

proving (7.1). Since β^{-1} is nonincreasing, $\epsilon\beta^{-1}(\epsilon) \leq B(\epsilon)$ and both the condition on a and the definition of $\delta_h(\epsilon)$ imply that $a \leq Q_h(\epsilon)$. This yields (7.2) and the proof of Lemma 7.1 is complete.

Next, we use Lemma 7.1 to prove the assertions of Section 4.3.

Lemma 7.2. *The function m defined in Section 4.3 by*

$$m(n, \delta, k) = \frac{8\delta}{q(H(k)/n)}\sqrt{\frac{n}{H(k)}}$$

satisfies Condition 2 of Theorem 3.3.

Proof. Apply Lemma 7.1 with

$$4a = m(n, \delta, k) \quad \text{and} \quad \epsilon = \frac{H(k)}{n} \max(1, q(H(k)/n) - 1).$$

By definition of $q(.)$, we obtain $\epsilon \leq \beta(q(H(k)/n))$ and therefore

$$\beta^{-1}(\epsilon) \geq q(H(k)/n) \geq \max(1, q(H(k)/n) - 1),$$

which implies that $m(n, \delta, k)\sqrt{\epsilon\beta^{-1}(\epsilon)} \geq 4\delta$. Now, take g in $\mathcal{L}_{2,\beta}(P)$ such that $\|g\|_{2,\beta} \leq \delta$. Since $\delta_g(1) \leq \|g\|_{2,\beta} \leq \delta$, inequality (7.2) holds yielding

$$\|g\mathbf{1}_{4g>m(n,\delta,k)}\|_1 \leq 2\delta\sqrt{\frac{\epsilon}{\beta^{-1}(\epsilon)}} \leq 2\delta\sqrt{\frac{H(k)}{n}},$$

and Condition 2 of Theorem 3.3 holds with $C_2 = 2$.

Lemma 7.3. *Let F be any nonnegative function in $\mathcal{L}_{2,\beta}(P)$. Let $\epsilon(n, k)$ be the unique solution of the equation $nx^2 = B(x)H(k)$. Then the function $M(n, k) = 4Q_F(\epsilon(n, k))$ satisfies Conditions 3 and 4 of Theorem 3.3.*

Proof. Let $\mathcal{G}$ be a finite subset of centered functions of $\mathcal{L}_{2,\beta}(P)$ with cardinality $|\mathcal{G}| \leq k$, whose elements satisfy $\|g\| \leq \delta$ and $\|g\|_\infty \leq 4Q_F(\epsilon)$. According to (4.11), we have

$$\mathbb{E}\left(\max_{g\in\mathcal{G}} |Z_n(g)|\right) \leq D\Big(\delta\sqrt{H(k)} + \frac{4Q_F(\epsilon)qH(k)}{\sqrt{n}} + 4\sqrt{n}Q_F(\epsilon)\beta(q)\Big).$$

Choose $q = \beta^{-1}(\epsilon)$. Since $\epsilon\beta^{-1}(\epsilon) \leq B(\epsilon)$, we have $qH(k) \leq n\epsilon$, which leads to

$$\mathbb{E}\left(\max_{g\in\mathcal{G}} |Z_n(g)|\right) \leq 4D\Big(\delta\sqrt{H(k)} + 2\sqrt{n}\epsilon Q_F(\epsilon)\Big).$$

To prove that Condition 3 holds, it remains to see that $\sqrt{n}\epsilon Q_F(\epsilon)$ tends to zero as n tends to infinity. Bearing in mind that $Q_F(\epsilon)\sqrt{B(\epsilon)} \leq \delta_F(\epsilon)$, we infer that

$$\sqrt{n}\epsilon Q_F(\epsilon) \leq \frac{\sqrt{n}\epsilon\delta_F(\epsilon)}{\sqrt{B(\epsilon)}} \leq \delta_F(\epsilon)\sqrt{H(k)}.$$

Since $\delta_F(\epsilon)$ tends to zero as n tends to infinity, we infer that Condition 3 is satisfied.

It remains to check Condition 4. By inequality (7.1) of Lemma 7.1, we have

$$\sqrt{n}\|F\mathbf{1}_{F>Q_F(\epsilon)}\|_1 \leq \frac{2\sqrt{n}\epsilon\delta_F(\epsilon)}{\sqrt{B(\epsilon)}} \leq 2\delta_F(\epsilon)\sqrt{H(k)},$$

and we conclude as above. The proof of Lemma 7.3 is complete.

References

[1] N. T. Andersen, E. Giné, M. Ossiander, and J. Zinn: The central limit theorem and the law of the iterated logarithm for empirical processes under local conditions. *Probability Theory and Related Fields* **77** (1988), 271–305.

[2] D. W. K. Andrews and D. Pollard: An introduction to functional central limit theorems for dependent stochastic processes. *International Statistical Review* **62** (1994), 119–132.

[3] M. A. Arcones and B. Yu: Central limit for empirical and U-processes of stationary mixing sequences. *Journal of Theoretical Probability* **7** (1994), 47–71.

[4] R. F. Bass: Law of the iterated logarithm for set-indexed partial sum processes with finite variance. *Zeitschrift für Wahrscheinlichkeitstheorie und verwandte Gebiete* **70** (1985), 591–608.

[5] H. C. P. Berbee: *Random Walks with Stationary Increments and Renewal Theory.* Mathematical Centre Tracts **112**, Amsterdam, 1979.

[6] P. Billingsley: *Convergence of Probability Measures.* John Wiley, New York, 1968.

[7] J. Dedecker: A central limit theorem for stationary random fields. *Probability Theory and Related Fields* **110** (1998), 397–426.

[8] J. Dedecker: Exponential inequalities and functional central limit theorems for random fields. *ESAIM, Probabilités et Statistique* **5** (2001), 77–104.

[9] M. D. Donsker: Justification and extension of Doob's heuristic approach to the Kolmogorov–Smirnov theorems. *Annals of Mathematical Statistics* **23** (1952), 277–281.

[10] P. Doukhan, J. Léon and F. Portal: Principe d'invariance faible pour la mesure empirique d'une suite de variables aléatoires dépendantes. *Probability Theory and Related Fields* **76** (1987), 51–70.

[11] P. Doukhan, P. Massart and E. Rio: Invariance principles for absolutely regular empirical processes. *Annales de l'Institut Henri Poincaré Probabilités et Statistiques* **31** (1995), 393–427.

[12] R. M. Dudley: Weak convergence of measures on nonseparable metric spaces and empirical measures on Euclidean spaces. *Illinois Journal of Mathematics* **10** (1966), 109–126.

[13] R. M. Dudley: Central limit theorems for empirical measures. *Annals of Probability* **6** (1978), 899–929.

[14] S. Goldstein: Maximal coupling. *Zeitschrift für Wahrscheinlichkeitstheorie und verwandte Gebiete* **46** (1979), 193–204.

[15] I. A. Ibragimov: Some limit theorems for stationary processes. *Theory of Probability and Its Applications* **7** (1962), 349–382.

[16] P. Massart: Invariance principle for empirical processes: The weakly dependent case, Quelques problèmes de vitesse de convergence pour des mesures empiriques. *Thèse d'état, Université de Paris-Sud,* 1987.

[17] Y. Nishiyama: Weak convergence of some classes of martingales with jumps. *Annals of Probability* **28** (2000), 685–712.

[18] M. Ossiander: A central limit theorem under metric entropy with $\mathbb{L}^2$-bracketing. *Annals of Probability* **15** (1987), 897–919.

[19] W. Phillip: Invariance principles for sums of mixing random elements and the multivariate empirical process. *Colloquia Mathematica Societatis János Bolyai* **36** (1982), 843–873.

[20] I. F. Pinelis: Optimum bounds for the distribution of martingales in Banach spaces. *Annals of Probability* **22** (1994), 1679–1706.

[21] D. Pollard: A central limit theorem for empirical processes. *Journal of the Australian Mathematical Society A* **33** (1982), 235–248.

[22] D. Pollard: *Empirical Processes: Theory and Applications. NSF-CBMS Regional Conference Series in Probabity and Statistics* **2**. Institute of Mathematical Statistics and American Statistical Association, 1990.

[23] E. Rio: Covariance inequalities for strongly mixing processes. *Annales de l'Institut Henri Poincaré Probabilités et Statistiques* **29** (1993), 587–597.

[24] E. Rio: Processus empiriques absolument réguliers et entropie universelle. *Probability Theory and Related Fields* **111** (1998), 585–608.

[25] E. Rio: *Théorie asymptotique des processus aléatoires faiblement dépendants. Collection Mathématiques & Applications* **31**. Springer, Berlin, 2000.

[26] Y. A. Rozanov and V. A. Volkonskii: Some limit theorems for random functions I. *Theory of Probability and Its Applications* **4** (1959), 178–197.

[27] A. W. van der Vaart and J. A. Wellner: *Weak Convergence and Empirical Processes.* Springer, Berlin, 1996.

Jérôme Dedecker
L.S.T.A. Université Pierre et Marie Curie
4 Place Jussieu
F-75252 Paris cedex 05, France
dedecker@ccr.jussieu.fr

Sana Louhichi
Laboratoire de Probabilités, Statistique et modélisation
Université de Paris-Sud
Bâtiment 425
F-91405 Orsay cedex, France
sana.louhichi@math.u-psud.fr

On Hoeffding's Inequality for Dependent Random Variables

SARA A. VAN DE GEER

ABSTRACT. Let $\{Z_\theta : \theta \in \Theta\}$ be a random process indexed by a parameter θ in some (metric) space Θ. Given an appropriate moment or probability inequality for each fixed θ (a *pointwise* inequality), one can often derive an inequality that holds *uniformly* in $\theta \in \Theta$ by applying the chaining technique. Therefore, pointwise inequalities are (apart from being of intrinsic interest) quite relevant within the theory of stochastic processes. We present a generalization of Hoeffding's inequality, and the related bounded difference inequality of McDiarmid [7]. We also state the corresponding uniform inequality. As an application, we consider estimation in the autoregression model.

1 Introduction

For a stochastic process $\{Z_\theta : \theta \in \Theta\}$, one is often interested in moment or probability inequalities, that hold *uniformly* in $\theta \in \Theta$. For instance, in some statistical applications, one may want to apply an inequality with θ randomly chosen (an estimator, say), which is possible indeed when the inequality is uniform. Uniform inequalities are in quite a few cases an immediate consequence of *pointwise* inequalities, that is, inequalities that hold for all $\theta \in \Theta$ fixed.

In Theorem 1.2 below we briefly review a result given in van der Vaart and Wellner [11] (their Corollary 2.2.5). This theorem serves as a major illustration that for many situations, it is enough to prove a pointwise inequality. The theorem is followed by a short discussion. We then establish in Section 2 a pointwise Hoeffding type inequality. The extension to a uniform inequality is presented in Section 3. Section 4 is a statistical application, namely on proving rates of convergence in autoregression.

The inequalities of Theorem 1.2 concern the Orlicz norm of the random variables involved, which is defined as follows.

Definition 1.1. *Let ψ be a convex, nondecreasing, nonzero function on $[0, \infty)$, with $\psi(0) = 0$ (an Orlicz function). The Orlicz norm of the random variable X is defined as*

$$\|X\|_\psi = \inf\{K > 0 : \mathbf{E}\psi(|X|/K) \leq 1\},$$

(with the usual conventions if the expectation is infinite).

For example, when $\psi(x) = x^p$, with $p \geq 1$, the Orlicz norm $\|X\|_\psi$ is equal to the L_p norm $(E|X|^p)^{1/p}$.

Condition (1.1) below is a pointwise inequality for the Orlicz norm (albeit that there are two parameters θ and ϑ involved), and (1.2) is the resulting uniform inequality.

Theorem 1.2. *Let ψ be an Orlicz function satisfying* $\limsup_{x,y\to\infty} \psi(x)\psi(y)/\psi(cxy) < \infty$, *for some constant c. Suppose $\{Z_\theta : \theta \in \Theta\}$ is a separable stochastic process indexed by θ in the (pseudo-)metric space (Θ, τ). Assume that*

$$\|Z_\theta - Z_\vartheta\|_\psi \le C\tau(\theta, \vartheta), \text{ for every } \theta, \vartheta, \tag{1.1}$$

where C is some constant. Then there exists a constant C' depending only on ψ and C, such that

$$\| \sup_{\theta,\vartheta} |Z_\theta - Z_\vartheta| \|_\psi \le C' \int_0^{\mathrm{diam}(\Theta)} \psi^{-1}(D(\delta))d\delta, \tag{1.2}$$

where $\mathrm{diam}(\Theta)$ *is the diameter of Θ and $D(\delta)$ is the δ-packing number of Θ (i.e., the maximum number of δ-separated points in Θ).*

When ψ is the L_p norm, the theorem may be applied to the case where Θ is a bounded subset of d-dimensional Euclidean space, with $d < p$. But for very large (infinite dimensional) Θ, Theorem 1.2 needs more than an L_p norm.

An important special case is

$$\psi(x) = \exp[x^2] - 1. \tag{1.3}$$

A random variable with finite Orlicz norm for this choice of ψ is called sub-Gaussian, because its tails behave like those of a Gaussian random variable, or are even slimmer. Theorem 1.2 with ψ given in (1.3) gives the uniform bound in terms of the square root of the log-packing number (or the *entropy*) of Θ. The resulting uniform bound can be used to prove, e.g., tightness of empirical processes based on independent observations, (using symmetrization techniques) or equicontinuity of certain Gaussian processes indexed by functions.

One may also consider

$$\psi(x) = \exp[x] - 1. \tag{1.4}$$

A random variable with finite Orlicz norm for this choice of ψ is called subexponential, because its tails decrease at least exponentially fast. In Theorem 1.2, the uniform bound in the subexponential case is again in terms of the log-packing number, but the square root which appeared in the sub-Gaussian case is now lost.

We remark however that in special cases one can prove that a bound of the form (1.2) holds in the subexponential case, with on the left-hand side ψ given by (1.4) and on the right-hand side a different ψ, namely the one given in (1.3). This is for example true when $\{Z_\theta\}$ is the empirical process based on i.i.d. observations with distribution P. One may then take Θ as a subset of $L_2(P)$, in which case the log-packing number(entropy), should be replaced by the stronger entropy with bracketing. The result is closely intertwined with Bernstein's inequality, which is roughly speaking sub-Gaussian in the near tails, and subexponential in the far tails.

Theorem 1.2 can be applied to sums of random variables, say $Z_\theta = \sum_{i=1}^n X_{i,\theta}$. It is then not necessary to assume independence of the $X_{i,\theta}$, as long as (1.1) is met. Also more complicated objects based on dependent random variables can be studied with the help of Theorem 1.2. See for example the paper of Dahlhaus and Polonik [4] (in this volume), where the empirical spectral process is examined.

In this paper, we consider an analogue of the sub-Gaussian case for dependent variables. Related results for the subexponential case have been treated in Nishiyama [8], and van de Geer [9].

Our main result is the derivation of a *pointwise* inequality, which will be a generalization of Hoeffding's inequality (and which will be further generalized to a bounded difference inequality; cf. [7]). The extension to a uniform inequality for weighted sums is straightforward. This extension makes use of a partitioning entropy condition, which relies on a generalization of the usual entropy concept.

2 Hoeffding's inequality and related results

Consider a probability triple $(\Omega, \mathcal{F}, \mathbf{P})$ and let $\emptyset = \mathcal{F}_0 \subset \mathcal{F}_1, \subset \cdots \subset \mathcal{F}$ be an increasing sequence of sigma-algebras. Let for each i, X_i be a real-valued $\mathcal{F}_i$-measurable random variable, satisfying

$$\mathbf{E}(X_i | \mathcal{F}_{i-1}) = 0 \quad \text{a.s.}$$

We will consider the martingale

$$S_n = \sum_{i=1}^{n} X_i, \ n \geq 1.$$

Throughout the rest of this paper, we let ψ be the Orlicz function

$$\psi(x) = \exp[x^2] - 1, \ x \geq 0.$$

Consider $\mathcal{F}_{i-1}$ measurable random variables $K_i > 0$, $i = 1, 2, \ldots$. Define $B_0^2 = 0$ and for $n \geq 1$,

$$B_n^2 = \sum_{i=1}^{n} K_i^2 \left(1 + \mathbf{E}\left(\psi\left(\frac{|X_i|}{K_i}\right) | \mathcal{F}_{i-1}\right)\right).$$

Lemma 2.1. *For all* β,

$$\{\xi_n(\beta) = \exp[\beta S_n - 2\beta^2 B_n^2]\}$$

is a supermartingale.

Proof. Let us use the notation

$$\tau_i^2 = K_i^2 \left(1 + \mathbf{E}\left(\psi\left(\frac{|X_i|}{K_i}\right) | \mathcal{F}_{i-1}\right)\right), \quad i \geq 1.$$

Then, by a slight extension of Lemma 8.1 in van de Geer [10], we have for all $i \geq 1$,

$$\mathbf{E}\left(\exp[\beta X_i - 2\beta^2 \tau_i^2] | \mathcal{F}_{i-1}\right) \leq 1 \quad \text{a.s.}$$

Thus,

$$\mathbf{E}(\exp[\beta S_n - 2\beta^2 B_n^2] | \mathcal{F}_{i-1}) \leq \exp[\beta S_{n-1} - 2\beta^2 B_{n-1}^2] \quad \text{a.s.,} \quad n \geq 1.$$

□

Theorem 2.2. *For all* $a > 0$, $b > 0$,

$$\mathbf{P}(S_n \geq a \text{ and } B_n^2 \leq b^2 \text{ for some } n) \leq \exp\left[-\frac{a^2}{8b^2}\right].$$

Proof. The argument is as in Freedman [5]. Let $\{\xi_n(\beta)\}$ be defined as in Lemma 2.1. Then, for any stopping time σ,

$$\mathbf{E}(\xi_\sigma(\beta)\{\sigma < \infty\}) \leq 1.$$

Take A as the set

$$A = \{S_n \geq a \text{ and } B_n^2 \leq b^2 \text{ for some } n\},$$

and take

$$\sigma = \inf\{n : S_n \geq a\}.$$

Because $A \subset \{\sigma < \infty\}$,

$$\mathbf{E}\xi_\sigma(\beta)1_A \leq 1.$$

On A,

$$\xi_\sigma(\beta) \geq \exp[\beta a - 2\beta^2 b^2].$$

So

$$\mathbf{P}(A) \leq \exp[-\beta a + 2\beta^2 b^2].$$

Now, take $\beta = a/(4b^2)$. □

An important special case is the one where S_n is a weighted sum of martingale differences, with predictable weights.

Corollary 2.3. *Let W_i be $\mathcal{F}_i$-measurable, and $\mathbf{E}(W_i|\mathcal{F}_{i-1}) = 0$, $i \geq 1$. Suppose that for some constant $c < \infty$,*

$$\mathbf{E}\left(\psi\left(\frac{|W_i|}{c}\right)|\mathcal{F}_{i-1}\right) \leq 1 \quad a.s., \quad i = 1, 2, \ldots.$$

Let g_i be $\mathcal{F}_{i-1}$-measurable weights, and take $X_i = g_i W_i$, $i = 1, 2, \ldots$. It follows that for all $a > 0$, $d > 0$,

$$\mathbf{P}\left(\sum_{i=1}^{n} g_i W_i \geq a \text{ and } \sum_{i=1}^{n} g_i^2 \leq d^2 \text{ for some } n\right) \leq \exp\left[-\frac{a^2}{16c^2d^2}\right]. \tag{2.1}$$

Clearly, Theorem 2.2 can also be applied to the case where the X_i are bounded. In that case, one arrives at a Hoeffding-type of inequality. However, the constants that appear in the exponential bound are then larger than those of Hoeffding [6]. We therefore present the improvement with Hoeffding's constants for the bounded case.

Lemma 2.4. *Suppose that*

$$L_i \leq X_i \leq U_i \quad a.s. \quad \text{for all } i \geq 1,$$

where $L_i < U_i$ are $\mathcal{F}_{i-1}$-measurable random variables, $i \geq 1$. Define $C_0^2 = 0$ and

$$C_n^2 = \sum_{i=1}^{n} (U_i - L_i)^2, \; n \geq 1.$$

Then for all β,

$$\{\zeta_n(\beta) = \exp[\beta S_n - \beta^2 C_n^2/8]\}$$

is a supermartingale.

Proof. This follows from a slight extension of Hoeffding [6]:

$$\mathbf{E}(\exp[\beta X_i]|\mathcal{F}_{i-1}) \leq \exp[\beta^2(U_i - L_i)^2/8] \quad \text{a.s.}$$

□

Theorem 2.5 (Hoeffding's inequality). *Suppose the condition of Lemmma 2.4 holds. Let $\{C_n\}$ be defined as there. Then for all $a > 0$, $c > 0$,*

$$\mathbf{P}(S_n \geq a \text{ and } C_n^2 \leq c^2 \text{ for some } n) \leq \exp\left[-\frac{2a^2}{c^2}\right].$$

Proof. In view of Lemma 2.4, this can be derived using the same arguments as in Theorem 2.2. □

Indeed, Theorem 2.5 is the inequality of Hoeffding [6], except that Hoeffding assumes non-random bounds $\{L_i, U_i\}$. The inequality is also known as Azuma's inequality (Azuma [1]).

Theorem 2.5 considers the $\mathcal{F}_n$-measurable random variables $S_n = \sum_{i=1}^n X_i$. We now examine general $\mathcal{F}_n$-measurable random variables. We state the result for fixed n.

Theorem 2.6 (Bounded difference inequality). *Fix $n \geq 1$. Let Z_n be an $\mathcal{F}_n$-measurable random variable, satisfying for each $i = 1, \ldots, n$,*

$$L_i \leq \mathbf{E}(Z_n|\mathcal{F}_i) \leq U_i \quad a.s.$$

where $L_i < U_i$ are $\mathcal{F}_{i-1}$-measurable, $i = 1, \ldots, n$. Define $C_n^2 = \sum_{i=1}^n (U_i - L_i)^2$. Then for all $a > 0$, $c > 0$,

$$\mathbf{P}(Z_n - \mathbf{E}Z_n \geq a \text{ and } C_n^2 \leq c^2) \leq \exp\left[-\frac{2a^2}{c^2}\right].$$

Proof. We may write

$$Z_n - \mathbf{E}Z_n = \sum_{i=1}^n \tilde{X}_i,$$

where $\tilde{X}_i = \mathbf{E}(Z_n|\mathcal{F}_i) - \mathbf{E}(Z_n|\mathcal{F}_{i-1})$, $i = 1, \ldots, n$. Clearly, $\tilde{X}_i$ is $\mathcal{F}_i$-measurable, and $\mathbf{E}(\tilde{X}_i|\mathcal{F}_{i-1}) = 0$, $i = 1, \ldots, n$. Moreover, for $i = 1, \ldots, n$,

$$\tilde{L}_i = L_i - \mathbf{E}(Z_n|\mathcal{F}_{i-1}) \leq \tilde{X}_i \leq U_i - \mathbf{E}(Z_n|\mathcal{F}_{i-1}) = \tilde{U}_i,$$

so that $\tilde{L}_i < \tilde{U}_i$ are $\mathcal{F}_{i-1}$-measurable random variables, with $\tilde{U}_i - \tilde{L}_i = U_i - L_i$. The result is thus a consequence of Theorem 2.5. □

Theorem 2.6 generalizes the bounded difference inequality of McDiarmid [7]. As an illustration, consider independent random variables $Y_1, \ldots, Y_n$, with values in some measurable space $\mathcal{Y}$. Let $\mathcal{F}_i = \sigma(Y_1, \ldots, Y_i)$, $i = 1, \ldots, n$. Let $g(y_1, \ldots, y_n)$ be some real-valued function on $\mathcal{Y}^n$ satisfying the bounded difference assumption

$$\begin{aligned}&|g(y_1, \ldots, y_{i-1}, y_i, y_{i+1}, \ldots, y_n) - g(y_1, \ldots, y_{i-1}, y_i', y_{i+1}, \ldots, y_n)|\\ &\quad \leq k_i, \text{ for all } y_i, y_i', \text{ and all } i,\end{aligned}$$

where $k_1, \ldots, k_n$ are constants. Take $Z_n = g(Y_1, \ldots, Y_n)$. Then one can take

$$\begin{aligned} L_i &= \inf_y \mathbf{E}(g(Y_1, \ldots, Y_{i-1}, y, Y_{i+1}, Y_n)|\mathcal{F}_i), \\ U_i &= \sup_y \mathbf{E}(g(Y_1, \ldots, Y_{i-1}, y, Y_{i+1}, \ldots, Y_n|\mathcal{F}_i), \end{aligned}$$

which satisfy $U_i - L_i \leq k_i$, $i = 1, \ldots, n$. The bounded difference inequality now reads: for all $a > 0$,

$$\mathbf{P}(g(Y_1, \ldots, Y_n) - \mathbf{E}g(Y_1, \ldots, Y_n) \geq a) \leq \exp\left[-\frac{2a^2}{\sum_{i=1}^n k_i^2}\right].$$

In the case $\mathcal{Y} = \mathbf{R}$, and $g(Y_1, \ldots, Y_n) = \sum_{i=1}^n Y_i$, this is Hoeffding's inequality for independent random variables.

3 A uniform inequality for weighted sums

In this section, we present an analogue of Theorem 1.2, with Orlicz function $\psi(x) = \exp[x^2]-1$. Consider some space of parameters Θ. Instead of using a metric τ on Θ, we will define a random quantity $D_n(\theta, \vartheta)$ to measure the closeness of two parameters θ and ϑ (see below). Consider $\mathcal{F}_{i-1}$-measurable real-valued random variables $g_{i,\theta}$, $\theta \in \Theta$. Let W_i be real-valued $\mathcal{F}_i$-measurable, with $\mathbf{E}(W_i|\mathcal{F}_{i-1}) = 0$, $i = 1, 2, \ldots$. Assume, moreover, that for all i, and some constant c,

$$\mathbf{E}\left(\psi\left(\frac{|W_i|}{c}\right)|\mathcal{F}_{i-1}\right) \leq 1. \tag{3.1}$$

Write

$$D_n^2(\theta, \vartheta) = \frac{1}{n}\sum_{i=1}^n (g_{i,\theta} - g_{i,\vartheta})^2. \tag{3.2}$$

Take a fixed $\theta_0 \in \Theta$. Let $F \in \mathcal{F}$ be some measurable set. This will be the set where the partitioning entropy defined below is supposed to be well behaved. It depends on the particular application, but generally one wants that $\mathbf{P}(F^c)$, where F^c is the complement of F, is small.

The log-packing number in (1.2) will be replaced by a partitioning entropy.

Assumption 3.1 (The partitioning entropy condition). For $0 \leq \delta \leq d$, let $\{\theta_j\}_{j\in\mathcal{J}}$ be such that for all $\theta \in \Theta$, there is a $j = j(\theta) \in \mathcal{J}$ such that $D_n(\theta, \theta_j) \leq \delta$ on $\{D_n(\theta, \theta_0) \leq d\} \cap F$. We assume that $\mathcal{J}$ can be chosen as a finite set $\mathcal{J} = \{1, \ldots, J\}$ for each δ and d. We then write $N(\delta, d) = J$, and let $H(\delta, d)$ be a continuous majorant of $\log(1 + J)$. We call $H(\delta, d)$ a partitioning entropy.

Theorem 3.2. *Suppose that on F, $\sum_{i=1}^n W_i^2/n \leq \sigma^2$ a.s. There exists a constant c_0 depending only on c such that for*

$$\sqrt{n}a \geq c_0\left(\int_{\frac{a}{\sigma c_0}}^d H^{1/2}(\delta, d)d\delta \vee d\right), \tag{3.3}$$

we have

$$\mathbf{P}\left(\left(\frac{1}{n}\sum_{i=1}^{n}(g_{i,\theta}-g_{i,\theta_0})W_i \geq a \text{ and } D_n(\theta,\theta_0)\leq d \text{ for some } \theta\in\Theta\right)\cap F\right)$$
$$\leq c_0\exp\left[-\frac{na^2}{c_0^2d^2}\right].$$

Proof. This follows by copying the proof of Lemma 3.2 in van de Geer [10], with its pointwise sub-Gaussian probability inequality replaced by the extension (2.1) of Corollary 2.3. □

4 Application to autoregression

Now that Theorem 3.2 is available, one can start the engines for a theory on M-estimation in autoregression, possibly using sieves and/or penalties. The results are very similar to the independent case. The only additional technicality is the handling of random metrics. In this section, we will briefly present a sample of the theory.

Consider n real-valued observations $Y_1,\ldots,Y_n$, with Y_i $\mathcal{F}_i$-measurable, $i\geq 1$. Let Θ be a given parameter space, and suppose that

$$\mathbf{E}(Y_i|\mathcal{F}_{i-1}) = g_{i,\theta_0} \quad \text{a.s.,} \quad i\geq 1,$$

where $\theta_0\in\Theta$. Define $W_i = Y_i - g_{i,\theta_0}$, $i\geq 1$. We assume that $\{W_i\}$ satisfies the sub-Gaussianity condition (3.1).

We let $\hat{\theta}_n$ be the least squares estimator

$$\hat{\theta}_n = \arg\min_{\Theta}\sum_{i=1}^{n}(Y_i-g_{i,\theta})^2.$$

Let $D_n(\theta,\vartheta)$ be given in (3.2). Consider a partitioning entropy $H(\delta,d)$ as introduced in Assumption 3.1. A rate of convergence for $D_n(\hat{\theta}_n,\theta_0)$ can now be obtained in exactly the same way as in Theorem 9.1 of van de Geer [10]. To facilitate the exposition, we only consider the case where the entropy-integral exists.

Theorem 4.1. *Define*

$$\phi(d) = \int_0^d H^{1/2}(\delta,d)d\delta \vee d,$$

and suppose that $\phi(d)/d^2$ is a nonincreasing function of d. Then there exists a constant c_0, such that for all $d\geq d_n$, where $\sqrt{n}d_n^2\geq c_0\phi(d_n)$,

$$\mathbf{P}(D_n(\hat{\theta}_n,\theta_0)\geq d)\leq c_0\exp\left[-\frac{nd^2}{c_0^2}\right]+\mathbf{P}(F^c).$$

Example 4.2 (Linear autoregression). Take $\Theta=\mathbf{R}^r$ and

$$g_{i,\theta}=\theta_1\psi_{i,1}+\cdots+\theta_r\psi_{i,r},\ \theta\in\Theta,$$

where for each i, the variables $\psi_{i,1},\ldots,\psi_{i,r}$ are $\mathcal{F}_{i-1}$-measurable. Note that now

$$D_n^2(\theta,\vartheta)=(\theta-\vartheta)'\Sigma_n(\theta-\vartheta),$$

with Σ_n a random $r \times r$ matrix. We assume that the range of Σ_n is nonrandom, and let $\Sigma \geq 0$ be a symmetric nonrandom $r \times r$ matrix with the same range. Take

$$F \subset \left\{ \frac{1}{4} \leq \|\Sigma^{-1/2} \Sigma_n \Sigma^{-1/2}\| \leq 4 \right\},$$

where $\Sigma^{-1/2}$ is the square root of the (generalized) inverse of Σ, and where $\|\cdot\|$ denotes the norm of a matrix as linear map. On F, the random metric D_n is equivalent to the nonrandom metric D given by

$$D^2(\theta, \vartheta) = (\theta - \vartheta)' \Sigma (\theta - \vartheta).$$

Therefore (invoke, e.g., Corollary 2.6 of van de Geer [10]),

$$H(\delta, d) \leq r \log \left(\frac{20d}{\delta} \right).$$

It follows that

$$\phi(d) \leq A d \sqrt{r},$$

with $A = \int_0^1 \sqrt{\log(\frac{20}{u})} du$. Thus, for $c_1 = c_0 A$ a constant depending only on c, and for all $T \geq c_1$,

$$\mathbf{P}(D(\hat{\theta}_n, \theta_0) \geq T \sqrt{\frac{r}{n}}) \leq c_1 \exp \left[-\frac{r^2 T^2}{c_1^2} \right] + \mathbf{P}(F^c).$$

For ways of handling $\mathbf{P}(F^c)$, see Baraud, Comte and Viennet ([2]). In fact, they study the much more involved problem of adaptation.

Example 4.3 (Smooth regression). One can easily extend Theorem 4.1 to the case of penalized least squares. Let us briefly consider a simple example. Suppose that $Y_i \in [0, 1]$ for all i, and that $g_{i,\theta} = \theta(Y_{i-1})$, $i \geq 1$ (with $Y_0 = 1/2$ (say)). Let $m \in \{1, 2, \ldots\}$ be given, and let

$$\Theta = \{\theta : I(\theta) < \infty\},$$

where $I^2(\theta) = \int_0^1 |\theta^{(m)}(y)|^2 dy$. Consider the penalized least squares estimator

$$\hat{\theta}_n = \arg \min_{\theta \in \Theta} \left\{ \frac{1}{n} \sum_{i=1}^{n} (Y_i - \theta(Y_{i-1}))^2 + \text{pen}^2(\theta) \right\},$$

with penalty

$$\text{pen}^2(\theta) = \lambda_n^2 I^2(\theta).$$

Here $0 < \lambda_n^2 \leq 1$ is a smoothing parameter. The partitioning entropy $H(\delta, d)$ should now be taken on the set where also $\text{pen}(\theta) \leq d$. Birman and Solomjak [3] show that the δ-entropy for the uniform norm of a set of functions θ, bounded by 1 and with $I(\theta) \leq L$, is of order $(L/\delta)^{1/m}$. Because in our case, $\text{pen}(\theta) \leq d$ reads $I(\theta) \leq d/\lambda_n$, we find

$$H(\delta, d) \leq \text{const.} \left(\frac{d}{\lambda_n \delta} \right)^{1/m}.$$

The condition $\sqrt{n} d_n^2 \geq c_0 (\int_0^{d_n} H^{1/2}(\delta, d) d\delta \vee d_n)$ leads to $d_n \geq c_0' n^{-1/2} \lambda_n^{-1/2m}$. As a consequence, one finds the rate $n^{-1/2} \lambda_n^{-1/2m} \vee \lambda_n I(\theta_0)$ for $D_n(\hat{\theta}_n, \theta_0)$.

References

[1] K. Azuma: Weighted sums of certain dependent random variables. *Tokuku Mathematical Journal* **19** (1967), 357–367.

[2] Y. Baraud, F. Comte and G. Viennet: Adaptive estimation in autoregression or β-mixing regression via model selection. Technical Report 566, Laboratoire de Probabilités et Modèles Aléatoires, Universités de Paris 6 et Paris 7, 2000. Available under www.dma.ens.fr/~baraud. *The Annals of Statistics* **29** (2001), 839–875.

[3] M.Š. Birman and M.Z. Solomjak: Piece-wise polynomial approximations of functions in the classes W_p^α. *Mathematics of the USSR Sbornik* **73** (1967), 295–317.

[4] R. Dahlhaus and W. Polonik: Empirical spectral process and nonparametric maximum likelihood estimation for time series. In this volume: H. Dehling, T. Mikosch, and M. Sørensen (Eds.) *Empirical Process for Dependent Data.* Birkhäuser, Boston, 2002, 275–298.

[5] D. Freedman: On tail probabilities for martingales. *The Annals of Probability* **3** (1975), 100–118.

[6] W. Hoeffding: Probability inequalities for sums of bounded random variables. *Journal of the American Statistical Association* **58** (1963), 13–30.

[7] C. McDiarmid: On the method of bounded differences. In *Surveys in Combinatorics* 148–188. Cambridge University Press, Cambridge, 1989.

[8] Y. Nishiyama: Weak convergence of some classes of martingales with jumps. *The Annals of Probability* **28** (2000), 685–712.

[9] S.A. van de Geer: Exponential inequalities for martingales, with application to maximum likelihood estimation for counting processes. *The Annals of Statistics* **23** (1995), 1779–1801.

[10] S.A. van de Geer: *Empirical Processes in M-Estimation.* Cambridge University Press, Cambridge, 2000.

[11] A.W. van der Vaart and J.A. Wellner: *Weak Convergence and Empirical Processes.* Springer, New York, 1996.

Sara van de Geer
Mathematical Institute
University of Leiden
P.O. Box 9512
2300 RA Leiden, The Netherlands
geer@math.leidenuniv.nl

On the Coupling of Dependent Random Variables and Applications

FLORENCE MERLEVÈDE AND MAGDA PELIGRAD[1]

ABSTRACT. In this paper we survey some results and further investigate the coupling of a sequence of dependent random variables with an independent one having the same marginal distributions. The upper bound of the distance between the variables with the same rank is given in terms of mixing coefficients. We shall apply the coupling methods to derive uniform laws of large numbers for the dependent random processes under various types of dependence. We shall also discuss the importance of coupling for obtaining the central limit theorem for strongly mixing sequences.

1 Introduction

One of the most useful techniques in obtaining limit theorems for weakly dependent random variables is the coupling of the initial sequence with an independent one. Generally speaking a coupling theorem enables us to replace the initial dependent sequence by an independent one, having the same marginals. Moreover, a variable in the newly constructed sequence is independent of the past of the initial sequence and it is closed, in some sense, to the variable having the same rank. This approximation has to be carried out in a manner that permits limit theorems for the independent random variables to carry over directly to the dependent ones.

The idea of coupling is going back at least to Doeblin (1937) for the case of the Markov chains with a countable state space and further extended for more general Markov processes by Vasershtein (1969) and Griffeath (1975). The problem of coupling was also considered in ergodic theory, in the context of Bernoulli shifts by Ornstein (1974), and Shields (1973).

This method gained popularity for obtaining limit theorems for dependent random variables since the publication of two articles by Berkes and Philipp (1977/78, 1979). Their approach is based on estimates of the Prohorov distance and the Strassen–Dudley theorem (Strassen (1965), Dudley (1968)). This method was successfully exploited by many authors including Berkes and Philipp (1977/78), Philipp (1979), Kuelbs and Philipp (1980), Dehling and Philipp (1982), Berger (1982), Dabrowski (1982), Eberlein (1983), Dudley and Philipp (1983), Borovkova, Burton, and Dehling (1999).

In parallel, various approaches to coupling have been developed by different authors.

One method can be called the "partition method" and is similar to the one used in the ergodic theory setting. It was developed in papers by Goldstein (1979), Berbee (1979), Schwarz (1980), Bryc (1982), Bradley (1983) and Peligrad (2001-a). A variant of this method will be presented in the next section.

Another approach, developed first by the Hungarian school for deriving an approximation theorem for independent random variables, is the quantile transform. Major (1978) extended this

[1] Supported in part by a Taft Research Grant and University of Paris X.

construction to the dependent case introducing the conditional quantile transform. The method was further exploited by Rio (1995) and Peligrad (2001-b).

The paper is organized as follows. In Section 2, we shall present some coupling results based on the last two methods. Estimates of the distance between a variable in an original sequence and the corresponding one in the newly constructed one will be given in terms of mixing coefficients. Let us mention that the estimate given at the point (f) of Theorem 2.9 is new. Section 3 is devoted to the application of the coupling results to derive uniform laws of large numbers for dependent processes (some results in this section are new), whereas in Section 4, we show the importance of such coupling results for obtaining the central limit theorem for dependent sequences.

Throughout the paper, the notation $\mathbb{I}(A)$ means the indicator of A and $[.]$ the integer part as usual.

2 Some results on coupling

In this paper we shall use the following coefficients of dependence. Let $(\Omega, \mathcal{T}, P)$ be a probability space and let $\mathcal{A}$ and $\mathcal{B}$ be two σ-algebras of $\mathcal{T}$.

Define the strong mixing coefficient by

$$\alpha(\mathcal{A}, \mathcal{B}) = \sup_{A\in\mathcal{A}, B\in\mathcal{B}} |\mathbb{P}(A \cap B) - \mathbb{P}(A)\mathbb{P}(B)|,$$

the version λ of the maximal coefficient of correlation by

$$\lambda(\mathcal{A}, \mathcal{B}) = \sup_{A\in\mathcal{A}, B\in\mathcal{B}} |\mathbb{P}((A \cap B) - \mathbb{P}(A)\mathbb{P}(B)|/\mathbb{P}^{1/2}(B),$$

the φ mixing coefficient by

$$\varphi(\mathcal{A}, \mathcal{B}) = \sup_{A\in\mathcal{A}, B\in\mathcal{B}} |\mathbb{P}(A \cap B) - \mathbb{P}(A)\mathbb{P}(B)|/\mathbb{P}(B)$$

and the absolute regular coefficient by

$$\beta(\mathcal{A}, \mathcal{B}) = \frac{1}{2} \sup \sum_{i=1}^{I} \sum_{j=1}^{J} |\mathbb{P}(A_i \cap B_j) - \mathbb{P}(A_i)\mathbb{P}(B_j)|,$$

where this latter sup is taken over all pairs of partitions $\{A_1, \ldots, A_I\}$ and $\{B_1, \ldots, B_J\}$ of Ω such that $A_i \in \mathcal{A}$ and each $B_j \in \mathcal{B}$.

Obviously $\alpha(\mathcal{A}, \mathcal{B}) \leq \lambda(\mathcal{A}, \mathcal{B}) \leq \varphi(\mathcal{A}, \mathcal{B})$ and $2\alpha(\mathcal{A}, \mathcal{B}) \leq \beta(\mathcal{A}, \mathcal{B}) \leq \varphi(\mathcal{A}, \mathcal{B})$.

Throughout the paper we shall make use of the short notation $\gamma(X, Y)$ to denote $\gamma(\sigma(X), \sigma(Y))$, γ being one of the dependence coefficients.

We shall now recall a well-known coupling result which characterizes the coefficient β of absolutely regularity.

Theorem 2.1. *Let X and Y be random variables defined on $(\Omega, \mathcal{T}, \mathbb{P})$ with values in a Polish space S. Let $\sigma(X)$ be a σ-algebra generated by X and U be a random variable uniformly distributed on $[0, 1]$ independent of (X, Y). Then there exists a random variable Y^* measurable with respect to $\sigma(X) \vee \sigma(Y) \vee \sigma(U)$, independent of X and distributed as Y, and such that*

$$\mathbb{P}(Y \neq Y^*) = \beta(X, Y).$$

This theorem is frequently named Berbee's Lemma (1979). Related results can be found in works by Goldstein (1979), Berkes and Philipp (1979), Schwarz (1980), Bryc (1982), Dehling and Philipp (1982) and, in the context of ergodic theory in Ornstein (1974) and Shields (1973). In all these papers, the constructions of the variable Y^* are equivalent. We shall sketch here the construction by Bryc (1982) which is versatile and leads to further interesting estimates for the distance between Y and Y^*.

Because the variables are S-valued, by completeness, according to the result in Bryc (1982), we can reduce the problem to the case of discrete S-valued random variables. Leaving the technical aspect apart, the main step of the construction is to define the following probabilistic structure:

With the notation $e_{x,y} := \mathbb{P}(X = x, Y = y) - \mathbb{P}(X = x)\mathbb{P}(Y = y)$, we have

$$\mathbb{P}(X = x, Y = y, Y^* = z) = \begin{cases} \dfrac{e^+_{x,y} e^-_{x,z}}{\sum_t e^+_{x,t}} & \text{if } z \neq y, \\ \mathbb{P}(X = x)\mathbb{P}(Y = y) & \text{if } z = y \text{ and } e_{x,y} \geq 0, \\ \mathbb{P}(X = x, Y = y) & \text{if } z = y \text{ and } e_{x,y} < 0. \end{cases}$$

It is easy to verify that with this construction, we have

1. Y^* distributed as Y,
2. Y^* independent of X,

and for all A, B and C Borel sets of S,

3. $\mathbb{P}(X \in A, Y \in B, Y \neq Y^*) = \sum_{x \in A} \sum_{y \in B} [\mathbb{P}(X = x, Y = y) - \mathbb{P}(X = x)\mathbb{P}(Y = y)]^+$,
4. $\mathbb{P}(X \in A, Y^* \in C, Y \neq Y^*) = \sum_{x \in A} \sum_{y \in C} [\mathbb{P}(X = x, Y = y) - \mathbb{P}(X = x)\mathbb{P}(Y = y)]^-$.

By this construction, one can easily obtain the following proposition.

Proposition 2.2. *Let Y be a random variable defined on $(\Omega, \mathcal{T}, \mathbb{P})$ with values in a Polish space (S, d). Let p be a positive number and s in S. Let Y^* be constructed as in Theorem 2.1. Then, for all $x \geq 0$,*

(i) $\mathbb{P}(Y \neq Y^* | \sigma(X)) \leq \sup_{B \in \sigma(Y)} |\mathbb{P}(B|\sigma(X)) - \mathbb{P}(B)|$ *a.s.*,

(ii) $\mathbb{P}(d(Y, Y^*) \geq x) \leq 2\varphi(X, Y)\mathbb{P}(d(Y, s) \geq x/2)$,

and if we assume that $d^p(Y, s)$ is integrable, then

(iii) $\mathbb{E}(d^p(Y, Y^*)) \leq 2^p \varphi(X, Y)\mathbb{E}(d^p(Y, s))$.

Point (ii) comes from Bryc (1982). Point (i) is probably well known (a proof can be found in Peligrad (2001-a)). Point (iii) is obtained by integrating with respect to x in (ii).

We shall establish now a new result for estimating the expectation of the distance between Y and Y^* in terms of the absolutely regular coefficient. Before stating it, we need to introduce the following definition: For any nonnegative random variable W, define the "upper tail" quantile function via

$$Q_W(u) = \inf\{t \geq 0 : \mathbb{P}(W > t) \leq u\}.$$

With this notation, we have the following.

Proposition 2.3. *Let X and Y be two random variables defined on $(\Omega, \mathcal{T}, \mathbb{P})$ with values in a Polish space S. Let p be a positive number and s in S and assume that $d^p(Y,s)$ is integrable. Let Y^* be constructed as in Theorem* 2.1. *Then for every positive p,*

$$\mathbb{E}\left(d^p(Y,Y^*)\right) \leq 2^{p+2} \int_0^{\beta(X,Y)} Q_{d^p(Y,s)}(u)du.$$

Remark 2.4. In the case when $p \leq 2$, with another approach the constant is sharper, namely, 2^{2p}.

Remark 2.5.

1. If $\mathbb{E}(d^{p+\delta}(Y,s)) < +\infty$ for a $\delta > 0$, then

$$\mathbb{E}(d^p(Y,Y^*)) \leq 2^{p+2}\beta^{\frac{\delta}{p+\delta}}(X,Y)(\mathbb{E}(d^{p+\delta}(Y,s)))^{\frac{p}{p+\delta}}.$$

2. Since

$$\begin{aligned}\mathbb{E}(d^p(Y,s)\mathbb{I}(d(Y,s) > Q_{d(Y,s)}(\beta(X,Y)))) &\leq \int_0^{\beta(X,Y)} Q_{d^p(Y,s)}(u)du \\ &\leq \mathbb{E}(d^p(Y,s)\mathbb{I}(d(Y,s) \geq Q_{d(Y,s)}(\beta(X,Y)))),\end{aligned}$$

we get

$$\mathbb{E}(d^p(Y,Y^*)) \leq 2^{p+2}\mathbb{E}(d^p(Y,s)\mathbb{I}(d(Y,s) \geq Q_{d(Y,s)}(\beta(X,Y)))).$$

Proof of Proposition 2.3. Set $T := Q_{d(Y,s)}(\beta(X,Y))$ and notice that the triangle inequality and the fact that Y and Y^* have the same distribution, yield, for all $s \in S$,

$$\begin{aligned}\mathbb{E}(d^p(Y,Y^*)) &= \mathbb{E}(d^p(Y,Y^*)\mathbb{I}(Y \neq Y^*)) \\ &\leq \mathbb{E}\{(d(Y,s) + d(Y^*,s))^p\mathbb{I}(Y \neq Y^*)\} \\ &\leq 2^p\mathbb{E}(d^p(Y,s)\mathbb{I}(Y \neq Y^*)) + 2^p\mathbb{E}(d^p(Y^*,s)\mathbb{I}(Y \neq Y^*)) \\ &\leq 2^p\mathbb{E}(d^p(Y,s)\mathbb{I}(d(Y,s) \leq T)\mathbb{I}(Y \neq Y^*)) + 2^p\mathbb{E}(d^p(Y,s)\mathbb{I}(d(Y,s) > T)) \\ &\quad + 2^p\mathbb{E}(d^p(Y^*,s)\mathbb{I}(d(Y^*,s) \leq T)\mathbb{I}(Y \neq Y^*)) + 2^p\mathbb{E}(d^p(Y^*s)\mathbb{I}(d(Y^*s) > T)) \\ &:= I_1 + I_2 + I_1^* + I_2^*.\end{aligned}$$

First we use Theorem 2.1, which yields

$$I_1 \leq 2^pT^p\beta(X,Y).$$

Now by using the equality: for all positive u

$$Q^p_{d(Y,s)}(u) = Q_{d^p(Y,s)}(u) \tag{2.1}$$

combined with the fact that $Q_{d^p(Y,s)}(u)$ is a nonincreasing function, we get

$$I_1 \leq 2^p \int_0^{\beta(X,Y)} Q_{d^p(Y,s)}(u)du. \tag{2.2}$$

In order to treat I_2, let us mention some results. Recall first that if U is a random variable uniformly distributed on the interval $[0,1]$ and if W is a nonnegative random variable, then the r.v. $Q_W(U)$ has the same distribution as the r.v. W, and then

$$\mathbb{E}(W) = \int_0^1 Q_W(u)du.$$

This last equality applied to the r.v. $d^p(Y,s)\mathbb{I}(d(Y,s) > T)$ entails that

$$\mathbb{E}\left(d^p(Y,s)\mathbb{I}(d(Y,s) > T)\right) = \int_0^1 Q_{d^p(Y,s)\mathbb{I}(d(Y,s)>T)}(u)du.$$

We also have

$$Q_{d^p(Y,s)\mathbb{I}(d(Y,s)>T)}(u) \leq \begin{cases} Q_{d^p(Y,s)}(u) & \text{if } u < \beta(X,Y) \\ 0 & \text{if } u \geq \beta(X,Y). \end{cases}$$

From these last considerations we derive that

$$I_2 \leq 2^p \int_0^{\beta(X,Y)} Q_{d^p(Y,s)}(u)du. \tag{2.3}$$

Gathering (2.2) and (2.3) and since similar bounds are valid for I_1^* and I_2^*, one gets the result. □

Let us mention now the following result of Bradley (1983): "For two σ-fields $\mathcal{A}$ and $\mathcal{B}$ such that one of them (say $\mathcal{A}$) is completely atomic with exactly N atoms, one has $\beta(\mathcal{A},\mathcal{B}) \leq N\alpha(\mathcal{A},\mathcal{B})$." By using this result, we can easily see that some coupling results involving the strong mixing coefficient can be obtained for Y taking its values in S and having a finite number of values, and the result will strongly depend on N. To derive more general strong approximation theorems for random variables taking their values in a Polish space, under the strong mixing condition, we would like to mention the two following results due to Bradley (1983).

Theorem 2.6. *Suppose Y is a random variable, defined on a probability space $(\Omega, \mathcal{T}, \mathbb{P})$ and taking its values in a Polish space $(S, \mathcal{S})$. Suppose $\mathcal{A}$ is a σ-field $\subset \mathcal{T}$. Suppose U is a uniform-*[0, 1] *random variable that is independent of the σ-field $\mathcal{A} \vee \sigma(Y)$.*

Suppose N is a positive integer, and $\mathcal{H} := \{H_1, H_2, \ldots, H_N\}$ is a partition of S, with $H_i \in \mathcal{S}$ for all $i = 1, 2, \ldots, N$. Then there exists a random variable Y^* which takes its values in $(S, \mathcal{S})$ and is measurable with respect to the σ-field $\mathcal{A} \vee \sigma(Y) \vee \sigma(U)$, such that Y^* has the following three properties:

(a) *Y^* is independent of the σ-field $\mathcal{A}$.*

(b) *Y^* has the same distribution (on $(S, \mathcal{S})$) as the random variable Y.*

(c) *One has that*

$$\mathbb{P}(Y^* \text{ and } Y \text{ are not elements of the same } H_i \in \mathcal{H}) \leq (8N)^{1/2}\alpha(\mathcal{A}, Y).$$

This theorem is taken from Bradley (1983, Theorem 2), and we can mention that it sharpens "codification" of arguments from the proofs of earlier strong approximation theorems in Berkes and Philipp (1977/78, 1979) and Bryc (1981).

The next theorem and comments that follow are due to Bradley (2001) in a personal communication.

Theorem 2.7. *Suppose (S, d) is a Polish space. Let $\mathcal{S}$ denote the σ-field on S generated by the open balls in the metric d.*

Suppose Y is a random variable, defined on a probability space $(\Omega, \mathcal{T}, \mathbb{P})$ and taking its values in $(S, \mathcal{S})$. Suppose $\mathcal{A}$ is a σ-field $\subset \mathcal{F}$. Suppose U is a uniform-[0, 1] *random variable which is independent of $\mathcal{A} \vee \sigma(Y)$.*

Suppose further that $\varepsilon > 0$, $\delta > 0$, N *is a positive integer,* $D \in \mathcal{S}$, $\mathbb{P}(X \in D) \geq 1 - \delta$, *and that there exist points* $a_1, a_2, \ldots, a_N \in D$ *such that* $\forall a \in D$, $\exists k \in \{1, 2, \ldots, N\}$ *satisfying* $d(a_k, a) \leq \varepsilon$.

Then there exists a random variable Y^* *which takes its values in* $(S, \mathcal{S})$ *and is measurable with respect to the* σ*-field* $\mathcal{A} \vee \sigma(Y) \vee \sigma(U)$, *such that* Y^* *has the following three properties:*

(a) Y^* *is independent of the* σ*-field* $\mathcal{A}$.

(b) Y^* *has the same distribution as the random variable* Y.

(c) $\mathbb{P}(d(Y, Y^*) > 2\varepsilon) \leq \delta + (8(N+1))^{1/2}\alpha(\mathcal{A}, Y)$.

This theorem evolved through several strong approximation theorems:

1. Berkes and Philipp (1977/78, Theorem 2 (special case)) involving the measure of dependence $\alpha(.,.)$ and random variables taking their values in $\mathbb{Z}^d$ for positive integers d.
2. Berkes and Philipp (1979, Theorem 2), involving the measure of dependence $\varphi(.,.)$ and random variables taking their values in complete separable metric spaces. The "structure" of the statement and proof of that result is in a certain sense imitated by the statement and proof of Theorem 2.7.
3. Bryc (1981), involving the measure of dependence $\alpha(.,.)$ and real-valued random variables.
4. Bradley (1983, Theorem 3), involving $\alpha(.,.)$ and real-valued random variables.
5. Novak (2000, Lemma 4), involving $\alpha(.,.)$ and random variables taking their values in $\mathbb{Z}^d$ for positive integers d.

In the context when Y is a real random variable, we would like to mention the following useful statement which is a corollary of Theorem 2.7 (see Theorem 3 of Bradley (1983)).

Theorem 2.8. *Suppose* (S, d) *is a Polish space. Suppose* X *and* Y *are random variables taking their values on* S *and* $\mathbb{R}$, *respectively. Suppose* U *is a uniform-*$[0, 1]$ *random variable which is independent of* (X, Y)*; and suppose* q *and* γ *are positive numbers such that* $q \leq \|Y\|_\gamma < \infty$. *Then there exists a random variable* Y^* *which takes its values in* $(S, \mathcal{S})$ *and is measurable with respect to the* σ*-field* $\sigma(X) \vee \sigma(Y) \vee \sigma(U)$, *such that* Y^* *has the following three properties:*

(a) Y^* *is independent of* X.

(b) Y^* *has the same distribution as the random variable* Y.

(c) $\mathbb{P}(|Y^* - Y| \geq q) \leq 18(\|Y\|_\gamma/q)^{\gamma/(2\gamma+1)}(\alpha(X, Y))^{2\gamma/(2\gamma+1)}$.

This result has been widely used in the case when Y is bounded (i.e., $\gamma = +\infty$) for instance in works by Bosq (1998) and Guillou and Merlevède (2001).

The next theorem gathers a variety of coupling results for S-valued sequences. The upper bounds for the discrepancy between the variables with the same rank are given in terms of mixing coefficients. The points (a), (b), (d), and (e) are proven in Bryc (1982). The point (c) is proven in Peligrad (2001-a). The point (f) is new.

Theorem 2.9. *Let* $(X_n, n \geq 1)$ *be a sequence of random variables with values in a Polish space* (S, d). *Denote by* $\mathcal{F}_1^n = \sigma(X_1, \ldots, X_n)$. *Then, we can redefine* $\{X_n, n \geq 1\}$ *onto a richer probability space (which supports a sequence* $(\delta_n, n \geq 1)$ *of independent uniformly* $[0, 1]$*-distributed random variables and independent of* $\{X_n, n \geq 1\}$*) together with a sequence* $\{X_n^*, n \geq 1\}$ *of independent random variables such that, for each* $1 \leq m \leq n$, *we have*

(a) X_m^* *is independent of* $\mathcal{F}_1^{m-1}$,

(b) X_m^* *has the same distribution as* X_m,

(c) $\mathbb{P}(X_m \neq X_m^* | \mathcal{F}_1^{m-1}) \leq \sup_{B \in \sigma(X_m)} |\mathbb{P}(X_m \in B|\mathcal{F}_1^{m-1}) - \mathbb{P}(X_m \in B)|$ *a.s.*,

(d) $\mathbb{P}(X_m \neq X_m^*) = \beta(\mathcal{F}_1^{m-1}, \sigma(X_m))$,

(e) $\mathbb{P}(d(X_m, X_m^*) \geq x) \leq 2\varphi(\mathcal{F}_1^{m-1}, \sigma(X_m))\mathbb{P}(d(X_m, s) \geq x/2)$ *for all* $x \geq 0$,

and if $\mathbb{E}\,(d^p(X_m, s)) < \infty$ *for a positive* p *and* $s \in S$,

(f) $\mathbb{E}(d^p(X_m, X_m^*)) \leq 2^{p+2} \int_0^{\beta(\mathcal{F}_1^{m-1}, \sigma(X_m))} Q_{d^p(X_m, s)}(u)du.$

Remark 2.10. Of course, now the following question arises: "Does the point (f) of the above-mentioned approximation theorem also hold in the context of strongly mixing sequences with values in an infinite dimensional space?"

Following Dehling (1983), it turns out that the answer is negative without any additional assumptions. In his paper he has constructed an example of strongly mixing sequences of ℓ^2-valued random variables X_n with values in an infinite dimensional space, which cannot be approximated by independent random variables X_n^* in such a way that $X_n - X_n^*$ converges to zero in probability.

By integrating (c) of Theorem 2.9 on the set $\bigcap_{i=1}^m (X_i \neq X_i^*)$, we easily deduce by an iterative procedure the following useful corollary which estimates the joint distributions of discrepancies between $(X_i)_{i\geq 1}$ and $(X_i^*)_{i\geq 1}$. Before stating it we need to introduce the following coefficient of dependence: let $(X_n, n \geq 1)$ be a sequence of random variables with values in a Polish space S. For $m \geq n$ we denote by $\mathcal{F}_n^m$ the sigma-algebra generated by the variables $(X_n, X_{n+1}, \ldots, X_m)$ and set

$$\tilde{\varphi}_k = \sup_{m\geq 1} \sup\{|\mathbb{P}(B|A) - \mathbb{P}(B)|; \mathbb{P}(A) \neq 0,\, A \in \mathcal{F}_1^m,\, B \in \mathcal{F}_{m+k}^{m+k}\}. \tag{2.4}$$

The definition of $\tilde{\varphi}_k$ uses the sigma-algebra generated by only one random variable in the future after k steps. This coefficient has already been used in many papers, including Deddens, Peligrad, and Yang (1987) and Rio (2000).

Corollary 2.11 (Peligrad (2001-a)). *Under the setting of Theorem* 2.9, *we obtain that: For all* $1 \leq i_1 < i_2 < \cdots < i_m$,

$$\mathbb{P}\left(\bigcap_{j=1}^m X_{i_j} \neq X_{i_j}^*\right) \leq \tilde{\varphi}_1^m. \tag{2.5}$$

Moreover, for $1 \leq \ell \leq n$,

$$\mathbb{P}\left(X_i \neq X_i^* \text{ at least } \ell \text{ times for } 1 \leq i \leq n\right) \leq \frac{4}{\sqrt{2\pi}} \frac{2^{n+1}}{\sqrt{n}} \frac{\tilde{\varphi}_1^\ell}{\sqrt{1 - \tilde{\varphi}_1^\ell}}. \tag{2.6}$$

Due to Remark 2.10, the case of strong mixing processes has to be analyzed in the real case and with a different approach that is called conditional quantile transform.

In order to construct Y^* we shall use, as in Rio (1995), the method based on the conditional quantile transform, introduced and studied by Major (1978):

Let us denote $F(y) = P(Y \le y)$, $F_{\mathcal{A}}(y) = P(Y \le y|\mathcal{A})$, and

$$V = F_{\mathcal{A}}(Y-0) + U(F_{\mathcal{A}}(Y) - F_{\mathcal{A}}(Y-0)),$$

where U is a random variable uniformly distributed on [0, 1] and independent of Y and $\mathcal{A}$.

By Lemma F.1 in Rio (2000), page 161, we can see that $P(V \le x|\mathcal{A}) = x$ a.s., and, as a consequence, by integrating this expression, we deduce that V is independent on $\mathcal{A}$ and uniformly distributed on [0, 1]. In addition,

$$F_{\mathcal{A}}^{-1}(V) = Y \quad \text{a.s.},$$

where we use the notation of generalized inverse: $h^{-1}(y) = \inf\{t \in R : h(t) \ge y\}$.

We now define

$$Y^* = F^{-1}(V).$$

According to Major (1978), this choice of X^* minimizes the distance in $\mathbb{L}_1$ between X and any other variable distributed as X and independent of $\mathcal{A}$.

By the above notation and considerations, since V is uniformly distributed on [0,1], clearly,

$$\mathbb{E}|Y^* - Y| = \mathbb{E}|F^{-1}(V) - F_{\mathcal{A}}^{-1}(V)| = \mathbb{E}(\int_0^1 |F^{-1}(u) - F_{\mathcal{A}}^{-1}(u)|du). \tag{2.7}$$

In order to obtain an upper bound for (2.7), we shall introduce the following weaker version of the coefficients of dependence: Let $(\Omega, \mathcal{T}, P)$ be a probability space and let X be a real random variable defined on Ω. Let $\mathcal{A}$ and $\mathcal{B}$ be two σ-algebras of $\mathcal{T}$ and let $\mathcal{T}(X)$ denote the family of events of the form $\{X < x\}$ and $\{X > x\}$ for $x \in \mathbb{R}$. Then define the strong mixing coefficient by

$$\overline{\alpha}(\mathcal{A}, X) = \sup_{A \in \mathcal{A}, x \in \mathbb{R}} |\mathbb{P}((X \ge x) \cap A) - \mathbb{P}(X \ge x)\mathbb{P}(A)|,$$

the λ-version of the maximal coefficient of correlation by

$$\overline{\lambda}(\mathcal{A}, X) = \sup_{A \in \mathcal{A}, B \in \mathcal{T}(X)} |\mathbb{P}(B \cap A) - \mathbb{P}(B)\mathbb{P}(A)|/\mathbb{P}^{1/2}(B),$$

the φ mixing coefficient by

$$\overline{\varphi}(\mathcal{A}, X) = \sup_{A \in \mathcal{A}, B \in \mathcal{T}(X)} |\mathbb{P}(B \cap A) - \mathbb{P}(B)\mathbb{P}(A)|/\mathbb{P}(B).$$

Theorem 2.12 (Peligrad (2001-b)). *Let Y be a real -valued integrable random variable that is defined on a probability space $(\Omega, \mathcal{T}, \mathbb{P})$. Suppose $\mathcal{A}$ is a sub σ-algebra of $\mathcal{T}$. Suppose U is a uniform random variable on [0, 1] that is independent on the σ-algebra generated by $\mathcal{A}$ and Y. Then, there is a random variable Y^*, which is measurable with respect to the σ-algebra $\mathcal{A} \vee \sigma(Y) \vee \sigma(U)$ with the following properties:*

(a) *Y^* is independent of the σ-algebra $\mathcal{A}$,*

(b) *Y^* has the same distribution as Y,*

(c) $\mathbb{E}|Y - Y^*| \le 4 \int_0^{\overline{\alpha}(\mathcal{A},Y)} Q_{|Y|}(u)du,$

(d) $\mathbb{E}|Y - Y^*| \le 2\sqrt{2}\,\overline{\lambda}(\mathcal{A}, Y) \int_0^{\infty} \mathbb{P}^{1/2}(|Y| > t)dt,$

(e) $\mathbb{E}|Y - Y^*| \le 2\overline{\varphi}(\mathcal{A}, Y)\mathbb{E}|Y|.$

Remark 2.13. Point (e) improves on the constant used in the proof of Corollary 2.4 of Bryc (1982). Moreover, point (c) extends a result of Rio (1995) who considered the case where Y is bounded. Indeed suppose that Y is bounded, say $a \le Y \le b$ a.s., where $a < b$ are real numbers. Let $C := (b-a)/2$. Subtracting the constant $(a+b)/2$ from Y (and from Y^*), we may assume without loss of generality that $-C \le Y \le C$ a.s., that is we can assume $a = -C$ and $b = C$. Now applying point (c), since $|Y| \le C$ a.s., and hence $Q_{|Y|}(u) \le C$ for all $u \in (0, 1)$, we get

$$\mathbb{E}|Y - Y^*| \le 4C\overline{\alpha}(\mathcal{A}, Y) = 2(b-a)\overline{\alpha}(\mathcal{A}, Y),$$

which is exactly the result of Rio (Theorem 4, 1995).

The idea of the proof of Theorem 2.12 is the following. For simplicity we shall assume that Y is almost surely positive. Let us only notice that the extension of class of sets $\mathcal{T}(Y)$ in the definition of $\overline{\lambda}(\mathcal{A}, Y)$ and $\overline{\phi}(\mathcal{A}, Y)$ is needed to treat the negative part, if Y takes its values in $\mathbb{R}$.

Equality (2.7) gives

$$\mathbb{E}|Y^* - Y| = \mathbb{E}\left(\int_0^\infty |\mathbb{P}(Y > u) - \mathbb{P}(Y > u|\mathcal{A})|du\right).$$

By classical arguments, we get

$$\mathbb{E}|\mathbb{P}(Y > u) - \mathbb{P}(Y > u|\mathcal{A})| \le 2 \sup_{A \in \mathcal{A}} |\mathbb{P}\left((Y > u) \cap A\right) - \mathbb{P}(Y > u)\mathbb{P}(A)|.$$

Now, by taking into account our definition of the mixing coefficients we obtain the following upper bound estimates:

$$\begin{aligned}
\mathbb{E}|\mathbb{P}(Y > u) - \mathbb{P}(Y > u|\mathcal{A})| &\le 2 \min\left(\overline{\alpha}(\mathcal{A}, Y), \mathbb{P}(Y > u)\right), \\
\mathbb{E}|\mathbb{P}(Y > u) - \mathbb{P}(X > u|\mathcal{A})| &\le 2\overline{\lambda}(\mathcal{A}, Y)\mathbb{P}^{1/2}(Y > u), \\
\mathbb{E}|\mathbb{P}(Y > u) - \mathbb{P}(Y > u|\mathcal{A})| &\le 2\overline{\varphi}(\mathcal{A}, Y)\mathbb{P}(Y > u),
\end{aligned}$$

and the results follow by integrating the above relations on $[0, \infty)$. □

Remark 2.14. The constants in the formulations (c), (d), and (e) of Theorem 2.12 are as sharp as possible in the sense that it is possible to exhibit an example which attains exactly the bounds. To be more precise, Bradley in a personal communication (2002), provided us with the following example: Let W and Y be random variables that take only the values 1 and -1 and such that

$$\begin{aligned}
&\mathbb{P}(W = Y = 1) = \mathbb{P}(W = Y = -1) = 3/10, \text{ and} \\
&\mathbb{P}(W = 1 \text{ and } Y = -1) = \mathbb{P}(W = -1 \text{ and } Y = 1) = 1/5.
\end{aligned}$$

Note that the distribution of (W, Y) is the same as that of (Y, W). Take $\mathcal{A} = \sigma(W)$, and suppose that Y^* is a random variable which is independent of $\mathcal{A}$ and has the same distribution as Y. Some calculations yield that $\mathbb{E}|Y - Y^*| \ge 1/5$, and that

$$\overline{\alpha}(\mathcal{A}, Y) = 0.05, \ \overline{\lambda}(\mathcal{A}, Y) = 0.05/\sqrt{1/2} \text{ and } \overline{\phi}(\mathcal{A}, Y) = 0.1.$$

From easy computations, it follows that the bounds in Theorem 2.12 are exactly attained in this particular case.

Concerning the case of a sequence of real random variables, Theorem 2.12 gives the following useful result.

Theorem 2.15 (Peligrad (2001-b)). *Let $(X_n, n \geq 1)$ be a sequence of real-valued random variables. Let $\mathcal{F}_1^n = \sigma(X_1, \ldots, X_n)$. Then, we can redefine $\{X_n, n \geq 1\}$ onto a richer probability space (which supports a sequence $(\delta_n, n \geq 1)$ of independent uniformly $[0, 1]$-distributed random variables and independent of $\{X_n, n \geq 1\}$) together with a sequence $\{X_n^*, n \geq 1\}$ of independent random variables such that, for each $1 \leq m \leq n$, we have*

(a) *X_m^* is independent of $\mathcal{F}_1^{m-1}$,*

(b) *X_m^* has the same distribution as X_m,*

(c) $\mathbb{E}|X_m - X_m^*| \leq 4 \int_0^{\overline{\alpha}(\mathcal{F}_1^{m-1}, X_m)} Q_{|X_m|}(u)du$,

(d) $\mathbb{E}|X_m - X_m^*| \leq 2\sqrt{2}\,\overline{\lambda}(\mathcal{F}_1^{m-1}, X_m) \int_0^\infty \mathbb{P}^{1/2}(|X_m| > t)dt$,

(e) $\mathbb{E}|X_m - X_m^*| \leq 2\overline{\varphi}(\mathcal{F}_1^{m-1}, X_m)\mathbb{E}|X_m|$.

The next remark makes a connection between Theorem 2.8 and point (c) of Theorem 2.15, and is useful when one makes use of truncation arguments.

Remark 2.16. Let us notice that if $(X_n, n \geq 1)$ is a sequence of real-valued random variables such that for all $n \geq 1$, $\mathbb{P}(a_n \leq X_n \leq b_n) = 1$, then, by using Theorem 2.8 and the relation (5.25) in Rio (2000), we can redefine $\{X_n, n \geq 1\}$ onto a richer probability space on which there exists a sequence $\{X_n^*, n \geq 1\}$ of independent random variables such that, for each $n \geq 1$, X_n and X_n^* have the same distribution and

$$\mathbb{E}|X_n - X_n^*| \leq 3\,(2(b_n - a_n))^{1/3}\,(\overline{\alpha}(\mathcal{F}_1^{n-1}, X_n))^{2/3}. \tag{2.8}$$

Compared to point (c) of Theorem 2.15 which gives in this case

$$\mathbb{E}|X_n - X_n^*| \leq 2(b_n - a_n)\overline{\alpha}(\mathcal{F}_1^{n-1}, X_n),$$

the inequality (2.8) seems to be less powerful. However, (2.8) can lead to better upper bound estimates when $(b_n - a_n)$ increases rapidly when n tends to infinity.

3 Application to the uniform laws of large numbers for dependent processes

We shall see now how the above-mentioned coupling results are very useful in proving asymptotic results and in this section we focus on the uniform laws of large numbers.

Uniform laws of averages extend the classical laws of large numbers from a single function to a collection of such functions. The uniform law addresses the question of finding a class $\mathcal{F}$ of functions such that

$$\lim_{n\to\infty} \sup_{f\in\mathcal{F}} \frac{1}{n} \left| \sum_{i=1}^n f(X_i) - \mathbb{E}f(X_i) \right| = 0, \tag{3.1}$$

in some stochastic sense.

If the convergence in (3.1) holds with probability 1, $\mathcal{F}$ is said to be a Glivenko–Cantelli class. Alternatively $\mathcal{F}$ is said to satisfy a uniform law of averages with respect to the sequence $(X_i)_{i\geq 1}$.

Vapnik and Cervonenkis (1981) found necessary and sufficient conditions under which a class of functions $\mathcal{F}$ is almost surely uniformly convergent with respect to an i.i.d. process $(X_i)_{i\geq 1}$.

The conditions involve the asymptotic behavior of covering numbers associated with the class $\mathcal{F}$ and sample sequences of the process $(X_i)_{i\geq 1}$. For more details, we refer the reader to Vapnik and Cervonenkis (1981) or Pollard (1984).

Concerning weak dependent sequences, the uniform convergence has been studied by many authors. Among them Philipp (1982), Yukich (1986), Massart (1988), Yu (1991), Doukhan, Massart, Rio (1995). All these results used specific rates of convergence to zero of the mixing coefficients.

In this paper we present recent and new results about the uniform law of large numbers for weak dependent sequences. This section is divided in several parts. First we deal with absolutely regular sequences, then with $\tilde{\varphi}$-mixing sequences and finally with strong mixing sequences. In these results no condition is imposed on the rates of convergence to zero of the mixing coefficients.

Since the proofs are based mainly on coupling results, we shall assume in what follows the probability space to be large enough to accommodate all the variables constructed.

Let $\mathcal{F}$ be a class of real-valued measurable functions on S. A measurable function $F : S \to R$ is said to be an envelope for $\mathcal{F}$, if $|f(x)| \leq F(x)$, for every $f \in \mathcal{F}$. The class $\mathcal{F}$ is said to be uniformly bounded if for some $0 \leq K < \infty$, $F = K$ is an envelope for $\mathcal{F}$. If the class $\mathcal{F}$ is uncountable, it must satisfy regularity conditions in order to insure that no problems of measurability will arise when considering quantities such as the supremum. Following Pollard, we use the term *permissible* to indicate that the class $\mathcal{F}$ satisfies such conditions (for more details, see Dudley (1978) or Pollard (1984)).

3.1 The case of absolutely regular sequences. Nobel and Dembo (1993) obtained a uniform law for the class of absolute regular sequences without imposing any condition on the rate of convergence to 0 of the dependence coefficients. In fact they obtained the almost sure uniform convergence for absolutely regular strictly stationary sequences as a direct consequence of uniform convergence for independent random variables without imposing any additional conditions on the class of functions. Their result is the following:

Theorem 3.1. *Let $(X_j)_{j\in\mathbb{Z}}$ be a sequence of strictly stationary S-valued variables defined on $(\Omega, \mathcal{T}, \mathbb{P})$ and such that $\beta(\mathcal{F}_{-\infty}^{0}, \mathcal{F}_{n}^{+\infty}) \to 0$ as $n \to \infty$. Let $(X_j^*)_{j\in\mathbb{Z}}$ be an independent sequence with values in S such that for all $j \in \mathbb{Z}$ X_j^* and X_j have the same distribution. Let $\mathcal{F}$ be a permissible class of real-valued functions on S, with an envelope F such that $F \in \mathbb{L}_1(\mathbb{P})$ and such that*

$$\sup_{f\in\mathcal{F}} \left| \frac{1}{n} \sum_{i=1}^{n} \left(f(X_i^*) - \mathbb{E} f(X_i^*) \right) \right| \to 0 \quad a.s.$$

Then

$$\sup_{f\in\mathcal{F}} \left| \frac{1}{n} \sum_{i=1}^{n} \left(f(X_i) - \mathbb{E} f(X_i) \right) \right| \to 0 \quad a.s.$$

Let us notice that the coefficient of absolute regularity used by Nobel and Dembo involves all the future of the process, which means that the process is ergodic and the proofs of their theorem benefitted by the results in ergodic theory.

3.2 The case of φ-mixing sequences. Concerning the case of $\tilde{\varphi}$-mixing sequences, Peligrad (2001-a) has obtained a uniform strong law without imposing any rates of convergence to zero on $\tilde{\varphi}_n$ and without requiring ergodicity. She has shown that the class of strictly stationary and

$\tilde{\varphi}$-mixing sequences satisfies a uniform law of averages if an independent sequence having the same marginals does. Therefore the result available for independent sequences translates to similar results for $\tilde{\varphi}$-mixing sequences. Her result can be formulated as follows:

Theorem 3.2. *Let $\mathcal{F}$ be a permissible class of real-valued functions on S having envelope $F \in \mathbb{L}_1(\mathbb{P})$. Then $\mathcal{F}$ satisfies a uniform strong law of averages with respect to a strictly stationary $\tilde{\varphi}$-mixing sequence of S-valued random variables if it satisfies the same law with respect to an i.i.d.*

The proof of the above theorem can be divided in two steps: the first one involves a class of functions $\mathcal{F}$ which is uniformly bounded and the second one involves a truncation argument. In fact by analyzing these two steps, we can give two new results on the uniform law of convergence for $\tilde{\varphi}$-mixing sequences without imposing stationarity. This is the aim of the next statements.

Proposition 3.3. *Let $(X_j)_{j\geq 1}$ be a sequence of S-valued variables such that $\tilde{\varphi}_n \to 0$ as $n \to \infty$. Let $(X_j^*)_{j\geq 1}$ be an independent sequence such that for all $j \geq 1$, X_j^* and X_j have the same distribution. Let $\mathcal{F}$ be a class of real-valued functions on S which is uniformly bounded and such that for a fixed $p \geq 1$ and each $0 \leq j \leq p-1$, we have*

(i) $\lim_{k\to\infty} \sup_{f\in\mathcal{F}} |\frac{1}{k}\sum_{i=0}^{k-1}(f(X_{ip+j}^*) - \mathbb{E}f(X_{ip+j}^*))| = 0$ *a.s., where* $X_0^* = 0$.

Then

(ii) $\lim_{n\to\infty} \sup_{f\in\mathcal{F}} |\frac{1}{n}\sum_{i=1}^{n}(f(X_i) - \mathbb{E}f(X_i))| = 0$ *a.s.*

Proof. For all $n \geq 1$, we shall set $f_n(X_n) := f(X_n) - \mathbb{E}f(X_n)$. However, for the sake of simplicity, we use the notation $f(X_n)$ instead of $f_n(X_n)$.

Let p be a natural number $p < n$. Let $k = [\frac{n}{p}]$, and extend the vector $(X_1, \ldots, X_n)$ to the sequence $(0, X_1, \ldots, X_n, 0, 0, \ldots)$. Define $Y_{i,j} = X_{ip+j}$ for $0 \leq i \leq k-1$, $0 \leq j \leq p-1$. For each $0 \leq j \leq p-1$, consider the vector $(Y_{0,j}^*, \ldots, Y_{k-1,j}^*)$ constructed in Theorem 2.9, and notice that

$$\sup_{f\in\mathcal{F}} \frac{1}{n}\left|\sum_{i=1}^{n} f(X_i)\right| \leq \sup_{f\in\mathcal{F}} \frac{1}{n}\left|\sum_{i=1}^{n} f(X_i) - \sum_{i=0}^{k}\sum_{j=0}^{p-1} f(Y_{i,j}^*)\right| + \sup_{f\in\mathcal{F}} \frac{1}{kp}\sum_{j=0}^{p-1}\left|\sum_{i=0}^{k-1} f(Y_{i,j}^*)\right|. \tag{3.2}$$

Under assumption (i), the second term in the right-hand side term in the above inequality tends to zero by letting n tend to infinity and then p. Then it is enough to show that

$$\lim_{n\to\infty} \sup_{f\in\mathcal{F}} \frac{1}{n}\left|\sum_{i=1}^{n} f(X_i) - \sum_{i=0}^{k-1}\sum_{j=0}^{p-1} f(Y_{i,j}^*)\right| = 0. \tag{3.3}$$

By the triangle inequality, we have

$$\frac{1}{n}\left|\sum_{i=1}^{n} f(X_i) - \sum_{i=0}^{k}\sum_{j=0}^{p-1} f(Y_{i,j}^*)\right| \leq \max_{0\leq j\leq p-1} \frac{1}{k}\left|\sum_{i=0}^{k-1}\Big(f(Y_{i,j}) - f(Y_{i,j}^*)\Big)\right|. \tag{3.4}$$

Fix j, $0 \leq j \leq p-1$, and let η be such that $\tilde{\varphi}_p < \frac{1}{2(1+\eta)}$. Let $\ell_k = [k \log \frac{1}{2(1+\eta)} / \log \tilde{\varphi}_p]$. We write

$$\frac{1}{k}\left(\sum_{i=0}^{k-1} f(Y_{i,j}) - f(Y_{i,j}^*)\right) =: I_1 + I_2, \tag{3.5}$$

where

$$I_1 = \frac{1}{k}\sum_{i=0}^{k-1}\left(f(Y_{i,j}) - f(Y_{i,j}^*)\right)\mathbb{I}\left(\sum_{i=0}^{k-1}\mathbb{I}(Y_{i,j} \neq \tilde{Y}_{i,j}) < \ell_k\right)$$

and

$$I_2 = \frac{1}{k}\sum_{i=0}^{k-1}\left(f(Y_{i,j}) - f(Y_{i,j}^*)\right)\mathbb{I}\left(\sum_{i=0}^{k-1}\mathbb{I}(Y_{i,j} \neq Y_{i,j}^*) \geq \ell_k\right)$$

Since $\mathcal{F}$ is bounded, we can find $C > 0$ such that for all $f \in \mathcal{F}$,

$$I_2 \leq CI\left(\sum_{i=0}^{k-1}\mathbb{I}(Y_{i,j} \neq \tilde{Y}_{i,j}) \geq \ell_k\right) \quad \text{a.s.} \tag{3.6}$$

and also

$$I_1 \leq C\frac{\ell_k}{k} = C\left(\log\frac{1}{2(1+\eta)}\right)/\log\tilde{\varphi}_p. \tag{3.7}$$

By combining (3.4)–(3.7) we can find a constant K such that

$$\frac{1}{n}\left|\sum_{i=1}^{n} f(X_i) - \sum_{i=0}^{k-1}\sum_{j=0}^{p-1} f(Y_{i,j}^*)\right| \leq K\left\{\left(\sum_{i=0}^{k-1}\mathbb{I}(Y_{i,j} \neq Y_{i,j}^*) \geq \ell_k\right)\right.$$
$$\left. + \left[\log\frac{1}{2(1+\eta)}\right]/\log\tilde{\varphi}_p\right\} \quad \text{a.s.}$$

By applying now (2.6) of Corollary 2.11, it is easy to obtain via Borel–Cantelli lemma that

$$\mathbb{P}\left(\sum_{i=0}^{k-1}\mathbb{I}(Y_{i,j} \neq Y_{i,j}^*) \geq \ell_k \ i.o.\right) = 0$$

which yields

$$\lim_{n\to\infty}\sup_{f\in\mathcal{F}}\frac{1}{n}\left|\sum_{i=1}^{n} f(X_i) - \sum_{i=0}^{k-1}\sum_{j=0}^{p-1} f(Y_{i,j}^*)\right| \leq K\left[\log\frac{1}{2(1+\eta)}\right]/\log\tilde{\varphi}_p \quad \text{a.s.}$$

The result follows by letting $p \to \infty$. □

Remark 3.4. In view of the proof, Proposition 3.3 holds with convergence in probability replacing almost sure convergence everywhere.

In the next theorem we remove the condition of boundedness of the class of function $\mathcal{F}$ and replace it with an integrability condition imposed on the envelope of the class. However, our result is now a convergence in probability instead of an almost sure result.

Theorem 3.5. *Assume that $(X_j)_{j\geq 1}$ and $(X_j^*)_{j\geq 1}$ are as in Proposition 3.3. Let $\mathcal{F}$ be a class of functions with an envelope F such that $\{F(X_i)\}_{i\geq 1}$ is a uniformly integrable family and*

$$\sup_{f\in\mathcal{F}}\left|\frac{1}{n}\sum_{i=1}^{n}\left(f(X_i^*) - \mathbb{E}f(X_i^*)\right)\right| \xrightarrow{\mathbb{P}} 0. \tag{i}$$

Then

$$\sup_{f\in\mathcal{F}}\left|\frac{1}{n}\sum_{i=1}^{n}(f(X_i)-\mathbb{E}f(X_i))\right| \overset{\mathbb{P}}{\to} 0. \tag{ii}$$

Proof of Theorem 3.5. Once again for all $n \geq 1$, we shall use the notation $f(X_n)$ to mean $f(X_n) - \mathbb{E}f(X_n)$.

Let C be a positive constant and let $f \in \mathcal{F}$. Let $f_1 = f\mathbb{I}(F \leq C)$ and $f_2 = f\mathbb{I}(F > C)$. We obviously have

$$\begin{aligned}\sup_{f\in\mathcal{F}}\frac{1}{n}\left|\sum_{i=1}^{n}f(X_i)\right| &= \sup_{f_1,f\in\mathcal{F}}\frac{1}{n}\left|\sum_{i=1}^{n}(f_1(X_i)-\mathbb{E}f_1(X_i))\right| \\ &+ \sup_{f_2,f\in\mathcal{F}}\frac{1}{n}\left|\sum_{i=1}^{n}(f_2(X_i)-\mathbb{E}f_2(X_i))\right|.\end{aligned} \tag{3.8}$$

By Proposition 3.3 and Remark 3.4, since the family $\{f_1, f \in \mathcal{F}\}$ has a bounded envelope, we will obtain that the first term in the right-hand side of equality (3.8) converges to zero in probability as soon as Assumption (i) of Proposition 3.3 holds in probability. Note that due to the uniform integrability of the envelope, Assumption (i) of Theorem 3.5 entails that the convergence still holds in $\mathbb{L}_1$. This last consideration combined with the following statement proves the desired convergence.

Claim 1. *Let $(Z_j)_{j\geq 1}$ be a sequence of independent S-valued random variables defined on $(\Omega, \mathcal{T}, \mathbb{P})$ and let $\mathcal{F}$ be a permissible class of real-valued functions on S such that*

$$\sup_{f\in\mathcal{F}}\left|\frac{1}{n}\sum_{i=1}^{n}(f(Z_i)-\mathbb{E}f(Z_i))\right| \overset{\mathbb{L}_1}{\to} 0.$$

Then for each p, $p \geq 1$, and all $1 \leq i_1 < i_2 < \cdots < i_k \leq n$, where $k = [\frac{n}{p}]$, we have

$$\sup_{f\in\mathcal{F}}\left|\frac{1}{k}\sum_{j=1}^{k}\left(f(Z_{i_j})-\mathbb{E}f(Z_{i_j})\right)\right| \overset{\mathbb{L}_1}{\to} 0.$$

Proof. The proof is based on the fact that if X and Y are two independent S-valued random variables such that for all $f \in \mathcal{F}$ we have $\mathbb{E}f(Y) = 0$, then

$$\mathbb{E}\left(\sup_{f\in\mathcal{F}}|f(X)|\right) \leq \mathbb{E}\left(\sup_{f\in\mathcal{F}}|f(X)+f(Y)|\right). \tag{3.9}$$

We give here a short proof of (3.9) based on a suggestion of Pollard (personal communication). Notice first that since $\mathbb{E}f(Y) = 0$ and X and Y are independent,

$$f(X) = \mathbb{E}(f(X) + f(Y)|X).$$

Then

$$\mathbb{E}\left(\sup_{f\in\mathcal{F}}|f(X)|\right)=\mathbb{E}\left(\sup_{f\in\mathcal{F}}|\mathbb{E}(f(X)+f(Y)|X)|\right)$$
$$\leq\mathbb{E}\left(\mathbb{E}\left(\sup_{f\in\mathcal{F}}|f(X)+f(Y)||X\right)\right)$$
$$\leq\mathbb{E}\left(\sup_{f\in\mathcal{F}}|f(X)+f(Y)|\right).$$

□

Now in order to treat the last term in the right-hand side of equality (3.8), it suffices to notice that

$$\mathbb{E}\sup_{f_2,f\in\mathcal{F}}\frac{1}{n}\left|\sum_{i=1}^{n}(f_2(X_i)-\mathbb{E}f_2(X_i))\right|\leq\frac{2}{n}\sum_{i=1}^{n}\mathbb{E}|F(X_i)|\mathbb{I}(|F(X_i)|>C).$$

The result follows by the uniform integrability of $\{F(X_i)\}_{i\geq 1}$, by letting $n\to\infty$ and then $C\to\infty$. □

3.3 The case of strong mixing sequences. Concerning the case of strong mixing sequences of real-valued random variables, some new results have been established recently by Peligrad (2001-b). The proofs are based mainly on the coupling result (c) of Theorem 2.15 and on a decomposition in "p-sequences" as in the proof of Proposition 3.3. The results are again obtained without imposing rates on the strong mixing coefficients and without requiring ergodicity. Peligrad has shown that this class of strong mixing sequences satisfies a uniform weak law of averages if an independent sequence having the same marginal does. Therefore the results available for independent sequences imply similar results for strong mixing sequences.

Let us also notice that, as we mentioned above, the proofs are based mainly on the coupling result for strongly mixing sequences. Due to Remark 2.10, results for strong mixing sequences taking their values in a Polish space cannot be derived by a similar approach.

For the sake of simplicity, we set $\overline{\alpha}_n=\overline{\alpha}(\mathcal{F}_1^{n-1},X_n)$, where $\overline{\alpha}$ is defined in the previous section.

Theorem 3.6 (Peligrad (2001-b)). *Assume that $\{X_n\}_{n\geq 1}$ is a sequence of identically distributed real-valued random variables in $\mathbb{L}_1$ such that $\overline{\alpha}_n\to 0$ as $n\to\infty$. Let $\{X_n^*\}_{n\geq 1}$ be a sequence of independent identically distributed real-valued random variables such that X_1^* has the same distribution as X_1. Let $\mathcal{F}$ be a class of Lipschitz functions with the Lipschitz constants bounded by C ($C>0$) such that*

(i) $\lim_{n\to\infty}\mathbb{E}\sup_{f\in\mathcal{F}}\frac{1}{n}|\sum_{i=1}^{n}(f(X_i^*)-\mathbb{E}f(X_i^*))|=0.$

Then

(ii) $\lim_{n\to\infty}\mathbb{E}\sup_{f\in\mathcal{F}}\frac{1}{n}|\sum_{i=1}^{n}(f(X_i)-\mathbb{E}f(X_i))|=0.$

In the statement of Theorem 3.6, let us notice that the assumption that the variables have the same distribution can be relaxed by imposing some weaker forms of stationarity. In this direction, we would like to mention the following result.

Proposition 3.7. *If we assume in Theorem 3.6 that $\{X_n\}_{n\geq 1}$ is uniformly integrable instead of stationary, then Theorem 3.6 holds.*

Proof of Proposition 3.7. Once again for all $n \geq 1$, we shall use the notation $f(X_n)$ to mean $f(X_n) - \mathbb{E}f(X_n)$. To prove the result, we consider the "p-sequences" as in the proof of Proposition 3.3, for each $1 \leq j \leq p-1$, the vector $(Y^*_{0,j}, \dots, Y^*_{k-1,j})$ being this time constructed as in Theorem 2.15. We notice that by the triangle inequality, we have

$$\begin{aligned}\mathbb{E} \sup_{f\in\mathcal{F}} \frac{1}{n}\left|\sum_{i=1}^{n} f(X_i)\right| &\leq \mathbb{E} \sup_{f\in\mathcal{F}} \frac{1}{n}\sum_{j=0}^{p-1}\left|\sum_{i=0}^{k-1}\left(f(Y_{i,j}) - f(Y^*_{i,j})\right)\right| \\ &\quad + \mathbb{E} \sup_{f\in\mathcal{F}} \frac{1}{kp}\sum_{j=0}^{p-1}\left|\sum_{i=0}^{k-1} f(Y^*_{i,j})\right|.\end{aligned} \tag{3.10}$$

By using Claim 1 together with Assumption (i), the last term in the right-hand side of inequality (3.10) tends to zero by letting n tend to infinity. Now, to treat the first term, we first notice that since f is Lipschitz, by Theorem 2.15, we derive that

$$\begin{aligned}\mathbb{E} \sup_{f\in\mathcal{F}} \frac{1}{n}\sum_{j=0}^{p-1}\left|\sum_{i=0}^{k-1}\left(f(Y_{i,j}) - f(Y^*_{i,j})\right)\right| &\leq \frac{4}{n}C\sum_{j=0}^{p-1}\sum_{i=0}^{k-1}\int_0^{\overline{\alpha}_p} Q_{|X_{ip+j}|}(u)du \\ &\leq 4C \sup_{1\leq i\leq n}\int_0^{\overline{\alpha}_p} Q_{|X_i|}(u)du.\end{aligned}$$

To obtain the desired result, we first let n tend to infinity and then we apply the following lemma.

Lemma 3.8. *Suppose that* $\{X_n\}_{n\geq 1}$ *is a uniformly integrable sequence of random variables, then*

$$\lim_{\varepsilon\to 0}\sup_{i\geq 1}\int_0^{\varepsilon} Q_{|X_i|}(u)du = 0. \tag{3.11}$$

Proof. Let us suppose that ess sup $\sup_{i\geq 1}|X_i| = +\infty$, otherwise (3.11) is trivial. Now notice that

$$\sup_{i\geq 1}\int_0^{\varepsilon} Q_{|X_i|}(u)du \leq \sup_{i\geq 1}\mathbb{E}\left\{|X_i|\mathbb{I}(|X_i| \geq Q_{|X_i|}(\varepsilon))\right\},$$

which converges to zero when $\varepsilon \to 0$ by using the fact that $\lim_{\varepsilon\to 0} Q_{|X_i|}(\varepsilon) = +\infty$ together with the uniform integrability condition. □

We can also formulate the following new result using strong mixing coefficients.

Theorem 3.9. *Assume* $\{X_n\}_{n\geq 1}$ *is a sequence of uniformly integrable, centered real-valued random variables and such that* $\overline{\alpha}_n \to 0$ *as* $n \to \infty$. *Then*

$$\lim_{n\to\infty}\mathbb{E}\max_{1\leq j\leq n}\frac{1}{n}\left|\sum_{i=1}^{j} X_i\right| = 0.$$

Proof of Theorem 3.9. Let us first truncate the random variables in the following way: Take A to be a positive number and set

$$X_i' = X_i\mathbb{I}(|X_i| \leq A) - \mathbb{E}X_i\mathbb{I}(|X_i| \leq A)$$

and

$$X_i'' = X_i\mathbb{I}(|X_i| > A) - \mathbb{E}X_i\mathbb{I}(|X_i| > A).$$

Now we set $S_k = \sum_{i=1}^k X_i$ and $S'_k = \sum_{i=1}^k X'_i$, and we notice that

$$\frac{1}{n}\mathbb{E}\max_{1\le j\le n}|S_j| \le \frac{\sum_{i=1}^n \mathbb{E}|X''_i|}{n} + \frac{1}{n}\mathbb{E}\max_{1\le j\le n}|S'_j|. \tag{3.12}$$

Under the condition of uniform integrability of the sequence $\{X_i\}$, it is clear that the first term in the right-hand side of the above inequality tends to zero by letting n and after A tend to infinity. Then it remains to treat the second term of inequality (3.12).

Now let p be a natural number $p < n$. Let $k = [\frac{n}{p}]$, and extend the vector $(X'_1, \dots, X'_n)$ to the sequence $(0, X'_1, \dots, X'_n, 0, 0, \dots)$. Define $Y'_{i,j} = X'_{ip+j}$ for $0 \le i \le k-1, 0 \le j \le p-1$. For each $0 \le j \le p-1$, consider the vector $(Y'^*_{0,j}, \dots, Y'^*_{k-1,j})$ constructed in Theorem 2.15, and notice that

$$\begin{aligned}\frac{1}{n}\mathbb{E}\max_{1\le j\le n}|S'_j| &= \frac{1}{n}\mathbb{E}\max_{1\le m\le p}\max_{1\le \ell\le k}\left|\sum_{j=0}^{m-1}\sum_{i=0}^{\ell-1}X'_{ip+j}\right| \\ &\le \frac{1}{n}\mathbb{E}\max_{1\le m\le p}\max_{1\le \ell\le k}\left|\sum_{j=0}^{m-1}\sum_{i=0}^{\ell-1}Y'^*_{i,j}\right| \\ &\quad + \frac{1}{n}\mathbb{E}\max_{1\le m\le p}\max_{1\le \ell\le k}\left|\sum_{j=0}^{m-1}\sum_{i=0}^{\ell-1}(Y'_{i,j} - Y'^*_{i,j})\right|.\end{aligned} \tag{3.13}$$

First we treat the last term in the right-hand side of the above inequality. By using Theorem 2.15, we derive that

$$\begin{aligned}\frac{1}{n}\mathbb{E}\max_{1\le m\le p}\max_{1\le \ell\le k}\left|\sum_{j=0}^{m-1}\sum_{i=0}^{\ell-1}(Y'_{i,j} - Y'^*_{i,j})\right| &\le \frac{1}{n}\sum_{j=0}^{p-1}\sum_{i=0}^{k-1}\mathbb{E}|Y'_{i,j} - Y'^*_{i,j}| \\ &\le 4\sup_{1\le i\le n}\int_0^{\bar\alpha_p} Q_{|X'_i|}(u)du \\ &\le 8A\bar\alpha_p,\end{aligned}$$

which converges to zero by letting, respectively, n and then p tend to infinity. To treat now the first term in the right-hand side of inequality (3.13), first we write

$$\frac{1}{n}\mathbb{E}\max_{1\le m\le p}\max_{1\le \ell\le k}\left|\sum_{j=0}^{m-1}\sum_{i=0}^{\ell-1}Y'^*_{i,j}\right| \le \frac{1}{n}\sum_{j=0}^{p-1}\mathbb{E}\max_{1\le \ell\le k}\left|\sum_{i=0}^{\ell-1}Y'^*_{i,j}\right|.$$

By using Rosenthal's inequality applied to the independent sequence (see, e.g., Hall and Heyde (1980), Theorem 2.11), there exists a constant C such that

$$\mathbb{E}\max_{1\le \ell\le k}\left|\sum_{i=0}^{\ell-1}Y'^*_{i,j}\right| \le C\left\{\left(\sum_{i=0}^{k-1}\mathbb{E}(Y'^*_{i,j})^2\right)^{1/2} + \mathbb{E}\max_{1\le i\le k-1}|Y'^*_{i,j}|\right\}.$$

It follows that

$$\frac{1}{n}\mathbb{E}\max_{1\le m\le p}\max_{1\le \ell\le k}\left|\sum_{j=0}^{m-1}\sum_{i=0}^{\ell-1}Y'^*_{i,j}\right| \le AC\left(\sqrt{\frac{p}{n}} + \frac{p}{n}\right),$$

which converges to zero by letting n tend to infinity.

Gathering all these considerations, we obtain the desired result. □

4 Application to the central limit theorem

In this section we shall see how the coupling results of Theorem 2.15 can be used to solve the central limit theorem question for stationary strongly mixing sequences, say $X := \{X_k, k \in \mathbb{Z}\}$. In this section, we mean by strongly mixing sequences, sequences such that

$$\alpha_n := \alpha\left(\mathcal{F}_{-\infty}^0, \mathcal{F}_n^{+\infty}\right) \to 0 \text{ as } n \to \infty$$

where $\mathcal{F}_m^p := \sigma(X_j, m \le j \le p)$.

Since 1956, a vast body of work has been devoted to study the limiting behavior of strongly mixing sequences. This is a large class of random variables which contains both weakly dependent sequences and sequences with long range dependence. Examples include time series, Gaussian processes and Markov processes. These processes appear in other branches of mathematics, as well as statistics and mathematical physics, giving rise to a great deal of interest in their asymptotic properties.

The question concerning the central limit theorem in this setting is the following: under what assumptions, besides that of strong mixing, do there exist real numbers $a_1, a_2, a_3, \ldots$ and positive numbers $b_1, b_2, b_3, \ldots$ with $b_n \to \infty$ as $n \to \infty$, such that

$$\frac{S_n - a_n}{b_n} \xrightarrow{\mathcal{D}} N \sim \mathcal{N}(0,1) \text{ as } n \to \infty, \tag{4.1}$$

where $S_n = \sum_{i=1}^n X_i$.

By answering a conjecture of Bradley (1997), Merlevède, and Peligrad (2000) obtained the following sharp central limit theorem that extends previous results by Ibragimov and Linnik (1971) and Doukhan, Massart, and Rio (1994).

Theorem 4.1. *Suppose that $\{X_k, k \in \mathbb{Z}\}$ is a strictly stationary, centered, strong mixing sequence with finite second moment. Assume that*

$$\liminf_{n\to\infty} \frac{\mathbb{E}(S_n^2)}{n} > 0 \tag{4.2}$$

and

$$\int_0^{\alpha_n} Q_{|X_0|}^2(u)du = o\left(\frac{1}{n}\right) \text{ as } n \to \infty. \tag{4.3}$$

Then

$$\frac{S_n}{\sqrt{\frac{\pi}{2}}\mathbb{E}|S_n|} \xrightarrow{\mathcal{D}} N \sim \mathcal{N}(0,1) \text{ as } n \to \infty. \tag{4.4}$$

The result of the above theorem was predicted by Bradley (1997). In his paper he formulated the following conjecture: Under the assumptions of Theorem 4.1, there exists a sequence $b_n \to \infty$ as $n \to \infty$, such that $\frac{S_n}{b_n}$ converges in distribution to a centered Gaussian random variable with variance 1.

The proof of Theorem 4.1 in Merlevède and Peligrad (2000) is based on the Bernstein-type blocking arguments and the coupling results. In an earlier version of this paper, the coupling was used only for identifying the normalizing sequence b_n. In a recent paper of Merlevède (2001), the proof does not require coupling. We would like to mention here the idea of proof based on coupling.

The proof of Theorem 4.1 involves several steps.

Step 1: As a consequence of (4.2) and (4.3), we proved that $\sigma_n^2 := \mathrm{Var}(S_n)$ has the representation

$$\sigma_n^2 = nh(n), \tag{4.5}$$

where $h(n)$ is a slowly varying function of n.

Step 2: To prepare the decomposition in "big" and "small" blocks, we constructed two sequences p_n and q_n which converge to infinity and such that $q_n = o(p_n)$. In fact the construction of p_n and q_n is the crucial part of the proof. The sequences are somehow implicit solutions of an equation involving the mixing coefficients and a continuous approximation of the quantile function.

Step 3: The proof continues with the following truncation: Set $T_n = Q_{|X_0|}(\alpha_{q_n})$ and

$$X_i' = \begin{cases} X_i\mathbb{I}(|X_i| \le T_n) - \mathbb{E}X_i\mathbb{I}(|X_i| \le T_n) & \text{if } X \text{ is an unbounded sequence,} \\ X_i & \text{if ess.sup} X = A \quad \text{a.s.} \end{cases}$$

and (4.6)

$$X_i'' = \begin{cases} X_i\mathbb{I}(|X_i| > T_n) - \mathbb{E}X_i\mathbb{I}(|X_i| > T_n) & \text{if } X \text{ is an unbounded sequence,} \\ 0 & \text{if ess.sup} X = A \quad \text{a.s.,} \end{cases}$$

and let $S_n' = \sum_{i=1}^n X_i'$ and $S_n'' = \sum_{i=1}^n X_i''$. Then $S_n = S_n' + S_n''$.

Step 4: At this step we reduce the proof to a central limit theorem for S_n' properly normalized by showing that $\frac{S_n''}{\sqrt{n}}$ converges to zero in $\mathbb{L}_1$.

Step 5: Here we partion the index set $1, 2, 3, \ldots$ into an alternating sequence of "big" blocks of size p_n and "small" blocks of size q_n (in fact we will obtain k_n "big" blocks and $k_n + 1$ "small" blocks, where $k_n = [\frac{n}{p_n+q_n}]$).

Let $Y_1', Y_2', Y_3', \ldots$ denote the respective sums of the X_k''s in the "big" blocks each having p_n summands, and let $Z_1', Z_2', Z_3', \ldots$ denote the respective sums of the X_k''s in the "small" blocks each having q_n summands.

Moreover, we use the selection of b_n

$$b_n^2 = k_n \sigma_{p_n}^2,$$

and prove that the asymptotic behavior of $\{\frac{\sum_{j=1}^{k_n+1} Z_j'}{b_n}\}_{n\ge1}$ is negligible for the convergence in distribution.

Step 6: At this step, we use coupling to study the asymptotic behavior of

$$\left\{\frac{\sum_{j=1}^{k_n} Y_j'}{b_n}\right\}_{n\ge1}.$$

To this aim, we consider a sequence $\{Y_j'^*\}_{1\le i\le k_n}$ of independent real random variables each, respectively, distributed as Y_j' and constructed as in Theorem 2.15. Since for all $1 \le j \le k_n$, $|Y_j'| \le 2p_n Q_{|X_0|}(\alpha_{q_n})$, point (c) of Theorem 2.15 yields

$$\frac{E\left|\sum_{j=1}^{k_n}\left(Y_j' - Y_j'^*\right)\right|}{b_n} \le \frac{8p_nk_n\alpha_{q_n}Q_{|X_0|}(\alpha_{q_n})}{b_n}, \tag{4.7}$$

which converges to 0 under (4.2) and our selection of p_n and q_n. This result implies that the proof of Theorem 4.1 is reduced to studying the limiting behavior of $\{\frac{\sum_{j=1}^{k_n} Y_j'^*}{b_n}\}_{n\ge1}$.

Step 7: We show that $\frac{\sum_{j=1}^{k_n} Y'^*_j}{b_n}$ converges in distribution to a standard normal random variable by checking the Lindeberg condition.

Step 8: To end the proof we identify the normalizing sequence b_n. To this end, we prove that $\{\frac{S_n}{b_n}\}_{n\geq 1}$ is a uniformly integrable family. Notice that (4.7) entails that $\{\frac{\sum_{j=1}^{k_n}(Y'_j - Y'^*_j)}{b_n}\}_{n\geq 1}$ is a uniformly integrable family. This consideration together with Steps 4, 5, and 7 imply the desired result. We finish the proof by using Theorem 5.4 in Billingsley (1968), which yields

$$\lim_{n\to\infty} \frac{b_n}{\sqrt{\frac{\pi}{2}}E|S_n|} = 1.$$

For the sake of applications, we give the following corollary of Theorem 4.1 in term of conditions imposed on mixing rates and on moments of individual summands. It extends the corresponding results of Ibragimov (1962).

Corollary 4.2. *Suppose that* $X := \{X_k, k \in \mathbb{Z}\}$ *is a strictly stationary, centered, strong mixing sequence which satisfies (4.2). In addition if*

1. *X has moments of order* $2 + \delta$ *finite, for* $\delta > 0$ *and*

$$n\alpha_n^{\frac{\delta}{2+\delta}} \to 0 \text{ as } n \to \infty \tag{4.8}$$

or if

2. *X is bounded and*

$$n\alpha_n \to 0 \text{ as } n \to \infty, \tag{4.9}$$

then (4.4) *holds.*

Let us notice that under the additional assumption $\lim_{n\to\infty} \frac{\mathbb{E}(S_n^2)}{n} = \sigma^2 > 0$, all the results of type (4.4) hold with the normalization $\sigma\sqrt{n}$ instead of $\sqrt{\frac{\pi}{2}}\mathbb{E}|S_n|$.

Moreover, we can mentioned that the functional version of Theorem 4.1 has also been established in the paper of Merlevède and Peligrad (2000), and it can be formulated as follows:

Theorem 4.3. *Assume that the conditions of Theorem* 4.1 *are satisfied. Then* W_n *converges in distribution to* W *in the Skorohod space* $D([0, 1])$*, where* $W_n(t) = \frac{\sum_{i=1}^{[nt]} X_i}{\sqrt{\frac{\pi}{2}}\mathbb{E}|S_n|}$, *and* W *is the standard Brownian motion on* [0, 1].

Acknowledgments. The authors would like to thank Richard Bradley for providing us with useful information regarding estimates for the coupling error under strong mixing conditions for multivariate random variables. We are also grateful to him for his help in clarifying the constants in Theorem 2.12, and, in particular, for having provided us the example of Remark 2.14 and also the connection exhibited in Remark 2.13. This paper was written when Magda Peligrad visited the University of Paris X.

References

[1] Arcones, M.A. and Yu, B., Central limit theorems for empirical and U-processes of stationary mixing sequences. *J. Theoret. Probab.* **7** (1994), 47–71.

[2] Berbee, H.P.C., Random walks with stationary increments and renewal theory, *Math. Center Tracts*, **112** (1979), Amsterdam.

[3] Berger, E., Fastsichere Approximation von Partialsummen unabhängiger und stationärer ergodischer Folgen von Zufallsvektoren, Dissertation, Univ. Gottingen, 1982.

[4] Berkes, I., Philipp, W., An almost sure invariance principle for the empirical distribution function of mixing random variables, *Z. Wahr. verw. Gebiete*, **41** (1977/78), 115–137.

[5] Berkes, I., Philipp, W., Approximation theorems for independent and weakly dependent random vectors, *Ann. Probability*, **7** (1979), 29–54.

[6] Billingsley, P., *Convergence of Probability Measures*, Wiley, New York, 1968.

[7] Borovkova, S., Burton,R., Dehling, H.G., Limit theorems for functionals of mixing processes with applications to U-Statistics and dimension estimation, Preprint 99-5-10, University of Groningen IWI, 1999.

[8] Bosq, D., *Nonparametric Statistics for Stochastic Processes, Estimation and Prediction*, Second Edition, Springer, New York, 1998.

[9] Bradley, R.C., Approximation theorems for strongly mixing random variables, *Michigan Math. J.*, **30** (1983), 69–81.

[10] Bradley, R.C., On quantiles and the central limit theorem question for strongly mixing sequences, *J. Theor. Probab.*, **10** (1997), 507–555.

[11] Bryc, W., Sequences of weakly dependent random variables, Ph.D. dissertation, 1981 (in Polish).

[12] Bryc, W., On the approximation theorem of Berkes and Philipp, *Demonstratio Matematica*, **15**-3 (1982), 807–816.

[13] Dabrowski, A. R., A note on a theorem of Berkes and Philipp for dependent sequences, *Statistic and Probability Letters*, **1** (1982), 53–55.

[14] Deddens, J., Yang, T. and Peligrad, M., On strong consistency of kernel estimators under dependence assumptions, in *Proceedings of the Sixth Pannonian Symposium of Mathematical Statistics*, Reidel, Dordrecht, 1987, 33–43.

[15] Dehling, H., and Philipp, W., Almost sure invariance principles for weakly dependent vector valued random variables, *Ann. Probability*, **10** (1982), 689–701.

[16] Dehling, H., A note on a theorem of Berkes and Philipp, *Z. Wahr. verw. Gebiete*, **62** (1983), 39–42.

[17] Doeblin, W., Exposé de la théorie des chaînes simples constantes de Markov à un nombre fini d'états, *Rev. Math. de l'Union Interbalkanique*, **2** (1937), 77–105.

[18] Doukhan, P., Massart, P., Rio, E., The functional central limit theorem for strongly mixing processes, *Ann. Inst. H. Poincaré Probab. Statist.*, **30** (1994), 63–82.

[19] Doukhan, P., Massart, P., Rio, E., Invariance principles for absolutely regular empirical processes, *Ann. Inst. H. Poincaré Probab. Statist.*, **31** (1995), 393–427.

[20] Dudley, R.M., Distances of probability measures and random variables, *Ann. Math. Stat.*, **39** (1968), 1563–1572.

[21] Dudley, R.M., Central limit theorems for empirical measures, *Ann. Probab.*, **6** (1978), 899–929.

[22] Dudley, R.M., and Philipp, W., Invariance principles for sums of Banach space valued random elements and empirical processes, *Z. Wahr. verw. Gebiete*, **62** (1983), 509–552.

[23] Eberlein, E., Strong approximation of very weak Bernoulli processes, *Z. Wahr. verw. Gebiete*, **62** (1983), 17–37.

[24] Goldstein, S., Maximal coupling, *Z. Wahr. verw. Gebiete*, **46** (1979), 193–204.

[25] Griffeath, D., A maximal coupling for Markov chains, *Z. Wahr. verw. Gebiete*, **31** (1975), 95–106.

[26] Guillou, A., and Merlevède, F., Second-order properties of the blocks of blocks bootstrap for density estimators for continuous time processes, submitted, 2001.

[27] Hall, P. and Heyde, C.C., *Martingale Limit Theory and Its Application*. North-Holland, New York, 1980.

[28] Ibragimov, I.A., Some limit theorems for stationary processes, *Teor. Verojatnost. i Primenen.*, **7** (1962), 361–392.

[29] Ibragimov, I.A. and Linnik, Yu. V., *Independent and Stationary Sequences of Random Variables*, Wolters-Noorhoff, Groningen, 1971.

[30] Kuelbs, J. and Philipp, W., Almost sure invariance principles for partial sums of mixing *B* valued random variables, *Ann. Probab.*, **8** (1980), 1003–1036.

[31] Major, P., On the invariance principle for sums of identically distributed random variables, *J. Multivariate Anal.*, **8** (1978), 487–517.

[32] Massart, P., Invariance principles for empirical processes: The weakly dependent case, Ph.D. thesis, University of Paris, 1988.

[33] Merlevède, F., and Peligrad, M., The functional central limit theorem for strong mixing sequences of random variables, *Ann. Probab.*, **28**-3 (2000), 1336–1352.

[34] Merlevède, F., On the central limit theorem and its weak invariance principle for strongly mixing sequences with values in a Hilbert space via martingale approximation, submitted, 2001.

[35] Nobel, A. Dembo, A., A note on uniform laws of averages for dependent processes, *Stat. and Prob. Letters*, **17** (1993), 169–172.

[36] Novak, S., A note on uniform laws of averages for dependent processes, *Stat. and Prob. Letters*, **17** (2000), 169–172.

[37] Ornstein, D., *Ergodic Theory, Randomness, and Dynamical Systems*, Yale University Press, New Haven, London, 1974.

[38] Peligrad, M., A note on the uniform laws for dependent processes via coupling, *J. Theoret. Probab.*, **14** (2001-a), 979–988.

[39] Peligrad, M., 2001-b. Some remarks on coupling of dependent random variables. *Stat. and Prob. Letters*, to appear.

[40] Philipp, W., Almost sure invariance principles for sums of B-valued random variables, in *Probability in Banach Spaces*, Proc. Conf. Oberwolfach 1978, Lecture Notes in Math. 709, Springer, New York, Heidelberg, Berlin, 1979, 171–193.

[41] Philipp, W., Invariance principles for sums of mixing random elements and the multivariate empirical process, in *Colloq. Math. Soc. János Bolyai* 36: *Limit Theorems in Probability and Statistics, Veszprém, Hungary*, North-Holland, Amsterdam, 1982, 843–873.

[42] Pollard, D., *Convergence of Stochastic Processes*, Springer, New York, 1984.

[43] Rio, E., The functional law of the iterated logarithm for stationary strongly mixing sequences, *Ann. Probability*, **23** (1995), 1188–1203.

[44] Rio, E., Théorie Asymtotique des Processus Aléatoires Faiblement Dépendants, Mathematiques et Applications de la SMAI 31, Springer, Berlin, Heidelberg, 2000.

[45] Schwarz, G., Finitely determined processes–an indiscrete approach, *J. Math. Anal. Appl.*, **76** (1980), 146–158.

[46] Shields, P., *The Theory of Bernoulli Shifts*, University of Chicago Press, Chicago, 1973.

[47] Strassen, V., The existence of probability measures with given marginals, *Ann. Math. Statist.*, **36** (1965), 423–439.

[48] Vapnik, V.N. and Cervonenkis, A. Ya., Necessary and sufficient conditions for the uniform convergence of means to their expectations, *Theory Probab. Appl.*, **26** (1981), 532–553.

[49] Vasershtein, L.N., Markov processes on countable product spaces describing large systems of automata, *Problemy Peredači Informacii*, **3** (1969), 64–72.

[50] Yu, B., Rates of convergence and central limit theorems for empirical processes of stationary mixing sequences, Technical Report 260, Department of Statistics, University of California at Berkeley, 1991.

[51] Yukich, J.E., Rates of convergence for classes of functions: The non-i.i.d. case, *J. Multivariate Anal.*, **20** (1986), 175–189.

Florence Merlevède
L.S.T.A. Université Pierre et Marie Curie
Université Paris VI
Tour 45-55, 4 place Jussieu
75252 Paris cedex 5, France
merleve@ccr.jussieu.fr

Magda Peligrad
Department of Mathematical Sciences
University of Cincinnati
P.O. Box 210025
Cincinnati, OH 45221-0025, U.S.A.
peligrm@math.uc.edu

Empirical Processes of Residuals

István Berkes[1] and Lajos Horváth

Abstract. We review the asymptotic behaviour of the empirical processes of residuals (parameter estimated variables) of independent, identically distributed random variables, ARMA(p, q), ARCH(p) and GARCH(p, q) sequences.

1 Independent, identically distributed observations

Let $X_1, X_2, \ldots$ be independent, identically distributed random variables with a common distribution function F. The empirical distribution function F_n of the sample $X_1, X_2, \ldots, X_n$, i.e.,

$$F_n(x) = \frac{1}{n} \sum_{1 \le i \le n} I\{X_i \le x\}, \qquad -\infty < x < \infty$$

is used to estimate F. Following Doob's (1949) heuristic, Donsker (1952) proved that

$$e_n(x) \overset{\mathcal{D}[-\infty,\infty]}{\longrightarrow} B(F(x)), \tag{1.1}$$

where $e_n(x) = n^{1/2}(F_n(x) - F(x))$ is the empirical process of $X_1, \ldots, X_n$ and $B(t)$ stands for a Brownian bridge. Here, and in the rest of the paper, the symbol $\mathcal{D}$ in a convergence relation means convergence in distribution in $\mathbf{R}$ and the symbols $\mathcal{D}[0, 1]$, $\mathcal{D}[-\infty, \infty]$ and $\mathcal{D}([-\infty, \infty] \times [0, 1])$ indicate convergence in the corresponding Skorokhod space. We say that the stochastic process $\{B(t),\ \ 0 \le t \le 1\}$ is a Brownian bridge if it is Gaussian with $EB(t) = 0$ and $EB(t)B(s) = \min(t, s) - ts,\ \ 0 \le t, s \le 1$. For a proof of (1.1) we refer to Billingsley (1968). An alternative proof, based on strong approximation, is in Csörgő and Révész (1981). For results on the properties of the Brownian bridge we refer to Csörgő and Révész (1981) and Shorack and Wellner (1986). Relation (1.1) and the continuous mapping theorem immediately yield that

$$\sup_{-\infty<x<\infty} |e_n(x)| \overset{\mathcal{D}}{\to} \sup_{-\infty<x<\infty} |B(F(x))| \tag{1.2}$$

(Kolmogorov (1933)),

$$\sup_{-\infty<x<\infty} e_n(x) \overset{\mathcal{D}}{\to} \sup_{-\infty<x<\infty} B(F(x)) \tag{1.3}$$

(Smirnov (1939)),

$$\int_{-\infty}^{\infty} e_n^2(x)dF(x) \overset{\mathcal{D}}{\to} \int_{-\infty}^{\infty} B^2(F(x))dF(x) \tag{1.4}$$

[1] Supported by the Hungarian National Foundation for Scientific Research grants T 29621 and T 37886.

(Smirnov (1937), Cramér (1946)),

$$\sup_{-\infty<x<\infty} e_n(x) - \inf_{-\infty<x<\infty} e_n(x) \xrightarrow{\mathcal{D}} \sup_{-\infty<x<\infty} B(F(x)) - \inf_{-\infty<x<\infty} B(F(x)) \tag{1.5}$$

(Kuiper (1960)),

$$\int_{-\infty}^{\infty} \frac{e_n^2(x)}{F(x)(1-F(x))} dF(x) \xrightarrow{\mathcal{D}} \int_{-\infty}^{\infty} \frac{B^2(F(x))}{F(x)(1-F(x))} dF(x) \tag{1.6}$$

(Anderson and Darling (1952)), and

$$\begin{aligned} &\int_{-\infty}^{\infty} e_n^2(x) dF(x) - \left(\int_{-\infty}^{\infty} e_n(x) dF(x) \right)^2 \\ &\qquad \xrightarrow{\mathcal{D}} \int_{-\infty}^{\infty} B^2(F(x)) dF(x) - \left(\int_{-\infty}^{\infty} B(F(x)) dF(x) \right)^2 \end{aligned} \tag{1.7}$$

(Watson (1961, 1967)).

We wish to point out that in case of a continuous F, the limits in (1.2)– (1.7) do not depend on F, i.e., $F(x)$ can be replaced by the distribution function of the uniform on $[0, 1]$ distribution. In this case the distribution functions of the limits are known and tables are available (cf. Shorack and Wellner (1986)).

The rate of convergence for strong approximations in (1.1) was investigated by several authors. The most complete result was obtained by Komlós, Major and Tusnády (1975, 1976). They constructed a sequence of Brownian bridges $\{B_n(t),\ \ 0 \le t \le 1\}$ such that

$$\sup_{-\infty<x<\infty} |e_n(x) - B_n(F(x))| \overset{\text{a.s.}}{=} O(n^{-1/2} \log n) \tag{1.8}$$

and proved that there are no Brownian bridges B_n^* satisfying

$$\sup_{-\infty<x<\infty} |e_n(x) - B_n^*(F(x))| = o_P(n^{-1/2} \log n) \tag{1.9}$$

(cf. also Chapter 3 in Csörgő and Horváth (1993)). For weighted versions of (1.8) we refer to Csörgő, Csörgő, Horváth, and Mason (1986) and Mason and van Zwet (1987).

In order to improve the power of asymptotic goodness of fit tests, weighted versions of (1.1)–(1.7) have been suggested. Surprisingly, weak convergence in a weighted metric does not imply the convergence in distribution of all functionals of the weighted empirical process. Let q be a positive weight function on $(0, 1)$, nondecreasing in a neighborhood of 0 and nonincreasing in a neighborhood of 1. Then there are Brownian bridges $\{B_n(t),\ \ 0 \le t \le 1\}$ such that

$$\sup_{-\infty<x<\infty} |e_n(x) - B_n(F(x))|/q(F(x)) = o_P(1) \tag{1.10}$$

if and only if

$$\int_0^1 \frac{1}{t(1-t)} \exp(-cq^2(t)/(t(1-t)))dt < \infty \ \text{ for all } \ c > 0.$$

For example,

$$\sup_{-\infty<x<\infty} |e_n(x)|/q(F(x)) \xrightarrow{\mathcal{D}} \sup_{-\infty<x<\infty} |B(F(x))|/q(F(x)) \tag{1.11}$$

if and only if

$$\int_0^1 \frac{1}{t(1-t)} \exp(-cq^2(t)/(t(1-t)))dt < \infty \text{ for some } c > 0$$

(cf. Csörgő, Csörgő, Horváth and Mason (1986), Mason and van Zwet (1987), and Csörgő and Horváth (1993)).

Clearly, (1.10) and (1.11) do not hold if we choose $q(t) = (t(1-t))^{1/2}$ as weight function. We note that $(t(1-t))^{1/2}$ is a natural choice since it is the standard deviation of $B(t)$. Eicker (1979) and Jaeschke (1979), assuming that F is continuous, proved that

$$\lim_{n\to\infty} P\{(2\log\log n)^{1/2} \sup_{-\infty<x<\infty} |e_n(x)|/(F(x)(1-F(x)))^{1/2} \le t$$
$$+ 2\log\log n + \frac{1}{2}\log\log\log n - \frac{1}{2}\log\pi\} = \exp(-2e^{-t}) \tag{1.12}$$

for all t.

In hypothesis testing, F is sometimes assumed to be in parametric form. This means that there is a given function $F_0(x;\boldsymbol{\theta})$, $\boldsymbol{\theta} = (\theta_1, \ldots, \theta_d) \in \Theta \subset \mathbf{R}^d$ such that

$$F(x) = F_0(x, \boldsymbol{\theta}_0) \quad \text{for some } \boldsymbol{\theta}_0 \in \Theta.$$

Of course, $\boldsymbol{\theta}_0$ is unknown. First, the unknown parameter is estimated from the sample $X_1, X_2, \ldots, X_n$ by an estimator $\hat{\boldsymbol{\theta}}_n$. Thus, instead of the empirical process $e_n(x)$, we have to investigate the asymptotic properties of the parameter estimated empirical process

$$\hat{e}_n(x) = n^{1/2}(F_n(x) - F_0(x;\hat{\boldsymbol{\theta}}_n)), \qquad -\infty < x < \infty.$$

Let

$$\mathbf{h}(x,\boldsymbol{\theta}) = \left(\frac{\partial}{\partial\theta_1}F_0(x;\boldsymbol{\theta}), \ldots, \frac{\partial}{\partial\theta_d}F_0(x;\boldsymbol{\theta})\right).$$

Using the mean value theorem (assuming some smoothness conditions on $F(x;\boldsymbol{\theta})$), we get

$$\sup_{-\infty<x<\infty} |\hat{e}_n(x) - \{e_n(x) - \mathbf{h}(x;\boldsymbol{\theta}_0)n^{1/2}(\hat{\boldsymbol{\theta}}_n - \boldsymbol{\theta}_0)^T\}| = o_P(1). \tag{1.13}$$

The next assumption says that the estimator is asymptotically equivalent to a partial sum of i.i.d. random variables, i.e.,

$$\hat{\boldsymbol{\theta}}_n - \boldsymbol{\theta}_0 = \frac{1}{n}\sum_{1\le i\le n} \boldsymbol{\ell}(X_i) + o_P(n^{-1/2}) \tag{1.14}$$

with some function $\boldsymbol{\ell}$ satisfying

$$E\boldsymbol{\ell}(X_1) = \mathbf{0} \quad \text{and} \quad E||\boldsymbol{\ell}(X_1)||^2 < \infty \tag{1.15}$$

($||\cdot||$ denotes the Euclidean norm). By (1.14) we have

$$n^{1/2}(\hat{\boldsymbol{\theta}}_n - \boldsymbol{\theta}_0) = \int_{-\infty}^{\infty} \boldsymbol{\ell}(x)de_n(x) + o_P(1). \tag{1.16}$$

Now (1.8), (1.13) and (1.16) imply that

$$\sup_{-\infty x<\infty} |\hat{e}_n(x) - \{B_n(F(x)) - \mathbf{h}(x;\boldsymbol{\theta}_0) \int_{-\infty}^{\infty} \boldsymbol{\ell}^T(u) dB_n(F(u))\}| = o_P(1). \tag{1.17}$$

Let

$$\Gamma(x) = B(F(x)) - \mathbf{h}(x;\boldsymbol{\theta}_0) \int_{-\infty}^{\infty} \boldsymbol{\ell}^T(u) dB(F(u))$$

with some Brownian bridge $\{B(t),\ 0 \le t \le 1\}$. By (1.17) we have

$$\hat{e}_n(x) \overset{\mathcal{D}[-\infty,\infty]}{\longrightarrow} \Gamma(x). \tag{1.18}$$

In general, the process $\Gamma(x)$ depends on the unknown $\boldsymbol{\theta}_0$ so (1.18) may not look useful for applications. Before we discuss the applicability of (1.18), we provide some references where proofs of (1.18) can be found. Durbin (1973) introduced the parameter estimated process and obtained the first results. Burke et al. (1979) proved (1.18) under very general conditions. For surveys we refer to Csörgő and Révész (1981) and Csörgő (1983).

We note that $\Gamma(x)$ does not depend on $\boldsymbol{\theta}_0$ in location, scale and location/scale families (cf. Lilliefors (1967), Margolin and Maurer (1976), Tyurin (1984) and Stephans (1986a, b).)

Khmaladze (1979) and Dzaparidze and Nikulin (1976) searched for a kernel $k(x, y)$ such that

$$\int_{-\infty}^{\infty}\int_{-\infty}^{\infty} k(x, y) d\hat{e}_n(x) d\hat{e}_n(y) \overset{\mathcal{D}}{\to} \xi \tag{1.19}$$

and the distribution function of ξ can be calculated explicitly. Unfortunately, no such kernel was found. The result in (1.19) was generalized by Csörgő (1983) who showed that

$$\int_{-\infty}^{\infty} k(x, y) d\hat{e}_n(y) \overset{\mathcal{D}[0,1]}{\longrightarrow} \int_{-\infty}^{\infty} k(x, y) d\Gamma(y). \tag{1.20}$$

Again, no transformation is known to give a prescribed distribution of any of the functionals of $\int k(x, y) d\Gamma(x)$. (Due to the probability integral transformation, it suffices to consider convergence in $\mathcal{D}[0, 1]$ in (1.20).) Horváth (1985) showed that (1.20) remains true if $k(x, y)$ is replaced by a sequence of suitable random functions $k_n(x, y)$ computed from the data such that

$$\int_{-\infty}^{\infty} k_n(x, y) d\hat{e}_n(x) \overset{\mathcal{D}[0,1]}{\longrightarrow} B(x),$$

where $\{B(x),\ 0 \le x \le 1\}$ stands for a Brownian bridge.

The distribution of the L_2-functionals of $\Gamma(x)$ has received special attention in the literature. Darling (1955, 1957), Kac, Kiefer and Wolfowitz (1955) and Sukhatme (1972) showed that

$$\int_{-\infty}^{\infty} \Gamma^2(x) dF_0(x;\boldsymbol{\theta}_0) = \sum_{1\le j<\infty} \lambda_j N_j^2, \tag{1.21}$$

where $N_1, N_2, \ldots$ are independent, standard normal random variables and $\lambda_1, \lambda_2, \ldots$ are the eigenvalues of the associated covariance function. They also provided a method to compute $\lambda_1, \lambda_2, \ldots$. Durbin, Knott, and Taylor (1975) observed that in (1.21) $\lambda_j \to 0$ as $j \to \infty$ very fast and therefore they suggested the computation of the principal components in (1.21). The analytical methods developed for the maximum of Gaussian processes can be used to get bounds for the distributions of $\sup \Gamma(x)$ and $\sup |\Gamma(x)|$ (cf. Piterbarg (1996)).

Khmaladze (1981, 1988, 1993) studied the properties of $\hat{e}_n(x)$ in the context of martingale theory and showed that there is a transformation $\mathcal{T}$ of $\hat{e}_n(x)$ such that $\mathcal{T}(\hat{e}_n(x)) \overset{\mathcal{D}(-\infty,\infty)}{\longrightarrow} W(x)$, where $\{W(x),\ 0 \le x < \infty\}$ is a Wiener process. Since $\mathcal{T}$ may depend on unknown parameters and functions, he constructed random (data-driven) transformations $\hat{\mathcal{T}}_n$ such that $\hat{\mathcal{T}}_n(\hat{e}_n(x)) \overset{\mathcal{D}(-\infty,\infty)}{\longrightarrow} W(x)$.

Rao (1972) and Durbin (1973) observed that

$$n^{1/2}(F_n(x) - F_0(x;\boldsymbol{\theta}_n^*)) \overset{\mathcal{D}[-\infty,\infty]}{\longrightarrow} B(F_0(x;\boldsymbol{\theta}_0)), \tag{1.22}$$

where B is a Brownian bridge and $\boldsymbol{\theta}_n^*$ is the maximum likelihood estimator of $\boldsymbol{\theta}_0$ based on a subsample of $[n/2]$ observations. The result in (1.22) is referred to as the "half-sample rule."

The review paper of Martynov (1992) contains several applications of parameter estimated processes. It also contains comparisons of the different techniques mentioned above.

2 Residuals of ARMA(p, q) processes

Consider the following ARMA(p, q) (autoregressive moving average of order (p, q)) time series model:

$$Y_k = \nu_1 Y_{k-1} + \nu_2 Y_{k-2} + \cdots + \nu_p Y_{k-p} + \epsilon_k + \gamma_1 \epsilon_{k-1} + \cdots + \gamma_q \epsilon_{k-q},$$

where $\{\epsilon_i,\ -\infty < i < \infty\}$ is a sequence of independent, identically distributed random variables with common distribution function J. Assume that the parameters ν_i and γ_j are chosen in such a way that the sequence $\{Y_k,\ -\infty < k < \infty\}$ is strictly stationary and invertible (cf. Brockwell and Davis (1987) for details). Denote the parameter of the process $\{Y_k,\ -\infty < k < \infty\}$ by $\boldsymbol{\theta} = (\nu_1, \nu_2, \ldots, \nu_p, \gamma_1, \ldots, \gamma_q)$. The residuals are defined by the recursion

$$\hat{\epsilon}_k = Y_k - \hat{\nu}_{1,n} Y_{k-1} - \hat{\nu}_{2,n} Y_{k-2} - \cdots - \hat{\nu}_{p,n} Y_{k-p} - \hat{\gamma}_{1,n} \hat{\epsilon}_{k-1} - \cdots - \hat{\gamma}_{q,n} \hat{\epsilon}_{k-q},$$

where $\hat{\boldsymbol{\theta}}_n = (\hat{\nu}_{1,n}, \hat{\nu}_{2,n}, \ldots, \hat{\nu}_{p,n}, \hat{\gamma}_{1,n}, \hat{\gamma}_{2,n}, \ldots, \hat{\gamma}_{q,n})$ is an estimator for $\boldsymbol{\theta}$ from $Y_1, Y_2, \ldots, Y_n$ and $\hat{\epsilon}_i$, $1 \le i \le \max(p, q)$ are arbitrarily chosen initial values. The sequential process of the residuals is

$$e_n^*(x, s) = n^{1/2} s (J_n^*(x, s) - J(x)),$$

where

$$J_n^*(x, s) = \frac{1}{ns} \sum_{1 \le i \le ns} I\{\hat{\epsilon}_i \le x\}$$

stands for the empirical distribution function of $\hat{\epsilon}_1, \ldots, \hat{\epsilon}_{[ns]}$.

Boldin (1982, 1989), Koul and Levental (1989), and Koul (1991) investigated the weak convergence of the empirical process of the residuals in some ARMA models under various conditions. The results were applied to goodness-of-fit testing.

Let

$$J_n(x, s) = \frac{1}{ns} \sum_{1 \le i \le ns} I\{\epsilon_i \le x\}$$

denote the empirical distribution function of the innovations $\epsilon_1, \ldots, \epsilon_{[ns]}$. Assuming only that $||\hat{\boldsymbol{\theta}}_n - \boldsymbol{\theta}|| = O_P(n^{-1/2})$, Bai (1994) proved that

$$\sup_{0 \le s \le 1} \sup_{-\infty < x < \infty} n^{1/2} s |J_n^*(x, s) - J_n(x, s)| = o_P(1). \tag{2.1}$$

The result in (2.1) with the weak convergence of the sequential empirical process

$$y_n(x, s) = n^{1/2} s (J_n(x, s) - J(x))$$

of $\epsilon_1, \epsilon_2, \ldots, \epsilon_n$ implies the main result in Bai (1994):

$$e_n^*(x, s) \overset{\mathcal{D}([-\infty,\infty]\times[0,1])}{\longrightarrow} K(J(x), s), \tag{2.2}$$

where $K(t, s)$ is a Kiefer process, i.e., a Gaussian process with $EK(t, s) = 0$ and $EK(t, s)K(t', s') = \min(s, s')(\min(t, t') - tt')$. Comparing Bai's (1994) result in (2.2) with the behavior of the parameter estimated empirical process in Section 1 and the residuals of ARCH and GARCH sequences in Sections 3 and 4, we see that the estimation of the parameter $\boldsymbol{\theta}$ does not have any effect on the limit behavior of the empirical process of the residuals.

The weak convergence in (2.2) was used by Bai (1994) to detect changes in the structure of time series models (cf. Chapter 4 in Csörgő and Horváth (1997)).

3 ARCH(p) residuals

The classical ARCH(p) (autoregressive conditionally heteroskedastic of order p) model of Engle (1982) is defined by the equations

$$y_k = \sigma_k \epsilon_k \tag{3.1}$$

and

$$\sigma_k^2 = \omega + \sum_{1 \le j \le p} \alpha_j y_{k-j}^2, \tag{3.2}$$

where $\{\epsilon_i, \ -\infty < i < \infty\}$ are independent, identically distributed random variables. The parameters $\omega, \alpha_1, \ldots, \alpha_p$ satisfy

$$\omega > 0 \text{ and } \alpha_j \ge 0, \ 1 \le j \le p. \tag{3.3}$$

Following Giraitis, Kokoszka and Leipus (2000) we assume that

$$E\epsilon_0 = 0, \ E\epsilon_0^2 = 1 \text{ and } E\epsilon_0^4 < \infty. \tag{3.4}$$

(We note that (3.4) is not necessary for the existence of a unique stationary solution of (3.1) and (3.2). More details will be given in the next section.) If

$$E\epsilon_0^4 \sum_{1 \le j \le p} \alpha_j < 1, \tag{3.5}$$

then $Ey_0^4 < \infty$ and the squares y_k^2 have a Volterra representation

$$y_k^2 = \epsilon_k^2 \left(\omega + \sum_{\ell=1}^{\infty} \sum_{j_1, \ldots, j_\ell = 1}^{p} \alpha_{j_1} \alpha_{j_2} \cdots \alpha_{j_\ell} \epsilon_{k-j_1}^2 \epsilon_{k-j_1-j_2}^2 \cdots \epsilon_{k-j_1-\cdots-j_\ell}^2 \right)$$

(cf. Giraitis, Kokoszka and Leipus (2000)). The Volterra representation immediately yields that y_k^2 is an ergodic sequence.

Suppose we observe a sample $y_1, \ldots, y_n$ and use an estimator $\hat{\boldsymbol{\theta}}_n$ to estimate $\boldsymbol{\theta} = (\omega, \alpha_1, \ldots, \alpha_p)$. The estimator $\hat{\boldsymbol{\theta}}_n$ in ARCH(p) models will not satisfy a decomposition of type (1.14) due to the dependence structure of ARCH(p). However, nearly all estimators for $\boldsymbol{\theta}$ considered so far satisfy

$$\hat{\boldsymbol{\theta}}_n - \boldsymbol{\theta} = \frac{1}{n} \sum_{1 \le j \le n} \boldsymbol{\ell}(\epsilon_j^2)\mathbf{A}(\epsilon_{j-1}, \epsilon_{j-2}, \ldots) + o_P(n^{-1/2}) \tag{3.6}$$

with vector $\boldsymbol{\ell}$ and matrix $\mathbf{A}$. Horváth, Kokoszka and Teyssière (2001a) proved that the pseudo maximum, conditional least squares and the conditional likelihood estimators satisfy (3.6). Gourièroux (1997) surveys several methods applicable to parameter estimation in ARCH models. Horváth and Liese (2001) study M-estimators. In general, if $n^{1/2}(\hat{\boldsymbol{\theta}}_n - \boldsymbol{\theta})$ is asymptotically normal, a linear representation, like (3.6), will hold under suitable smoothness conditions.

Now we define the residuals $\hat{\epsilon}_k^2$, $p < k \le n$. The squared residuals are defined as

$$\hat{\epsilon}_k^2 = \frac{y_k^2}{w_k(\hat{\boldsymbol{\theta}}_n)}, \qquad p < k \le n, \tag{3.7}$$

where

$$w_k(\mathbf{u}) = u_0 + \sum_{1 \le j \le p} u_j y_{k-j}^2, \qquad \mathbf{u} = (u_0, u_1, \ldots, u_p). \tag{3.8}$$

Horváth, Kokoszka, and Teyssière (2001a) introduced the sequential (or two-time parameter) empirical process of the squared residuals

$$\tilde{e}_n(x, s) = n^{1/2} s(\tilde{H}_n(x, s) - H(x)),$$

where

$$\tilde{H}_n(x, s) = \frac{1}{ns} \sum_{p < k \le ns} I\{\hat{\epsilon}_k^2 \le x\}, \qquad p/n < s \le 1,$$

$\tilde{H}_n(x, s) = 0$ if $0 < s \le p/n$ and $H(x)$ denotes the distribution function of ϵ_0^2. We note that $\tilde{e}_n(x, 1)$ is the empirical process of the squared residuals $\hat{\epsilon}_k^2$, $p < k \le n$. Conditionally on $y_{k-1}, y_{k-2}, \ldots, \sigma_k^2$ is the scale parameter for y_k^2 and since $w_k(\hat{\boldsymbol{\theta}}_n)$ is an estimator for σ_n^2, one can look at $\tilde{e}_n(x, s)$ as a parameter estimated empirical process. Indeed, this approach was utilized by Horváth, Kokoszka and Teyssière (2001a). Using the mean value theorem one can show that

$$
\begin{aligned}
\tilde{e}_n(x,s) &= n^{1/2}s\left(\frac{1}{ns}\sum_{p<k\le ns}\left(I\left\{\epsilon_k^2 \le x\frac{w_k(\hat{\boldsymbol\theta}_n)}{\sigma_k^2}\right\} - H\left(x\frac{w_k(\hat{\boldsymbol\theta}_n)}{\sigma_k^2}\right)\right)\right.\\
&\quad \left. + xh(x)\frac{1}{ns}\sum_{p<k\le ns}\frac{w_k(\hat{\boldsymbol\theta}_n)-\sigma_k^2}{\sigma_k^2}\right) + o_P(1)\\
&= n^{-1/2}\sum_{p<k\le ns}\left(I\left\{\epsilon_k^2 \le x\frac{w_k(\hat{\boldsymbol\theta}_n)}{\sigma_k^2}\right\} - H\left(x\frac{w_k(\hat{\boldsymbol\theta}_n)}{\sigma_k^2}\right)\right)\\
&\quad + xh(x)n^{1/2}(\hat{\boldsymbol\theta}_n - \boldsymbol\theta)s\frac{1}{ns}\sum_{p<k\le ns}\frac{(w_k'(\boldsymbol\theta))^T}{\sigma_k^2} + o_P(1)
\end{aligned}
\tag{3.9}
$$

where $h(x) = H'(x)$ and $w_k'(\mathbf{u}) = \left(\frac{\partial}{\partial u_0}w_k(\mathbf{u}), \ldots, \frac{\partial}{\partial u_p}w_k(\mathbf{u})\right)$. Next one needs to show

$$
\begin{aligned}
&\left\{\left(n^{-1/2}\sum_{p<k\le ns}(I\{\epsilon_k^2 \le x\} - H(x), n^{1/2}(\hat{\boldsymbol\theta}_n - \boldsymbol\theta))\right), 0 \le x < \infty, 0 \le s \le 1\right\}\\
&\quad \overset{\mathcal{D}([0,\infty]\times[0,1])}{\longrightarrow} \{(K(H(x),s),\boldsymbol\eta),\ 0 \le x < \infty,\ 0 \le s \le 1\},
\end{aligned}
\tag{3.10}
$$

where $(K(H(x),s),\boldsymbol\eta)$ is Gaussian. The ergodic theorem gives that

$$
\frac{1}{n}\sum_{p<k\le ns}\frac{w_k'(\boldsymbol\theta)}{\sigma_k^2} \overset{P}{\to} \mathbf{c}. \tag{3.11}
$$

Heuristically, $w_k(\hat{\boldsymbol\theta}_n)/\sigma_k^2$ is essentially 1, so by (3.9)–(3.11) we have

$$
\tilde{e}_n(x,s) \overset{\mathcal{D}([0,\infty]\times[0,1])}{\longrightarrow} \tilde{\Gamma}(x,s),
$$

where

$$
\tilde{\Gamma}(x,s) = K(H(x),s) + sxh(x)\boldsymbol\eta\mathbf{c}^T.
$$

The process $\tilde{\Gamma}(x,1)$ looks like $\Gamma(x)$ in Section 1. However, $\boldsymbol\eta$ is not an integral with respect to $K(H(x),1)$, and therefore it is not clear if the computations of the distribution functions of functionals of $\Gamma(x)$ can be applied to functionals of $\tilde{\Gamma}(x,s)$. The bootstrap method in Horváth, Kokoszka and Teyssière (2001b) provides an alternative way to get critical values.

The squared residuals have also been used to estimate the correlation between ϵ_0^2 and ϵ_k^2. Let

$$
\hat{r}_k = \frac{1}{\hat{r}_0}\sum_{\max(k,p)<i\le n}(\hat{\epsilon}_i^2 - 1)(\hat{\epsilon}_{i-k}^2 - 1), \qquad 1 \le k < n
$$

with

$$
\hat{r}_0 = \sum_{p<i\le n}(\hat{\epsilon}_i^2 - 1)^2.
$$

Results on the asymptotic properties of $\hat{r}_k$ can be found in Li and Mak (1994) and Horváth and Kokoszka (2001).

4 GARCH(p, q) residuals

Bollerslev (1986) introduced the following generalizations of ARCH models:

$$y_k = \sigma_k \epsilon_k \tag{4.1}$$

and

$$\sigma_k^2 = \omega + \sum_{1 \le i \le p} \alpha_i y_{k-i}^2 + \sum_{1 \le j \le q} \beta_j \sigma_{k-j}^2, \tag{4.2}$$

where $\{\epsilon_i,\ -\infty < i < \infty\}$ are independent, identically distributed random variables. The parameter vector of this GARCH(p, q) (generalized ARCH(p, q)) sequence will be denoted by $\boldsymbol{\theta} = (\omega, \alpha_1, \ldots \alpha_p, \beta_1, \ldots, \beta_q)$. Nelson (1990) and Bougerol and Picard (1992a, b) found necessary and sufficient conditions for the existence of a unique stationary solution of (4.1) and (4.2). Davis, Mikosch, and Basrak (1999) pointed out that $\{y_k,\ -\infty < k < \infty\}$ is strongly mixing with an exponential mixing rate if $E|\epsilon_0|^\delta < \infty$ with some $\delta > 0$ and ϵ_0^2 has a density on $(0, \infty)$. Berkes and Horváth (2001a) obtained strong approximations for the empirical process of $y_1, \ldots, y_n$. Representations for GARCH(p, q) are given in Berkes, Horváth and Kokoszka (2001a).

Lee and Hansen (1994) and Lumsdaine (1996) obtained asymptotic results for the quasi-maximum likelihood estimator of $\boldsymbol{\theta}$ in GARCH(1, 1). These results were extended by Berkes, Horváth, and Kokoszka (2001a) to the general case.

We continue with the formal definition of the quasi-maximum likelihood estimator for $\boldsymbol{\theta}$. First we define a sequence of functions by recursion. Let $\mathbf{u} = (x, \mathbf{s}, \mathbf{t}) \in \mathbf{R}^{p+q+1}$, $x \in \mathbf{R}$, $\mathbf{s} \in \mathbf{R}^p$ and $\mathbf{t} \in \mathbf{R}^q$. We start with the initial condition: If $q \ge p$, then

$$
\begin{aligned}
c_0(\mathbf{u}) &= x/(1 - (t_1 + \cdots + t_q)),\\
c_1(\mathbf{u}) &= s_1,\\
c_2(\mathbf{u}) &= s_2 + t_1 c_1(\mathbf{u}),\\
&\vdots\\
c_p(\mathbf{u}) &= s_p + t_1 c_{p-1}(\mathbf{u}) + \cdots + t_{p-1} c_1(\mathbf{u}),\\
c_{p+1}(\mathbf{u}) &= t_1 c_p(\mathbf{u}) + \cdots + t_p c_1(\mathbf{u}),\\
&\vdots\\
c_q(\mathbf{u}) &= t_1 c_{q-1}(\mathbf{u}) + \cdots + t_{q-1} c_1(\mathbf{u}),
\end{aligned}
$$

and if $q < p$, the equations above are replaced with

$$
\begin{aligned}
c_0(\mathbf{u}) &= x/(1 - (t_1 + \cdots + t_q)),\\
c_1(\mathbf{u}) &= s_1,\\
c_2(\mathbf{u}) &= s_2 + t_1 c_1(\mathbf{u}),\\
&\vdots\\
c_{q+1}(\mathbf{u}) &= s_{q+1} + t_1 c_q(\mathbf{u}) + \cdots + t_q c_1(\mathbf{u}),\\
&\vdots\\
c_p(\mathbf{u}) &= s_p + t_1 c_{p-1}(\mathbf{u}) + \cdots + t_q c_{p-q}(\mathbf{u}).
\end{aligned}
$$

In general, if $i > \max(p, q)$, then

$$c_i(\mathbf{u}) = t_1 c_{i-1}(\mathbf{u}) + t_2 c_{i-2}(\mathbf{u}) + \cdots + t_q c_{i-q}(\mathbf{u}).$$

One can show that

$$\sigma_k^2 = c_0(\boldsymbol{\theta}) + \sum_{1 \le i < \infty} c_i(\boldsymbol{\theta}) y_{k-i}^2.$$

The logarithm of the quasi-likelihood function of the sample $y_1, y_2, \ldots, y_n$ is given by

$$\hat{L}_n(\mathbf{u}) = \sum_{1 \le k \le n} -\frac{1}{2}\left\{\log \hat{w}_k(\mathbf{u}) + \frac{y_k^2}{\hat{w}_k(\mathbf{u})}\right\},$$

where

$$\hat{w}_k(\mathbf{u}) = c_0(\mathbf{u}) + \sum_{1 \le i \le k-1} c_i(\mathbf{u}) y_{k-i}^2.$$

The form of the likelihood function is motivated by the likelihood function of independent standard normal ϵ's. The quasi-maximum likelihood estimator is defined as

$$\hat{\boldsymbol{\theta}}_n = \arg\max_{\mathbf{u} \in U} \hat{L}_n(\mathbf{u})$$

where U is given by

$$U = \{\mathbf{u} : t_1 + t_2 + \cdots + t_q \le \rho_0 \text{ and } u_1 \le \min(x, s_1, s_2, \ldots, s_p, t_1, t_2, \ldots t_q)$$

$$\le \max(x, s_1, s_2, \ldots, s_p, t_1, t_2, \ldots, t_q) \le u_2\},$$

with some $0 < u_1 < u_2$, $0 < \rho_0 < 1$ and $qu_1 < \rho_0$, $\mathbf{u} = (x, s_1, \ldots s_p, t_1, \ldots, t_q)$. The choice of U guarantees the convergence of $\hat{\boldsymbol{\theta}}_n$. We note that Lee and Hansen (1994) and Lumsdaine (1996) assume that U is a small neighborhood of the unknown parameter of $\boldsymbol{\theta}$. Berkes, Horváth and Kokoszka (2001a) removed this assumption. The asymptotic normality of $n^{1/2}(\hat{\boldsymbol{\theta}}_n - \boldsymbol{\theta})$ was proven by Lee and Hansen (1994) and Lumsdaine (1996) under very strong conditions for GARCH(1, 1). Those regularity conditions were removed and the asymptotic normality of $n^{1/2}(\hat{\boldsymbol{\theta}}_n - \boldsymbol{\theta})$ was established for GARCH(p, q) models.

If the innovations $\epsilon_1^2, \ldots, e_n^2$ were observable, the sample autocovariances of the sequence would be

$$\rho_n(k) = \frac{1}{n} \sum_{k < i \le n} (\epsilon_i^2 - 1)(\epsilon_{i-k}^2 - 1).$$

Since $\epsilon_1^2, \ldots, \epsilon_n^2$ are not observable, we replace them with the squared residuals

$$\hat{\epsilon}_i^2 = y_i^2 / \hat{w}_i(\hat{\boldsymbol{\theta}}_n), \qquad 1 \le i \le n$$

and we shall use

$$\hat{\rho}_n(k) = \frac{1}{n} \sum_{k < i \le n} (\hat{\epsilon}_i^2 - 1)(\hat{\epsilon}_{i-k}^2 - 1).$$

Ling and Li (1997) introduced $\hat{\rho}_n(k)$ and obtained some asymptotic results assuming that ϵ_k, $-\infty < k < \infty$ are standard normal random variables. Berkes, Horváth and Kokoszka

(2001b) showed that $\hat{\rho}_n(i_1), \ldots, \hat{\rho}_n(i_k)$ are jointly asymptotically normal and they used this result to study weighted χ^2-statistics for the goodness of fit of GARCH processes.

Applications require the knowledge of the number of moments of the observations $y_1, y_2, \ldots, y_n$. The results in Nelson (1990), Goldie (1991) and Mikosch and Stărică (2000) connect, in the GARCH(1, 1) case, the number of moments of $y_1, \ldots, y_n$ with the solution of the equation $E(\beta_1 + \alpha_1\epsilon_0^2)^t = 1$. It can be shown under general conditions on the distribution of ϵ_0 that

$$P\{y_1 > x\} \sim cx^{-t} \quad \text{as} \quad x \to \infty.$$

Since α_1, β_1 are unknown and ϵ_i^2, $-\infty < i < \infty$ are unobservable, they must be estimated from the sample. This program was carried out by Berkes, Horváth, and Kokoszka (2000) who obtained an estimator of the tail index t in GARCH(1, 1) and showed its asymptotic normality. Similar results for GARCH(p, q) sequences have not been derived yet.

Similarly to ARCH sequences one can define the empirical process of the squared residuals in GARCH(p, q) sequences. The asymptotic results in Horváth, Kokoszka and Teyssière (2001a) assume the existence of Ey_0^4. However, empirical evidence shows that this may not hold in practice. Also, the limit of the empirical process of the squared residuals in ARCH(p) exists under very mild moment conditions on y_0. The main result on the empirical process of squared residuals in Horváth, Kokoszka and Teyssière (2001a) was extended to GARCH(p, q) square residuals by Berkes and Horváth (2001b).

References

[1] Anderson, T. W. and Darling, D. A., Asymptotic theory of certain "goodness of fit" criteria based on stochastic processes. *Ann. Math. Statist.*, **23** (1952), 193–212.

[2] Bai, J., Weak convergence of the sequential empirical processes of residuals in ARMA models. *Ann. Statist.*, **22** (1994), 2051–2061.

[3] Berkes, I. and Horváth, L., Strong approximation of the empirical process of GARCH sequences. *Ann. Appl. Probab.*, **11** (2001a), 789–809.

[4] Berkes, I. and Horváth, L., Limit results for the empirical process of squared residuals in GARCH models. Preprint, 2001b.

[5] Berkes, I., Horváth, L. and Kokoszka, P., Estimation of the maximal moment exponent of a GARCH(1, 1) sequence. Preprint, 2000.

[6] Berkes, I., Horváth, L. and Kokoszka, P., GARCH processes: structure and estimation. *Bernoulli*, to appear, 2001a.

[7] Berkes, I., Horváth, L. and Kokoszka, P., Asymptotics for GARCH squared residual correlations. Preprint, 2001b.

[8] Billingsley, P., *Convergence of Probability Measures*. Wiley, New York, 1968.

[9] Boldin, M. V., Estimation of the distribution of noise in an autoregression scheme. *Theory Probab. Appl.*, **27** (1982), 866–871.

[10] Boldin, M. V., On testing hypothesis in sliding average scheme by Kolmogorov–Smirnov and ω^2 tests. *Theory Probab. Appl.*, **34** (1989), 699–704.

[11] Bollerslev, T., Generalized autoregressive conditional heteroskedasticity. *J. Econometrics*, **31** (1986), 307–327.

[12] Bougerol, P. and Picard, N., Strict stationary of generalized autoregressive processes. *Ann. Probab.*, **20** (1992a), 1714–1730.

[13] Bougerol, P. and Picard, N., Stationarity of GARCH processes and of some non-negative time series. *J. Econometrics*, **52** (1992b), 115–127.

[14] Brockwell, P. J. and Davis, R. A., *Times Series: Theory and Method.* Springer-Verlag, New York, 1987.

[15] Burke, M. D., Csörgő, M., Csörgő, S. and Révész, P., Approximations of the empirical process when parameters are estimated. *Ann. Prob.*, **7** (1979), 790–810.

[16] Cramér, H., *Mathematical Methods of Statistics.* Princeton University Press, Princeton, NJ, 1946.

[17] Csörgő, M., *Quantile Processes with Statistical Applications.* SIAM, Philadelphia, 1983.

[18] Csörgő, M., Csörgő, S., Horváth, L. and Mason, D. M., Weighted empirical and quantile processes. *Ann. Prob.*, **14** (1986), 86–118.

[19] Csörgő, M. and Horváth, L., *Weighted Approximations in Probability and Statistics.* Wiley, Chichester, 1993.

[20] Csörgő, M. and Horváth, L., *Limit Theorems in Change-Point Analysis.* Wiley, Chichester, 1997.

[21] Csörgő, M. and Révész, P., *Strong Approximations in Probability and Statistics.* Academic Press, New York, Akadémiai Kiadó, Budapest, 1981.

[22] Csörgő, S., Kernel-transformed empirical processes. *J. Multivariate Analysis*, **13** (1983), 511–533.

[23] Darling, D. A., The Cramér–Smirnov test in the parametric case. *Ann. Math. Statist.*, **26** (1955), 1–20.

[24] Darling, D. A., The Kolmogorov–Smirnov, Cramér–von Mises tests. *Ann. Math. Statist.*, **28** (1957), 823–838.

[25] Davis, R. A., Mikosch, T. and Basrak, B., Sample ACF of multivariate stochastic recurrence equations with applications to GARCH. Preprint, 1999.

[26] Donsker, M., Justification and extension of Doob's heuristic approach to the Kolmogorov–Smirnov theorems. *Ann. Math. Statist.*, **23** (1952), 277–283.

[27] Doob, J. L., Heuristic approach to the Kolmogorov–Smirnov theorems. *Ann. Math. Statist.*, **20** (1949), 393–403.

[28] Durbin, J., Weak convergence of the sample distribution function when parameters are estimated. *Ann. Statist.*, **1** (1973), 279–290.

[29] Durbin, J., Knott, M. and Taylor, C., Components of Cramér–von Mises statistics II. *J. Royal Statist. Soc. B*, **37** (1975), 216–237.

[30] Dzaparidze, K. O. and Nikulin, M. S., The probability distributions of the Kolmogorov and ω^2 statistics for continuous distributions with shift and location parameters. *Zap. Naučn. Sem. Leningrad. Otdel. Mat. Inst. Steklov (LOMI)*, **85** (1976), 46–74 (in Russian).

[31] Eicker, F., The asymptotic distribution of the suprema of standardized empirical processes. *Ann. Statist.*, **7** (1979), 116–138.

[32] Engle, R. F., Autoregressive conditional heteroskedasticity with estimates of the variance of the United Kingdom inflation. *Econometrica*, **50** (1982), 987-1007.

[33] Giraitis, L., Kokoszka, P. and Leipus, R., Stationary ARCH models: dependence structure and central limit theorem. *Econometric Theory*, **16** (2000), 3–22.

[34] Goldie, C.M., Implicit renewal theory and tails of solutions of random equations. *Ann. Appl. Probab.*, **1** (1991), 126–166.

[35] Gourièroux, C., *ARCH Models and Financial Applications.* Springer-Verlag, Heidelberg, 1997.

[36] Horváth, L., Empirical kernel transforms of parameter-estimated empirical processes. *Acta Sci. Math. (Szeged)*, **48** (1985), 201–213.

[37] Horváth, L., Kokoszka, P., Large sample distribution of weighted sums of ARCH(p) square residual correlations. *Econometric Theory*, **17** (2001), 283–295.

[38] Horváth, L., Kokoszka, P. and Teyssière, G., Empirical process of the squared residuals of an ARCH sequence. *Ann. Statist.*, **29** (2001a), 445–469.

[39] Horváth, L., Kokoszka, P. and Teyssière, G., Bootstrap specification tests for ARCH. Preprint, 2001b.

[40] Horváth, L. and Liese, F., L_p-estimators in ARCH models. *J. Stat. Plan. Inf.*, to appear, 2001.

[41] Jaeschke, D., The asymptotic distribution of the supremum of the standardized empirical distribution function on subintervals. *Ann. Statist.*, **7** (1979), 108–115.

[42] Kac, M., Kiefer, J. and Wolfowitz, J., On tests of normality and other tests of goodness of fit based on the distance methods. *Ann. Math. Statist.*, **26** (1955), 189–211.

[43] Khmaladze, E. V., The use of ω^2 for testing parameteric hypothesis. *Theory Probab. Appl.*, **24** (1979), 283–301.

[44] Khmaladze, E. V., Martingale approach to the theory of goodness of fit tests. *Theory Probab. Appl.*, **26** (1981), 240–257.

[45] Khmaladze, E. V., An innovation approach in goodness-of-fit tests in R^m. *Ann. Statist.*, **16** (1988), 1503–1516.

[46] Khmaladze, E. V., Goodness of fit problem and scanning innovation martingales. *Ann. Statist.*, **21** (1993), 798–829.

[47] Kolmogorov, A. N., Sulla determinazione empirica di una legge di distribuzione. *Giorn. Inst. Ital. Attuari*, **4** (1933), 83–91.

[48] Komlós, J., Major, P. and Tusnády, G., An approximation of partial sums of independent R.V.'s and the sample DF. *I. Z. Wahrschein. verw. Gebiete*, **32** (1975), 111–131.

[49] Komlós, J., Major, P. and Tusnády, G., An approximation of partial sums of independent R.V.'s and the sample DF. II. *Z. Wahrschein. verw. Gebiete*, **34** (1976), 33–58.

[50] Koul, H. L., A weak convergence result useful in robust autoregression. *J. Statist. Plann. Inference*, **29** (1991), 1291–1308.

[51] Koul, H. L. and Levental, S., Weak convergence of the residual empirical process in explosive autoregression. *Ann. Statist.*, **17** (1989), 1784–1794.

[52] Kuiper, N. H., Tests concerning random points on a circle. *Proc. Nederl. Akad. Wetensch. Indag. Math. (A)*, **63** (1960), 38–47.

[53] Lee, S.-W. and Hansen, B. E., Asymptotic theory for the GARCH(1, 1) quasi-maximum likelihood estimator. *Econometric Theory*, **10** (1994), 29–52.

[54] Li, W. K. and Mak, T. K., On the squared residual autocorrelations in non-linear time series with conditional heteroskedasticity. *J. Time Series Analysis*, **15** (1994), 627–636.

[55] Lilliefors, H. W., On the Kolmogorov–Smirnov test for normality with mean and variance unknown. *J. Amer. Stat. Assoc.*, **62** (1967), 399–402.

[56] Ling, S. and Li, W. K., On fractionally integrated autoregressive moving-average time series models with conditional heteroskedasticity. *J. Amer. Stat. Assoc.*, **92** (1997), 1184–1194.

[57] Lumsdaine, R. L., Consistency and asymptotic normality of the quasi-maximum likelihood estimator in GARCH(1, 1) and covariance stationary GARCH(1, 1) models. *Econometrica*, **6** (1996), 575–596.

[58] Margolin, B. H. and Maurer, W., Tests of the Kolmogorov–Smirnov type for exponential data with unknown scale and related problems. *Biometrika*, **63** (1976), 149–160.

[59] Martynov, G. V., Statistical tests based on empirical processes and related questions. *J. Soviet Mathematics*, **61** (1992), 3–112.

[60] Mason, D. M. and van Zwet, W., A refinement of the KMT inequality for the uniform empirical process. *Ann. Prob.*, **15** (1987), 871–884.

[61] Mikosch, T. and Stărică, C., Limit theory for the sample autocorrelation and extremes of a GARCH(1, 1) process. *Ann. Statist.*, **28** (2000), 1427–1451.

[62] Nelson, D. B., Stationarity and persistence in GARCH(1, 1) model. *Econometric Theory*, **6** (1990), 318–334.

[63] Piterbarg, V. I., *Asymptotic Methods in the Theory of Gaussian Processes and Fields.* Translation of Mathematical Monographs 148, AMS, Providence, RI, 1996.

[64] Rao, K. C., The Kolmogorov, Cramér–von Mises, chi-square statistics for goodness of fit tests in the parametric case (Abstract). *Bulletin of the IMS*, **1** (1972), 87.

[65] Shorack, G. R. and Wellner, J. A., *Empirical Processes with Applications to Statistics.* Wiley, New York, 1986.

[66] Smirnov, N. V., On the distribution of the von Mises ω^2-criterion. *Matem. Sbornik*, **5** (1937), 973–993 (in Russian).

[67] Smirnov, N. V., On the estimation of the discrepancy between empirical curves of distribution for two independent samples. *Bull. Math. de l'Université de Moscou*, **2** (1939), Fasc. 2

[68] Stephans, M.A., Tests based on EDF statistics. In: *Goodness-of-Fit Techniques*, Marcel Dekker, New York, Basel, 1986a, 97–123.

[69] Stephans, M. A., Tests for exponential distribution. In: *Goodness-of-Fit Techniques*, Marcel Dekker, New York, Basel, 1986b, 421–459.

[70] Sukhatme, S., Fredholm determinant of a positive definite kernel of a special type and its application. *Ann. Math. Statist.*, **43** (1972), 1914–1926.

[71] Tyurin, Yu. V., On the limit distribution of Kolmogorov–Smirnov statistics for a composite hypothesis. *Izv. Akad. Nauk SSSR Ser. Mat.*, **48** (1984), 1314–1343.

[72] Watson, G. S., Goodness-of-fit tests on a circle. *Biometrika*, **48** (1961), 109–114.

[73] Watson, G. S., Another test for the uniformity of a circular distribution. *Biometrika*, **54** (1967), 675–676.

István Berkes
Rényi Institute of Mathematics
Hungarian Academy of Sciences
P.O. Box 127
H-1364 Budapest, Hungary
berkes@renyi.hu

Lajos Horváth
Department of Mathematics
University of Utah
155 South 1440 East
Salt Lake City, UT 84112-0090, U.S.A.
horvath@math.utah.edu

Part III

The Empirical Process of Long Range Dependent Processes

Asymptotic Expansion of the Empirical Process of Long Memory Moving Averages

HIRA L. KOUL[1] AND DONATAS SURGAILIS

ABSTRACT. Moving averages in i.i.d. variables form one of the most important classes of long memory time series. The paper reviews various results on the asymptotic distribution of empirical processes of long memory moving averages with finite and infinite variance. It also discusses some interesting applications to goodness-of-fit testing for the marginal stationary error distribution in linear regression models and M-estimation in the one sample location model.

1 Introduction

Let $X_t, t = 1, \ldots, N$, be the observed sample of size N from a strictly stationary and ergodic time series $\{X_t;\ t \in \mathbb{Z}\}$, where $\mathbb{Z} = \{0, \pm 1, \pm 2, \ldots\}$. The classical Glivenko–Cantelli theorem says that the empirical distribution function $F_N(x) = N^{-1} \sum_{t=1}^{N} I(X_t \le x)$ converges a.s. to the marginal distribution function $F(x) = P(X_0 \le x)$ uniformly in $x \in \mathbb{R}$, i.e.,

$$\sup_x |F_N(x) - F(x)| \longrightarrow 0, \quad \text{a.s.}$$

The magnitude and the character of the "fluctuations" $F_N(x) - F(x)$ can be very different depending on the dependence properties and the structure of the underlying process $\{X_t\}$. Roughly speaking, if $\{X_t\}$ is weakly dependent, then $N^{1/2}(F_N(x) - F(x))$ approaches a nondegenerate Gaussian process, while in the case of long range dependent (long memory) observations, the fluctuations of $F_N(x) - F(x)$ are of order $N^{-\theta}, 0 < \theta < 1/2$, and the weak limit of $N^{\theta}(F_N(x) - F(x))$, if it exists, is a degenerate process in x. Furthermore, in the long memory case, an asymptotic expansion of $F_N(x)$ that is similar in spirit to a Taylor expansion can be derived. This expansion is important for obtaining higher order approximations of certain statistical functionals of the empirical process.

One of the most important classes of long memory processes form *moving averages*

$$X_t = \sum_{j=1}^{\infty} b_j \zeta_{t-j}, \quad t \in \mathbb{Z}, \tag{1.1}$$

where $\zeta_s, s \in \mathbb{Z}$ are i.i.d. random variables with zero mean and finite variance, and $b_j, j \ge 1$, are square summable weights whose sum of absolute values diverges: $\sum_{j=1}^{\infty} |b_j| = \infty$. Moving averages (1.1) include several important parametric classes such as ARFIMA$(p, d, q), d \in (0, 0.5)$, as well as a large class of stationary Gaussian processes with nonsummable autocovariances $\sum_{t \in \mathbb{Z}} |r_t| = \infty$.

[1] Research of this author was partly supported by the NSF grant DMS 0071619.

The large sample asymptotic behavior of empirical processes of long memory moving averages (1.1) was studied in Ho and Hsing [18], [19], Giraitis, Koul, and Surgailis [13], Koul and Surgailis [33], [35], Giraitis and Surgailis [16], Marinucci [39], Doukhan, Lang, and Surgailis [10], and other papers. The case of Gaussian and Gaussian subordinated processes $\{X_t\}$ was discussed by Dehling and Taqqu [8], Koul and Mukherjee [32], and Csörgő and Mielniczuk [7]. The important paper by Ho and Hsing [18] obtained an asymptotic expansion of the empirical process of (1.1) and developed a new approach to the study of nonlinear functionals of non-Gaussian moving averages which in some sense replaces the method of Hermite expansions in the Gaussian case.

In this paper we present an exposition of the method of Ho and Hsing and various asymptotic results for empirical processes of long memory moving averages (1.1). Section 2 starts by discussing the case of Gaussian processes $\{X_t\}$. In this case, a complete asymptotic expansion of the empirical process is derived via the standard techniques of Hermite expansions and using the ideas of Dehling and Taqqu [8] and Csörgő and Mielniczuk [7]. In Section 3 we formulate and discuss the asymptotic expansion of Ho and Hsing [18], Koul and Surgailis [33], as well as some further results on empirical processes of long memory moving averages (1.1). This expansion is heuristically "derived" in Section 4. Sections 5 and 6 are devoted to the proof of the main estimate on the covariance of the "remainder term" of the expansion of the indicator function $I(X_t \le x)$ (Lemma 6.1). This result is crucial in obtaining the above mentioned asymptotic expansion. Section 7 is devoted to some applications of these asymptotic results to some inference problems. Finally, in Section 8 we discuss the empirical process of long memory moving averages with *infinite variance*, with ζ_0 in the domain of attraction of a symmetric α-stable law ($1 < \alpha < 2$).

2 Asymptotic expansion: The Gaussian case

Let $\{X_t\}$ be a stationary Gaussian process with zero mean, unit variance, and covariance function $r_t = EX_0X_t$. It is well known that any square integrable function H, $EH^2(X_t) < \infty$ can be expanded in a Hermite series:

$$H(X_t) = \sum_{k=0}^{\infty} \frac{c_{k,H}}{k!} H_k(X_t), \tag{2.1}$$

where $c_{k,H} = EH(X_0)H_k(X_0)$, and where H_k, $k \ge 0$, are the standard Hermite polynomials; $H_0(y) = 1$, $H_1(y) = y$, $H_2(y) = y^2 - 1$ etc. For the indicator function $H(y) \equiv I(y \le x)$, the expansion (2.1) is very explicit:

$$I(X_t \le x) = \sum_{k=0}^{\infty} \frac{(-1)^k}{k!} F^{(k)}(x) H_k(X_t), \tag{2.2}$$

where $F(x) = (2\pi)^{-1} \int_{-\infty}^{x} e^{-y^2/2} dy$ is the standard normal c.d.f. Assume that the covariance function r_t decays as

$$r_t \sim c_1 t^{-\theta}, \qquad t \to \infty, \tag{2.3}$$

where $0 < \theta < 1$ and $c_1 \ne 0$ is some constant. Note that (2.3) implies that $\{r_t\}$ is not summable: $\sum_{t=1}^{\infty} |r_t| = \infty$, in other words, the process $\{X_t\}$ has long memory. By the well-known property of Hermite polynomials,

$$\mathrm{Cov}(H_k(X_0), H_k(X_t)) = k! r_t^k,$$

and therefore

$$\sum_{t=1}^{\infty} |\mathrm{Cov}(H_k(X_0), H_k(X_t))| = k! \sum_{t=1}^{\infty} |r_t|^k \begin{cases} = \infty, & \text{if } k\theta < 1, \\ < \infty, & \text{if } k\theta > 1. \end{cases}$$

Accordingly, the r.h.s. of (2.2) can be split into two parts—the long memory part,

$$\sum_{0 \le k < 1/\theta} \frac{(-1)^k}{k!} F^{(k)}(x) H_k(X_t),$$

and the short memory part,

$$R_t(x) = I(X_t \le x) - \sum_{0 \le k < 1/\theta} \frac{(-1)^k}{k!} F^{(k)}(x) H_k(X_t).$$

For simplicity, we assume that $1/\theta$ is not an integer. The behavior of the sums of the long memory terms in (2.2) is described by *noncentral limit theorems* (Dobrushin and Major [9], Taqqu [47], [49]). Put

$$Z_N^{(k)} := N^{k\theta/2-1} \sum_{t=1}^{N} H_k(X_t)/k!, \quad k = 1, 2, \ldots .$$

As shown in [9], [47], [49], for each $1 \le k < 1/\theta$,

$$Z_N^{(k)} \Longrightarrow Z^{(k)}, \tag{2.4}$$

in distribution, where the limit random variable $Z^{(k)}$ can be written as the multiple Itô–Wiener integral

$$Z^{(k)} = c_k(\theta) \int_{\mathbb{R}^k} \left\{ \int_0^1 \prod_{i=1}^{k} (\tau - s_i)_+^{-(1+\theta)/2} d\tau \right\} W(ds_1) \cdots W(ds_k) \tag{2.5}$$

w.r.t. a Gaussian white noise $W(ds)$, $E(W(ds))^2 = ds$; $s_+^{-\lambda} := s^{-\lambda}$ if $s > 0$, $:= 0$ otherwise, and where

$$c_k(\theta) := (c_1/B(\theta, (1-\theta)/2))^{k/2}, \quad 0 < \theta < 1,$$

with $B(\cdot, \cdot)$ denoting the beta-function.

Theorem 2.1. *Let $\{X_t\}$ be a stationary Gaussian process satisfying the above conditions including* (2.3) *for some $0 < \theta < 1$. Assume $1/\theta$ is not an integer. Then the following asymptotic expansion of the empirical c.d.f. holds:*

$$F_N(x) - F(x) = \sum_{1 \le k < 1/\theta} (-1)^k N^{-k\theta/2} F^{(k)}(x) Z_N^{(k)} + N^{-1/2} Q_N(x), \tag{2.6}$$

where $Z_N^{(k)}$, $1 \le k < 1/\theta$, satisfy (2.4), *and $\{Q_N(x), x \in \mathbb{R}\}$ converges weakly, in the Skorohod space $D(\bar{\mathbb{R}})$ with the sup-topology, to a Gaussian process $\{Q(x), x \in \mathbb{R}\}$ with zero mean and covariance*

$$\mathrm{Cov}(Q(x), Q(y)) = \sum_{t \in \mathbb{Z}} \mathrm{Cov}(R_0(x), R_t(y)).$$

Proof. Write $Q_N(x) = N^{-1/2}\sum_{t=1}^N R_t(x) = N^{-1/2}\sum_{t=1}^N h_x(X_t)$, where the function

$$\begin{aligned} h_x(z) &:= I(z \le x) - \sum_{0\le k<1/\theta} \frac{(-1)^k}{k!} F^{(k)}(x) H_k(z) \\ &= \sum_{k>1/\theta} \frac{(-1)^k}{k!} F^{(k)}(x) H_k(z), \qquad z \in \mathbb{R}, \end{aligned}$$

is in $L^2(F)$ and has Hermite rank $> 1/\theta$, for each $x \in \mathbb{R}$. The last fact together with (2.3) implies the asymptotic normality of $Q_N(x)$:

$$Q_N(x) \Longrightarrow Q(x), \tag{2.7}$$

where $Q(x) \sim N(0, \sigma_x^2)$, $\sigma_x^2 := \sum_{t\in\mathbb{Z}} \mathrm{Cov}(h_x(X_0), h_x(X_t)) = \sum_{t\in\mathbb{Z}} \mathrm{Cov}(R_0(x), R_t(x))$, see Breuer and Major [4], Giraitis and Surgailis [14]. A similar conclusion applies to any linear combination $\sum_{i=1}^m a_i Q_N(x_i)$, $a_i, x_i \in \mathbb{R}$, implying the finite dimensional convergence $\{Q_N(x)\} \Longrightarrow \{Q(x)\}$.

It remains to show the tightness. To that end, it suffices to check the well-known tightness criterion (Shao and Yu [43]):

$$E\big(Q_N(y) - Q_N(x)\big)^4 \le C\big((\mu(x,y))^{1+\delta} + N^{-\delta}\mu(x,y)\big), \qquad x < y, \tag{2.8}$$

where μ is a bounded continuous measure on $\mathbb{R}$, and $\delta > 0$.

The proof of (2.8) uses the ideas of Csörgő and Mielniczuk [7]. Its main ingredient is the following inequality. Let $\langle a \rangle = a - [a] \in [0, 1)$ be the fractional part of $a \in \mathbb{R}$.

Lemma 2.2 (Taqqu [48]). *Let $N, m, p \ge 2$, $1 \le q \le p$ be integers, and suppose $\bar r_N := \max\{|r_t| : 1 \le t \le N\} < 1/(p-1)$. If $G_1, \dots, G_p$ are functions in $L^2(F)$ such that at least q of them have Hermite rank greater than or equal to m, then*

$$\sum{}' |E(G_1(X_{t_1}) \cdots G_p(X_{t_p})| \le K_p N^{p-q/2} \left(\sum_{t=1}^N |r_t|^m\right)^{q/2} \left(N^{-1}\sum_{t=1}^N |r_t|^m\right)^{\langle mq/2\rangle/m}, \tag{2.9}$$

where the summation $\sum{}'$ extends over all different indices $1 \le t_1, \dots, t_p \le N$ $(t_i \ne t_j,\ i \ne j)$, and

$$K_p \le C_p \left\{E(G_1^2(X)) \cdots E(G_p^2(X))\right\}^{1/2}, \tag{2.10}$$

where $C_p := 2^p/(1 - \{(p-1)\bar r_N\}^{1/2})^p$.

In order to apply the above lemma, assume first $\bar r := \sup_{t\ne 0} |r_t| < 1/3$, and write $R_t(y) - R_t(x) = h_{x,y}(X_t)$, where $h_{x,y}(z) = h_y(z) - h_x(z)$. Then

$$\left(\sum_{t=1}^N h_{x,y}(X_t)\right)^4 = \Sigma_1 + \Sigma_{21} + \Sigma_{22} + \Sigma_3 + \Sigma_4,$$

where

$$\begin{aligned}\Sigma_1 &= \sum{}' h^4_{x,y}(X_t),\\ \Sigma_{21} &= \sum{}' h^3_{x,y}(X_{t_1})h_{x,y}(X_{t_2}), \qquad \Sigma_{22} = \sum{}' h^2_{x,y}(X_{t_1})h^2_{x,y}(X_{t_2}),\\ \Sigma_3 &= \sum{}' h^2_{x,y}(X_{t_1})h_{x,y}(X_{t_2})h_{x,y}(X_{t_3}),\\ \Sigma_4 &= \sum{}' h_{x,y}(X_{t_1})h_{x,y}(X_{t_2})h_{x,y}(X_{t_3})h_{x,y}(X_{t_4}).\end{aligned}$$

Note for any even integer $p \geq 2$,

$$\begin{aligned}E(h_{x,y}(X_0))^p &\leq C\left(P(x < X_0 \leq y) + \sum_{1\leq k<1/\theta} (F^{(k)}(x,y))^p E H_k^p(X_0)\right)\\ &\leq C\mu(x,y), \qquad (2.11)\end{aligned}$$

where $\mu(x,y) := \sum_{1\leq k<1/\theta} \int_x^y |f^{(k-1)}(z)|dz$ defines a finite continuous measure on $\mathbb{R}$.

To estimate $E|\Sigma_4|$, use Lemma 2.2 with $p = q = 4$, $m > 1/\theta$. As $\sum_{t\in\mathbb{Z}} |r_t|^m < \infty$, we obtain by (2.9), (2.10), (2.11),

$$E|\Sigma_4| \leq CN^2(Eh^2_{x,y}(X_0))^2 \leq CN^2(\mu(x,y))^2.$$

Next, consider $E|\Sigma_3|$. In this case, use Lemma 2.2 with $p = 3$, $q = 2$, $m > 1/\theta$, to obtain

$$E|\Sigma_3| \leq CN^2 Eh^2_{x,y}(X_0)(Eh^4_{x,y}(x,y))^{1/2} \leq CN^2(\mu(x,y))^{3/2}.$$

For $E|\Sigma_{21}|$, take $p = 2, q = 1, m > 1/\theta$ and obtain

$$E|\Sigma_{21}| \leq CN^{3/2}(Eh^2_{x,y}(X))^{1/2}(Eh^6_{x,y}(X))^{1/2} \leq CN^{3/2}\mu(x,y).$$

Consider $E|\Sigma_{22}|$. We have as in [7] with $\tilde{h}_{x,y}(z) = I(x < z \leq y) - F(x,y)$,

$$\begin{aligned}&Eh^2_{x,y}(X_{t_1})h^2_{x,y}(X_{t_2})\\ &= E\tilde{h}_{x,y}(X_{t_1})\tilde{h}_{x,y}(X_{t_2})(1-2F(x,y))^2 + (F(x,y))^2(1-2F(x,y)+3(F(x,y))^2)\\ &\leq |E\tilde{h}_{x,y}(X_{t_1})\tilde{h}_{x,y}(X_{t_2})| + C(\mu(x,y))^2.\end{aligned}$$

Next, by the orthogonality of the Hermite expansion (2.2),

$$\begin{aligned}\left|E\tilde{h}_{x,y}(X_{t_1})\tilde{h}_{x,y}(X_{t_2})\right| &= \left|Eh_{x,y}(X_{t_1})h_{x,y}(X_{t_2}) + \sum_{1\leq k<1/\theta} (F^{(k)}(x,y))^2 r_t^k/k!\right|\\ &\leq |Eh_{x,y}(X_{t_1})h_{x,y}(X_{t_2})| + C(\mu(x,y))^2.\end{aligned}$$

Therefore, applying Lemma 2.2 with $p = q = 2$, $m > 1/\theta$, we get

$$\begin{aligned}E|\Sigma_{22}| &\leq CN^2(\mu(x,y))^2 + \sum_{t_1,t_2=1}^{N} |E\tilde{h}_{x,y}(X_{t_1})\tilde{h}_{x,y}(X_{t_2})|\\ &\leq CN^2(\mu(x,y))^2 + CN\mu(x,y).\end{aligned}$$

Finally,

$$E|\Sigma_1| \le CN\mu(x, y).$$

These bounds clearly imply (2.8) with $\delta = 1/2$.

The general case $\bar{r} \in (0, 1)$ can be reduced to the case $\bar{r} < 1/3$, by writing, similarly as in Csörgő and Mielniczuk [7],

$$Q_N(x) = \sum_{j=1}^{\ell} Q_{N,j}(x),$$

where

$$Q_{N,j}(x) = N^{-1/2} \sum_{\tau=0}^{[(N-j)/\ell]} h_x(X_{j+\tau\ell})$$

and where the covariance function $r_{\tau,\ell}$ of the "decimated" Gaussian process $\{X_{j+\tau\ell}, \tau \in \mathbb{Z}\}$ satisfies $\sup_{\tau \ne 0} |r_{\tau,\ell}| < 1/3$ by choosing ℓ sufficiently large. See [7] for details. Theorem 2.1 is proved. □

3 Asymptotic expansion: The moving average case

Let $\{X_t\}$ be the moving average sequence of (1.1) in i.i.d. random variables $\{\zeta_s\}$ with zero mean and variance 1. The moving average is well defined if and only if $\sum_{j=1}^{\infty} b_j^2 < \infty$, and has zero mean and autocovariance

$$r_t = \sum_{j=1}^{\infty} b_j b_{t+j}.$$

We shall write $F(x) = P(X_t \le x)$ for the marginal c.d.f. of (1.1) and $f(x) = F'(x)$ for the density (provided it exists). The decay of b_j, $j \ge 1$, determine the degree of dependence between the past and the future of the time series (1.1). We shall assume that the weights decay hyperbolically:

$$b_j \sim c_0 j^{-(1+\theta)/2}, \qquad j \to \infty, \tag{3.1}$$

where $c_0 \ne 0$ is some constant, and $0 < \theta < 1$ is the long memory parameter. It easily follows from (3.1) that the covariance function

$$r_t \sim c_1 t^{-\theta}, \qquad t \to \infty, \tag{3.2}$$

and therefore it is not summable; the constant $c_1 = c_0^2 B(\theta, (1-\theta)/2)$. The class of moving averages (1.1) satisfying (3.1), (3.2) include ARFIMA(p, d, q), $d = (1-\theta)/2 \in (0, 0.5)$; cf. Brockwell and Davis [5]. In some cases, (3.1) can be replaced by the weaker condition

$$b_j = O(j^{-(1+\theta)/2}), \qquad j \to \infty, \tag{3.3}$$

implying $r_t = O(t^{-\theta})$, which is typical for "oscillating long memory" in *seasonal ARFIMA* processes; cf. Leipus and Viano [37].

The asymptotic expansion of the empirical process of non-Gaussian long memory moving average observations (1.1) is very similar to the expansion (2.6) in the Gaussian case except for the definition of $Z_N^{(k)}$. There are two possibilities to define the $Z_N^{(k)}$'s. The first one is to use the formal *Appell expansion* of the indicator function (Koul and Surgailis [33])

$$I(X_t \le x) = \sum_{k=0}^{\infty} \frac{(-1)^k}{k!} F^{(k)}(x) A_k(X_t),$$

which is similar to (2.2) for the Gaussian case. Recall that the nth Appell polynomial A_n relative to the c.d.f. F_X of a random variable X with $E|X|^n < \infty$ is defined by

$$A_n(x) = (-i)^n \frac{\partial^n}{\partial z^n} \left(\frac{e^{izx}}{Ee^{izX}} \right)_{z=0}.$$

See [1], [15], [46] for more on Appell polynomials. The corresponding $Z_N^{(k)}$'s can be defined as

$$Z_N^{(k)} := N^{k\theta/2-1} \sum_{t=1}^{N} A_k(X_t)/k!. \tag{3.4}$$

As shown in Surgailis [45] and Avram and Taqqu [1], under conditions (3.1) and $E\zeta_0^{2k} < \infty$, for any $0 \le k < 1/\theta$,

$$Z_N^{(k)} \Longrightarrow Z^{(k)},$$

where the limiting random variable is the same multiple integral (2.5) as in the Gaussian case.

However, although Appell polynomials coincide with Hermite polynomials when $X \sim N(0,1)$, they lack the orthogonality property of the latter ([15]), and $Z_N^{(k)}, Z_N^{(\ell)}$ for $k \ne \ell$ are not orthogonal. Moreover, the use of (3.4) in the asymptotic expansion of the empirical processes requires the existence of $2[1/\theta]$th moments of ζ_0 ([33]). For the above reasons, it is more convenient to use instead of (3.4) the sums

$$\tilde{Z}_N^{(k)} := N^{k\theta/2-1} \sum_{t=1}^{N} X_t^{(k)}, \tag{3.5}$$

where $X_t^{(0)} := 1$, and

$$X_t^{(k)} := \sum_{s_k < \cdots < s_1 < t} b_{t-s_1} \cdots b_{t-s_k} \zeta_{s_1} \cdots \zeta_{s_k}, \tag{3.6}$$

is a polynomial form of order k in the i.i.d. random variables ζ_s, $s \in \mathbb{Z}$, $k \ge 1$. Note that the series (3.6) converges in mean square for each $k \ge 1$, under the conditions $\sum_{j=1}^{\infty} b_j^2 < \infty$, $E\zeta_0^2 < \infty$ alone, and (3.6) defines a strictly stationary process in $t \in \mathbb{Z}$ with zero mean and covariance

$$EX_t^{(k)} X_0^{(k)} = \sum_{0 < i_1 < \cdots < i_k} b_{t+i_1} b_{i_1} \cdots b_{t+i_k} b_{i_k}. \tag{3.7}$$

Moreover, (3.6) are *orthogonal* in the sense that

$$EX_t^{(k)} X_s^{(\ell)} = 0 \qquad (k \ne \ell,\ k, \ell = 0, 1, \ldots). \tag{3.8}$$

It follows from (3.3) and (3.7), that for each $k \geq 1$,

$$|EX_t^{(k)}X_0^{(k)}| \leq (k!)^{-1}\left(\sum_{i=1}^{\infty}|b_{t+i}b_i|\right)^k = O(t^{-k\theta}), \qquad t \to \infty, \tag{3.9}$$

and

$$E\left(\sum_{t=1}^{N}X_t^{(k)}\right)^2 = \begin{cases} O(N^{2-k\theta}), & \text{if } k\theta < 1, \\ O(N), & \text{if } k\theta > 1, \end{cases} \qquad N \to \infty. \tag{3.10}$$

Properties (3.7)–(3.10) are very similar to the properties of the normalized Hermite polynomials $(k!)^{-1}H_k(X_t)$ when $\{X_t\}$ is Gaussian. In fact, the polynomial form (3.6) is the "off-diagonal" part (modulus the factor $(k!)^{-1}$) of the Appell polynomial $A_k(X_t)$. For example, in the case $k = 2$, one has

$$A_2(X_t) = X_t^2 - EX_t^2 = 2X_t^{(2)} + \sum_{j=1}^{\infty} b_j^2(\zeta_{t-j}^2 - E\zeta_{t-j}^2),$$

where the difference $\{A_2(X_t) - 2X_t^{(2)}\}$ is the moving average in the i.i.d. random variables $\{A_2(\zeta_s) = \zeta_s^2 - E\zeta_s^2\}$ with summable weights b_j^2, $j \geq 1$, and therefore this difference is a short memory process. For general k, the relation between $A_k(X_t)$ and $X_t^{(k)}$ follows from the multinomial formula for Appell polynomials of Avram and Taqqu [1]. The limit distribution of $\tilde{Z}_N^{(k)}$ given in (3.5) is the same as that of $Z_N^{(k)}$ given in (3.4), but its existence does not require additional moments of ζ_0. Namely, under the conditions (3.1) and $E\zeta_0^2 < \infty$, for any $0 \leq k < 1/\theta$,

$$\tilde{Z}_N^{(k)} \Longrightarrow Z^{(k)}. \tag{3.11}$$

Now we are ready to formulate the main result of this section. Write $G(x) = P(\zeta_0 \leq x)$ for the c.d.f. of ζ_0.

Theorem 3.1 (Ho and Hsing [18]). *Let $\{X_t\}$ be a long memory moving average satisfying* (1.1) *and* (3.1). *Suppose the c.d.f. G of ζ_0 is $k_* + 3$ times differentiable with bounded, continuous and integrable derivatives, where k_* is the maximal positive integer $< 1/\theta$, and*

$$E|\zeta_0|^4 < \infty. \tag{3.12}$$

Then,

$$F_N(x) - F(x) = \sum_{1 \leq k < 1/\theta} (-1)^k N^{-k\theta/2} F^{(k)}(x)\tilde{Z}_N^{(k)} + N^{-1/2}Q_N(x), \tag{3.13}$$

where, for any $\delta > 0$,

$$\sup_x |Q_N(x)| = O_p(N^{\delta}). \tag{3.14}$$

Although similar to (2.6), Theorem 3.1 falls short of the statement of Theorem 2.1 in the sense that the limit of the "remainder term" $Q_N(x)$ in (3.13) is not established. In fact, using the results of [19] or [33], one can show the finite dimensional convergence of $Q_N(x)$ to a Gaussian process; however, the question of the tightness is still open. The conditions on G in Theorem 3.1 can be weakened; see Lemma 6.1 below.

From Theorem 3.1, the *uniform reduction principle* for the empirical process follows: there exists $\kappa > 0$ such that

$$\sup_{x\in\mathbb{R}} N^{\theta/2}|F_N(x) - F(x) + f(x)\bar{X}_N| = O_p(N^{-\kappa}), \tag{3.15}$$

where $\bar{X}_N = N^{-1}\sum_{t=1}^N X_t$ (cf. Dehling and Taqqu [8]). Giraitis and Surgailis [16] proved (3.15) under weak moment and smoothness conditions on the c.d.f. G.

A uniform reduction principle is also valid for *weighted residual empirical processes:*

$$\sup_{x\in\mathbb{R}} \left| N^{-1+\theta/2} \sum_{t=1}^N \gamma_{Nt}\left[I(X_t \le x + \xi_{Nt}) - F(x+\xi_{Nt}) + f(x+\xi_{Nt})X_t\right]\right| = o_p(1), \tag{3.16}$$

where $\gamma_{Nt}, \xi_{Nt}, 1 \le t \le N$, are nonrandom and such that $\max_{1\le t\le N}|\gamma_{Nt}| = O(1)$, $\max_{1\le t\le N}|\xi_{Nt}| = o(1)$ (Giraitis, Koul, and Surgailis [13]). For Gaussian and/or Gaussian subordinated long memory errors, an analogous result was obtained in Koul and Mukherjee [32]. Doukhan, Lang and Surgailis [10] extended (3.15) to weighted empirical processes of linear random fields $X_t, t \in \mathbb{Z}^d (d \ge 1)$ with long memory. As noted in Koul [28], weighted residual empirical processes arise naturally in multiple linear regression models. The result (3.16) proves to be a crucial tool in developing a unified theory for obtaining limiting distributions of several classes of so-called M-, R- and minimum distance estimators of the regression parameters when errors have long memory as is evidenced in [32] and [13]. It is also useful in developing asymptotic distributions of various goodness-of-fit tests based on residual empirical processes when fitting the marginal error distribution, as will be seen in Section 7 below.

The result (3.15) easily implies the Functional Central Limit Theorem

$$N^{\theta/2}(F_N(x) - F(x)) \Longrightarrow_{D(\bar{\mathbb{R}})} c^{1/2}(\theta)f(x)Z, \tag{3.17}$$

where $Z \sim N(0,1)$ is standard normal variable and the constant $c(\theta) > 0$ is given by

$$c(\theta) = \frac{2c_1}{(1-\theta)(2-\theta)}, \tag{3.18}$$

where c_1 is the asymptotic constant in (3.2) also depending on θ.

Theorem 3.1 extends to sums of instantaneous functionals

$$S_{N,H} = N^{-1}\sum_{t=1}^N (H(X_t) - EH(X_t)),$$

where $H(x), x \in \mathbb{R}$, is a nonlinear function with $E(H(X_0))^2 < \infty$. The corresponding asymptotic expansion can be formally obtained from (3.13), by noting $S_{N,H} = \int_{\mathbb{R}} H(x)d(F_N(x) - F(x))$ and integrating the terms on the right-hand side of (3.13). This results in

$$S_{N,H} = \sum_{1\le k<1/\theta} H_\infty^{(k)}(0)N^{-k\theta/2}\tilde{Z}_N^{(k)} + N^{-1/2}Q_{N,H}, \tag{3.19}$$

where

$$H_\infty(x) := EH(x+X_0) = \int_{\mathbb{R}} H(x+y)dF(y), \quad x \in \mathbb{R},$$

is a smooth function, see Lemma 5.1 below, $H_\infty^{(k)}(0) = (-1)^k \int_{\mathbb{R}} H(x) f^{(k)}(x)dx$. Under some regularity and boundedness conditions on H and further conditions on moments of ζ_0, the "remainder term" $Q_{N,H}$ in (3.19) is asymptotically normal (Ho and Hsing [19], Koul and Surgailis [33]):

$$Q_{N,H} \Longrightarrow N(0, \sigma^2), \tag{3.20}$$

where $\sigma^2 := \sum_{t\in\mathbb{Z}} \mathrm{Cov}(H_{0,k_*}, H_{t,k_*})$, $H_{t,k} := H(X_t) - \sum_{j=0}^k H_\infty^{(j)}(0) X_t^{(j)}$. It follows from (3.19) that the limiting distribution of $S_{N,H}$ is determined by the integer

$$k_{*,H} := \min\{k \geq 1 : H_\infty^{(k)}(0) \neq 0\},$$

called the *Appell rank of* H ([33], [46]), or the *power rank of* H ([19]). If $X_0 \sim N(0,1)$, then $H_\infty^{(k)}(0) = k! E H(X_0) H_k(X_0)$ are the coefficients of the Hermite expansion of H and $k_{*,H}$ coincides with the Hermite rank of H. One may conclude from the above discussion that the asymptotic behavior of sums $S_{N,H}$ for long memory moving averages $\{X_t\}$ with finite 4th moment is very similar as in the Gaussian case. However, the behavior of $S_{N,H}$ changes dramatically if the variance of $\{X_t\}$ is infinite; see Section 8.

4 A heuristic "proof"

Following Ho and Hsing [18], consider the decomposition

$$X_t = \sum_{i=1}^{j} b_i \zeta_{t-i} + \sum_{i=j+1}^{\infty} b_i \zeta_{t-i} =: X_{t,j} + \tilde{X}_{t,j}. \tag{4.1}$$

For fixed t and j, the summands $X_{t,j}$ and $\tilde{X}_{t,j}$ in (4.1) are independent; moreover, given j, $\{X_{t,j}\}$ and $\{\tilde{X}_{t,j}\}$ are strictly stationary moving averages in t, although they are not independent as processes. When j is large, the second term in (4.1) is small, due to $E\tilde{X}_{t,j}^2 = \sum_{i=j+1}^{\infty} b_j^2 \to 0 (j \to \infty)$. Moreover, under mild assumptions on the distribution of ζ_0,

$$F_j(x) := P(X_{t,j} \leq x) \to P(X_t \leq x) = F(x), \tag{4.2}$$

uniformly in $x \in \mathbb{R}$ as $j \to \infty$, together with all derivatives $F_j^{(p)}(x)$, $p = 1, 2, \dots$; see Lemma 5.1 below. This means that for large j, $X_{t,j}$ is the dominating term of the decomposition (4.1) in a distributional sense. On the other hand, $\{\tilde{X}_{t,j}\}$ is the dominating term in the sense of "memory," which follows from the fact that, for fixed j, the process $\{X_{t,j} = X_t - \tilde{X}_{t,j}\}$ is finitely dependent, while $\{\tilde{X}_{t,j}\}$ has long memory:

$$\mathrm{Cov}(\tilde{X}_{0,j}, \tilde{X}_{t,j}) \sim \mathrm{Cov}(X_0, X_t) \sim c_1 t^{-\theta}, \qquad t \to \infty, \tag{4.3}$$

where (4.3) follows from (1.1), (3.1) and c_1 is the same as in (3.2). Let $\mathcal{F}_t$ be the "past" σ-field generated by $\zeta_s : s \leq t$. The starting point of the Ho and Hsing approach is the telescoping identity:

$$I(X_t \leq x) - F(x) = \sum_{j=1}^{\infty} (P(X_t \leq x|\mathcal{F}_{t-j}) - P(X_t \leq x|\mathcal{F}_{t-j-1})), \tag{4.4}$$

which converges in $L^2(\Omega)$ as the summands on the r.h.s. of (4.4) are orthogonal by the well-known property of nested conditional expectations. Observe because of the independence of $X_{t,j}$ and $\tilde{X}_{t,j}$,

$$P(X_t \le x|\mathcal{F}_{t-j-1}) = P(X_{t,j} + \tilde{X}_{t,j} \le x|\mathcal{F}_{t-j-1}) = F_j(x - \tilde{X}_{t,j}).$$

Therefore,

$$\begin{aligned}
I(X_t \le x) - F(x) &= \sum_{j=1}^{\infty}(F_{j-1}(x - \tilde{X}_{t,j-1}) - F_j(x - \tilde{X}_{t,j})) \\
&\approx \sum_{j=1}^{\infty}(F_j(x - \tilde{X}_{t,j-1}) - F_j(x - \tilde{X}_{t,j})) && \text{(by (4.2))} \\
&= \sum_{j=1}^{\infty} F_j(x - b_j\zeta_{t-j} - \tilde{X}_{t,j}) - F_j(x - \tilde{X}_{t,j})) \\
&\approx -\sum_{j=1}^{\infty} b_j\zeta_{t-j}F_j'(x - \tilde{X}_{t,j}) && \text{(by Taylor formula)} \\
&= -F'(x)X_t + \sum_{j=1}^{\infty} b_j\zeta_{t-j}[F'(x) - F_j'(x - \tilde{X}_{t,j})]. && (4.5)
\end{aligned}$$

The last sum on the r.h.s. of (4.5) is negligible as compared to the "main term" $-F'(x)X_t$ because of (4.2). Therefore $I(X_t \le x) - F(x) \approx -F'(x)X_t$ in the "first approximation," yielding $F_N(x) \approx F(x) + f(x)\bar{X}_N$, or the first order expansion of $F_N(x)$. In a similar way, one can formally obtain the second term of the expansion (3.15), by noting $EF_j'(x - \tilde{X}_{t,j}) = (EF_j(x - \tilde{X}_{t,j}))' = F'(x)$, and expanding $F_j'(x - \tilde{X}_{t,j})$ in (4.5) exactly similarly:

$$\begin{aligned}
&F_j'(x - \tilde{X}_{t,j}) - F'(x) \\
&= \sum_{j_2=1}^{\infty}(F'_{j+j_2-1}(x - \tilde{X}_{t,j+j_2-1}) - F'_{j+j_2}(x - \tilde{X}_{t,j+j_2})) \\
&\approx \sum_{j_2=1}^{\infty}(F'_{j+j_2}(x - \tilde{X}_{t,j+j_2-1}) - F'_{j+j_2}(x - \tilde{X}_{t,j+j_2})) \\
&= \sum_{j_2=1}^{\infty}(F'_{j+j_2}(x - b_{j+j_2}\zeta_{t-j-j_2} - \tilde{X}_{t,j+j_2}) - F'_{j+j_2}(x - \tilde{X}_{t,j+j_2})) \\
&\approx -\sum_{j_2=1}^{\infty} b_{j+j_2}\zeta_{t-j-j_2}F''_{j+j_2}(x - \tilde{X}_{t,j+j_2}) \\
&= -F''(x)\sum_{j_2=1}^{\infty} b_{j+j_2}\zeta_{t-j-j_2} \\
&\qquad + \sum_{j_2=1}^{\infty} b_{j+j_2}\zeta_{t-j-j_2}[F''(x) - F''_{j+j_2}(x - \tilde{X}_{t,j+j_2})]. \qquad (4.6)
\end{aligned}$$

Dropping the last sum in (4.6), we obtain by (4.5) the second order expansion of the indicator

$$I(X_t \le x) - F(x) \approx -F'(x) \sum_{j_1=1}^{\infty} b_{j_1} \zeta_{t-j_1} + F''(x) \sum_{j_1, j_2=1}^{\infty} b_{j_1} b_{j_2} \zeta_{t-j_1} \zeta_{t-j_1-j_2} + \cdots$$

and the two first terms of the expansion of F_N :

$$N(F_N(x) - F(x)) = -F'(x) \sum_{t=1}^{N} X_t^{(1)} + F''(x) \sum_{t=1}^{N} X_t^{(2)} + \cdots .$$

Proceeding in a similar way, one can derive heuristically further terms of the expansion (3.13), until the last expansion term is "short memory," or $(k+1)\theta > 1$.

However, to make the above heuristic argument work, one needs to justify rigorously the approximate relations in (4.5), (4.6), in particular, the approximation of F_{j-1} by F_j, the use of the Taylor expansion, and the iteration step $k-1 \to k$, i.e., the possibility to obtain the next expansion term by induction on k. Some of these steps are explained in detail in Sections 5 and 6 below.

5 Some auxiliary bounds

Similarly to [13], [33], [16], we shall assume below that there exist constants C, $\delta > 0$, such that for all $u \in \mathbb{R}$,

$$|Ee^{iu\zeta_0}| = \left| \int e^{iux} dG(x) \right| \le C(1 + |u|)^{-\delta}. \tag{5.1}$$

Condition (5.1) is a weak restriction on the smoothness of the distribution function G, and implies its uniform δ-Hölder condition. It guarantees, however, that the distribution function F is infinitely differentiable. The following lemma is a version of the statements given in [13], [33], [16].

Lemma 5.1. *For any $p = 0, 1, \ldots,$ there exist $j_0 \ge 1$ and a constant C such that for any $j \ge j_0$, the distribution functions $F(x)$, $F_j(x)$ are $(p+1)$-times continuously differentiable and for any $j \ge j_0, x \in \mathbb{R}$,*

$$|F^{(p+1)}(x)| + |F_j^{(p+1)}(x)| \le C(1 + x^2)^{-1}, \tag{5.2}$$

$$|F_j^{(p+1)}(x) - F_{j+1}^{(p+1)}(x)| \le Cb_j^2(1 + x^2)^{-1}. \tag{5.3}$$

Proof. We shall merely sketch the argument. The densities $f(x) = F'(x)$, $f_j(x) = F_j'(x)$ exist if the characteristic functions $\hat{f}(u) = Ee^{iuX_0}$, $\hat{f}_j(u) = Ee^{iuX_{0,j}}$ are integrable and (5.2) follows by integration by parts if we show that for any $p \ge 0$ there exist j_0 and a constant C such that for any $j \ge j_0$, $u \in \mathbb{R}$,

$$|(u^p \hat{f}(u))^{(k)}| + |(u^p \hat{f}_j(u))^{(k)}| \le C/(1 + |u|)^2, \quad k = 0, 1, 2. \tag{5.4}$$

We shall prove (5.4) for $\hat{f}_j$ and $k = 2$. We have

$$(u^p \hat{f}_j(u))'' = p(p-1)u^{p-2} \hat{f}_j(u) + 2pu^{p-1} \hat{f}_j'(u) + u^p \hat{f}_j''(u). \tag{5.5}$$

Put $\phi(u) = Ee^{iu\zeta_0}$, $\Phi_I(u) = \prod_{r\in I}\phi(ub_r) = E\exp\{iu\sum_{r\in I} b_r\zeta_r\}$ $(I \subset \{1, 2, \dots\})$. Then $\hat{f}_j(u) = \Phi_J(u)$ $(J = \{1, \dots, j\})$ and

$$\hat{f}_j'(u) = \sum_{r\in J}\phi'(ub_r)b_r\Phi_{J\setminus\{r\}}(u), \tag{5.6}$$

$$\begin{aligned}\hat{f}_j''(u) &= \sum_{r\in J}\phi''(ub_r)b_r^2\Phi_{J\setminus\{r\}}(u) \\ &+ \sum_{r_1,r_2\in J, r_1\neq r_2}\phi'(ub_{r_1})\phi'(ub_{r_2})b_{r_1}b_{r_2}\Phi_{J\setminus\{r_1,r_2\}}(u).\end{aligned} \tag{5.7}$$

Here,

$$\sum_{r\in J}|\phi'(ub_r)b_r| \le C|u|, \qquad \sum_{r\in J}|\phi''(ub_r)|b_r^2 \le C, \tag{5.8}$$

where the constant does not depend on u and j. On the other hand, by (5.1) the characteristic functions $\Phi_J(u)$, $\Phi_{J\setminus\{r\}}(u)$, $\Phi_{J\setminus\{r_1,r_2\}}(u)$ decay polynomially fast with $|u| \to \infty$, uniformly in r, r_1, r_2, provided j is large enough; more precisely, there exist j_0 and a constant $C < \infty$, such that for all $j \ge j_0$, $u \in \mathbb{R}$, and any $r, r_1, r_2 \in J$,

$$|\Phi_J(u)| + |\Phi_{J\setminus\{r\}}(u)| + |\Phi_{J\setminus\{r_1,r_2\}}(u)| \le C(1+u^2)^{-p-4}. \tag{5.9}$$

Statements (5.5)–(5.9) imply (5.4), hence also (5.2), while (5.3) follows analogously, by showing a similar inequality to (5.4) for the differences $\hat{f}_j(u) - \hat{f}_{j+1}(u)$; see [33] for details. □

Let

$$g_\gamma(x) := (1+|x|)^{-\gamma}, \qquad \mu_\gamma(x,y) := \int_x^y g_\gamma(z)dz, \qquad x<y,\ \gamma>1,\ x,y\in\mathbb{R}.$$

Note μ_γ defines a finite measure on $\mathbb{R}$. The bounds given in Lemma 5.2 below are elementary; their proofs can be found in [36], [10].

Lemma 5.2. *Let $h(x)$, $x \in \mathbb{R}$, be a real-valued function such that*

$$|h(x)| \le Cg_\gamma(x), \qquad |h(x)-h(y)| \le C|x-y|g_\gamma(x), \tag{5.10}$$

hold for any $x, y \in \mathbb{R}$, $|x-y| \le 1$ and some $C < \infty$, $1 < \gamma \le 2$. Then there exists a constant C_γ depending only on γ and C in (5.10), *such that for any $x, y, v, z \in \mathbb{R}$*

$$|h(x+y)| \le C_\gamma g_\gamma(x)(1 \vee |y|^\gamma), \tag{5.11}$$

$$\left|\int_0^y h(x+w)dw\right| \le C_\gamma g_\gamma(x)(|y| \vee |y|^\gamma), \tag{5.12}$$

$$\left|\int_x^y (h(\xi+v)-h(\xi))d\xi\right| \le C_\gamma \mu_\gamma(x,y)(|v| \vee |v|^\gamma), \tag{5.13}$$

$$\left|\int_0^v dw\int_x^y (h(\xi+w-z)-h(\xi-z))d\xi\right| \le C_\gamma\mu_\gamma(x,y)|v|^\gamma(1\vee|z|^\gamma). \tag{5.14}$$

6 Proof of Theorem 3.1

Essentially, Theorem 3.1 reduces to the statement (3.14) about $Q_N(x) \equiv Q_{N,k_*}(x)$, where $k_* := \max\{k \ge 1 : k\theta < 1\}$ and where

$$Q_{N,k}(x) := N^{-1/2}\sum_{t=1}^{N} R_{t,k}(x),$$

$$R_{t,k}(x) := I(X_t \le x) - \sum_{j=0}^{k}(-1)^j F^{(j)}(x)X_t^{(j)}.$$

The control of $\sup_{x\in\mathbb{R}} |Q_{N,k}(x)|$, $1 \le k \le k_*$, is largely facilitated by the fact that the indicator $I(X_t \le x)$ is a monotone function of x, and reduces by the well-known chaining argument of Dehling and Taqqu [8] to an estimate for the variance of increments $Q_{N,k}(y) - Q_{N,k}(x)$, $x < y$. To that end, one needs to estimate the decay rate of the covariance $\mathrm{Cov}(R_{0,k}(y) - R_{0,k}(x), R_{t,k}(y) - R_{t,k}(x))$ as $t \to \infty$. The relevant bound is provided by Lemma 6.1 below, which is the most difficult part of the proof of Theorem 3.1. Lemma 6.1 is of independent interest and is presented and proved here in more generality than necessary for the proof of Theorem 3.1, as it refers to any $1 \le k \le k_*$ and not only to $k = k_*$. Moreover, Lemma 6.1 is proved under weaker conditions than those of Theorem 3.1, in particular, we aim to weaken the 4th moment condition (3.12) of Theorem 3.1, by assuming only

$$E|\zeta_0|^{2r} < \infty, \qquad \text{for some } r > 1. \tag{6.1}$$

If $r < 2$, the number of terms in the asymptotic expansion of $F_N(x)$ may be less than k_* and depends on both r and θ; in particular, for the first order asymptotic expansion $F_N(x) = F(x) - f(x)\bar{X}_N + o_p(N^{-\theta/2})$, it suffices to have condition (6.1) with $r > 1$, but arbitrarily close to 1 (see Giraitis and Surgailis [16]). The last result follows from Lemma 6.1 via the chaining argument as a particular case.

Let $X_{t,j}$, $\tilde{X}_{t,j}$, $F_j(x)$ be defined as in (4.1), (4.2), respectively, and let

$$\tilde{X}_{t,j}^{(k)} := \sum_{i_k > \cdots > i_1 > j} b_{i_1}\cdots b_{i_k}\zeta_{t-i_1}\cdots\zeta_{t-i_k}, \quad k \ge 1, \quad \tilde{X}_{t,j}^{(0)} = 1.$$

Note, for each $k = 1, 2, \ldots$,

$$X_t^{(k)} = \sum_{j=1}^{\infty} b_j\zeta_{t-j}\tilde{X}_{t,j}^{(k-1)}. \tag{6.2}$$

Also, put,

$$\tilde{R}_{t,j,k}(x) := F_j(x - \tilde{X}_{t,j}) - \sum_{q=0}^{k}(-1)^q F^{(q)}(x)\tilde{X}_{t,j}^{(q)}, \qquad x \in \mathbb{R}. \tag{6.3}$$

The formulation of Lemma 6.1 includes the bound (6.7) for the variance of an increment of (6.2) which is used in the proof of the main bound (6.5) by induction on k. For any function $g(x)$, $x \in \mathbb{R}$, and any $x, y \in \mathbb{R}$, put $g(x, y) := g(y) - g(x)$.

Lemma 6.1. *Let the distribution of ζ_0 satisfy conditions* (5.1) *and* (6.1), *with*

$$r > \max\left(1, \frac{1 + k\theta}{1 + \theta}\right), \tag{6.4}$$

where $0 \le k < 1/\theta$, is a given integer. Then, one can find a constant $C < \infty$ such that for any $t \ge 1$, $x < y$,

$$\left|\mathrm{Cov}(R_{t,k}(x,y), R_{0,k}(x,y))\right| \le C\mu_{r_*}(x,y) \begin{cases} t^{-k\theta-\kappa}, & \text{if } k\theta+\kappa<1, \\ t^{-(1+k\theta+\kappa)/2}, & \text{if } k\theta+\kappa>1, \end{cases} \tag{6.5}$$

where $r_* = r \wedge 2$ and

$$\kappa = \min(\theta, (1+\theta)r - 1 - k\theta) > 0. \tag{6.6}$$

Moreover, for any integer $p \ge 0$, there exist i_0 and a constant $C < \infty$ such that for any $i \ge i_0$, $x < y$,

$$\mathrm{Var}(\tilde{R}^{(p)}_{0,i,k}(x,y)) \le C\mu_{r_*}(x,y) \begin{cases} i^{-k\theta-\kappa}, & \text{if } k\theta+\kappa<1, \\ i^{-(1+k\theta+\kappa)/2}, & \text{if } k\theta+\kappa>1. \end{cases} \tag{6.7}$$

Proof. Note that the r.h.s. of (6.4) is less than 2, and $\kappa = \theta$ for $r \ge 2$. Therefore in the sequel we may assume $r \le 2$ and $r_* = r$ without loss of generality.

The proof below uses induction on $k \ge 0$. Accordingly, we need to verify the lemma for $k = 0$ and then to prove the induction step $k-1 \to k$. However, the two arguments are very similar and we shall start by discussing the general case $k \ge 0$ and later shall distinguish between the cases $k = 0$ and $k \ge 1$.

By the telescoping identity (4.4) and (6.2) one can rewrite $R_{t,k}(x)$ as

$$R_{t,k}(x) = \sum_{i=1}^{\infty} U_{t,i}(x), \tag{6.8}$$

where

$$U_{t,i}(x) := F_{i-1}(x - \tilde{X}_{t,i-1}) - F_i(x - \tilde{X}_{t,i}) - b_i\zeta_{t-i}\sum_{q=1}^{k}(-1)^q F^{(q)}(x)\tilde{X}^{(q-1)}_{t,i}. \tag{6.9}$$

(For $k = 0$, the last sum is zero). We claim that, for r, k, as in the statement of the lemma, there exists a constant $C < \infty$ such that for all $i \ge 1$, $x < y$,

$$E(U_{t,i}(x,y))^2 \le C\mu_r(x,y)i^{-(1+\theta)\rho}, \tag{6.10}$$

where

$$\rho := \min\left(r, \frac{1+(k+1)\theta}{1+\theta}\right) = \frac{1+k\theta+\kappa}{1+\theta}. \tag{6.11}$$

The bound (6.10) and the orthogonality of the expansion (6.8) easily imply (6.5):

$$\begin{aligned} &\left|\mathrm{Cov}(R_{t,k}(x,y), R_{0,k}(x,y)\right| \\ &= \left|\sum_{i=1}^{\infty} EU_{t,t+i}(x,y)U_{0,i}(x,y)\right| \\ &\le \sum_{i=1}^{\infty}\Big(E(U_{t,t+i}(x,y))^2\Big)^{1/2}\Big(E(U_{0,i}(x,y))^2\Big)^{1/2} \\ &\le C\,\mu_r(x,y)\sum_{i=1}^{\infty} i^{-(1+\theta)\rho/2}(t+i)^{-(1+\theta)\rho/2} \\ &\le C\,\mu_r(x,y)\begin{cases} t^{-k\theta-\kappa}, & \text{if } k\theta+\kappa<1, \\ t^{-(1+k\theta+\kappa)/2}, & \text{if } k\theta+\kappa>1. \end{cases} \end{aligned}$$

In the last line above we used the elementary inequality: for any $\alpha > 0,\ \beta > 0,\ \alpha + \beta > 1$,

$$\sum_{i=1}^{\infty} i^{-\alpha}(t+i)^{-\beta} \le C \begin{cases} t^{-\alpha-\beta+1}, & \text{if } 0 < \alpha < 1, \\ t^{-\beta}, & \text{if } \alpha > 1. \end{cases} \tag{6.12}$$

To prove the claim (6.10), assume first $k \ge 1$ and $i \ge i_0$ sufficiently large so that F_i is differentiable and the conclusions of Lemma 5.1 hold. Write $U_{t,i}(x) = \sum_{\tau=1}^{3} U_{t,i}^{(\tau)}(x)$, where

$$\begin{aligned} U_{t,i}^{(1)}(x) &:= F_i(x - \tilde{X}_{t,i-1}) - F_i(x - \tilde{X}_{t,i}) + b_i \zeta_{t-i} F_i'(x - \tilde{X}_{t,i}), \\ U_{t,i}^{(2)}(x) &:= F_{i-1}(x - \tilde{X}_{t,i-1}) - F_i(x - \tilde{X}_{t,i}), \\ U_{t,i}^{(3)}(x) &:= -b_i \zeta_{t-i} \left(F_i'(x - \tilde{X}_{t,i}) - \sum_{q=1}^{k} (-1)^{q-1} F^{(q)}(x) \tilde{X}_{t,i}^{(q-1)} \right). \end{aligned}$$

Clearly, (6.10) follows from

$$E(U_{t,i}(x,y))^2 \le C\mu_r(x,y), \quad 1 \le i \le i_0, \tag{6.13}$$

and

$$E(U_{t,i}^{(\tau)}(x,y))^2 \le C\mu_r(x,y) i^{-(1+\theta)\rho}, \qquad i \ge i_0, \quad \tau = 1, 2, 3. \tag{6.14}$$

To show (6.14), by the integral representation of the remainder,

$$U_{t,i}^{(1)}(x,y) = \int_0^{-b_i\zeta_{t-i}} du \int_x^y [f_i'(w + u - \tilde{X}_{t,i}) - f_i'(w - \tilde{X}_{t,i})] dw.$$

According to Lemma 5.1 (5.2) and Lemma 5.2 (5.14), for any $i \ge i_0,\ x < y,\ \gamma \in (1, 2]$, with probability 1,

$$|U_{t,i}^{(1)}(x,y)| \le C\mu_\gamma(x,y)|b_i|^\gamma |\zeta_{t-i}|^\gamma (1 \vee |\tilde{X}_{t,i}|)^\gamma. \tag{6.15}$$

From (5.2), (5.11), we also have the bound

$$|U_{t,i}^{(1}(x,y)| \le \int_{-|b_i\zeta_{t-i}|}^{|b_i\zeta_{t-i}|} \int_{\mathbb{R}} (|f_i'(w + u - \tilde{X}_{t,i})| + |f_i'(w - \tilde{X}_{t,i})|) dw \le C|b_i\zeta_{t-i}|. \tag{6.16}$$

Combining (6.15) and (6.16) and using the independence of ζ_{t-i} and $\tilde{X}_{t,i}$ as well as $E|\tilde{X}_{t,i}|^\gamma \le (E\tilde{X}_{t,i}^2)^{\gamma/2} \le C$ and taking $\gamma = r$, one obtains

$$E(U_{t,i}^{(1)}(x,y))^2 \le C\mu_\gamma(x,\ y)|b_i|^{2r} \le C\mu_r(x,y) i^{-(1+\theta)\rho}. \tag{6.17}$$

Consider $U_{t,i}^{(2)}$. We have again by Lemma 5.1, (5.3) and Lemma 5.2, (5.11), for $i \ge i_0$,

$$\begin{aligned} |U_{t,i}^{(2)}(x,y)| &= \left| \int_x^y [f_{i-1}(u - \tilde{X}_{t,i-1}) - f_i(u - \tilde{X}_{t,i-1})] du \right| \\ &\le Cb_i^2 \int_x^y (1 + |u - \tilde{X}_{t,i-1}|)^{-2} du \\ &\le C\mu_r(x,y) b_i^2 (1 \vee |\tilde{X}_{t,i}|)^r. \end{aligned}$$

Together with $|U^{(2)}_{t,i}(x,y)| \le Cb_i^2$, this yields

$$E(U^{(2)}_{t,i}(x,y))^2 \le C\mu_r(x,y)b_i^4. \tag{6.18}$$

Finally, note $U^{(3)}_{t,i}(x)$ can be rewritten as

$$U^{(3)}_{t,i}(x) = -b_i\zeta_{t-i}\tilde{R}^{(1)}_{t,i,k-1}(x).$$

Therefore, by the inductive assumption (6.7) with k replaced by $k-1$,

$$\begin{aligned} E(U^{(3)}_{t,i}(x,y))^2 &= b_i^2 E(\tilde{R}^{(1)}_{0,i,k-1}(x,y))^2 \\ &\le C\mu_r(x,y)b_i^2 i^{-(k-1)\theta-\kappa} \le C\mu_r(x,y)i^{-(1+\theta)\rho}. \end{aligned} \tag{6.19}$$

The claim (6.14) follows from (6.17)–(6.19). Relation (6.13) follows from

$$E(F_i(x-\tilde{X}_{t,i}, y-\tilde{X}_{t,i}))^2 \le C\mu_r(x,y), \quad 0 \le i \le i_0, \tag{6.20}$$

and

$$(F^{(q)}(x,y))^2 E(X_0^{(q)}))^2 \le C\mu_r(x,y), \quad 1 \le q \le k. \tag{6.21}$$

Here, (6.21) is immediate by Lemma 5.1 (5.2), while (6.20) follows from the inequality

$$\begin{aligned} E(F_i(x-\tilde{X}_{t,i}, y-\tilde{X}_{t,i}))^2 &\le EF_i(x-\tilde{X}_{t,i}, y-\tilde{X}_{t,i}) \\ &= F(x,y) = \int_x^y f(z)dz \le C\mu_2(x,y), \end{aligned}$$

see Lemma 5.1 (5.2).

This proves (6.10) and (6.5) for $k \ge 1$.

The above argument applies also to the case $k=0$, with some changes. Namely, the definition of $U^{(\tau)}_{t,i}(x)$ has to be modified to

$$U^{(1)}_{t,i}(x) := F_i(x-\tilde{X}_{t,i-1}) - F_i(x-\tilde{X}_{t,i}), \quad U^{(2)}_{t,i}(x) := F_{i-1}(x-\tilde{X}_{t,i-1}) - F_i(x-\tilde{X}_{t,i}),$$

$U^{(3)}_{t,i}(x) := 0$. Writing $U^{(1)}_{t,i}(x,y) = \int_x^y (f_i(u-b_i\zeta_{t-i}-\tilde{X}_{t,i}) - f_i(u-\tilde{X}_{t,i}))du$ and using Lemma 5.1 (5.2) and Lemma 5.2 (5.13), (5.11), for any $\gamma \in (1,2]$ we have the bound

$$|U^{(1)}_{t,i}(x,y)| \le C\mu_\gamma(x,y)(|b_i\zeta_{t-i}| + |b_i\zeta_{t-i}|^\gamma)(1+|\tilde{X}_{t,i}|^\gamma).$$

Clearly, $|U^{(1)}_{t,i}(x,y)| \le C|b_i\zeta_{t-i}|$ and, by combining the above bounds and taking $\gamma = r$, we obtain $E(U^{(1)}_{t,i}(x,y))^2 \le C\mu_\gamma(x,y)|b_i|^2$. This yields (6.14) for $k=0$ and $\tau=1$, with $\rho=1$, see the definition (6.11). The proof of (6.14) for $k=0$, $\tau=2$ is analogous as in the case $k \ge 1$ discussed above.

Next, let us prove (6.7). The proof mimics that of (6.5) given above. Fix $p \ge 0$ and $j \ge j_0$, where j_0 is such that the inequalities (5.2), (5.3) hold for all $j \ge j_0$, with the constant C independent of j. Note that $EF_j^{(p)}(x-\tilde{X}_{t,j}) = F^{(p)}(x)$, $E[F_j^{(p)}(x-\tilde{X}_{0,j})|\mathcal{F}_{-i-j}] = F^{(p)}_{i+j-1}(x-\tilde{X}_{0,i+j-1})$. Write

$$\begin{aligned} F_j(x-\tilde{X}_{0,j}) &= \sum_{i=1}^{\infty}(E[F_j(x-\tilde{X}_{0,j})|\mathcal{F}_{-i-j}] - E[F_j(x-\tilde{X}_{0,j})|\mathcal{F}_{-i-j-1}]) \\ &= \sum_{i=1}^{\infty}(F_{i+j-1}(x-\tilde{X}_{0,i+j-1}) - F_{i+j}(x-\tilde{X}_{0,i+j})). \end{aligned}$$

Hence we obtain the following expansion for $\tilde{R}^{(p)}_{0,j,k}(x-\tilde{X}_{0,j})$ analogous to (6.8):

$$\tilde{R}^{(p)}_{0,j,k}(x-\tilde{X}_{0,j}) = \sum_{i=1}^{\infty} \tilde{U}_{0,i}(x), \tag{6.22}$$

where

$$\tilde{U}_{0,i}(x) = F^{(p)}_{i+j-1}(x-\tilde{X}_{0,i+j-1}) - F^{(p)}_{i+j}(x-\tilde{X}_{0,i+j}) - b_{i+j}\zeta_{-i-j}\sum_{q=1}^{k}(-1)^q F^{(q+p)}(x)\tilde{X}^{(q-1)}_{0,-i-j}.$$

Then (6.7) follows from

$$E(\tilde{U}^{(\tau)}_{0,i}(x,y))^2 \le C\mu_r(x,y)(i+j)^{-(1+\theta)\rho}, \qquad i\ge 1,\ \ \tau=1,2,3, \tag{6.23}$$

where $\tilde{U}_{0,i}(x) = \sum_{\tau=1}^{3}\tilde{U}^{(\tau)}_{0,i}(x)$ is the decomposition analogous to the decomposition of $U_{t,i}(x)$ above:

$$\begin{aligned}
\tilde{U}^{(1)}_{0,i}(x) &:= F^{(p)}_{i+j}(x-\tilde{X}_{0,i+j-1}) - F^{(p)}_{i+j}(x-\tilde{X}_{0,i+j}) + b_{i+j}\zeta_{-i-j}F^{(p+1)}_{i+j}(x-\tilde{X}_{0,i+j}),\\
\tilde{U}^{(2)}_{0,i}(x) &:= F^{(p)}_{i+j-1}(x-\tilde{X}_{0,i+j-1}) - F^{(p)}_{i+j}(x-\tilde{X}_{0,i+j-1}),\\
\tilde{U}^{(3)}_{0,i}(x) &:= -b_{i+j}\zeta_{-i-j}\left(F^{(p+1)}_{i+j}(x-\tilde{X}_{0,i+j}) - \sum_{q=1}^{k}(-1)^{q-1}F^{(p+q)}(x)\tilde{X}^{(q-1)}_{0,-i-j}\right)\\
&= -b_{i+j}\zeta_{t+j}\tilde{R}^{(p+1)}_{0,i,k-1}(x).
\end{aligned}$$

The proof of (6.23) is completely analogous to that of (6.10) and uses the inductive assumption of (6.7) with k and p replaced by $k-1$ and $p+1$, respectively. It is important to note, however, that (6.23) holds for all $i\ge 1$, and that, contrary to (6.10), the bounds of Lemma 5.1 on $F^{(p)}_{i+j}$ etc., needed to prove (6.23), do not depend on $i\ge 1$, as they are guaranteed alone by choosing $j\ge j_0$ sufficiently large. □

7 Some applications

7.1 Goodness-of-fit testing. Here, we shall focus on some goodness-of-fit testing problems. Suppose F_0 is a specified d.f. and we wish to test the hypothesis $H_0:\ F=F_0$, i.e., the marginal d.f. of the given stationary process is F_0. Consider three test statistics for this problem: the Kolmogorov–Smirnov statistic, and the L_p, $p=1,2$, statistics:

$$D_N := \sup_{x\in\mathbb{R}}|F_N(x)-F_0(x)|, \qquad C_{Np} := \int |F_N(x)-F_0(x)|^p dx.$$

Also, recall from Ibragimov and Linnik [24], Theorem 18.6.5, that $(\mathrm{Var}(\bar{X}_N))^{-1/2}\bar{X}_N \Longrightarrow Z\sim N(0,1)$, which is true if $\mathrm{Var}(\bar{X}_N)\to\infty$, as is the case here. Use this together with (3.15), to obtain the following weak convergence under H_0:

$$N^{\theta/2}D_N \Longrightarrow c^{1/2}(\theta)|Z|\sup_{x\in\mathbb{R}} f_0(x), \tag{7.1}$$

$$N^{p\theta/2}C_{Np} \Longrightarrow c^{p/2}(\theta)|Z|^p\int f_0^p(x)dx, \tag{7.2}$$

where the constant $c(\theta)$ is determined by

$$c(\theta) = \lim_{N\to\infty} N^{\theta} E\bar{X}_N^2, \tag{7.3}$$

and is given in (3.18). Observe that it is a continuous function on (0, 1). We mention that (7.1) holds under the conditions (3.3), (5.1), (7.3) and $E|\zeta_0|^{2+\delta} < \infty$, for some $\delta > 0$, alone. In particular, the regular decay condition (3.1) of the coefficients b_j in the moving average representation (1.1) is not needed. The limit (7.3) is strictly positive under (3.1) and may also be zero if (3.1) is not satisfied.

From (7.1) and (7.2), we readily obtain the following facts:

$$T_{N1}(\theta) := \frac{N^{\theta/2} D_N}{c^{1/2}(\theta) \sup_{x\in\mathbb{R}} f_0(x)} \Longrightarrow |Z|,$$

$$T_{N2}(\theta) := \frac{N^{\theta/2} C_{N1}}{c^{1/2}(\theta)} \Longrightarrow |Z|, \qquad T_{N3}(\theta) := \frac{N^{\theta} C_{N2}}{c(\theta) \int f_0^2(x)dx} \Longrightarrow Z^2.$$

Thus implementation of any one of these tests requires only an estimator $\hat{\theta}_N$ of θ satisfying

$$\log(N)(\hat{\theta}_N - \theta) = o_p(1). \tag{7.4}$$

Fox and Taqqu [11] and Robinson [40], among others, give some such estimators. Notice that the determination of the critical values of tests based on any of the three statistics $T_{Nj}(\hat{\theta}_N)$, $j = 1, 2, 3$, is trivial, relative to their analog in the i.i.d. setup. Another point worth mentioning is that the limiting null distribution of $N^{\theta/2} C_{N1}$ does not depend on F_0. To remove the dependence of the other two statistics on F_0, one may replace f_0 in there by a consistent density estimator. Csörgő and Mielniczuk [6] and Hall, Lahiri, and Truang [17] study some estimators of the marginal density of a long memory process.

As noted in [18], p. 998, (7.1) holds with $\sup_{x\in\mathbb{R}}$ replaced by $\sup_{x\in B}$, where $B \subset \mathbb{R}$ is an arbitrary Borel set. In particular, for any fixed point $x_0 \in \mathbb{R}$,

$$N^{\theta/2} |F_N(x_0) - F(x_0)| \Longrightarrow c^{1/2}(\theta) f(x_0) |Z|. \tag{7.5}$$

If $f(x_0) = 0$, one can use the asymptotic expansion (3.13) to make (7.5) more precise. Assuming $f'(x_0) = 0$, $f''(x_0) > 0$ and $\theta < 1/3$, from Theorem 3.1 one obtains

$$N^{3\theta/2} |F_N(x_0) - F(x_0)| \Longrightarrow f''(x_0) |Z^{(3)}|.$$

Next, consider the linear regression model:

$$Y_{Nt} = \beta' v_{Nt} + X_t, \qquad 1 \le t \le N,\ \beta \in \mathbb{R}^q,$$

where v'_{Nt} is the tth row of the known nonrandom $N \times q$ design matrix V_N, and X_t is as in (1.1). Assume

$$\mathcal{V}_N := V_N' V_N \quad \text{is nonsingular for all } N \ge q, \text{ and that}$$
$$N^{1/2} \max_{1\le t\le N} \|v'_{Nt} \mathcal{V}_N^{-1/2}\| = O(1).$$

Again, consider the above H_0. Let $\hat{\beta}_N$ be the least square estimator of β. Observe that

$$\Delta_N := \hat{\beta}_N - \beta = \mathcal{V}_N^{-1} \sum_{t=1}^{N} v_{Nt} X_t.$$

An appropriate process here to base tests of H_0 on is the residual empirical process:

$$\tilde{F}_N(x) := N^{-1}\sum_{t=1}^{N} I\left(Y_{Nt} - \hat{\beta}' v_{Nt} \le x\right) = N^{-1}\sum_{t=1}^{N} I\left(X_t \le x + (\hat{\beta} - \beta)' v_{Nt}\right).$$

Then using (3.16), (3.15), and the smoothness of F_0 (implied by (5.1)), one obtains that

$$\begin{aligned} & N^{\theta/2}[\tilde{F}_N(x) - F_0(x)] \\ &= \sup_{x\in\mathbb{R}} \left| N^{\theta/2}[F_N(x) - F_0(x)] - N^{\theta/2}\bar{v}_N' \Delta_N f_0(x) \right| + u_p(1) \\ &= N^{\theta/2}\left(c_\theta \bar{X}_N - \bar{v}_N' \Delta_N\right) f_0(x) + u_p(1), \end{aligned}$$

where $u_p(1)$ denotes a sequence of processes converging to zero, uniformly on $\mathbb{R}$, in probability, and $\bar{v}_N := N^{-1}\sum_{t=1}^{N} v_{Nt}$. The random variables on the r.h.s. of the above fact can be shown to converge in distribution to a normal random variable. See Giraitis, Koul, and Surgailis [13] for relevant details. Again, it follows that the tests of H_0 based on analogs of D_N or C_{Np} in the present case are also relatively easy to implement compared to the i.i.d. setup. Here, one may use the Whittle estimator of θ based on the residuals $Y_{Nt} - \hat{\beta}' v_{Nt}$; $1 \le t \le N$, in carrying out these tests. The $\log(N)$ consistency (7.4) of this estimator follows from the works of Giraitis and Koul [12], Koul and Surgailis [34], and Koul [31]. See Ho and Hsing [18] for some other applications of Theorem 3.1.

7.2 U-statistics (see Dehling and Taqqu [8], Ho and Hsing [18], and Hsing [23]). Let $H(x_1, x_2)$ be a symmetric function in $x_1, x_2 \in \mathbb{R}$. The (nonnormalized) U-statistic of order 2 and kernel H over observations $X_1, \ldots, X_N$ is defined by

$$U_{N,H} = \sum{}' H(X_i, X_j),$$

where the sum $\sum'$ is taken over all $1 \le i, j \le N$, $i \ne j$. A kernel H or the statistic $U_{N,H}$ is called *degenerate* if

$$\int_{\mathbb{R}} H(x, y) dF(x) = 0, \qquad \forall\, y \in \mathbb{R}.$$

Put

$$\begin{aligned} \mu &:= \int\int H(x, y) dF(x) dF(y), \qquad H_1(x) := \int H(x, y) dF(y) - \mu, \\ H_2(x, y) &:= H(x, y) - H_1(x) - H_1(y) - \mu, \qquad x, y \in \mathbb{R}. \end{aligned}$$

It is well known (Hoeffding [20]) that any square integrable U-statistic $U_{N,H}$ of order 2 can be decomposed into a sum $U_{N,H} = U_{N,H_2} + 2N^2 S_{N,H_1} + N^2\mu$, where U_{N,H_2} is a degenerate U-statistic and $S_{N,H_1} = N^{-1}\sum_{i=1}^{N} H_1(X_i)$ is a sum of zero mean random variables. The Hoeffding decomposition is the starting point in the study of the limit distribution of U-statistics in the case of i.i.d. or weakly dependent observations; see, e.g., Serfling [42]. In this case, it is well known that the limit distribution of a U-statistic is normal if the statistic is nondegenerate, and nonnormal (given by a double Itô–Wiener integral) if the statistic is degenerate.

The asymptotic distribution of degenerate U-statistics from Gaussian or Gaussian subordinated observations with long memory was studied by Dehling and Taqqu [8]. Ho and Hsing [18] extended their results to linear processes (1.1) with long memory. Let the U statistic be degenerate. Then

$$\begin{aligned} U_{N,H} &= N^2 \int_{\mathbb{R}^2} H(x_1, x_2) dF_N(x_1) dF_N(x_2) - N\tilde{S}_{N,H} \\ &= N^2 \int_{\mathbb{R}^2} H(x_1, x_2) d(F_N(x_1) - F(x_1)) d(F_N(x_2) - F(x_2)) - N\tilde{S}_{N,H}, \end{aligned}$$

where $\tilde{S}_{N,H} = N^{-1} \sum_{i=1}^{N} H(X_i, X_i) = \int_{\mathbb{R}} H(x, x) dF_N(x)$ is the "diagonal part."

Clearly, if $E|H(X_0, X_0)| < \infty$ and $0 < \theta < 1$, then $N^{\theta-1} S_{N,H} = o_P(1)$, by the Ergodic Theorem. Thus, an application of (3.17) yields, under some regularity conditions on H,

$$N^{\theta-2} U_{N,H} \Longrightarrow c_\theta H_\infty^{(1,1)}(0,0) Z^2, \tag{7.6}$$

where $Z \sim N(0, 1)$, and

$$\begin{aligned} H_\infty(z_1, z_2) &:= \int_{\mathbb{R}^2} H(x_1 + z_1, x_2 + z_2) dF(x_1) dF(x_2), \qquad z \in \mathbb{R}, \\ H_\infty^{(1,1)}(0,0) &= \partial^2 H_\infty(z_1, z_2) / \partial z_1 \partial z_2 |_{z_1 = z_2 = 0} = \int_{\mathbb{R}^2} H(x_1, x_2) f'(x_1) f'(x_2) dx_1 dx_2. \end{aligned}$$

If the coefficient $H_\infty^{(1,1)}(0,0) = 0$ in (7.6), the question of the limiting distribution of appropriately normalized $U_{N,H}$ becomes more complicated. For a class of (possibly nondenegerate) U-statistics, this problem was recently studied by Hsing [23]. As it turned out, the Hoeffding decomposition is not necessarily the right tool for obtaining the limiting distributions of these statistics under long memory, as different terms of this decomposition may contribute to the limit distribution of $U_{N,H}$. The above remarks apply to the general case of kth order U-statistics also.

7.3 M-estimators (see [3], [13], [29], [32], [30], [33], [35]). Consider the one sample location model where one observes $Y_1, \ldots, Y_N$, satisfying

$$Y_t = m + X_t, \qquad 1 \le t \le N, \tag{7.7}$$

for an unknown location parameter m, and the "errors" $\{X_t\}$ follow the long memory moving average of (1.1), (3.1). Given a monotone weight function H, $EH(X_0) = 0$, the corresponding M-estimator T_N of the unknown parameter m is defined as the solution of the equation

$$0 = \sum_{t=1}^{N} H(Y_t - T_N) = N \int_{\mathbb{R}} H(x - T_N) dF_N(x).$$

In particular, the choices $H(x) = \mathrm{sgn}(x)$ and $H(x) = x$ correspond to the median and sample mean (least squares estimator), respectively. As Beran [3] observed, under Gaussian long memory errors M-estimators do not lose efficiency compared to the least squares estimator, in the sense that $T_N = \bar{Y}_N + o_p(\bar{Y}_N)$. Similar findings were reported in Koul [29] and Koul and Mukherjee [32] for certain classes of M- and R-estimators of the slope parameter in multiple linear regression models. Giraitis, Koul, and Surgailis [13] extended these results to non-Gaussian moving average errors of (1.1). This raised the natural question whether there exist any differences among M-estimators asymptotically at any stage.

To answer this question, Koul and Surgailis [33] obtained the asymptotic expansion of a class of M-estimators in the location model (7.7):

$$T_N - \bar{Y}_N = \sum_{2 \le k < 1/\theta} N^{-k\theta/2} Q_k(Z_N^{(k)}, \ldots, Z_N^{(k)}) + N^{-1/2} W_N.$$

Here, $Q_k(z_1, \ldots, z_k)$ is a polynomial in $z_1, \ldots, z_k$ with coefficients given by $\lambda_i := H_\infty^{(i)}(-m)$, $1 \le i \le k$; $Z_N^{(k)}$ are given by (3.4) and

$$Q_k(Z_N^{(k)}, \ldots, Z_N^{(k)}) \Longrightarrow Q_k(Z_1, \ldots, Z_k), \quad 2 \le k < 1/\theta,$$

while W_N is asymptotically Gaussian: $W_N \Longrightarrow N(0, \lambda_1^{-2}\sigma^2)$, where σ^2 and H_∞ are the same as in (3.19) and (3.20). The two lower order polynomials Q_k are

$$Q_2(z_1, z_2) = \frac{\lambda_2}{\lambda_1}(z_1^2 - 2z_2), \quad Q_3(z_1, z_2, z_3) = \frac{2\lambda_3}{\lambda_1}(z_1^3 - 3z_1z_2 + 3z_3).$$

See also Koul and Surgailis [35] for a similar result in a one-parameter linear regression model. The proofs in these papers are quite involved. It would be desirable to develop a less involved proof and extend the result of the latter paper to multiple linear regression models.

8 The infinite variance case

The results discussed in the previous sections apply to the finite variance case $E\zeta_0^2 < \infty$, or $EX_t^2 < \infty$ only. Long memory moving averages with *infinite variance* are of considerable interest, too. ARFIMA(p, d, q) series with i.i.d. innovations $\{\zeta_s\}$ belonging to the domain of attraction of α-stable law were investigated by Samorodnitsky and Taqqu [41], Kokoszka and Taqqu [26], [27]. The asymptotic behavior of empirical functionals of long memory moving averages with infinite variance was discussed by Hsing [22] and Koul and Surgailis [36].

In the sequel, we assume that $\{X_t\}$ is the moving average process of (1.1), where $\{\zeta_s\}$ are i.i.d. random variables with zero mean, satisfying the tail regularity condition

$$\lim_{x\to-\infty} |x|^\alpha G(x) = \lim_{x\to\infty} x^\alpha(1 - G(x)) = c_1, \tag{8.1}$$

where $1 < \alpha < 2$ and $c_1 > 0$ is some constant. In such case, the moving average (1.1) is well defined if and only if $\sum_{j=1}^\infty |b_j|^\alpha < \infty$, see [41]. Similarly as in Section 3, we shall assume that the weights decay hyperbolically:

$$b_j \sim c_0 j^{-\beta}, \qquad j \to \infty, \tag{8.2}$$

where $1/\alpha < \beta < 1$ and $c_0 \neq 0$ is some constant. It is well known ([25], [41]) that under (8.1) and (8.2), the sample mean is asymptotically stable:

$$N^{\beta-1/\alpha}\bar{X}_N \Longrightarrow c_{\alpha,\beta} Z, \tag{8.3}$$

where

$$c_{\alpha,\beta} = c_0 \left(2c_1 \frac{\Gamma(2-\alpha)\cos(\alpha\pi/2)}{1-\alpha} \int_{-\infty}^1 \int_0^1 (t-s)_+^{-\beta}\, dt\, ds \right)^{1/\alpha},$$

and $Z \sim S\alpha S$ is a standard symmetric α-stable random variable, $Ee^{iuZ} = e^{-|u|^\alpha}$, $u \in \mathbb{R}$. Note the normalization in (8.3) is different from the usual normalization $N^{1-1/\alpha}$ of the sample mean of weakly dependent random variables in the domain of attraction of α-stable law. This is an indication of the fact that the moving average (1.1) with (8.1), (8.2) exhibits long memory. While the usual characterization of long memory in terms of covariance function is not applicable in the infinite variance case, other related characterizations exist, see [41], [26], [36].

Theorem 8.1. (Koul and Surgailis [36]) *Let $\{X_t\}$ be the moving average sequence satisfying* (8.1), (8.2), *with* $1 < \alpha < 2$, $1/\alpha < \beta < 1$. *Moreover, we assume that G is twice differentiable and there is a constant* $C < \infty$ *such that for all* $x, y \in \mathbb{R}$, $|x-y| \le 1$,

$$|G''(x)| \le C(1+|x|)^{-\alpha}, \qquad |G''(x) - G''(y)| \le C|x-y|(1+|x|)^{-\alpha}. \tag{8.4}$$

Then,

$$N^{\beta-1/\alpha}(F_N(x) - F(x)) \Longrightarrow_{D(\bar{\mathbb{R}})} c_{\alpha,\beta} f(x) Z,$$

where $f(x) = F'(x)$ *and* Z *is a standard* $S\alpha S$ *random variable.*

From this theorem, one can easily obtain the limiting distribution of sums $S_{N,H} = N^{-1}\sum_{t=1}^{N}(H(X_t) - EH(X_t))$ for a large class of bounded measurable functions H.

Corollary 8.2. *For any measurable function H of bounded variation,*

$$N^{\beta-1/\alpha} S_{N,H} \Longrightarrow c_{\alpha,\beta} H'_\infty(0) Z, \tag{8.5}$$

where $Z \sim S\alpha S$ *and* $H_\infty(x) := EH(x + X_0)$.

Corollary 8.2 is *surprising* in the sense it shows that an α-stable ($1 < \alpha < 2$) limit law may arise from sums of *bounded* random variables $H(X_t)$. This is unlike the case of i.i.d. or weakly dependent summands, where such limit laws require a *long tailed* distribution of the summands and the contribution of the maximal summand to be comparable to the sum itself. It also reconfirms the deep differences between long and short memory.

Theorem 8.1 and Corollary 8.2 contradict the recent result of Hsing [22], which claims, under similar assumptions on $\{X_t\}$ and H, that

$$N^{(\alpha\beta-1)/2} S_{N,H} \Longrightarrow \sigma_H Z, \tag{8.6}$$

where $Z \sim N(0,1)$ is a standard normal variable. Note that for $1/\alpha < \beta < 1$, the normalization $N^{(\alpha\beta-1)/2} \ll N^{\beta-1/\alpha}$ in (8.6). The proof of (8.6) in [22] uses an approximation of $S_{N,H}$ by a sum $T_{N,H}$ of *independent* random variables whose normal limiting behavior is deduced by the classical Lindeberg–Feller central limit theorem. However, Lindeberg's condition for $T_{N,H}$ actually does not hold, see [36], and the proof of (8.6) fails.

If the asymptotic constant $H'_\infty(0) = -\int_{\mathbb{R}} H(u) f'(u) du = 0$, then (8.5) implies

$$N^{\beta-1/\alpha} S_{N,H} \Longrightarrow 0.$$

The natural question arises as to whether in this case it is possible to obtain a nondegenerate limit for $S_{N,H}$ with some other normalization which grows faster than $N^{\beta-1/\alpha}$. This question is clearly related to the possibility of asymptotic expansions in the infinite variance case similar to (3.13) and (3.19). According to Hsing [22], in the infinite variance case "asymptotic expansions are not feasible." On the other hand, $S\alpha S$ analogs of the multiple integrals $Z^{(k)}$ (2.5) exist (Surgailis [44]), and similar weak convergence results to such integrals as in (3.13) of polynomial forms in α-stable random variables can be proved. In any case, the question of asymptotic expansions for infinite variance moving averages remains open.

The proof of Theorem 8.1 follows from (8.3) and the *uniform reduction principle* similar to (3.15): there exists $\kappa > 0$, such that

$$\sup_{x\in\mathbb{R}} N^{\beta-1/\alpha}|F_N(x) - F(x) + f(x)\bar{X}_N| = O_p(N^{-\kappa}). \tag{8.7}$$

Contrary to the finite variance case, the covariance of the process $R_t(x) = I(X_t \le x) - F(x) + f(x)X_t$ or its increment $R_t(x, y)$ does not exist and cannot be estimated as in Lemma 6.1. Instead, the proof of Theorem 8.1 in [36] is based on the estimate (8.8) below of the rth moment ($r < \alpha$) of the increment of the empirical process in (8.7). Namely, there exist $1 < r < \alpha$, $\kappa > 0$, and a finite measure μ on the real line such that for any $x < y$, $N \ge 1$,

$$E\Big(N^{\beta - 1/\alpha}|F_N(x, y) - F(x, y) + f(x, y)\bar{X}_N|\Big)^r \le \mu(x, y)N^{-\kappa}. \tag{8.8}$$

The claim (8.7) can be deduced from (8.8) by a chaining argument as in the finite variance case. The starting point of the proof of (8.8) is the telescoping identity similar to (4.4), (6.8):

$$N(F_N(x) - F(x) + f(x)\bar{X}_N) = \sum_{s=-\infty}^{N} M_{s,N}(x),$$

where $M_{s,N}(x) := \sum_{t=1\vee(s+1)}^{N} U_{t,t-s}(x)$ and where $U_{t,i}(x)$ are defined as in (6.9) (with $k = 1$). Although the $U_{t,i}(x)$'s have infinite variance and the argument of Section 6 is not applicable, the random variables $M_{s,N}(x)$, $s \in \mathbb{Z}$ form *martingale differences*

$$E[M_{s,N}(x)|\mathcal{F}_{s-1}] = 0,$$

w.r.t. the σ-fields $\mathcal{F}_t = \sigma\{\zeta_s : s \le t\}$. By the inequality due to von Bahr and Esséen [2], for any $1 < r < \alpha$,

$$N^r E|F_N(x, y) - F(x, y) + f(x, y)\bar{X}_N|^r \le 2 \sum_{s=-\infty}^{N} E|M_{s,N}(x, y)|^r.$$

Then, (8.8) follows from the bound

$$E|M_{s,N}(x, y)|^r \le \mu(x, y)\left(\sum_{t=1\vee(s+1)}^{N} |b_{t-s}|^{1+\kappa}\right)^r, \qquad x < y; \tag{8.9}$$

see [36], Lemma 4.3. The proof of (8.9) is broadly similar to that of Lemma 6.1 above, where the exponent $r < \alpha$ has to be chosen sufficiently close to α.

References

[1] F. Avram and M.S. Taqqu: Noncentral limit theorems and Appell polynomials, *Annals of Probability* **15** (1987), 767–775.

[2] B. von Bahr and C.-G. Esséen: Inequality for rth absolute moment of the sum of random variables, *Annals of Mathematical Statistics* **36** (1965), 299–303.

[3] J. Beran: M-estimators for location for Gaussian and related processes with slowly decaying serial correlations, *Journal of American Statistical Association* **86** (1991), 704–708.

[4] P. Breuer and P. Major: Central limit theorem for non-linear functionals of Gaussian fields, *Journal of Multivariate Analysis* **13** (1983), 425–441.

[5] P.J. Brockwell and R.A. Davis: *Time Series: Theory and Methods*, Springer-Verlag, New York, 1991.

[6] M. Csörgő and J. Mielniczuk: Density estimation under long range dependence, *Annals of Statistics* **23** (1995), 990–999.

[7] M. Csörgő and J. Mielniczuk: The empirical process of a short-range dependent stationary sequence under Gaussian subordination, *Probability Theory and Related Fields* **104** (1996), 15–25.

[8] H. Dehling and M.S. Taqqu: The empirical process of some long-range dependent sequences with an application to U-statistics, *Annals of Statistics* **17** (1989), 1767–1783.

[9] R.L Dobrushin and P. Major: Non-central limit theorems for non linear functionals of Gaussian fields, *Zeitschrift für Wahrscheinlichkeitstheorie und verwandte Gebiete* **50** (1979), 27–52.

[10] P. Doukhan, G. Lang and D. Surgailis: Asymptotics of empirical processes of linear random fields with long range dependence, *Annales de l'Institut Henri Poincaré* (B) *Probabilités et Statistiques*, to appear.

[11] R. Fox and M.S. Taqqu: Large sample properties of parameter estimates for strongly dependent stationary Gaussian time series, *Annals of Statistics* **14** (1986), 517–532.

[12] L. Giraitis and H.L. Koul: Estimation of the dependence parameter in linear regression with long-range dependent errors, *Stochastic Processes and Their Applications* **71** (1997), 207–224.

[13] L. Giraitis, H.L. Koul and D. Surgailis: Asymptotic normality of regression estimators with long memory errors, *Statistics and Probability Letters* **29** (1996), 317–335.

[14] L. Giraitis and D. Surgailis: CLT and other limit theorems for functionals of Gaussian sequences, *Zeitschrift für Wahrscheinlichkeitstheorie und verwandte Gebiete* **70** (1985), 191–212.

[15] L. Giraitis and D. Surgailis: Multivariate Appell polynomials and the central limit theorem. In: E. Eberlein and M.S. Taqqu (eds.), *Dependence in Probability and Statistics*, Birkhäuser, Boston, 1986, 21–71.

[16] L. Giraitis and D. Surgailis: Central limit theorem for the empirical process of a linear sequence with long memory, *Journal of Statistical Planning and Inference* **80** (1999), 81–93.

[17] P. Hall, S.N. Lahiri and Y.K. Truang: On bandwidth choice for density estimation with dependent data, *Annals of Statistics* **23** (1995), 2241–2263.

[18] H.-C. Ho and T. Hsing: On the asymptotic expansion of the empirical process of long memory moving averages, *Annals of Statistics* **24** (1996), 992–1024.

[19] H.-C. Ho and T. Hsing: Limit theorems for functionals of moving averages, *Annals of Probability* **25** (1997), 1636–1669.

[20] W. Hoeffding: A class of statistics with asymptotically normal distribution, *Annals of Mathematical Statistics* **19** (1948), 239–325.

[21] J.R.M. Hosking: Fractional differencing, *Biometrika* **68** (1981), 165–176.

[22] T. Hsing: On the asymptotic distributions of partial sums of functionals of infinite variance moving averages. *Annals of Probability* **27** (1999), 1579–1599.

[23] T. Hsing: A decomposition of the generalized U-statistic for long-memory linear process, preprint, 2000.

[24] I.A. Ibragimov and Yu. Linnik: *Independent and Stationary Sequences of Random Variables*, Walters-Noordhoff, Groningen, 1971.

[25] Y. Kasahara and M. Maejima: Weighted sums of i.i.d. random variables attracted to integrals of stable processes, *Probability Theory and Related Fields* **78** (1988), 75–96.

[26] P. Kokoszka and M.S. Taqqu: Fractional ARIMA with stable innovations, *Stochastic Processes and Applications* **60** (1995), 19–47.

[27] P. Kokoszka and M.S. Taqqu: Parameter estimation for infinite variance fractional ARIMA, *Annals of Statistics* **24** (1996), 1880–1913.

[28] H.L. Koul: *Weighted Empirical Processes and Linear Models*, Lecture Notes and Monograph Series 21. Institute of Mathematical Statistics, Hayward, CA, 1992.

[29] H.L. Koul: M-estimators in linear models with long range dependent errors, *Statistics and Probability Letters* **14** (1992), 153–164.

[30] H.L. Koul: Asymptotics of M-estimators in non-linear regression with long-range dependent errors. In: P.M. Robinson and M. Rosenblatt (eds.), *Athens Conference on Applied Probability and Time Series*, Lecture Notes in Statistics 115, Springer-Verlag, New York, 1996, 272–290.

[31] H.L. Koul: Estimation of the dependence parameter in non-linear regression with random design and long memory errors. In: A.K. Basu, J.K. Ghosh, P.K. Sen, and B.K. Sinha (eds.), *Perspectives in Statistical Sciences*, Oxford University Press, Oxford, 2000, 191–208.

[32] H.L. Koul and K. Mukherjee: Asymptotics of R-, MD- and LAD-estimators in linear regression models with long range dependent errors, *Probability Theory and Related Fields* **95** (1993), 535–553.

[33] H.L. Koul and D. Surgailis: Asymptotic expansion of M-estimators with long memory errors, *Annals of Statistics* **25** (1997), 818–850.

[34] H.L. Koul and D. Surgailis: Asymptotic normality of the Whittle estimator in linear regression models with long memory errors, *Statistical Inference for Stochastic Processes* **3** (2000), 129–147.

[35] H.L. Koul and D. Surgailis: Second order behavior of M-estimators in linear regression with long memory errors, *Journal of Statistical Planning and Inference* **91** (2000), 399–412.

[36] H.L. Koul and D. Surgailis: Asymptotics of the empirical processes of long memory moving averages with infinite variance, *Stochastic Processes and Their Applications* **91** (2001), 309–336.

[37] R. Leipus and M.-C. Viano: Modeling long-memory time series with finite or infinite variance: A general approach, *Journal of Time Series Analysis*, **21** (1997), 61–74.

[38] P. Major: *Multiple Wiener–Itô integrals*, Lecture Notes in Mathematics 849, Springer-Verlag, New York, 1981.

[39] D. Marinucci: The empirical process for bivariate sequences with long memory, preprint, 2000.

[40] P.M. Robinson: Semiparametric analysis of long-memory time series, *Annals of Statistics* **22** (1994), 515–539.

[41] G. Samorodnitsky and M.S.Taqqu: *Stable Non-Gaussian Random Processes*, Chapman and Hall, New York, 1994.

[42] R. Serfling: *Approximation Theorems of Mathematical Statistics*, Wiley, New York, 1980.

[43] Q.M. Shao and H. Yu: Weak convergence for weighted empirical processes of dependent sequences, *Annals of Probability* **24** (1996), 2052–2078.

[44] D. Surgailis: On L^2 and non-L^2 multiple stochastic integration. In: M. Arató, D. Vermes and A.V. Balakrishnan (eds.), *Stochastic Differential Systems.* Lecture Notes in Control and Information Sciences 36, Springer-Verlag, Berlin, 1981, 212–226.

[45] D. Surgailis: Zones of attraction of self-similar multiple integrals, *Lithuanian Mathematical Journal* **22** (1982), 327–340.

[46] D. Surgailis: Long-range dependence and Appell rank, *Annals of Probability* **28** (2000), 478–497.

[47] M.S. Taqqu: Weak convergence to fractional Brownian motion and to the Rosenblatt process, *Zeitschrift für Wahrscheinlichkeitstheorie und verwandte Gebiete* **31** (1975), 287–302.

[48] M.S. Taqqu: Law of the iterated logarithm for sums of non-linear functions of Gaussian variables that exhibit a long range dependence, *Zeitschrift für Wahrscheinlichkeitstheorie und verwandte Gebiete* **40** (1977), 203–238.

[49] M.S. Taqqu: Convergence of integrated processes of arbitrary Hermite rank, 1 *Zeitschrift für Wahrscheinlichkeitstheorie und verwandte Gebiete* **50** (1979), 53–83.

Hira L. Koul
Department of Statistics and Probability
Michigan State University
East Lansing, MI 48824-1027, U.S.A.
koul@stt.msu.edu

Donatas Surgailis
Vilnius Institute of Mathematics and Informatics
Akademijos 4
2600 Vilnius, Lithuania
sdonatas@ktl.mii.lt

The Reduction Principle for the Empirical Process of a Long Memory Linear Process

Liudas Giraitis[1] and Donatas Surgailis

Abstract. We discuss the uniform reduction principle for the empirical process of a long memory moving average process $X_t, t \in \mathbb{Z}$ with long memory, which generalizes the corresponding reduction principle of Dehling and Taqqu [9]. The proof is based on the expansion of the bivariate probability density $f_t(x_1, x_2)$ of (X_0, X_t):

$$f_t(x_1, x_2) \;=\; f(x_1)f(x_2) + r_t f'(x_1)f'(x_2) + o(r_t), \qquad t \to \infty,$$

uniformly in x_1, x_2, where $r_t = \operatorname{cov}(X_0, X_t)$, and $f(x)$ is the marginal probability density. An easy consequence of the reduction principle is the functional CLT for the empirical process. An application of the last result to the change-point problem of the marginal c.d.f. is discussed.

1 Introduction

Let

$$X_t \;=\; \sum_{s\in\mathbb{Z}} b_{t-s}\zeta_s, \qquad t \in \mathbb{Z}, \tag{1.1}$$

be a moving average process, where $\zeta_s, s \in \mathbb{Z}$ are i.i.d. random variables with zero mean and variance 1, and $b_s, s \in \mathbb{Z}$ are square summable weights

$$\sum_{s\in\mathbb{Z}} b_s^2 < \infty.$$

The process (1.1) is said to have *long memory* if $\sum_s |b_s| = \infty$, and *short memory* if $\sum_s |b_s| < \infty$ (Kokoszka and Mikosch [24]). (1.1) is said to be *causal* (one-sided) if $b_s = 0$ $(s < 0)$, and *non-causal* (two-sided) otherwise. It is well known (e.g., Brockwell and Davis [6]) that any zero mean stationary Gaussian process with absolutely continuous spectral measure $\Psi(d\lambda) = \psi(\lambda)d\lambda$ admits a (two-sided) moving average representation (1.1). It also admits a causal representation (1.1) if and only if its spectral density satisfies $\int_{-\pi}^{\pi} \log \psi(\lambda)d\lambda > -\infty$ ([6], Chapter 5).

One of the most interesting problems of long memory processes (1.1) concerns the limit distribution of the partial sums processes

$$S_{N,h}(t) = \sum_{s=1}^{[Nt]} h(X_s), \qquad t \in [0, 1], \tag{1.2}$$

[1]Research supported by the ESRC grant R000238212.

where $h(\cdot)$ is a square integrable measurable function, i.e., $Eh^2(X_0) < \infty$. A particular case of (1.2) is the *empirical process*

$$F_N(x,t) := N^{-1} \sum_{s=1}^{[Nt]} I(X_s \le x), \qquad (x,t) \in \mathbb{R} \times [0,1], \tag{1.3}$$

where $F(x) := P(X_t \le x)$ is the (marginal) probability distribution function which does not depend on $t \in \mathbb{Z}$, due to strict stationarity of (1.1). For general linear processes (1.1) with long memory, the study of the limit distribution of (1.3) seems quite difficult. In the causal case, one usually assumes that the weights b_s, $s \ge 0$ decay regularly as $s \to \infty$ in the sense that

$$b_s = L(s)s^{-(1+\theta)/2} \tag{1.4}$$

for some $0 < \theta < 1$, where $L(s)$ is a function slowly varying at infinity. In particular, one can have $L(s) = (\log s)^a$ $(a \in \mathbb{R})$, $L(s) = \log\log s$, or any converging sequence $L(s)$ with $\lim_{s\to\infty} L(s) \neq 0$. The parameter θ characterizes the decay rate of the dependence ("memory") between the past and the future of the series (1.1). A particular case of (causal) long memory processes (1.1), (1.4) are the fractional ARIMA processes (see Granger and Joyeux [20]; cf. Brockwell and Davis [6]). (1.4) implies the hyperbolic decay of the covariance

$$r_t := EX_sX_{t+s} = \sum_{s=0}^{\infty} b_s b_{t+s} = L_2(t)t^{-\theta}, \tag{1.5}$$

where $L_2(t)$ is a slowly varying function such that

$$\lim_{t\to\infty} L_2(t)/L^2(t) = \int_0^\infty (u(1+u))^{-(1+\theta)/2} du = B(\theta, (1-\theta)/2),$$

where $B(\cdot,\cdot)$ is the beta-function. Note in the case of fractional ARIMA, both sequences $L(t)$ and $L_2(t)$ tend (as $t \to \infty$) to finite constants not equal to 0. The above property of L_2 holds generally under (1.4), by the well-known properties of slowly varying functions (see, e.g., [34], [32]).

For Gaussian long memory processes, the above problems were studied in numerous publications, e.g., [10], [34], [27], [5], [16]. Most of these studies use Hermite expansions as the major tool. Recall that any function $h \in L^2(F)$, where F is the standard normal c.d.f., admits the convergent expansion

$$h(x) = \sum_{k=0}^{\infty} c_k H_k(x)/k! \tag{1.6}$$

in Hermite polynomials $H_k(x) = (-1)^k e^{x^2/2}(e^{-x^2/2})^{(k)}$, $k = 0, 1, \ldots$; with coefficients given by $c_k = Eh(X_0)H_k(X_0)$. The smallest $k \ge 1$ with $c_k \neq 0$ is called the *Hermite rank of h*. The well-known reduction principle of Taqqu [34] asserts that in the long memory case (1.5), the limit distribution of $S_{N,h}(t) - ES_{N,h}(t)$ coincides with that of $c_{k_H^*} S_{N,H_{k_H^*}}(t)$, provided the Hermite rank k_H^* of h satisfies $\theta k_H^* < 1$. The reduction principle is a direct consequence of the formula

$$\operatorname{cov}\Big(h(X_0) - \sum_{j=0}^{k} c_j H_j(X_0)/j!,\ h(X_t) - \sum_{j=0}^{k} c_j H_j(X_t)/j!\Big) = \sum_{j=k+1}^{\infty} r_t^j c_j^2/j!, \tag{1.7}$$

whose r.h.s. is $O(r_t^{k+1})$ $(t \to \infty)$. This follows from the well-known orthogonality properties of the Hermite polynomials. The weak limit of the partial sums $S_{N,H_k}(t)$ of the Hermite polynomials is well known; see, e.g., Taqqu [34]. Note that the Hermite expansion of the indicator function $h_x(y) = I(y \le x)$ is very explicit:

$$I(y \le x) = \sum_{k=0}^{\infty} (-1)^k F^{(k)}(x) H_k(y)/k!, \tag{1.8}$$

and therefore the Hermite rank of $I(\cdot \le x)$ equals 1 for each $x \in \mathbb{R}$. Whence it easily follows that the weak limit (of the finite dimensional distributions) of the normalized empirical process $d_N^{-1}(F_N(t,x) - ([Nt]/N)F(x))$ is the degenerate Gaussian process $c_\theta f(x) Z_t$, where

$$d_N = L(N) N^{-\theta/2} \tag{1.9}$$

is the normalization determined by $E(N^{-1}\sum_{t=1}^N X_t)^2 = c_\theta^2 d_N^2 (1 + o(1))$, c_θ is some constant depending only on θ, and $Z_t, 0 \le t \le 1$ is fractional Brownian motion, i.e., a continuous Gaussian process with zero mean and covariances

$$E Z_t Z_s = (1/2)(t^{2-\theta} + s^{2-\theta} - |t-s|^{2-\theta}).$$

The main result of Dehling and Taqqu [9] asserts that the above convergence holds in a two-parameter Skorohod space, with the sup-topology, and is based on the *uniform reduction principle*

$$\max_{n \le N} \sup_{x \in \mathbb{R}} (N d_N)^{-1} \left| \sum_{j=1}^{n} (I(X_j \le x) - F(x) + f(x) X_j) \right| = o_P(1). \tag{1.10}$$

If the linear long memory process of (1.1), (1.4) is non-Gaussian, the above problems become more complicated, because of the lack of an analog of the Hermite expansions in the non-Gaussian case. A generalization of the Dehling–Taqqu theorem was obtained independently in Ho and Hsing [21] and Giraitis, Koul and Surgailis [14] (an earlier version of the last result appeared in the preprint Giraitis and Surgailis [18] in 1994). The method of Ho and Hsing was later applied in several papers and is discussed in the review paper Koul and Surgailis [26] in this volume. In our paper we present the approach introduced in [18] and [14] which is more analytic as compared to Ho and Hsing's approach, and is based on the study of the asymptotics of the bivariate density and the bivariate characteristic function of the linear process (1.1). Namely, under some additional conditions on the c.d.f. G of the noise, we show that the bivariate p.d.f. $f_t(x_1, x_2) = P(X_0 \in dx_1, X_t \in dx_2)/dx_1 dx_2$ exists, for sufficiently large t, and satisfies the asymptotic relation

$$f_t(x_1, x_2) = f(x_1) f(x_2) + r_t f'(x_1) f'(x_2) + o(r_t), \qquad t \to \infty \tag{1.11}$$

uniformly in $x_1, x_2 \in \mathbb{R}$ (see Lemma 3.2 below). Put

$$p_t(x_1, x_2) := f_t(x_1, x_2) - f(x_1) f(x_2) - r_t f'(x_1) f'(x_2). \tag{1.12}$$

Let h be a (bounded, measurable) function, and let $Eh(X_0) = \int_{\mathbb{R}} h(x) f(x) dx = 0$, $a_1 := -\int_{\mathbb{R}} h(x) f'(x) dx$. Integration by parts yields

$$\begin{aligned} &\operatorname{cov}(h(X_0) - a_1 X_0, h(X_t) - a_1 X_t) \\ &= \int_{\mathbb{R}^2} (h(x_1) - a_1 x_1)(h(x_2) - a_1 x_2) f_t(x_1, x_2) dx_1 dx_2 \\ &= \int_{\mathbb{R}^2} (h(x_1) - a_1 x_1)(h(x_2) - a_1 x_2) p_t(x_1, x_2) dx_1 dx_2, \end{aligned} \tag{1.13}$$

formally implying, by (1.11),

$$\mathrm{cov}(h(X_0) - a_1 X_0, h(X_t) - a_1 X_t) = o(r_t),$$

or the first order reduction principle for $S_{N,h}$, similar as in the Gaussian case. Of course, additional effort is needed to make the above argument work. The number a_1 is nothing else but the first nonzero (as $a_0 = Eh(X_0) = 0$) coefficient in the expansion

$$h(x) = \sum_{j=0}^{\infty} a_j A_j(x)/j! \tag{1.14}$$

in *Appell polynomials* $A_j(x)$, $j = 0, 1, \ldots$, which are defined by means of the generating function $\sum_{j=0}^{\infty}(iz)^j A_j(x)/j! = e^{izx}/Ee^{izX_0}$ (see [17]). Although Appell polynomials coincide with Hermite polynomials in the Gaussian case and the partial sums $S_{N,A_k}(t)$ have a similar limit distribution as $S_{N,H_k}(t)$, (Surgailis [32], Avram and Taqqu [1]), they lack the orthogonality property of the latter and the expansion (1.14) is purely formal and cannot be directly applied to obtain the analog of the Taqqu reduction principle except for the case when h itself is a polynomial. However, the coefficients

$$a_k = (-1)^k \int_{\mathbb{R}} h(x) f^{(k)}(x)dx = \left.\frac{\partial^{(k)} Eh(y + X_0)}{\partial^k y}\right|_{y=0}$$

make sense under fairly mild regularity conditions on h and G, and the *Appell rank* $k_A^* = \min\{j \geq 1 : a_j \neq 0\}$ *of* h (also called the *power rank of* h in [22]) determines the limit distribution of $S_{N,h}(t)$, similarly as in the Gaussian case, see Ho and Hsing [21], [22], Koul and Surgailis [25], Surgailis [33] and other papers.

It is interesting to note that for Gaussian X_t, (1.11) is a consequence of the classical *Mehler's formula* (Erdélyi et al. [13], Chapter 10.13 (22)):

$$\begin{aligned} f_t(x_1, x_2) &= f(x_1)f(x_2)\sum_{j=0}^{\infty} r_t^j H_j(x_1)H_j(x_2)/j! \\ &= \sum_{j=0}^{\infty} r_t^j f^{(j)}(x_1)f^{(j)}(x_2)/j!, \end{aligned} \tag{1.15}$$

where $f(x) = (2\pi)^{-1/2}e^{-x^2/2}$ is the normal p.d.f. Formula (1.7) and the reduction principle in the Gaussian case follow from (1.15) and integration by parts. The first two terms $f(x_1)f(x_2) + r_t f'(x_1)f(x_2)$ of (1.15) appear in the asymptotic formula (1.11), too, without the assumption of Gaussianity. This raises the question of obtaining further terms in the expansion (1.11) which would imply a higher order reduction principle in the non-Gaussian case. In Section 5 (Appendix) we present a heuristic derivation of such a generalization of Mehler's formula, whose rigorous justification is an interesting open problem. Another interesting open problem is to generalize the asymptotics (1.11) to *multivariate* p.d.f.'s of the linear long memory process X_t of (1.1), (1.4).

The rest of the paper is organized as follows. In Section 2 we formulate the main result (Theorem 2.1), which is the analog of the Dehling–Taqqu uniform reduction principle (1.10) for linear processes. The main steps of the proof of Theorem 2.1 (Lemmas 2.3 and 3.3) are explained. An application of the functional convergence result of the two-parameter empirical process (1.3) to the change-point problem of the marginal c.d.f. is discussed in Section 4. Further applications and asymptotic results for empirical processes of linear processes with long memory can be found in the review paper Koul and Surgailis [26].

2 The reduction principle

Below, X_t is a (possibly two-sided) linear process (1.1) with long memory, where

$$b_s = L(s)|s|_+^{-(1+\theta)/2}, \qquad s \in \mathbb{Z}, \tag{2.1}$$

$0 < \theta < 1$; here, $x_+ = x \vee 1(x \geq 0)$, and the function $L(x), x \in \mathbb{R}$ is bounded on bounded intervals and varies slowly at $\pm\infty$, i.e., $L(tx)/L(t) \to 1 (t \to \infty)$ for any $x \neq 0$. Moreover, we assume in addition that there exists a constant ℓ (possibly zero) such that

$$\lim_{t\to\infty} L(-t)/L(t) = \ell. \tag{2.2}$$

The above assumptions include the causal case $L(s) = 0 (s < 0)$, in which case the constant $\ell = 0$. Relations (2.1) and (2.2) imply

$$r_t = EX_0X_t = \sum_s b_s b_{t+s} = L_2(t)t^{-\theta},$$

c.f. (1.5), where L_2 is slowly varying and

$$\lim_{t\to\infty} L_2(t)/L^2(t) = a^2_{\theta,\ell} := (1+\ell^2)B(\theta, (1-\theta)/2) + \ell B((1-\theta)/2, (1-\theta)/2).$$

Then the variance of the sample mean $\bar{X}_N = N^{-1}\sum_{t=1}^N X_t$ satisfies the following asymptotic relation:

$$\operatorname{var}\bar{X}_N = N^{-2}\sum_{t,s=1}^N r_{t-s} \sim c_\theta^2 L_2^2(N)N^{-\theta} \sim c_\theta^2 a^2_{\theta,\ell} d_N^2, \tag{2.3}$$

where $c_\theta^2 = 2/(1-\theta)(2-\theta)$, d_N is given by (1.9), and where $\sim$ means that the ratio of both sides tends to 1. It is well known that normalized partial sums of the linear process (1.1) converge weakly, as random elements of the Skorohod space $D[0,1]$, to fractional Brownian motion (Davydov [8]):

$$(Nd_N)^{-1}\sum_{s=1}^{[Nt]} X_s \Longrightarrow_{D[0,1]} a_{\theta,\ell}c_\theta Z_t. \tag{2.4}$$

Theorem 2.1. *Assume that the characteristic function $\phi(u) = Ee^{iu\zeta_0}$ of the noise in (1.1) satisfies the following condition: there exist constants $C > 0, \delta > 0$ such that*

$$|\phi(u)| \leq C(1+|u|)^{-\delta}, \qquad u \in \mathbb{R}. \tag{2.5}$$

Moreover, assume $E|\zeta_0|^{n*(\theta)} < \infty$, where $n^*(\theta) \in [12, \infty)$ is estimated in Remark 3.3 below. Then there are constants $C, \gamma > 0$ such that for any $0 < \epsilon < 1$,

$$P\left[\max_{n\leq N}\sup_{x\in\mathbb{R}} (Nd_N)^{-1}\left|\sum_{j=1}^n (I(X_j \leq x) - F(x) + f(x)X_j)\right| > \epsilon\right] \leq CN^{-\gamma}(1+\epsilon^{-3}). \tag{2.6}$$

Introduce the (two-parameter Skorohod) space $D(\bar{\mathbb{R}} \times [0,1])$ of real functions $g = g(x,t)$, $(x,t) \in \bar{\mathbb{R}} \times [0,1]$, $\bar{\mathbb{R}} := [-\infty, +\infty]$, with the sup-topology, and write $\Longrightarrow_{D(\bar{\mathbb{R}}\times[0,1])}$ for weak convergence of random elements in $D(\bar{\mathbb{R}} \times [0,1])$; see, e.g., Dehling and Taqqu [9]. (See also [4], [29] for more information on multiparameter generalizations of the Skorohod space $D[0,1]$.) Theorem 2.1 and (2.4) readily imply the following.

Corollary 2.2. *Under the conditions of Theorem* 2.1,

$$d_N^{-1}(F_N(x,t) - ([Nt]/N)F(x)) \Longrightarrow_{D(\bar{\mathbb{R}}\times[0,1])} a_{\theta,\ell} c_\theta f(x) Z_t. \tag{2.7}$$

Some comments on Theorem 2.1 and its assumptions are in order. Condition (2.5) is a weak assumption on the smoothness of the c.d.f. G which guarantees nevertheless that the marginal c.d.f. F is infinitely differentiable; see Lemma 3.1 below. The same condition on the noise distribution was imposed in several subsequent papers ([12], [19], [25]). It was further weakened in Wu [35]. The high moment condition $E|\zeta_0|^{n^*(D)} < \infty$ looks rather unpleasant but could be weakened by more accurate computations. Actually, for causal processes (1.1) and using the method of Ho and Hsing, a similar reduction principle has been proved under a minimal $(2+\delta)$-moment assumption on the noise (Giraitis and Surgailis [19]). It should be noted that the causality assumption plays a very important role in the latter method, which uses a martingale difference decomposition of the indicator $I(X_t \le x)$ with respect to the past σ-fields generated by $\zeta_u, u \le s$, for $s \le t$. (The method of Ho and Hsing, including an asymptotic expansion of the empirical process, is discussed in the review paper [26]). We also mention here related works on the empirical process of linear random fields (Doukhan, Lang and Surgailis [12]), seasonal ARFIMA processes (Ould Haye [30], Ould Haye and Philippe [31]) and the bivariate empirical processes (Marinucci [28]).

Theorem 2.1 follows from the chaining argument of Dehling and Taqqu [9] and Lemma 2.3 below. Put

$$S_n(x) := \sum_{t=1}^{n} (I(X_t \le x) - F(x) + f(x)X_t).$$

For any function $g(x)$, and any $x < y$, let $g(x, y) := g(y) - g(x)$ denote the increment. Everywhere below, we assume that the conditions of Theorem 2.1 are satisfied.

Lemma 2.3. *There exist $\gamma > 0$ and a finite continuous measure μ on $\mathbb{R}$ such that for any $N \ge 1$, $1 \le n \le N$, and any $x < y$,*

$$(Nd_N)^{-2} E S_n^2(x, y) \le \frac{n}{N^{1+\gamma}} \mu(x, y). \tag{2.8}$$

While (2.8) and (2.4) readily yield the convergence of the finite dimensional distributions in (2.7), the chaining argument provides the control of the $\sup_{x\in\bar{\mathbb{R}}}$ in (2.6) which is crucial for the tightness in $D(\bar{\mathbb{R}} \times [0, 1])$. The idea of chaining is the following. Consider the sequence $\Pi_k = \{\pi_{i,k}; i = 0, \ldots, 2^k - 1\}$, $k = 0, 1, \ldots, K$ of "dyadic" partitions of the real line, with the property that each interval from the partition Π_k has μ measure $\mu(\mathbb{R})2^{-k}$. Every point $x \in \mathbb{R}$ can be "chained" to the fixed point $-\infty$ along some "dyadic" points $\pi_{i_k^x,k}$, $k = 0, 1, \ldots, K$. The number K grows with N but not very fast and so that $\mu(\mathbb{R})2^{-K}$(= the "μ-length" of intervals of the finest partition $\Pi_K) = O(d_N N^{-\gamma})$. The $S_N(x)$ can be decomposed into the sum of its increments over the "dyadic" points of the chain connecting x to $-\infty$, and the increment $S_N(\pi_{j_K^x,K}, x)$, where $\pi_{j_K^x,K}$ is the nearest to x (from the left) point of Π_K. Using the monotonicity of the indicator function, one has a rough bound $O_P(N2^{-K}) = O_P((Nd_N)N^{-\gamma})$ for the supremum over $x \in \mathbb{R}$ of the last increment, which is sufficient to bound the contribution (to $(Nd_N)^{-1}\sup_x |S_N(x)|$) of this increment by $O_P(N^{-\gamma})$. On the other hand, the maximal contribution over the "dyadic" points does not exceed the sum of all absolute increments of $(Nd_N)^{-1}|S_N(\pi_{i,k}, \pi_{i+1,k})|$, $i = 0, \ldots, 2^{k-1}$, $k = 0, 1, \ldots, K$ which can be easily estimated by Lemma 2.3. Here, it is important that the sum of the increments of μ converges for fixed k (as μ is bounded), and that the number K grows logarithmically with N, i.e., much slower than N^γ.

3 Proof of Lemma 2.3

In this section we explain the main ideas of the proof of Lemma 2.3. The details can be found in [14] or [18]. Although the proofs given in [14], [18] are restricted to the causal case $b_s = 0$ $(s < 0)$, they easily generalize to the case of two-sided weights b_s as in (2.1).

Lemma 3.1.

(i) *Let the moving average process of* (1.1), (2.1) *satisfy the condition* (2.5). *Then the marginal density f exists on $\mathbb{R}$ and is infinitely differentiable everywhere.*

(ii) *Assume in addition that $E|\zeta_0|^m < \infty$ for some integer $m \geq 2$. Then there exists a constant $C < \infty$ such that for all $x \in \mathbb{R}$,*

$$f(x) \leq C(1+|x|)^{-m+1}, \tag{3.1}$$

$$|f'(x)| \leq C(1+|x|)^{-m}. \tag{3.2}$$

Generalizations and refinements of the above lemma can be found in [12], [25], [26]. The proof uses the form and properties of the characteristic function

$$\hat{f}(u) := Ee^{iuX_0} = \prod_{j \in \mathbb{Z}} \phi(ub_j), \qquad u \in \mathbb{R}. \tag{3.3}$$

In particular, (2.5) implies $|\hat{f}(u)| \leq \prod_j |\phi(ub_j)| \leq C \prod_J (1+|ub_j|)^{-\delta}$, for any finite set $J \subset \mathbb{Z}$ which can be chosen large enough so that $\delta|J| > k$, for a given $k \geq 0$. If $b_i \neq 0 \; \forall i \in J$, this implies $|\hat{f}(u)| \leq C(1+|u|)^{-k}$, or the integrability of $(1+|u|)^k \hat{f}(u)$, and the existence and continuity of $f^{(k)}$. (Here and below, C is a generic constant which may change from line to line.) Noting $x^m f'(x) = (-i)^{m+1} \int_{\mathbb{R}} e^{-ixu} (u\hat{f}(u))^{(m)} du$, it suffices for (3.2) to show the inequality $|(u\hat{f}^{(m)}(u)| \leq C/(1+|u|)^2$, which follows by differentiating the product (3.3) and using $|\phi'(u)| \leq C|u|$, $|\phi^{(k)}(u)| \leq C (2 \leq k \leq m)$. Of course, Lemma 3.1 does not require the asymptotics (1.4) and is valid under very general assumptions on b_j, namely, $\sum_j b_j^2 < \infty$ and $|\{i \in \mathbb{Z} : b_i \neq 0\}| = \infty$.

A similar argument allows one to show the existence and regularity of the bivariate density $f_t(x_1, x_2) = P(X_0 \in dx_1, X_t \in dx_2)/dx_1 dx_2$ for $|t|$ sufficiently large. Write $f_t(x_1, x_2) = (2\pi)^{-2} \int_{\mathbb{R}^2} e^{-i(u_1x_2+u_2x_2)} \hat{f}_t(u_1, u_2) du_1 du_2$, where $\hat{f}_t(u_1, u_2)$ is the bivariate characteristic function

$$\hat{f}_t(u_1, u_2) = Ee^{i(u_1X_0+u_2X_t)} = \prod_j \phi(u_1b_j + u_2b_{t+j}), \quad u_1, u_2 \in \mathbb{R}. \tag{3.4}$$

Given $\delta > 0$, for all sufficiently large $|t|$ there exist disjoint finite sets J_1, $J_2 \subset \mathbb{Z}$, "centered" at 0 and $-t$, respectively, each containing at least $[3/\delta]$ points, and such that $|b_j| > 2|b_{t+j}|$ $(j \in J_1)$, $|b_{t+j}| > 2|b_j|$ $(j \in J_2)$. Then by (2.5) one observes

$$\begin{aligned}|\hat{f}_t(u_1, u_2)| &\leq C \prod_{J_1 \cup J_2} (1+|u_1b_j + u_2b_{t+j}|)^{-\delta} \\ &\leq C \prod_{J_1} (1+|u_1 + u_2a_{1j}|)^{-\delta} \prod_{j \in J_2} (1+|u_2 + u_1a_{2j}|)^{-\delta}\end{aligned}$$

for some $|a_{1j}|, |a_{2j}| \leq 1/2$. Whence and from the definition of J_1, J_2, it easily follows that $|\hat{f}_t(u_1, u_2)| \leq C(1+|u_1|+|u_2|)^{-3}$, or the integrability of $\hat{f}_t(u_1, u_2)$ and the existence of

$f_t(x_1, x_2)$. By choosing the sets J_1, J_2 sufficiently large, the above fact can be strengthened to

$$\int_{\mathbb{R}^2} (|u_1| + |u_2|)^k |\hat{f}_t(u_1, u_2)| du_1 du_2 \le C, \tag{3.5}$$

for arbitrary $k \ge 0$ and all sufficiently large $|t|$.

Lemma 3.2. *Assume conditions* (1.4), (2.5) *and* $E|\zeta_0|^3 < \infty$. *Then*

$$f_t(x_1, x_2) = f(x_1) f(x_2) + r_t f'(x_1) f'(x_2) + O(t^{-\theta-\kappa}), \tag{3.6}$$

uniformly in $x_1, x_2 \in \mathbb{R}$ *as* $t \to \infty$, *where* $0 < \kappa < \min(\theta, (1-\theta)/2)$, *arbitrary otherwise.*

The upper bound $\kappa < \bar{\kappa} := \min(\theta, (1-\theta)/2)$ in Lemma 3.2 seems close to optimal, as suggested by the heuristic higher order expansion of $f_t(x_1, x_2)$ of Section 5. It would be desirable to obtain a *nonuniform* bound in x_1, x_2 of the remainder term in (3.6), as this would allow one to reduce the number of required moments of ζ_0 in Theorem 2.1.

Let us give a sketch of the proof of Lemma 3.2. Let $p_t(x_1, x_2)$ be defined as in (1.12), and $\hat{p}_t(u_1, u_2)$ be its Fourier transform. Clearly, (3.6) follows from

$$\int_{\mathbb{R}^2} |\hat{p}_t(u_1, u_2)| du_1 du_2 \le C t^{-\theta-\bar{\kappa}+\eta}, \qquad t \to \infty, \tag{3.7}$$

for arbitrary $\eta > 0$. Fix a small $\epsilon > 0$, and split the integral on the l.h.s. of (3.7) into the sum of two integrals, the first being taken over $|u_1| + |u_2| > t^\epsilon$, and the second over $|u_1| + |u_2| \le t^\epsilon$. Then by (3.5), the first integral is $O(t^{-k\epsilon})$ where $k \ge 0$ can be chosen arbitrarily large, in particular, such that $k\epsilon \ge \theta + \bar{\kappa}$. On the other hand, the required bound $O(t^{-\theta-\bar{\kappa}+\eta})$ for the second integral clearly follows from

$$\sup_{|u_1|+|u_2| \le t^d} |\hat{p}_t(u_1, u_2)| = O(t^{-\theta-\bar{\kappa}+4\epsilon}), \tag{3.8}$$

because of the arbitrariness of $\epsilon > 0$. The proof of (3.8) uses the representation (3.4) of the bivariate characteristic function and the Taylor expansion of the logarithm of the product on the r.h.s. of (3.4). More precisely, we split the product (3.4) into three parts:

$$\hat{f}_t(u_1, u_2) = \prod_{j \in \mathbb{Z}} \phi(u_1 b_j + u_2 b_{t+j}) = \prod_{I_1} \cdots \prod_{I_2} \cdots \prod_{I_3} \cdots =: a_1 \cdot a_2 \cdot a_3,$$

where the sets I_1 and I_2 are (disjoint) "neighborhoods" of 0 and $-t$, respectively, of radius $t^{2\epsilon}$, and $I_3 := \mathbb{Z} \backslash (I_1 \cup I_2)$. In a similar way,

$$\hat{f}(u_1) \hat{f}(u_2) = \prod_{j \in \mathbb{Z}} \phi(u_1 b_j) \phi(u_2 b_{t+j}) = \prod_{I_1} \cdots \prod_{I_2} \cdots \prod_{I_3} \cdots =: a_1' \cdot a_2' \cdot a_3'.$$

Then,

$$\begin{aligned} \hat{f}_t(u_1, u_2) - \hat{f}(u_1) \hat{f}(u_2) &= a_1 a_2 a_3 - a_1' a_2' a_3' \\ &= (a_1 - a_1') a_2 a_3 + a_1' (a_2 - a_2') a_3 + a_1' a_2' (a_3 - a_3'). \end{aligned}$$

Hence (3.8) will follow from $|a_i| \le 1$, $|a_i'| \le 1$, $i = 1, 2, 3$, and the facts

$$a_i - a_i' = O(t^{-((1+\theta)/2)+4\epsilon}), \qquad i = 1, 2, \tag{3.9}$$

$$a_3 - a_3' = -a_3' u_1 u_2 r_t + O(t^{-\theta-\bar{\kappa}+4\epsilon}), \tag{3.10}$$

uniformly in $|u_1|+|u_2| \le t^\epsilon$. (3.9) for $i = 1$ follow by the telescoping inequality $|\prod c_j - \prod c'_j| \le \sum_j |c_j - c'_j|$, $|c_j| \le 1$, $|c'_j| \le 1$, and $|\phi(u_1 b_j + u_2 b_{t+j}) - \phi(u_1 b_j)\phi(u_2 b_{t+j})| \le C|u_2 b_{t+j}| \le Ct^{-((1+\theta)/2)+2\epsilon}$, as $|u_2| \le t^\epsilon$ and $|b_{t+j}| \le Ct^{-((1+\theta)/2)+\epsilon}$ $(j \in I_1)$; the case (3.9), $i = 2$ is analogous. The main contribution to the l.h.s. of (3.8) comes from (3.10), whose l.h.s. can be written as

$$a_3 - a'_3 = a'_3(\exp\{Q_t(u_1, u_2)\} - 1),$$

where

$$Q_t(u_1, u_2) := \sum_{I_3} [\log(u_1 b_j + u_2 b_{t+j}) - \log\phi(u_1 b_j) - \log\phi(u_2 b_{t+j})].$$

In order to use the Taylor expansion of the above logarithms, we need both $|u_1 b_j|$ and $|u_2 b_{t+j}|$ to be small, which is guaranteed by the above choice of $I_3 = \{j : |j| > t^{2\epsilon}, |t + j| > t^{2\epsilon}$ and $|u_1| + |u_2| \le t^\epsilon$ as in (3.8). Then by the Taylor formula,

$$\begin{aligned}
&\log(u_1 b_j + u_2 b_{t+j}) - \log\phi(u_1 b_j) - \log\phi(u_2 b_{t+j}) \\
&\quad = \int_0^{u_1 b_i} \int_0^{u_2 b_{t+j}} (\log\phi)''(x+y)\,dx\,dy \\
&\quad = -u_1 u_2 b_j b_{t+j} + O((u_1 b_j)^2 |u_2 b_{t+j}| + |u_1 b_j|(u_2 b_{t+j})^2),
\end{aligned}$$

implying $Q_t(u_1, u_2) = -u_1 u_2 r_t + O(t^{-((1+\theta)/2)+4\epsilon})$ uniformly in $|u_1| + |u_2| \le t^\epsilon$. Hence by expanding $e^{Q_t(u_1,u_2)} - 1$, (3.10) easily follows; see [14] for details.

Now we come to the final step in the proof of Lemma 2.3. It suffices to consider the case $n = N$. Write $S_N(x, y) = \sum_{t=1}^N h(X_t)$, where $h(z) \equiv h(z; x, y) := I(x < z \le y) - F(x, y) + f(x, y)z$. Then

$$ES_N^2(x, y) = \sum_{t,s=1}^N \rho_{t-s},$$

where

$$\begin{aligned}
\rho_t := Eh(X_0)h(X_t) &= \int_{\mathbb{R}^2} h(z_1)h(z_2) f_t(z_1, z_2)\,dz_1 dz_2 \\
&= \int_{\mathbb{R}^2} h(z_1)h(z_2) p_t(z_1, z_2)\,dz_1 dz_2,
\end{aligned}$$

as in (1.13). By Lemma 3.2, for sufficiently large t,

$$|\rho_t| \le Ct^{-\theta-\gamma} \int_{\mathbb{R}^2} |h(z_1)h(z_2)|\,|p_t(z_1, z_2)|^\epsilon dz_1 dz_2, \tag{3.11}$$

where $\gamma := (\theta+\kappa)(1-\epsilon)-\theta > 0$, provided $\epsilon > 0$ is chosen sufficiently small $(\epsilon < \kappa/(\kappa+\theta))$. Write $i(t)$ for the integral on the r.h.s. of (3.11); it suffices to show that there exists a finite measure μ on $\mathbb{R}$ such that for all $t \in \mathbb{Z}$, $x < y$,

$$i(t) \le \mu(x, y). \tag{3.12}$$

To show (3.11), split the integral $i(t) = \int_{B_-} + \int_{B_+} =: i_-(t) + i_+(t)$, where

$$B_- := \{(z_1, z_2) \in \mathbb{R}^2 : |p_t(z_1, z_2)|^\epsilon \le (1 + |z_1|)^{-3}(1 + |z_2|)^{-3}\}, \qquad B_+ := \mathbb{R}^2 \setminus B_-.$$

Then $i_-(t) \le (\int_{\mathbb{R}} |h(z)|(1+|z|)^{-3}dz)^2$, whence one easily has the bound $i_-(t) \le C\mu_-(x,y)$, with $\mu_-(A) = \int_A (f(z) + |f'(z)| + (1+|z|)^{-3})dz$ being a finite measure on the real line. Next, to evaluate $i_+(t)$, note

$$|p_t(z_1, z_2)|^{\epsilon - 1} \le (1+|z_1|)^{m_*}(1+|z_2|)^{m_*}, \qquad (z_1, z_2) \in B_+,$$

by the definition of the set B_+, where $m_* := [3(1-\epsilon)/\epsilon] + 1$. This implies

$$\begin{aligned} i_+(t) &\le \int\int |p_t||h(z_1)h(z_2)|(1+|z_1|)^{m_*}(1+|z_2|)^{m_*} dz_1 dz_2 \\ &\le \int\int |p_t| \Big(h^2(z_1)(1+|z_1|)^{2m_*} + h^2(z_2)(1+|z_2|)^{2m_*} \Big) dz_1 dz_2 \\ &\le 2\int\int |p_t(z_1,z_2)| \Big(|h(z_1)|(1+|z_1|)^{2m_*+1} + |h(z_2)|(1+|z_2|)^{2m_*+1} \Big) dz_1 dz_2 \\ &\le \int (4f(z) + 2|f'(z)|)\, |h(z)|(1+|z|)^{2m_*+1} dz, \end{aligned}$$

where the last inequality follows by the definition of p_t. Hence, taking into account Lemma 3.1 and the form of the function $h(z) = h(z; x, y)$, one obtains $i_+(t) \le C\mu_+(x,y)$, with μ_+ a finite measure on the real line, provided the moment $E|\zeta_0|^{2m_*+4} < \infty$. Details can be found in [18] or [14].

Remark 3.3. From the proof of Lemma 2.3, one obtains the following estimate of the number $n^*(\theta)$ of finite noise moments in Theorem 2.1:

$$n^*(\theta) = \begin{cases} 12, & \text{if } 0 < \theta \le 1/3, \\ 6(1+\theta)/(1-\theta), & \text{if } 1/3 < \theta < 1. \end{cases}$$

4 Application to the change-point problem

The two-parameter empirical process $F_N(x,t)$ can be applied to test for changes in the stationary distribution. Assume that, at some unknown point $1 \le [N\tau] \le N$, the marginal distribution of the observed sequence X_t, $1 \le t \le N$ changes, i.e.,

$$X_j = \begin{cases} X_j^{(1)}, & \text{if } 1 \le j \le [N\tau], \\ X_j^{(2)}, & \text{if } [N\tau] < j \le N, \end{cases} \tag{4.1}$$

where $0 < \tau < 1$ is fixed, and $X_t^{(1)}, X_t^{(2)}$ are two stationary sequences with marginal c.d.f.'s $P(X_j^{(i)} \le x) = F_i(x), i = 1, 2, F_1 \ne F_2$. To test the null hypothesis H_0 that no such change occurs, i.e., that all observations come from a single stationary sequence, it is natural to use the Kolmogorov–Smirnov-type statistic

$$T_N := \sup\{|V_N(x,t)| : (x,t) \in \mathbb{R} \times [0,1]\},$$

where

$$V_N(x,t) := \frac{[Nt]}{N}\Big(1 - \frac{[Nt]}{N}\Big)(F_{[Nt]}(x) - F^*_{N-[Nt]}(x)),$$

and where

$$F_{[Nt]}(x) = \frac{1}{[Nt]}\sum_{j=1}^{[Nt]} I(X_j \le x), \quad F^*_{N-[Nt]}(x) = \frac{1}{N-[Nt]}\sum_{j=[Nt]+1}^{N} I(X_j \le x)$$

are the empirical distribution functions based on partial samples $X_1, \dots, X_{[Nt]}$ and $X_{[Nt]+1}, \dots, X_N$, respectively. If the two-parameter empirical process (1.3) converges, under some normalization $\tilde{d}_N \to 0$, to a random field $W(x, t)$:

$$\tilde{d}_N^{-1}(F_N(x, t) - ([Nt]/N)F(x)) \Longrightarrow_{D(\bar{\mathbb{R}}\times[0,1])} W(x, t), \tag{4.2}$$

this implies the convergence in distribution of the statistic T_N under the null hypothesis:

$$\tilde{d}_N^{-1} T_N \Longrightarrow \sup\{|W(x, t) - tW(x, 1)| : (x, t) \in \mathbb{R} \times [0, 1]\}, \tag{4.3}$$

see Giraitis, Leipus, and Surgailis [15]. For i.i.d. observations, the convergence (4.1) is well known, $\tilde{d}_N = N^{-1/2}$, and the limit process $W(x, t) = K(x, t)$ is the *Kiefer process*, i.e., a zero mean Gaussian process with covariances

$$EK(x, t)K(x', t') = t \wedge t' \big(F(x \wedge x') - F(x)F(x')\big).$$

In the case of weakly dependent stationary processes X_t satisfying certain mixing assumptions, (4.2) holds with $\tilde{d}_N = N^{-1/2}$ and $W(x, t)$ a Gaussian random field (the generalized Kiefer process); see, e.g., Berkes and Philipp [3]. For causal short memory linear processes (1.1), Doukhan and Surgailis [11] obtained the functional CLT for the one-parameter empirical process $F_N(x, 1), x \in \mathbb{R}$:

$$N^{1/2}(F_N(x, 1) - F(x)) \Longrightarrow_{D(\bar{\mathbb{R}})} W(x, 1),$$

by assuming only $\sum_{j=1}^{\infty} |b_j| < \infty$, $E\zeta_0^4 < \infty$ and (2.5) with $\delta > 1/2$, and not assuming any mixing conditions. We expect that the last result can be extended to the two-parameter empirical process as in (4.2).

In the long memory case, (4.2) follows from Corollary 2.2, with $\tilde{d}_N = d_N$ given by (1.9) and

$$W(x, t) = a_{\theta,\ell} c_\theta f(x) Z_t$$

being a degenerate Gaussian field (Z_t is fractional Brownian motion). From (4.3) and Corollary 2.2 one immediately obtains the asymptotic probability of Type I error:

$$\lim_{N\to\infty} P(d_N^{-1} T_N > c) = P(\sup_{t\in[0,1]} |Z_t - tZ_1| > c/(a_{\theta,\ell} c_\theta \|f\|_\infty)) \qquad (\forall c > 0),$$

where $\|f\|_\infty = \sup_{x\in\mathbb{R}} f(x)$.

If the change (4.1) occurs, the test T_N will reject H_0 with power tending to 1:

$$\lim_{N\to\infty} P(\tilde{d}_N^{-1} T_N > c\} \to 1 \quad (\forall c > 0)$$

for any normalization $\tilde{d}_N \to 0$ such that for some fixed $x \in \mathbb{R}$, $F_1(x) \neq F_2(x)$,

$$\tilde{d}_N^{-1}(F_N^{(i)}(x, 1) - F_i(x)) = O_P(1), \qquad i = 1, 2,$$

where $F_N^{(i)}(x, t)$ is the empirical process (1.3) of $X_t^{(i)}$, $i = 1, 2$. More exact asymptotic results for the Type II error probability are usually discussed in the context of *contiguous* alternatives. For Kolmogorov–Smirnov type statistics and independent observations, related results were studied by several authors; see, e.g., Csörgő and Horváth [7]. However, the contiguity assumption is rather difficult to verify for certain dependent models, in particular, for the linear process (1.1). In [15], a related notion of *converging* alternatives is introduced and the corresponding asymptotic distribution of the test statistic T_N is obtained. See also the review paper Kokoszka and Leipus [23]. One should mention that in the long memory case, testing the change-point hypothesis requires knowledge of the long memory parameter θ, which, if unknown, has to be preliminarily estimated; see Beran [2].

5 Appendix

In this section we heuristically derive further terms of the expansion (1.11) of the bivariate p.d.f. $f_t(x_1, x_2)$. "Heuristically" means here that we perform formal algebraic transformations of infinite series, ignoring the questions of their convergence, integration, etc., and dropping all "short memory" terms of order $O(t^{-1})$. For simplicity, we assume the slowly varying function $L(t) = O(1)$ in (1.4). We start with the representation

$$\hat{f}_t(u_1, u_2) - \hat{f}(u_1)\hat{f}(u_2) = \hat{f}(u_1)\hat{f}(u_2)\Big(\exp\Big\{\sum_j \psi(u_1 b_j, u_2 b_{t+j})\Big\} - 1\Big),$$

see (3.3), (3.4), where $\psi(u_1, u_2) = \log\phi(u_1 + u_2) - \log\phi(u_1) - \log\phi(u_2)$. Note

$$\log\phi(u) = \sum_{k=2}^{\infty} \chi_k (iu)^k / k!$$

is the cumulant generating function; i.e., χ_k is the k-th cumulant of ζ_0, $\chi_1 = E\zeta_0 = 0$, $\chi_2 = E\zeta_0^2 = 1$. Therefore,

$$\sum_j \psi(u_1 b_j, u_2 b_{t+j}) = \sum_{k=2}^{\infty} i^k / k! \sum_j \chi_k((u_1 b_j + u_2 b_{t+j})^k - (u_1 b_j)^k - (u_2 b_{t+j})^k).$$

We have

$$\sum_j \chi_2((u_1 b_j + u_2 b_{t+j})^2 - (u_1 b_j)^2 - (u_2 b_{t+j})^2) = 2u_1 u_2 \sum_j b_j b_{t+j} = 2u_1 u_2 r_t,$$

$$\begin{aligned}\sum_j \chi_k((u_1 b_j + u_2 b_{t+j})^k - (u_1 b_j)^k - (u_2 b_{t+j})^k) &= \sum_{p=1}^{k-1} \binom{k}{p} u_1^p u_2^{k-p} \sum_j \chi_k b_j^p b_{t+j}^{k-p} \\ &= \sum_{p=1}^{k-1} \binom{k}{p} u_1^p u_2^{k-p} \chi_{p,k-p}(t),\end{aligned}$$

$k \geq 3$, where

$$\chi_{p,q}(t) := \mathrm{cum}(\underbrace{X_0, \ldots, X_0}_{p}, \underbrace{X_t, \ldots, X_t}_{q}) = \chi_{p+q} \sum_j b_j^p b_{t+j}^q$$

is the joint cumulant; the last equality follows by multilinearity of the cumulant and the independence of ζ_s, $s \in \mathbb{Z}$. It is easy to show that under (1.4),

$$\chi_{p,q}(t) = \begin{cases} O(t^{-(1+\theta)/2}), & \text{if } p = 1, q \geq 2 \text{ or } p \geq 2, q = 1, \\ O(t^{-1}), & \text{if } p \geq 2, q \geq 2. \end{cases} \tag{5.1}$$

Therefore

$$\begin{aligned}\sum_j \psi(u_1 b_j, u_2 b_{t+j}) = -u_1 u_2 r_t &+ \sum_{k\geq 2} (iu_1)^k (iu_2) \chi_{k,1}(t)/k! \\ &+ \sum_{k\geq 2} (iu_1)(iu_2)^k \chi_{1,k}(t))/k! + O(t^{-1}).\end{aligned}$$

Taking the Taylor expansion of $\exp\{\sum_j \psi(u_1 b_j, u_2 b_{t+j})\}$ and ignoring the terms $O(t^{-1})$, we obtain

$$\begin{aligned}\hat{f}_t(u_1,u_2) &= \hat{f}(u_1)\hat{f}(u_2)\Big\{\sum_{0\le k<1/\theta}(iu_1 iu_2)^k r_t^k/k! \\ &+ \sum_{0\le q<(1-\theta)/2\theta}\sum_{k\ge 2}\Big((iu_1)^{q+k}(iu_2)^{q+1} r_t^q \chi_{k,1}(t) \\ &\qquad +(iu_1)^{q+1}(iu_2)^{q+k} r_t^q \chi_{1,k}(t)\Big)\Big)/(q!k!) + O(t^{-1})\Big\}.\end{aligned}$$

Finally, by taking the inverse Fourier transform, we obtain

$$\begin{aligned}f_t(x_1,x_2) &= \sum_{0\le k<1/\theta} r_t^k f^{(k)}(x_1) f^{(k)}(x_2)/k! \\ &+ \sum_{0\le q<(1-\theta)/2\theta}\sum_{k\ge 2}\Big(r_t^q \chi_{k,1}(t) f^{(q+k)}(x_1) f^{(q+1)}(x_2) \\ &\qquad + r_t^q \chi_{1,k}(t) f^{(q+k)}(x_2) f^{(q+1)}(x_1)\Big)/(q!k!) + O(t^{-1}).\end{aligned} \tag{5.2}$$

Formula (5.2) can be regarded as the asymptotic analog of Mehler's formula (1.15) for the linear long memory process (1.1). Note the first sum on the r.h.s. of (5.2) is the same as in Mehler's formula for "long memory" terms of order $O(r_t^k)$, $k\theta < 1$. The second (double) sum vanishes in the Gaussian case as the cumulants $\chi_{k,1}(t) = \chi_{1,k}(t) = 0$, and accounts for non-Gaussianity of X_t. Also note that the next term after $f(x_1)f(x_2)+r_t f'(x_2)f'(x_2)$ in (5.2) (w.r.t. the decay rate in t) is of the order $O(t^{-\min(2\theta,(1+\theta)/2)})$, and equals $r_t^2 f'(x_1)f'(x_2)/2$ in the case $2\theta < (1+\theta)/2$, or $0 < \theta < 1/3$, and $\sum_{k=2}^{\infty}(\chi_{k,1}(t)f'(x_1)f^{(k)}(x_2) + \chi_{1,k}(t)f^{(k)}(x_1)f'(x_2))/k!$ in the case $1/3 < \theta < 1$. A rigorous justification of the formula (5.2) and the implications of it to limit theorems are the subjects for future investigations.

References

[1] F. Avram and M.S. Taqqu: Noncentral limit theorems and Appell polynomials. *Annals of Probability* **15** (1987), 767–775.

[2] J. Beran: *Statistics for Long-Memory Processes.* Chapman and Hall, New York, 1994.

[3] I. Berkes and W. Philipp: An almost sure invariance principle for the empirical distribution function of mixing random variables. *Zeitschrift für Wahrscheinlichkeitstheorie und verwandte Gebiete* **41** (1977), 115–137.

[4] P.J. Bickel and M.J. Wichura: Convergence criteria for multiparameter stochastic processes and some applications. *Annals of Mathematical Statistics* **42** (1971), 1656–1670.

[5] P. Breuer and P. Major: Central limit theorem for non-linear functionals of Gaussian fields. *Journal of Multivariate Analysis* **13** (1983), 425–441.

[6] P.J. Brockwell and R.A. Davis: *Time Series: Theory and Methods.* Springer-Verlag, New York, 1991.

[7] M. Csörgő and L. Horváth: *Limit Theorems in Change-Point Analysis.* Wiley, 1997.

[8] Yu.A. Davydov: An invariance principle for random processes. *Probability Theory and Its Applications* **15** (1970), 487–498.

[9] H. Dehling and M.S. Taqqu: The empirical process of some long-range dependent sequences with an application to U-statistics. *Annals of Statistics* **17** (1989), 1767–1783.

[10] R.L. Dobrushin and P. Major: Non-central limit theorems for non-linear functionals of Gaussian fields. *Zeitschrift für Wahrscheinlichkeitstheorie und verwandte Gebiete* **50** (1979), 27–52.

[11] P. Doukhan and D. Surgailis: Functional central limit theorem for the empirical process of short memory linear processes. *C. R. Acad. Sci. Paris, SerieI* **326** (1998), 87–92.

[12] P. Doukhan, G. Lang and D. Surgailis: Asymptotics of empirical processes of linear random fields with long range dependence. *Annales de l'Institut Henri Poincaré* (B) *Probabilités et Statistiques*, to appear.

[13] A. Erdélyi, W. Magnus, F. Oberhettinger and F.G. Tricomi: *Higher transcendental functions*, vol. 2. McGraw-Hill, New York, 1953.

[14] L. Giraitis, H.L. Koul and D. Surgailis: Asymptotic normality of regression estimators with long memory errors. *Statistics and Probability Letters* **29** (1996), 317–335.

[15] L. Giraitis, R. Leipus and D. Surgailis: The change-point problem for dependent observations. *Journal of Statistical Planning and Inference* **53** (1996), 297–310.

[16] L. Giraitis and D. Surgailis: CLT and other limit theorems for functionals of Gaussian sequences. *Zeitschrift für Wahrscheinlichkeitstheorie und verwandte Gebiete* **70** (1985), 191–212.

[17] L. Giraitis and D. Surgailis: Multivariate Appell polynomials and the central limit theorem. In: E. Eberlein and M.S. Taqqu (Eds.) *Dependence in Probability and Statistics*, pp. 21–71, Birkhäuser, Boston, 1986.

[18] L. Giraitis and D. Surgailis: Asymptotics of the empirical process of a long memory linear sequence. Preprint 24, "Beiträge für Statistik," Universität Heidelberg, 1994.

[19] L. Giraitis and D. Surgailis: Central limit theorem for the empirical process of a linear sequence with long memory. *Journal of Statistical Planning and Inference* **80** (1999), 81–93.

[20] C.W.J. Granger and R. Joyeux: An introduction to long memory time series and fractional differencing. *Journal of Time Series Analysis* **1** (1980), 15–30.

[21] H.-C. Ho and T. Hsing: On the asymptotic expansion of the empirical process of long memory moving averages. *Annals of Statistics* **24** (1996), 992–1024.

[22] H.-C. Ho and T. Hsing: Limit theorems for functionals of moving averages. *Annals of Probability* **25** (1997), 1636–1669.

[23] P. Kokoszka and R. Leipus: Detection and estimation of changes in regime. In: P. Doukhan, G. Oppenheim and M.S. Taqqu (Eds.) *Long-Range Dependence: Theory and Applications*, Birkhäuser, Boston, 2001.

[24] P. Kokoszka and T. Mikosch: The integrated periodogram for long memory processes with finite and infinite variance. *Stochastic Processes and Their Applications* **66** (1997), 55–78.

[25] H.L. Koul and D. Surgailis: Asymptotic expansion of M-estimators with long memory errors. *Annals of Statistics* **25** (1997), 818–850.

[26] H.L. Koul and D. Surgailis: Asymptotic expansion of the empirical process of long memory moving averages. In: H.G. Dehling, T. Mikosch, and M. Sørensen (Eds.) *Empirical Processes for Dependent Data*, Birkhäuser, Boston, 2002.

[27] P. Major: *Multiple Wiener-[Itô Integrals*. Lecture Notes in Mathematics, volume 849, Springer-Verlag, New York, 1981.

[28] D. Marinucci: The empirical process for bivariate sequences with long memory. Preprint, 2000.

[29] G. Neuhaus: On weak convergence of stochastic processes with multidimensional time parameter. *Annals of Mathematical Statistics* **42** (1971), 1285–1296.

[30] M. Ould Haye: Asymptotical behavior of the empirical process for Gaussian data presenting seasonal long-memory. Preprint, Pub. IRMA Lille, 53-V, 2000.

[31] M. Ould Haye and A. Philippe: A non central limit theorem for the empirical process of linear sequences with seasonal long-memory. Preprint, Pub. IRMA Lille, 55-VII, 2001.

[32] D. Surgailis: Zones of attraction of self-similar multiple integrals. *Lithuanian Mathematical Journal* **22** (1982), 327–340.

[33] D. Surgailis: Long range dependence and Appell rank. *Annals of Probability* **28** (2000), 478–497.

[34] M.S. Taqqu: Convergence of integrated processes of arbitrary Hermite rank. *Zeitschrift für Wahrscheinlichkeitstheorie und verwandte Gebiete* **50** (1979), 53–83.

[35] W.B. Wu: Central limit theorems for functionals of linear processes and their applications. Preprint, 2000.

Liudas Giraitis
London School of Economics
Department of Economics
Houghton Street, London WC2A 2AE, U.K.
L.Giraitis@lse.ac.uk

Donatas Surgailis
Vilnius Institute of Mathematics and Informatics
Akademijos 4
2600 Vilnius, Lithuania
sdonatas@ktl.mii.lt

Distributional Limit Theorems for Empirical Processes Generated by Functions of a Stationary Gaussian Process

MIGUEL A. ARCONES

ABSTRACT. We review the literature about limit theorems for sums of a nonlinear function of a stationary sequence of Gaussian r.v.'s. We also consider limit theorems for empirical processes in the previous setup.

1 Introduction

In many cases, the data is far from being independent. This dependence of the data could make senseless the inferences made assuming independence. There are several mathematical models studying dependent data. One of these models is the long range dependence model. Under long range dependence, the rate of convergence of the considered estimators increases. Thus, the precision of the estimators decreases.

In this paper, we review the literature on limit theorems for nonlinear functions of Gaussian random variables. Let $\{X_j\}_{j=1}^{\infty}$ be a stationary sequence of Gaussian random variables with $E[X_j] = 0$ and $E[X_j^2] = 1$. Let X be a standard normal random variable. Let $G : \mathbb{R} \to \mathbb{R}$ be a measurable function with $E[G^2(X)] < \infty$. We assume that the data is $G(X_1), \ldots, G(X_n)$. This type of dependence has been studied by several authors: Sun [18], [19]; Taqqu [21], [22], [23]; Dobrushin and Major [9]; Major [13]; Breuer and Major [6]; Rosenblatt [17]; Beran [3], [4].

We are interested in limit theorems for a convenient normalization of

$$n^{-1}\sum_{j=1}^{n}(G(X_j) - E[G(X_j)]). \tag{1.1}$$

$G(x)$ is a function in $L_2(\mathbb{R}, \phi(x)\,dx)$, where ϕ is the density of a standard normal distribution. A natural orthonormal basis of $L_2(\mathbb{R}, \phi(x)\,dx)$ is given by the Hermite polynomials. The Hermite polynomials $H_n(x)$, $n \geq 0$, are defined by $e^{ux-2^{-1}u^2} = \sum_{n=0}^{\infty} \frac{u^n}{n!} H_n(x)$. Doing the expansion

$$\sum_{n=0}^{\infty} \frac{u^n}{n!} H_n(x) = e^{ux-2^{-1}u^2} = \sum_{k=0}^{\infty} \frac{u^k x^k}{k!} \cdot \sum_{m=0}^{\infty} \frac{u^{2m}(-1)^m}{m!2^m},$$

we get that

$$H_n(x) = \sum_{0\leq m\leq[\frac{n}{2}]} \frac{x^{n-2m}(-1)^m}{(n-2m)!m!2^m}. \tag{1.2}$$

The first Hermite polynomials are given by

$$H_0(x)=1,\ H_1(x)=x,\ H_2(x)=x^2-1,\ H_3(x)=x^3-x,\ H_4(x)=x^4-6x^2+3.$$

Using that $e^{ux-\frac{u^2}{2}-\frac{x^2}{2}}=\sum_{n=0}^{\infty}\frac{u^n}{n!}H_n(x)e^{-\frac{x^2}{2}}$, we get that

$$H_n(x)e^{-\frac{x^2}{2}}=\frac{d^n}{du}e^{-\frac{(x-u)^2}{2}}|_{u=0}=\left(\frac{-d}{dx}\right)^n e^{-\frac{x^2}{2}}. \tag{1.3}$$

Since $e^{ux-2^{-1}u^2}e^{vx-2^{-1}v^2}=\sum_{n,m=0}^{\infty}\frac{u^n v^m}{n!m!}H_n(x)H_m(x)$, we have that

$$\begin{aligned}\sum_{n,m=0}^{\infty}\frac{u^n v^m}{n!m!}E[H_n(X)H_m(X)]&=E[e^{uX-2^{-1}u^2}e^{vX-2^{-1}v^2}]\\&=E[e^{(u+v)X-2^{-1}(u^2+v^2)}]=e^{uv}=\sum_{n=0}^{\infty}\frac{u^n v^n}{n!}.\end{aligned}$$

Consequently,

$$E[H_n(X)H_m(X)]=\begin{cases}n! & \text{if } m=n,\\ 0 & \text{if } m\neq n.\end{cases} \tag{1.4}$$

This proves that $\{H_n(x) : n\geq 0\}$ is a collection of orthonormal functions in the space $L_2(\mathbb{R},\phi(x)\,dx)$. It is possible to prove that $\{H_n(x) : n\geq 0\}$ is a complete orthonormal system of $L_2(\mathbb{R},\phi(x)\,dx)$. So, given a function $G\in L_2(\mathbb{R},\phi(x)\,dx)$, we have that

$$G(x)=\sum_{l=0}^{\infty}\frac{c_l}{l!}H_l(x), \tag{1.5}$$

where $c_l=E[G(X)H_l(X)]$ and the convergence is in $L_2(\mathbb{R},\phi(x)\,dx)$. Using that

$$e^{(rx+\sqrt{1-r^2}y)u-2^{-1}u^2}=e^{rux-2^{-1}r^2u^2}e^{\sqrt{1-r^2}uy-2^{-1}(1-r^2)u^2}$$

we get that

$$\sum_{n=0}^{\infty}\frac{u^nH_n(rx+\sqrt{1-r^2}y)}{n!}=\sum_{j=0}^{\infty}\frac{(ru)^jH_j(x)}{j!}\sum_{k=0}^{\infty}\frac{(\sqrt{1-r^2}u)^kH_k(y)}{k!}$$

and

$$H_n(rx+\sqrt{1-r^2}y)=\sum_{k=0}^{n}\binom{n}{k}r^kH_k(x)(1-r^2)^{(n-k)/2}H_{n-k}(y). \tag{1.6}$$

If X and Y are two jointly normal standard random variables, we claim that

$$E[H_n(X)H_m(Y)]=\begin{cases}n!(E[XY])^n & \text{if } m=n,\\ 0 & \text{if } m\neq n.\end{cases} \tag{1.7}$$

Let $r = E[XY]$. If $|r| = 1$, (1.7) follows from (1.4) and (1.6). Assume that $|r| < 1$, let $U = \frac{Y-rX}{\sqrt{1-r^2}}$, then X and U are independent standard normal random variables. So, by (1.4) and (1.6),

$$E[H_n(X)H_m(Y)] = E[H_n(X)H_m(rX + (1-r^2)^{1/2}U)]$$
$$= \sum_{k=0}^{m} \binom{m}{k} E[H_n(X)r^k H_k(X)(1-r^2)^{(m-k)/2} H_{m-k}(U)] = n! r^n \delta_{m,n}.$$

By (1.5) and (1.7),

$$\mathrm{Var}\left(n^{-1}\sum_{j=1}^{n}(G(X_j) - E[G(X_j)])\right) = n^{-2}\sum_{l=1}^{\infty}\frac{c_l^2}{l!}\sum_{j,k=1}^{n}(r(j-k))^l \tag{1.8}$$
$$= n^{-2}\sum_{l=1}^{\infty}\frac{c_l^2}{l!}\sum_{k=-(n-1)}^{n-1}(n-|k|)(r(k))^l.$$

If the sequence of random variables is close to be independent (the sequence of covariances $r(k)$ is close to zero), we have that for each function G with $E[G^2(X)] < \infty, n^{-1/2}\sum_{j=1}^{n}(G(X_j) - E[G(X_j)])$ converges in distribution to a normal distribution (Sun [18], [19]).

Theorem 1.1 (Sun [19]). *Suppose that $E[G^2(X_1)] < \infty$ and that the following limits exist:*

$$\lim_{n\to\infty}\frac{1}{n}\sum_{j,k=1}^{n} r(j-k) \tag{1.9}$$

and

$$\lim_{n\to\infty}\frac{1}{n}\sum_{j,k=1}^{n}(r(j-k))^2. \tag{1.10}$$

Then,

$$n^{-1/2}\sum_{j=1}^{n}(G(X_j) - E[G(X_j)]) \xrightarrow{d} N(0,\sigma^2), \tag{1.11}$$

where

$$\sigma^2 = \lim_{n\to\infty}\mathrm{Var}\left(n^{-1/2}\sum_{j=1}^{n}(G(X_j) - E[G(X_j)])\right) = \sum_{l=1}^{\infty}\frac{c_l^2}{l!}\lim_{n\to\infty}\frac{1}{n}\sum_{j,k=1}^{n}(r(j-k))^l.$$

Observe that (1.10) is equivalent to $\sum_{k=-\infty}^{\infty} r^2(k) < \infty$ which implies that for $l \geq 2$,

$$\lim_{n\to\infty}\frac{1}{n}\sum_{j,k=1}^{n}(r(j-k))^l = \lim_{n\to\infty}\sum_{k=-(n-1)}^{n-1}\frac{(n-|k|)}{n}(r(k))^l = \sum_{k=-\infty}^{\infty}(r(k))^l.$$

In this case the dependence between the random variables is called short range dependence. If $|r(k)| \leq \frac{c}{k^\alpha}$, for some $\alpha > 1$, then (1.9) and (1.10) hold. If $r(k) \sim \frac{c(-1)^k}{k^\alpha}$, for some $\alpha > 1/2$, then (1.9) and (1.10) hold.

If the sequence of covariances $r(k)$ is far from being zero, then the rate of convergence of the sequence in (1.1) is slower.

Theorem 1.2 (Taqqu [21], Theorem 5.1). *Suppose that there are* $\frac{1}{2} < H < 1$ *and a slowly varying function* L *such that*

$$\sum_{j,k=1}^{n} r(k-j) \sim n^{2H} L(n) \tag{1.12}$$

and

$$\sum_{j,k=1}^{n} |r(k-j)| = O(n^{2H} L(n)), \tag{1.13}$$

then

$$\left\{ n^{-H}(L(n))^{-1/2} \sum_{j=1}^{[nt]} (G(X_j) - E[G(X_j)]) : 0 \le t \le 1 \right\} \tag{1.14}$$
$$\xrightarrow{\text{w}} \{c_1 B_H(t) : 0 \le t \le 1\},$$

where $\{B_H(t) : 0 \le t\}$ *is mean zero Gaussian process with covariance given by*

$$E[B_H(s)B_H(t)] = t^{2H} + s^{2H} - |t-s|^{2H}, 0 \le s, t.$$

Moreover,

$$\sup_{0 \le t \le 1} n^{-H}(L(n))^{-1/2} \left| \sum_{j=1}^{[nt]} (G(X_j) - E[G(X_j)] - c_1 X_j) \right| \xrightarrow{\text{Pr}} 0.$$

The Gaussian process $\{B_H(t) : 0 \le t\}$ in the previous theorem is called a fractional Brownian motion of order H. If $r(k) \sim \frac{c}{k^{\alpha}}$, for some $c > 0$ and some $0 < \alpha < 1$, then the theorem above applies and

$$\sum_{j,k=1}^{n} r(k-j) \sim \frac{c2n^{2-\alpha}}{(1-\alpha)(2-\alpha)}.$$

In the previous theorem, only the first element of the series expansion of G contributes to the limit.

The previous results say under long range dependence the convergence of the sample mean to the population mean is slower than $n^{1/2}$. The implication to statistics of this is that under long range dependence estimators are less precise than in the i.i.d. situation. Thus, to get a determined precision of an estimator, under long range dependence a larger sample size is required than in the i.i.d. case.

Alternately, we may use spectral densities to motivate long range dependence. By the Herglotz theorem (see, for example, Section 4.2 in Brockwell and Davis [7]), there exists a right-continuous nondecreasing, bounded function F on $[-\pi, \pi]$ with $F(-\pi) = 0$ such that

$$r(k) = \int_{(-\pi,\pi]} e^{ikt} \, dF(t). \tag{1.15}$$

F is called the spectral distribution function of $\{X_j\}_{j=1}^{\infty}$. If F is absolutely continuous with density $f(x)$, the latter function is called the spectral density of $\{X_j\}_{j=1}^{\infty}$. If $r(k) \sim k^{-d} L(k)$, where $0 < d < 1$ and L is a slowly varying function, satisfying some additional conditions, then the spectral density has a singularity at zero: $f(x) \sim \text{cst}|x|^{d-1}$ as $x \to 0$ (see for example Zygmund [24], Chapter V.2).

Equation (1.7) gives the formula for the expectation of a product of two Hermite polynomials. The general formula giving the expectation of the product of several Hermite polynomials is known as the Diagram Formula. This formula was given by Breuer and Major [6] (p. 433). It appeared in the literature earlier in Lemma 3.2 in Taqqu [22]. A diagram G of order $(l_1, \dots, l_p)$ is a set of points $\{(j, l) : 1 \le j \le p, 1 \le l \le l_j\}$, called vertices, and a set pair of these points $\{((j, l), (k, m)) : 1 \le j < k \le p, 1 \le l \le l_j, 1 \le m \le l_k\}$, called edges, such that every vertex appears in one and only one edge. We denote by $\Gamma(l_1, \dots, l_p)$ the set of diagrams of order $(l_1, \dots, l_p)$. Notice that $\Gamma(l_1, \dots, l_p)$ is empty if $l_1 + \dots + l_p$ is an odd number. The set $L_j = \{(j, l) : 1 \le l \le l_j\}$ is called the j-th level of the graph G. Edges connect vertices of different levels. We will denote the set of edges of the diagram G by $E(G)$. Given an edge $w = ((j, l), (k, m))$, let $d_1(w) = j$ and $d_2(w) = k$. With this notation the diagram formula is as follows:

Diagram formula. Let $(X_1, \dots, X_p)$ be a Gaussian vector with

$$E[X_j] = 0,\ E[X_j^2] = 1 \quad \textit{and} \quad E[X_j X_k] = r(j, k) \quad \text{for each } 1 \le j, k \le p.$$

Then

$$E\left[\prod_{j=1}^{p} H_{l_j}(X_j)\right] = \sum_{G \in \Gamma(l_1, \dots, l_p)} \prod_{w \in E(G)} r(d_1(w), d_2(w)). \tag{1.16}$$

More properties on Hermite polynomials and their moments can be found in Neveu [14], Kwapień and Woyczyński [12], and de la Peña and Giné [15]. In particular, it is shown in these references that if Q is a polynomial of degree d in Gaussian r.v.'s, then for each $1 < p < q < \infty$,

$$\|Q\|_q \le \left(\frac{q-1}{p-1}\right)^{d/2} \|Q\|_p. \tag{1.17}$$

This implies that if Q_n is a sequence of polynomials of degree d in Gaussian r.v.'s which converges in distribution, then we have convergence of moments as well.

A possible extension of these limit theorems is to a stationary sequence $\{X_j\}_{j=1}^{\infty}$ of random vectors. Limit theorems in this setup can be found in Arcones [1]. A related problem concerns the limit theorems of nonlinear functions of a linear process. Several important results on this problem were obtained by Surgailis [20], Giraitis [10], and Ho and Hsing [11]. In this paper, we only review the case of functions of a Gaussian sequence of r.v.'s.

2 Limit theorems for sums of r.v.'s

In this section, we review the literature on limit theorems for

$$a_n^{-1} \sum_{j=1}^{n} (G(X_j) - E[G(X_j)]), \tag{2.1}$$

where $\{a_n\}$ is a sequence of real numbers converging to ∞.

Theorem 1.1 gives sufficient conditions for the asymptotic normality of the sequence in (2.1) under the normalization $a_n = n^{1/2}$. These conditions are optimal. If for each function $G \in L_2(\mathbb{R}, \phi(x)\,dx)$,

$$n^{-1/2}\sum_{j=1}^{n}(G(X_j) - E[G(X_j)])$$

converges in distribution to a normal distribution, then (1.9) and (1.10) hold. Notice that the convergence in distribution of $n^{-1/2}\sum_{j=1}^{n} X_j$ and $n^{-1/2}\sum_{j=1}^{n} H_2(X_j)$ implies convergence of moments and

$$E\left[(n^{-1/2}\sum_{j=1}^{n} X_j)^2\right] = \frac{1}{n}\sum_{j,k=1}^{n} r(j-k)$$

and

$$E\left[(n^{-1/2}\sum_{j=1}^{n} H_2(X_j))^2\right] = \frac{1}{n}\sum_{j,k=1}^{n} (r(j-k))^2.$$

Theorems 1.1 and 1.2 do not cover all possible types of sequences. For example if $r(k) = \frac{(-1)^k}{(1+k^2)^{\alpha/2}}$ where $0 < \alpha < 1/2$, then

$$\sum_{j,k=1}^{n} r(k-j) \sim n$$

and

$$\sum_{j,k=1}^{n} (r(k-j))^2 \sim n^{2-\alpha}.$$

So, neither Theorem 1.1 nor Theorem 1.2 apply. The following is a general theorem on the distributional limits for the sequence of r.v.'s in (2.1).

Theorem 2.1 (Arcones [2]). *Let $\{X_j\}_{j=1}^{\infty}$ be a stationary sequence of standard Gaussian r.v.'s with covariances $r(j) = E[X_1X_{1+j}]$. Let G be a function with $E[G^2(X)] < \infty$. Let*

$$A_{n,k} = a_n^{-k}\sum_{l_1,\dots,l_k=1}^{n} r(l_2-l_1)\cdots r(l_k-l_{k-1}),$$

$$B_{n,k} = a_n^{-k}\sum_{l_1,\dots,l_k=1}^{n} r(l_2-l_1)\cdots r(l_k-l_{k-1})r(l_1-l_k).$$

Suppose that the following limits exist:

$$\lim_{n\to\infty} r(n) = 0,$$

$$\lim_{n\to\infty} A_{n,k} =: A_k \quad \textit{for } k \geq 2;$$

$$\lim_{n\to\infty} B_{n,k} =: B_k \quad \textit{for } k \geq 2.$$

Then,

$$a_n^{-1}\sum_{j=1}^{n}(G(X_j)-E[G(X_j)]) \xrightarrow{\text{d}} \xi.$$

Moreover, we have the following:

(i) *If* $\sum_{k=1}^{\infty} r^2(k) < \infty$, *then* ξ *is a normal r.v. with mean zero and variance*

$$\lim_{n\to\infty} \operatorname{Var}\left(a_n^{-1}\sum_{j=1}^{n}(G(X_j)-E[G(X_j)])\right).$$

(ii) *If* $\sum_{k=1}^{\infty} r^2(k) = \infty$, *then* ξ *has characteristic function*

$$E[\exp(it\xi)] = \exp\left(\sum_{k=2}^{\infty}(it)^k\left(2^{-1}c_1^2c_2^{k-2}A_k + (2k)^{-1}c_2^kB_k\right)\right).$$

Moreover,

$$a_n^{-1}\left|\sum_{j=1}^{n}(G(X_j)-E[G(X_j)]-c_1X_j-2^{-1}c_2H_2(X_j))\right| \xrightarrow{\text{Pr}} 0.$$

In the case (ii),

$$\left(\frac{1}{a_n}\sum_{j=1}^{n}X_j, \frac{1}{a_n}\sum_{j=1}^{n}H_2(X_j)\right) \xrightarrow{\text{d}} (\xi_1,\xi_2)$$

with

$$E[\exp(is\xi_1+it\xi_2)] = \exp\left(\sum_{k=2}^{\infty}\left(\frac{(is)^2(it)^{k-2}}{2}A_k + \frac{(it)^k}{2k}B_k\right)\right).$$

The main step in the proof is to find the characteristic function

$$E\left[\exp\left(\frac{itc_1}{a_n}\sum_{j=1}^{n}X_j + \frac{itc_2}{a_n}\sum_{j=1}^{n}H_2(X_j)\right)\right] = \exp\left(\sum_{k=2}^{\infty}(it)^k\left(\frac{c_1^2c_2^{k-2}}{2}A_{n,k} + \frac{c_2^k}{2k}B_{n,k}\right)\right)$$

and then to take limits.

Next, we consider several sequences of covariances to which Theorem 2.1 applies. If $r(k) = \frac{(-1)^k}{(1+k^2)^{\alpha/2}}$, then there exists a stationary Gaussian sequence of r.v.'s with that covariance. Moreover, if $0<\alpha<1/2$ and $a_n = n^{1-\alpha}$, then

$$a_n^{-1}\sum_{j=1}^{n}(G(X_j)-E[G(X_j)]) \xrightarrow{\text{d}} \xi,$$

where

$$E[\exp(it\xi)] = \exp\left(\sum_{k=2}^{\infty}(it)^k(2k)^{-1}c_2^k B_k\right)$$

and

$$B_k = \int_0^1 \cdots \int_0^1 |x_2 - x_1|^{-\alpha} \cdots |x_k - x_{k-1}|^{-\alpha}|x_1 - x_k|^{-\alpha}\, dx_1 \cdots dx_k,$$

for $k \geq 2$. The distribution limit is not normal. Observe that only c_2 contributes to the limit, where

$$G(x) - E[G(X_1)] = \sum_{l=1}^{\infty}\frac{c_l}{l!}H_l(x).$$

Let $r(2k) = \frac{(-1)^k}{(1+4k^2)^{\alpha/2}}$ and $r(2k+1) = 0$, where $0 < \alpha < 1/2$. Let $a_n = n^{1-\alpha}$, then

$$a_n^{-1}\sum_{j=1}^{n}(G(X_j) - E[G(X_j)]) \xrightarrow{\text{d}} \xi,$$

where

$$E[\exp(it\xi)] = \exp\left(\sum_{k=2}^{\infty}(it)^k(2k)^{-1}c_2^k B_k\right)$$

and

$$B_k = 2^{1-k}\int_0^1 \cdots \int_0^1 |x_2 - x_1|^{-\alpha} \cdots |x_k - x_{k-1}|^{-\alpha}|x_1 - x_k|^{-\alpha}\, dx_1 \cdots dx_k,$$

for $k \geq 2$. Again only c_2 contributes to the limit.

If $r(k) = \frac{(-1)^k}{2(1+k^2)^{\alpha/2}} + \frac{1}{2(1+k^2)^{\alpha}}$, then there exists a stationary Gaussian sequence of r.v.'s with that covariance. Moreover, if $0 < \alpha < 1/2$ and $a_n = n^{1-\alpha}$, then

$$a_n^{-1}\sum_{j=1}^{n}(G(X_j) - E[G(X_j)]) \xrightarrow{\text{d}} \xi,$$

where

$$E[\exp(it\xi)] = \exp\left(\sum_{k=2}^{\infty}(it)^k\left(2^{-1}c_1^2 A_k + (2k)^{-1}c_2^k B_k\right)\right),$$

where

$$A_2 = 2^{-2}\int_0^1\int_0^1 |x_2 - x_1|^{-2\alpha}\, dx_1\, dx_2,$$
$$A_k = 0 \quad \text{for } k \geq 3,$$

and

$$B_k = 2^{-k}\int_0^1 \cdots \int_0^1 |x_2 - x_1|^{-\alpha} \cdots |x_k - x_{k-1}|^{-\alpha}|x_1 - x_k|^{-\alpha}\, dx_1 \cdots dx_k,$$

for $k \geq 2$. Here, both c_1 and c_2 contribute to the limit.

The conditions in Theorem 2.1 are sufficient in the following sense.

Theorem 2.2 (Arcones [2]). *Suppose that* $\lim_{n\to\infty} r(n) = 0$. *Then, the following conditions are equivalent:*

(i) *For each function* $G(x)$ *with* $E[G^2(X_1)] < \infty$, $a_n^{-1}\sum_{j=1}^n (G(X_j) - E[G(X_j)])$ *converges in distribution.*

(ii) *For each* $k \geq 2$, *the following limits exist and they finite:*

$$\lim_{n\to\infty} a_n^{-k} \sum_{l_1,\ldots,l_k=1}^{n} r(l_2 - l_1)\cdots r(l_k - l_{k-1})$$

and

$$\lim_{n\to\infty} a_n^{-k} \sum_{l_1,\ldots,l_k=1}^{n} r(l_2 - l_1)\cdots r(l_k - l_{k-1}) r(l_1 - l_k).$$

To get convergence of the partial sums process, conditions a little stronger than the ones before are needed.

Theorem 2.3. *With the usual notation, let* $\{a_n\}$ *be a sequence of real numbers converging to infinity. Supppose the following:*

(i) $\lim_{n\to\infty} r(n) = 0$.

(ii) *For each* $k \geq 2$ *and each* $0 \leq t_1, \ldots, t_k \leq 1$, *the following limits exist:*

$$\lim_{n\to\infty} a_n^{-k} \sum_{l_1=1}^{[nt_1]} \cdots \sum_{l_k=1}^{[nt_k]} r(l_2 - l_1) r(l_3 - l_2)\cdots r(l_k - l_{k-1})$$

and

$$\lim_{n\to\infty} a_n^{-k} \sum_{l_1=1}^{[nt_1]} \cdots \sum_{l_k=1}^{[nt_k]} r(l_2 - l_1) r(l_3 - l_2)\cdots r(l_k - l_{k-1}) r(l_1 - l_k).$$

(iii) $\{a_n\}$ *is regularly varying of order* α, *with* $2^{-1} < \alpha \leq 1$.

Then,

$$\left\{ a_n^{-1} \sum_{j=1}^{[nt]} (G(X_j) - E[G(X_j)]) : 0 \leq t \leq 1 \right\}$$

converges weakly. Moreover,

$$a_n^{-1} \sup_{0\leq t\leq 1} \left| \sum_{j=1}^{[nt]} (G(X_j) - E[G(X_j)] - c_1 X_j - 2^{-1} c_2 H_2(X_j)) \right| \overset{\Pr}{\to} 0.$$

Proof. The convergence of the finite-dimensional distributions follows from the methods in the proof of Theorem 2.1. Let

$$Z_n(t) := a_n^{-1} \sum_{j=1}^{[nt]} (G(X_j) - E[G(X_j)]).$$

By Theorem 15.6 in Billingsley [5], it suffices to show that for $0 \leq t_1 \leq t \leq t_2 \leq 1$,

$$E[|Z_n(t) - Z_n(t_1)||Z_n(t_2) - Z_n(t)|] \leq ct^{2\alpha-\epsilon},$$

where c is a constant and $2\alpha - \epsilon > 1$. But, by regular variation, for $0 \leq s < t \leq 1$,

$$E[|Z_n(t) - Z_n(s)|^2] \leq ca_n^{-2} a_{[nt]-[ns]}^2 \leq c|t - s|^{2\alpha-\epsilon}.$$

□

There are several theorems on the convergence in distribution of the sequence in (2.1) which depend on the notion of Hermite rank. Taqqu [21] defined the rank of a function $G(x)$ as

$$\text{rank}(G) := \inf\{k : E[(G(X) - EG(X))H_k(X)] \neq 0\}.$$

If $\text{rank}(G) = m$, then

$$G(x) - E[G(X_1)] = \sum_{k=m}^{\infty} \frac{c_k}{k!} H_k(x),$$

and $c_m \neq 0$. The following deep result gives a limit theorem for higher rank functions.

Theorem 2.4 (Taqqu [21], [23]; Dobrushin and Major [9]). *Assume $0 < d < m$ and a slowly varying function $L(k)$ such that $\frac{r(k)}{k^{-d}L(k)}$ converges. Let G be a real function with rank m. Then,*

$$\left\{ n^{md-2}(L(n))^{-m/2} \sum_{j=1}^{[nt]} (G(X_j) - E[G(X_j)]) : 0 \leq t \leq 1 \right\}$$

converges weakly. Moreover, the distributional limit of

$$n^{md-2}(L(n))^{-m/2} \sum_{j=1}^{[nt]} (G(X_j) - E[G(X_j)])$$

has the distribution of an m-dimensional Îto–Wiener integral and

$$a_n^{-1} \sum_{j=1}^{n} (G(X_j) - E[G(X_j)] - (m)^{-1} c_m H_m(X_j)) \xrightarrow{P} 0.$$

A way to obtain the previous theorem is through the limit of multiple stochastic integrals with respect to the spectral density. A review of these techniques and a nice proof of the previous theorem can be found in Major [13]. The limit distribution of

$$n^{md-2}(L(n))^{-m/2} \sum_{j=1}^{n} (G(X_j) - E[G(X_j)])$$

is

$$D^{-m/2} c_m \int_{-\infty}^{\infty} \cdots \int_{-\infty}^{\infty} e^{i(x_1+\cdots+x_m)} \frac{e^{i(x_1+\cdots+x_m)} - 1}{i(x_1+\cdots+x_m)} \times |x_1|^{(\alpha-1)/2} \cdots |x_m|^{(\alpha-1)/2} \, dW(x_1) \cdots dW(x_k),$$

where W is the random spectral measure of the white noise process and

$$D = \int_{-\infty}^{\infty} e^{ix} |x|^{\alpha-1} \, dx = 2\Gamma(\alpha) \cos\left(\frac{\alpha\pi}{2}\right).$$

The sufficient conditions for the CLT of the processes in (2.1) depend on the rank of G. The next theorem shows that the conditions needed to get asymptotic normality in (2.1) are less stringent for functions of higher rank.

Theorem 2.5 (Breuer and Major [9], Theorem 1). *Let G be a function with $E[G^2(X)] < \infty$ and rank m. Suppose that $\sum_{k=1}^{\infty} |r(k)|^m < \infty$. Then*

$$n^{-1/2} \sum_{i=1}^{n} (G(X_i) - E[G(X_i)]) \xrightarrow{d} N(0, \sigma^2),$$

where

$$\sigma^2 = \lim_{n\to\infty} \mathrm{Var}(n^{-1/2} \sum_{j=1}^{n} (G(X_j) - E[G(X_j)])) = \sum_{l=m}^{\infty} \frac{c_l^2}{l!} \sum_{k=-\infty}^{\infty} (r(k))^l.$$

Observe that the two previous theorems cover the convergence for functions with rank m and sequences of covariances $r(k) \sim ck^{-d}$ for $0 < d \neq 1/m$. The next theorem applies to functions with rank m and sequences of covariances $r(k) \sim ck^{-1/m}$.

Theorem 2.6 (Breuer and Major [9], Theorem 1′). *Let G be a function with $E[G^2(X)] < \infty$ and rank m. Suppose that*

$$L(n) := \sum_{j=1}^{n} |r(j)|^m$$

is a slowly varying function and for each $l \geq m$,

$$\lim_{n\to\infty} (L(n))^{-1} \sum_{j=1}^{n} (r(j))^l$$

exists. Then,

$$n^{-1/2} (L(n))^{-1/2} \sum_{j=1}^{n} (G(X_j) - E[G(X_j)]) \xrightarrow{d} N(0, \sigma^2),$$

where

$$\sigma^2 = \lim_{n\to\infty} \mathrm{Var}\left(n^{-1/2} (L(n))^{-1/2} \sum_{j=1}^{n} (G(X_j) - E[G(X_j)]) \right).$$

When the rank is 2, the methods in Theorem 2.1 give the following.

Theorem 2.7 (Arcones [2]). *Let $\{X_j\}_{j=1}^{\infty}$ be a a stationary Gaussian sequence of r.v.'s with mean zero. Let G be a real function with $E[G^2(X_1)] < \infty$ and rank 2 or larger. Let $\{a_n\}$ be a sequence of real numbers, $a_n \to \infty$. Suppose that the following conditions are satisfied:*

(i) $\lim_{n\to\infty} r(n) = 0$.

(ii) *For each $k \geq 2$, $\lim_{n\to\infty} B_{n,k} =: B_k$, where*

$$B_{n,k} = a_n^{-k} \sum_{l_1,\ldots,l_k=1}^{n} r(l_2 - l_1) \cdots r(l_k - l_{k-1}) r(l_1 - l_k).$$

Then,

$$a_n^{-1} \sum_{j=1}^{n} (G(X_j) - E[G(X_j)]) \xrightarrow{\text{d}} \xi.$$

Moreover, we have the following:

(i) *If $\sum_{k=1}^{\infty} r^2(k) < \infty$, then ξ is a normal r.v. with mean zero and variance*

$$\lim_{n\to\infty} \text{Var}\left(a_n^{-1} \sum_{j=1}^{n} (G(X_j) - E[G(X_j)]) \right).$$

(ii) *If $\sum_{k=1}^{\infty} r^2(k) = \infty$, then ξ has characteristic function*

$$E[\exp(it\xi)] = \exp\left(\sum_{k=2}^{\infty} (it)^k (2k)^{-1} c_2^k B_k \right).$$

3 Limit theorems for empirical processes

There are very few papers on the weak convergence of empirical processes under long range dependence.

Theorem 3.1 (Dehling and Taqqu [8]). *Suppose that $r(k) \sim k^{-d} L(k)$, where $0 < d < 1/m$ and L is a slowly varying function at infinity. Let G be a measurable function. Suppose that the minimum rank of the functions $I(G(x) \leq t)$, $t \in \mathbb{R}$, is m. Then,*

$$\{n^{2^{-1}md-1} (L(n))^{-2^{-1}m} \sum_{j=1}^{[nt]} (I(G(X_j) \leq x) - P\{G(X_j) \leq x\}) : x \in \mathbb{R}, 0 \leq t \leq 1\}$$

converges weakly.

Under the conditions of the previous theorem, for each $x \in R$, we have that

$$I(G(X) \leq x) - P\{G(X_j) \leq x\} = \sum_{k=m}^{\infty} (k!)^{-1} c_k(x) H_k(x),$$

where $c_k(x) = E[H_k(X)I(G(X) \leq x)]$. By Theorem 2.4,

$$\{n^{2^{-1}md-1}(L(n))^{-2^{-1}m}c_m(x)\sum_{j=1}^{[nt]} H_m(X_j) : x \in \mathbb{R}, 0 \leq t \leq 1\}$$

converges weakly. Theorem 3.1 is obtained by proving that

$$\sup_{x\in\mathbb{R},0\leq t\leq 1} n^{2^{-1}md-1}(L(n))^{-2^{-1}m}|R_{[nt]}(x)| \stackrel{\Pr}{\to} 0, \tag{3.1}$$

where

$$R_n(x) = \sum_{j=1}^{n}(I(G(X_j) \leq x) - P\{G(X_j) \leq x\} - (m!)^{-1}c_m(x)H_m(X_j)).$$

In order to get this, Dehling and Taqqu [8] proved that for each $x, y \in \mathbb{R}$,

$$n^{md-2}(L(n))^{-m}E[(R_n(y) - R_n(x))^2] \leq cn^{-1-\gamma}|F(y) - F(x)|,$$

where $c, \gamma > 0$. This is enough to prove (3.1).

It is easy to see that the arguments in the previous theorem give the following:

Theorem 3.2. *With the usual notation, let $\{a_n\}$ be a sequence of real numbers converging to infinity. Supppose the following:*

(i) $\lim_{n\to\infty} r(n) = 0$.

(ii) *For each $k \geq 2$ and each $0 \leq t_1, \ldots, t_k \leq 1$, the following limits exist:*

$$\lim_{n\to\infty} a_n^{-k}\sum_{l_1=1}^{[nt_1]}\cdots\sum_{l_k=1}^{[nt_k]} r(l_2-l_1)r(l_3-l_2)\cdots r(l_k-l_{k-1})$$

and

$$\lim_{n\to\infty} a_n^{-k}\sum_{l_1=1}^{[nt_1]}\cdots\sum_{l_k=1}^{[nt_k]} r(l_2-l_1)r(l_3-l_2)\cdots r(l_k-l_{k-1})r(l_1-l_k).$$

(iii) *$\{a_n\}$ is regularly varying of order α, with $2^{-1} < \alpha \leq 1$.*

(iv) *For some $\gamma > 0$,*

$$a_n^{-2}\sum_{j,k=1}^{n} |r(k-j)|^3 = O(n^{-\gamma}).$$

Then,

$$\left\{a_n^{-1}\sum_{j=1}^{[nt]}(I(G(X_j) \leq x) - P\{G(X_j) \leq x\}) : x \in \mathbb{R}, 0 \leq t \leq 1\right\}$$

converges weakly.

It is an open problem whether conditions (1.9) and (1.10) suffice to show that the empirical process

$$\left\{ n^{-1/2} \sum_{j=1}^{n} (I(G(X_j) \le x) - \Pr\{G(X_j) \le x\}) : x \in \mathbb{R} \right\}$$

converges weakly.

References

[1] M. A. Arcones: Limit theorems for nonlinear functionals of a stationary Gaussian sequence of vectors. *Ann. Probab.* **22** (1994), 2242–2274.

[2] M. A. Arcones: Distributional limit theorems over a stationary Gaussian sequence of random variables. *Stoch. Proc. Appl.* **88** (2000), 135–159.

[3] J. Beran: Statistical methods for data with long–range dependence. *Statistical Science.* **7** (1992), 404–427.

[4] J. Beran: *Statistics for Long Memory Processes.* Chapman & Hall, New York, 1994.

[5] P. Billingsley: *Convergence of Probability Measures.* Wiley, New York, 1969.

[6] P. Breuer and P. Major: Central limit theorems for non-linear functionals of Gaussian fields. *J. Mult. Anal.* **13** (1983), 425–441.

[7] P. J. Brockwell and R. A. Davis: *Time Series: Theory and Methods.* Springer, New York, 1991.

[8] H. Dehling and M. S. Taqqu: The empirical process of some long-range dependent sequences with an application to U-statistics. *Ann. Statist.* **17** (1989), 1767–1783.

[9] R. L. Dobrushin and P. Major: Non-central limit theorems for non-linear functionals of Gaussian fields. *Z. Wahrsch. verw. Geb.* **50** (1979), 27–52.

[10] L. Giraitis: Central limit theorem for functions of a linear process. *Lith. Math. J.* **25** (1985), 25–35.

[11] H. C. Ho and T. Hsing. Limit theorems for functional of moving averages. *Ann. Probab.* **25** (1997), 1636–1669.

[12] S. Kwapień and W. A. Woyczyński: *Random Series and Stochastic Integrals: Single and Multiple.* Birkhäuser, Boston, 1992.

[13] P. Major: *Multiple Wiener–Itô Integrals. Lect. Notes Math.* **849**. Springer–Verlag, New York, 1981.

[14] J. Neveu: *Processus Aléatoires Gaussianes.* Presses de l'Université de Montréal, 1968.

[15] V. H. de la Peña, and E. Giné: *Decoupling: From Dependence to Independence.* Springer, New York, 1999.

[16] M. Rosenblatt: Independence and dependency. In: *Proc. Fourth Berkeley Symp. Math. Stat. and Prob. II,* 431–443. Univ. of California Press, Berkeley, 1961.

[17] M. Rosenblatt: Stochastic processes with short-range and long-range dependence. In: *Statistics: An Appraisal.* Editors: H. A. David and H. T. David, 509–520, 1984.

[18] T. C. Sun: A central limit theorem for non-linear functions of a normal stationary process. *J. Math. Mech.* **12** (1963), 945–977.

[19] T. C. Sun: Some further results on central limit theorems for non-linear functions of a normal stationary process. *J. Math. Mech.* **14** (1965), 71–85.

[20] D. Surgailis. Zones of attraction of self-similar multiple integrals. *Lith. Mathem. J.* **22** (1982), 327–340.

[21] M. S. Taqqu: Weak convergence to fractional Brownian motion and to the Rosenblatt process. *Z. Wahrsch. verw. Geb.* **31** (1975), 287–302.

[22] M. S. Taqqu: Law of the iterated logarithm for sums of non-linear functions of Gaussian variables that exhibit long-range dependence. *Z. Wahrsch. verw. Geb.* **40** (1977), 203–238.

[23] M. S. Taqqu: Convergence of the integrated process of arbitrary Hermite rank. *Z. Wahrsch. verw. Geb.* **50** (1979), 53–83.

[24] A. Zygmund: *Trigonometric Series.* Cambridge University Press, London, 1953.

Miguel A. Arcones
Department of Mathematical Sciences
Binghamton University
Binghamton, NY 13902, U.S.A.
arcones@math.binghamton.edu

Part IV

Empirical Spectral Process Techniques

Empirical Spectral Processes and Nonparametric Maximum Likelihood Estimation for Time Series

RAINER DAHLHAUS AND WOLFGANG POLONIK

ABSTRACT. We survey recent developments in the theory of empirical spectral processes indexed by functions and their applications to nonparametric maximum Whittle likelihood estimation for times series. The exposition covers stationary and nonstationary time series as well as Gaussian and non-Gaussian cases. Previously unpublished results are added in order to complete the overall picture.

1 Introduction

During the last decade the theory of empirical processes has developed considerably and the number of statistical problems that are approached utilizing concepts from empirical process theory is steeply increasing. For instance, very recently Bühlmann and Yu [6] used the modern theory to derive some theory for classification and regression trees (CART) (see also Breiman et al. [3]), Polonik [31, 32] applied it to testing for unimodality, and testing for goodness-of-fit. Applications to prediction of stochastic processes can be found in Polonik and Yao [33, 34]. Van de Geer [40] applied it extensively to least squares, and nonparametric maximum likelihood estimation. Other examples of applications include learning theory, like methods based on support vector machines (SVM), as for instance methods for classification, regression, or image recognition. Probabilistic theory of SVM-methods is heavily based on the Vapnik–Chervonenkis (VC) theory for empirical processes; see, e.g., the text books by Christiani and Shawe-Taylor [7], and Vapnik [43, 44]. Another application of empirical process theory can be found in Beran and Dümbgen [2].

This short list of recent publications is given to underline the broad range of potential statistical applications of modern empirical process theory. By no means is it meant to be a comprehensive or even representative list.

In this paper we show how large parts of the existing methodology can fruitfully be used for time series analysis. In our setup the role of the empirical distribution of i.i.d. data is taken over by the empirical spectral measure. We study the empirical spectral process indexed by classes of functions, derive its convergence and show how maximal exponential inequalities can be used for nonparametric likelihood estimation.

Since the empirical spectral measure is utilized here, the spectral density of a stationary process plays a major role. Although our results are not restricted to Gaussian processes the method works particularly well for such processes, since a stationary Gaussian process is uniquely determined by its spectral density. For non-Gaussian processes the method is limited to those parameters or parameter curves which can be uniquely identified from the spectral density (or from the time-varying spectral density—see Section 6).

Although our concept is based on analysis in the frequency domain, the focus might still be the analysis of certain time-domain models. For example, Theorem 3.2 leads, in particular, to the asymptotic distribution of a test statistic of the composite hypothesis that the observed process is an autoregressive process of order p. The estimation of certain time-varying AR parameters is another example.

The paper is organized as follows. In Section 2 we set up the notation and give an introduction to those time series problems which are related to our approach. In particular, we introduce the Whittle likelihood. In Section 3 we present some properties of the empirical spectral process, such as its weak convergence to a Gaussian process. These results use conditions based on the metric entropy with bracketing which is introduced in Section 3 also. Section 4 presents a Bernstein-type exponential inequality and a maximal inequality for the empirical spectral process. Section 5 contains the results for nonparametric maximum likelihood estimation, and in Section 6 we briefly discuss some extensions, such as concepts for locally stationary processes.

2 Time Series Analysis based on the spectral domain

In this section we give a brief overview of some spectral domain methods—in particular, of methods related to the empirical spectral process treated in Sections 3 and 4. A more comprehensive treatment of spectral domain methods in time series can, e.g., be found in the monographs by Brillinger [4] and Shumway and Stoffer [38].

The starting point of our considerations is the classical spectral representation theorem of Cramér [9] (cf. Brillinger, [4]; Taniguchi and Kakizawa [39]) which says that a real-valued stochastic process $X_t, t \in \mathbb{Z}$, is stationary if and only if it can be written in the form

$$X_t = \mu + \int_{-\pi}^{\pi} \exp(i\lambda t) A(\lambda) d\xi(\lambda), \tag{2.1}$$

where $\mu = EX_t$, $A(\lambda) : [-\pi, \pi] \to \mathbb{C}$ is a deterministic function, and $\xi(\lambda)$ is some complex-valued stochastic process with orthonormal increments. This representation is a decomposition of the process into periodic components (sine and cosine) of frequencies λ and random amplitudes with variance $f(\lambda) := |A(\lambda)|^2$. The function $f(\lambda)$ is sometimes called power spectrum or energy spectrum of the process, because it measures the "power" of the component of frequency λ. We call $f(\lambda)$ the spectral density of the process. Throughout this paper we assume that $\mu = 0$.

It is easy to see that $f(\lambda)$ is the Fourier transform of the covariance function $c(u) = \mathrm{E}X_{t+u}X_t$, that is,

$$f(\lambda) = \frac{1}{2\pi} \sum_{u=-\infty}^{\infty} c(u) \exp(-i\lambda u) \tag{2.2}$$

and by the inverse Fourier transform

$$c(u) = \int_{-\pi}^{\pi} f(\lambda) \exp(i\lambda u) d\lambda. \tag{2.3}$$

We may also take (2.2) as the definition of the spectral density (avoiding the notion of the stochastic integral in (2.1)) but we feel that (2.1) is more intuitive. It is easy to see that $f(\lambda) = f(-\lambda)$ and $f(\lambda) \geq 0$. The integrated spectral density

$$\begin{aligned} F(\lambda) &:= \int_0^{\lambda} f(\alpha) d\alpha, \ \lambda \in [0, \pi], \\ F(-\lambda) &:= -F(\lambda) \end{aligned} \tag{2.4}$$

is called the spectral measure of the process. If X_t is Gaussian, then $F(\lambda)$ (or $f(\lambda)$) uniquely determines the distribution of the process, which means that all parameters of interest can be written as functionals of F. This is the basis for the following considerations. Note that F is monotone in λ with $F(\pi) = c(0)/2$.

Suppose we observe the process at times $t = 1, \ldots, n$ and we want to make some statistical inference on $f(\lambda)$ or $F(\lambda)$. The representation (2.1) suggests using as an estimate of $f(\lambda)$ the squared inverse Fourier transform of the data, that is,

$$I_n(\lambda) := \frac{1}{2\pi n} \left| \sum_{t=1}^{n} X_t \exp(-i\lambda t) \right|^2 \tag{2.5}$$
$$= \frac{1}{2\pi} \sum_{u=-(n-1)}^{n-1} \left(\frac{1}{n} \sum_{1 \le t, t+u \le n} X_t X_{t+u} \right) \exp(-i\lambda u)$$

with the last expression being the empirical counterpart of (2.2). The function $I_n(\lambda)$ is the well-known periodogram introduced by Schuster [37]. It is usually calculated via the Fast Fourier Algorithm (Cooley and Tukey [8]) at the Fourier frequencies $\lambda_s := \frac{2\pi s}{n}$ ($s = 0, \ldots, [n/2]$). It can be shown under some regularity conditions that

$$\mathrm{E} I_n(\lambda) = f(\lambda) + O(n^{-1}), \tag{2.6}$$
$$\mathrm{var} I_n(\lambda) = f(\lambda)^2 + O(n^{-1}), \tag{2.7}$$
$$\mathrm{cov}(I_n(\lambda), I_n(\mu)) = O(n^{-1}) \quad \text{for } \lambda \pm \mu \not\equiv 0 \bmod 2\pi \tag{2.8}$$

(cf. Brillinger [4], Chapter 5), which means that the periodogram is asymptotically unbiased but not consistent for estimating $f(\lambda)$. Moreover, it can be shown for $\lambda \neq 0, \pi$ that $I_n(\lambda)$ converges in distribution to an $\frac{f(\lambda)}{2} \chi_2^2$ variable. In the case where the process X_t consists of independent Gaussian variables, (2.8) can be strengthened in that $I_n(\lambda_s)$, $s = 0, \ldots, [n/2]$ are also i.i.d. variables.

A consistent estimator of $f(\lambda)$ can be obtained by smoothing $I_n(\lambda_s)$ in the neighborhood of λ, i.e., by some kernel estimator (cf. Brillinger [4], Chapter 5). However, we do not want to discuss the resulting spectral estimates in detail in this paper.

The sharp cut-off of the Fourier transform in (2.5) at times $t = 1$ and $t = n$ may cause some severe bias problems—in particular, if the spectral density contains strong peaks (i.e., the original series contains strong periodicities) or when the sample length is rather small. This effect is usually called leakage effect (cf. Dahlhaus [13]). It is worth mentioning that this bias problem also afflicts all estimators which can be written as functionals of the periodogram such as the empirical covariances, Yule–Walker estimates or Whittle estimates—see below (cf. Dahlhaus [12]). A simple possibility to reduce this bias is to multiply the data with some data taper which decays smoothly to zero at the edges of the observation domain prior to the Fourier transform, that is, to use

$$I_n^{(h)}(\lambda) := \frac{1}{2\pi H_{2,n}} \left| \sum_{t=1}^{n} h_{t,n} X_t \exp(-i\lambda t) \right|^2, \tag{2.9}$$

where $h_{t,n}$ for example is the cosine taper $h_{t,n} = h_\rho(\frac{t-0.5}{n})$ with

$$h_\rho(x) = \begin{cases} \frac{1}{2}(1 - \cos\frac{2\pi x}{\rho}), & \text{if } 0 \le x \le \rho/2, \\ 1, & \text{if } \rho/2 \le x \le 1/2, \\ h_\rho(1-x), & \text{if } 1/2 \le x \le 1, \end{cases} \tag{2.10}$$

and

$$H_{2,n} = \Sigma_{t=1}^{n} h_{t,n}^2.$$

If ρ is constant, then it can be shown under some regularity conditions that

$$\mathrm{E}I_n^{(h)}(\lambda) = f(\lambda) + O(n^{-2}) \tag{2.11}$$

demonstrating the bias-improvement in comparison to (2.6). It should be noted, however, that the use of data tapers usually leads to some small increase of the variance of the estimates which can be avoided if $\rho = \rho_T \to 0$ slowly (which is a reasonable assumption).

Equation (2.4) now immediately leads to an estimator for the spectral measure, namely,

$$F_n(\lambda) := \int_0^{\lambda} I_n^{(h)}(\alpha)d\alpha, \quad \lambda \in [0, \pi]. \tag{2.12}$$

More generally, we define for some arbitrary function ϕ,

$$F_n(\phi) := \int_{-\pi}^{\pi} \phi(\alpha) I_n^{(h)}(\alpha)d\alpha, \tag{2.13}$$

so that $F_n(\lambda)$ can be understood as a short-hand notation for $F_n(\mathbf{1}_{[0,\lambda]})$ where $\mathbf{1}_{[0,\lambda]}$ denotes the indicator function of the interval $[0, \lambda]$. In practice one usually prefers the sum over the Fourier frequencies instead of the integral—however, both estimators usually have the same asymptotic properties since the difference is small, e.g., of order $o_p(n^{-1/2})$. We also define

$$F(\phi) := \int_{-\pi}^{\pi} \phi(\alpha) f(\alpha)d\alpha$$

with the convention that $F(\lambda) = F(\mathbf{1}_{[0,\lambda]})$. We mention that under weak conditions $F_n(\phi)$ is a consistent estimator of $F(\phi)$ (cf. Proposition 3.1), even though the periodogram itself is not a consistent estimator of the spectral density.

The empirical spectral process is now defined as

$$E_n(\phi) = \sqrt{n}(F_n(\phi) - F(\phi)), \tag{2.14}$$

where ϕ is from some function class Φ. In Section 3 we consider asymptotic properties of this process. In particular, we will state below (see (3.7)) the weak convergence of the process $E_n(\phi), \phi \in \Phi$ to some Gaussian limit process indexed by $\phi \in \Phi$.

Examples for interesting choices of functions ϕ are the following.

(i) for $\phi(\alpha) = \mathbf{1}_{[0,\lambda]}(\alpha)$ we obtain $F_n(\phi) = F_n(\lambda)$, i.e., the above estimator for $F(\lambda)$.

(ii) for $\phi(\alpha) = \phi_u(\alpha) = \exp(i\alpha u)$ we obtain $F(\phi_u) = c(u)$ and

$$F_n(\phi_u) = c_n(u) = \frac{1}{H_{2,n}} \sum_{1 \le t, t+u \le n} h_{t,n} X_t \, h_{t+u,n} X_{t+u}, \tag{2.15}$$

i.e., an estimator for the covariance function.

(iii) for $\phi(\alpha) = \phi_{n,\lambda}(\alpha) = \frac{1}{b_n} K(\frac{\lambda - \alpha}{b_n})$ with $\int K(\gamma)d\gamma = 1$ and $b_n \to 0$, $nb_n \to \infty$ we obtain a kernel estimator of the spectral density $f(\lambda) = \lim_{n\to\infty} F(\phi_{n,\lambda})$, namely,

$$F_n(\phi_{n,\lambda}) = \frac{1}{b_n} \int_{-\pi}^{\pi} K\left(\frac{\lambda - \alpha}{b_n}\right) I_n^{(h)}(\alpha)d\alpha.$$

Another important application is the minimization of the Whittle likelihood (Whittle [46, 47]). Suppose we have some parametric Gaussian time series model X_t^θ usually given in the time domain. An example is the classical ARMA(p, q)-model

$$\sum_{j=0}^{p} a_j X_{t-j} = \sum_{j=0}^{q} b_j \varepsilon_{t-j} \quad \text{with } a_0 = b_0 = 1 \tag{2.16}$$

with $\varepsilon_t \sim_{\text{i.i.d.}} \mathcal{N}(0, \sigma^2)$. The exact Gaussian likelihood estimate for this model is difficult to calculate. An often-used alternative is the minimization of the Whittle likelihood

$$\mathcal{L}_n(\theta) = \frac{1}{4\pi} \int_{-\pi}^{\pi} \left\{ \log f_\theta(\lambda) + \frac{I_n(\lambda)}{f_\theta(\lambda)} \right\} d\lambda \tag{2.17}$$

with $f_\theta(\lambda)$ being the parametric spectral density of the model, which in this ARMA(p, q)-case becomes

$$f_\theta(\lambda) = \frac{\sigma^2}{2\pi} \frac{|\sum_{j=0}^{q} b_j \exp(i\lambda j)|^2}{|\sum_{j=0}^{p} a_j \exp(i\lambda j)|^2},$$

where $\theta = (a_1, \dots, a_p, b_1, \dots, b_q, \sigma^2)'$. The minimization also has to be done numerically - but the effort is smaller, in particular, since it is not difficult to calculate the derivatives of the above likelihood. In the pure AR(p)-case $(q = 0)$ it can be shown that the minimization of (2.17) leads exactly to the Yule–Walker equations. We also mention that Kolmogorov's formula (cf. Brockwell and Davis [5], Theorem 5.8.1) gives

$$\frac{1}{4\pi} \int_{-\pi}^{\pi} \log f_\theta(\lambda) d\lambda = \frac{1}{2} \log \frac{\tau^2}{2\pi},$$

where τ^2 is the optimal one-step ahead prediction error, which in the above ARMA-case is identical to σ^2, i.e., the likelihood simplifies considerably.

The Whittle likelihood is an approximation of "$-\frac{1}{n}$ log Gaussian likelihood" and it can be shown that the approximation error is uniformly of order $O_p(n^{-1})$ (the same holds for the derivatives—cf. Azencott and Dacunha-Castelle [1]). It may be interpreted as a minimum distance functional between $I_n(\lambda)$ as a nonparametric spectral estimate and the parametric spectral density $f_\theta(\lambda)$.

If $\hat{\theta}_n = \operatorname{argmin}_{\theta \in \Theta} \mathcal{L}_n(\theta)$ denotes the resulting Whittle estimator, then heuristically $\hat{\theta}_n$ converges to $\theta_o = \operatorname{argmin}_{\theta \in \Theta} \mathcal{L}(\theta)$, where

$$\mathcal{L}(\theta) = \lim_{n \to \infty} \mathrm{E}\mathcal{L}_n(\theta) = \frac{1}{4\pi} \int_{-\pi}^{\pi} \left\{ \log f_\theta(\lambda) + \frac{f(\lambda)}{f_\theta(\lambda)} \right\} d\lambda \tag{2.18}$$

and $f(\lambda)$ is the spectral density of the true process. If the model is correctly specified, i.e., $f(\lambda) = f_{\theta^*}(\lambda)$ with some $\theta^* \in \Theta$ then it is easy to see that $\theta_o = \theta^*$. In passing we mention that $\mathcal{L}(\theta)$ is up to some constant the asymptotic Kullback–Leibler information divergence between two Gaussian stationary processes—cf. Parzen [29].

Under suitable regularity conditions we now get from the properties of the empirical spectral process (cf. Proposition 3.1)

$$\sup_{\theta \in \Theta} |\mathcal{L}_n(\theta) - \mathcal{L}(\theta)| = \sup_{\theta \in \Theta} \frac{1}{\sqrt{n}} \left| E_n \left(\frac{1}{4\pi} f_\theta^{-1} \right) \right| = o_P(1). \tag{2.19}$$

The result (2.19) implies

$$\mathcal{L}_n(\hat{\theta}_n) \le \mathcal{L}_n(\theta_o) \xrightarrow{P} \mathcal{L}(\theta_o) \le \mathcal{L}(\hat{\theta}_n).$$

Since $\mathcal{L}_n(\hat{\theta}_n) - \mathcal{L}(\hat{\theta}_n) \xrightarrow{P} 0$ we obtain $\mathcal{L}(\hat{\theta}_n) \xrightarrow{P} \mathcal{L}(\theta_o)$. If Θ is compact, $\mathcal{L}$ is continuous, and θ_o is unique, we get $\hat{\theta}_n \xrightarrow{P} \theta_o$. The usual Taylor argument now yields, with $|\bar{\theta} - \theta_o| \le |\hat{\theta}_n - \theta_o|$,

$$\nabla\mathcal{L}_n(\hat{\theta}_n) - \nabla\mathcal{L}_n(\theta_o) = \nabla^2\mathcal{L}_n(\bar{\theta})(\hat{\theta}_n - \theta_o).$$

We have $\nabla\mathcal{L}_n(\hat{\theta}_n) = o_P(1)$. Since

$$\sqrt{n}\nabla\mathcal{L}_n(\theta_o) = E_n\left(\frac{1}{4\pi}\nabla f_{\theta_o}^{-1}\right)$$

is asymptotically normal (see Theorem 3.2 ff.), and

$$\sup_{\theta\in\theta} |\nabla^2\mathcal{L}_n(\theta) - \nabla^2\mathcal{L}(\theta)| = \sup_{\theta\in\Theta} \frac{1}{\sqrt{n}} \left|\frac{1}{4\pi} E_n(\nabla^2 f_\theta^{-1})\right| = o_p(1),$$

we finally obtain the asymptotic normality of $\sqrt{n}(\hat{\theta}_n - \theta_o)$.

As a more sophisticated application of the theory of empirical spectral processes, we will investigate nonparametric Whittle estimates of the spectral density obtained by minimizing

$$\mathcal{L}_n(\phi) = \frac{1}{4\pi}\int_{-\pi}^{\pi} \left\{\log\phi(\lambda) + \frac{I_n(\lambda)}{\phi(\lambda)}\right\} d\lambda \tag{2.20}$$

over some suitable class of spectral densities $\phi \in \Phi$ (see Section 5).

At the end of this section we add some comments that are not directly related to the topics of the next sections:

(i) The relation to the empirical quantile function: if one sets $x_{(i)} := \frac{2\pi}{n}\sum_{0\le j\le i} I_n^{(h)}(\lambda_j) = \tilde{F}_n(\lambda_i)$ and regards $x_{(1)}, \ldots, x_{(n/2)}$ as the order statistic of some observations, then it is easy to see that $\tilde{F}_n(\lambda_i)$ is up to some shift the empirical quantile function of these observations.

(ii) Processes with long range dependence and heavy tails are not covered by our assumptions. For an overview on methods for long range dependent processes see Giraitis and Robinson [21] or Moulines and Soulier [26], and for processes with heavy tails see Mikosch [24].

(iii) Bootstrap in the frequency domain: The similarities between the spectral measure and the empirical quantile function indicate the possibility of constructing a nonparametric bootstrap in the frequency domain. Frequency domain bootstrap methods are reviewed in the paper by Paparoditis [28] in the present book.

3 The empirical spectral process

The function indexed empirical spectral process based on observations from a strictly stationary process has been introduced in (2.14) as $E_n(\phi) = \sqrt{n}(F_n(\phi) - F(\phi))$, $\phi \in \Phi$ where F and F_n denote theoretical and empirical spectral measure, respectively, and Φ is some indexing class of functions which can be viewed as a model for the spectral density of the underlying process.

In Dahlhaus [11] convergence in distribution of $(E_n(\phi))_{\phi\in\Phi}$ to a Gaussian process is proved under the assumption of an exponential moment condition. Mikosch and Norvaiša [25] prove a similar result for linear processes with existing fourth order moments. Fay and Soulier [20] prove results for nonlinear functionals of the periodogram and include the case of long range dependence.

In this paper we relax the entropy conditions on the indexing class Φ in comparison to the above-mentioned results. Parts of the results are special cases of Dahlhaus and Polonik [18] where the locally stationary case is treated (see also Section 6.2). We focus on Gaussian processes—the non-Gaussian case is briefly discussed in Section 6.1.

Let us first formulate a set of assumptions that we will use below without further mention:

(A0) All the random quantities under consideration are measurable,

(A1) $(X_t, t \in \mathbb{Z})$ is a strictly stationary Gaussian process,

(A2) $f \in Lip(\alpha)$, $\alpha > 1/2$, and

(A3) $\sup_n \sum_{t=1}^n |h_{t,n} - h_{t-1,n}| < \infty$ and $0 < H_2, H_4 < \infty$, where $H_k := \lim_{n\to\infty} \frac{1}{n} \sum_{t=1}^n h_{t,n}^k$.

We do not want to discuss measurability issues here, hence assumption (A0) is made. Measurability questions in fact can be avoided (almost) completely in this context, by using outer/inner measures and integrals and the Hoffmann-Jørgensen approach to weak convergence (Hoffmann-Jørgensen [22], van der Vaart and Wellner [42]). Another possibility would be to adopt Pollards approach and assume Φ to be "permissible" (Pollard [30]). Gaussianity in (A1) can be dropped, but then we obtain weaker results (see Section 6.1). Further we mention that some results can also be obtained under weaker conditions on the function f, however the price to pay is stronger conditions on Φ. Since in this paper we aim at greater generality on Φ we have assumed (A2). Assumption (A3) is standard for data tapers, and it is particularly fulfilled in the nontapered case $h_{t,n} = 1, t = 1, \dots, n$.

Let further

$$\rho_\infty(\phi) = \sum_{s=-\infty}^{\infty} |\widehat{\phi}(s)| \quad \text{and} \tag{3.1}$$

$$\rho_2(\phi) = \left(\int_{-\pi}^{\pi} |\phi(\lambda)|^2 d\lambda \right)^{1/2}, \tag{3.2}$$

denote two norms defined on Φ, where $\widehat{\phi}$ denotes the Fourier transform of ϕ, and let $\rho_k(\phi, \psi) := \rho_k(\phi - \psi)$, $k = \infty, 2$ denote the induced metrics. We assume that

(A4) $\sup_{\phi\in\Phi} \rho_\infty(\phi) < \tau_\infty < \infty$, and

(A5) $\sup_{\phi\in\Phi} \rho_2(\phi) < \tau_2 < \infty$.

Note that $\sup_\lambda |\phi(\lambda)| \le \frac{1}{2\pi} \rho_\infty(\phi)$.

Further assumptions on the indexing class Φ are formulated in terms of metric entropy with bracketing. For each $\epsilon > 0$, the bracketing covering number of Φ with respect to some metric ρ is defined as

$$\begin{aligned} N_B(\epsilon, \Phi, \rho) = \; & \inf\{m \ge 1 : \exists\, \psi_1, \dots, \psi_m \in \Phi \;\text{ such that} \\ & \forall\, \phi \in \Phi \;\exists\, 1 \le i, j \le m \text{ with } \psi_i \le \phi \le \psi_j \;\text{ and }\; \rho(\psi_j, \psi_i) < \epsilon\}. \end{aligned} \tag{3.3}$$

The quantity $H_B(\epsilon, \Phi, \rho) = \log N_B(\epsilon, \Phi, \rho)$ is called the *metric entropy with bracketing* of Φ with respect to ρ. By identifying sets with their indicator functions, this concept can be applied to classes of sets also. In this case bracketing reduces to *inclusion*. Several examples for classes with finite metric entropy are given below.

The "usual" covering number without the bracketing condition, $N(\epsilon, \Phi, \rho)$, equals the smallest number of balls of radius $\leq \epsilon$ (w.r.t. the metric ρ) necessary to cover Φ. By definition $N(\epsilon, \Phi, \rho) \leq N_B(\epsilon, \Phi, \rho)\ \forall \epsilon > 0$. In fact, often both covering numbers are of the same order, and can be bounded by similar quantities. Using the bracketing condition has some technical advantages, and we formulate all our results by using assumptions on the metric entropy with bracketing, like finiteness, or integrability conditions. Since the metric entropy is nonincreasing in ϵ, validity of these conditions solely depends on the behaviour of $H_B(\epsilon, \Phi, \rho)$ for *small* $\epsilon > 0$.

The following important property is often exploited for deriving bounds for covering numbers or metric entropies, respectively. The proof is straightforward. Let ρ and ν be two metrics; then we have the implication

$$\exists c > 0 \quad \text{s.th.} \quad \rho(f, g) \leq c\, \nu(f, g)\ \ \forall f, g \in \Phi \quad \Rightarrow \quad N_B(c\,\epsilon, \Phi, \rho) \leq N_B(\epsilon, \Phi, \nu). \tag{3.4}$$

Now we formulate two results on the asymptotic behavior of the empirical spectral process. Let $\rho_{1,F}$ denote the $L_1(F)$-norm, i.e., $\rho_{1,F}(g, h) = \int_{-\pi}^{\pi} |g(z) - h(z)|\, dF(z)$, then we have the following Glivenko–Cantelli-type result.

Proposition 3.1. *If*

$$N_B(\epsilon, \Phi, \rho_{1,F}) < \infty \quad \forall\, \epsilon > 0, \tag{3.5}$$

then $\sup_{\phi \in \Phi} |(F_n - F)(\phi)| \to 0$ *in probability as* $n \to \infty$.

Remarks. (i) Assumption (3.5) implies total boundedness of Φ as a subset of $L_1(F)$.

(ii) The assertion of Proposition 3.1 actually holds under considerably weaker conditions than (A0)–(A5), but finding the weakest conditions is not of our interest here.

The proof of this result is a straightforward adaptation of analogous results for the function-indexed empirical process based on i.i.d. data (see Pollard [30]). There are two reason that the proof can be transferred easily. One is given by the fact that F_n and F are positive integrals (and hence, linear and monotone) just like their counterparts from the empirical process based on i.i.d. data, and the second reason is the availability of a Bernstein-type exponential inequality for $E_n(\phi)$ (see Section 4.)

The perhaps easiest example for a class Φ satisfying (3.5) is the class of intervals $\mathcal{I}_0$ starting at 0, i.e., $\mathcal{I}_0 = \{[0, \lambda], 0 \leq \lambda \leq \pi\}$ (identified with their indicator functions, i.e., $[0, \lambda] = 1_{[0,\lambda]}$). A finite cover of $\mathcal{I}_0$ of balls (in $L_1(F)$) that also satisfies the bracketing condition is constructed easily. For $\epsilon > 0$ let $k_j = F^{-1}(j\,\epsilon)$ for $j = 0, 1, \ldots, N - 1$, where N denotes the smallest integer such that $N\,\epsilon \geq F(\pi) = c(0)/2$, and put $k_N = \pi$. Then $\{[0, k_j],\ j = 0, \ldots, N\}$ defines such a cover, and it follows that $N_B(\epsilon, \mathcal{I}_0, \rho_{1,F}) \leq 2\,N \leq A\frac{1}{\epsilon}$ for some constant $A > 0$.

More examples are presented below. Now recall the definition of ρ_2 given in (3.2).

Theorem 3.2. *Suppose that*

$$\int_0^1 \Big(H_B(\epsilon, \Phi, \rho_2) \Big)^{1/2} d\epsilon < \infty. \tag{3.6}$$

Then $(E_n(\phi))_{\phi \in \Phi}$ *is asymptotically equicontinuous, i.e., for all* $\eta, \epsilon > 0$ *there exists a* $\delta > 0$ *such that*

$$\limsup_{n \to \infty} P(\sup_{[\delta]} |E_n(\phi - \psi)| > \eta) < \epsilon,$$

where $[\delta] = \{(\phi, \psi) \big| \phi, \psi \in \Phi \quad$ *and* $\quad \rho_2(\phi - \psi) < \delta\}$.

It is known that Assumptions (A1)–(A5) imply convergence of the finite dimensional distributions of the empirical spectral process. (This can be seen by using arguments similar to the one of the proof of Theorem 2 in Dahlhaus [11].) Thus Theorem 3.2 entails weak convergence (for instance, in the Hoffmann-Jørgensen sense) of the empirical spectral process $(E_n(\phi))_{\phi\in\Phi}$ to a mean zero Gaussian process $(E(\phi))_{\phi\in\Phi}$, i.e.,

$$(E_n(\phi))_{\phi\in\Phi} \xrightarrow{\mathcal{D}} (E(\phi))_{\phi\in\Phi} \quad \text{as } n \to \infty, \tag{3.7}$$

where $\mathrm{E}(E(\phi)) = 0\ \forall \phi \in \Phi$ and

$$\operatorname{cov}(E(\phi), E(\psi)) = 2\pi \frac{H_4}{H_2^2} \int_{-\pi}^{\pi} \phi(\lambda) \left(\overline{\psi(\lambda)} + \overline{\psi(-\lambda)}\right) f^2(\lambda) d\lambda.$$

It is worth pointing out that assumption (3.6) in Theorem 3.2 is weaker than the entropy assumptions in Dahlhaus [11] and Mikosch and Norvaiša [25]. Finiteness of a covering integral, like (3.6), is also used as an assumption in Dahlhaus and Mikosch & Norvaiša, respectively, but there the metric entropy enters with powers larger than $1/2$. The reason that we can use the power $1/2$ in (3.6) comes from the fact that we have a good exponential inequality for the empirical spectral process (cf. Section 6.1 below). We would like to mention that, although the use of metric entropy with bracketing has some technical advantages, it is not always necessary to do so. In fact, Mikosch and Norvaiša use entropy without bracketing.

Upper bounds for covering numbers. Often entropy bounds are given w.r.t. an $L_2(Q)$-metric, where Q is a probability measure. These bounds entail an entropy bound w.r.t. the $L_1(Q)$-metric (see (3.4)).

(i) First we consider a class that is *continuously parametrized*. The following result can be found in van de Geer [40], Lemma 3.10. Suppose $\mathcal{G}_\Theta = \{g_\theta : \theta \in \Theta\}$ where (Θ, τ) is a compact metric space, such that the map $\theta \to g_\theta$ is continuous with respect to the sup-norm. If in addition the envelope $G = \sup_{\theta\in\Theta} |g_\theta|$ is in $L_1(F)$, then $H_B(\epsilon, \mathcal{G}_\Theta, \rho_{1,F}) < \infty,\ \forall \epsilon > 0$. As a result, Proposition 3.1 holds for such continuously parametrized classes. If in addition to the above we know that the number of balls (in the metric τ) of radius at most ϵ necessary to cover Θ is of the order $\epsilon^{-\alpha}$ for some $\alpha > 0$, then there exists a constant $A > 0$ such that

$$H_B(\epsilon, \mathcal{G}_\Theta, \rho_{1,F}) \le A \log \frac{1}{\epsilon}. \tag{3.8}$$

Examples for parameter spaces satisfying these conditions are for instance intervals on the real line, as well as rectangles, balls, and ellipsoids in R^d, which consequently all satisfy the conditions of Theorem 3.2.

(ii) Let $\mathcal{G}_m$ denote the Sobolev classes of bounded, smooth functions, defined as

$$\mathcal{G}_m = \left\{ g : [0,1] \to [0,1],\ \int_0^1 \left(g^{(m)}(z)\right)^2 dz < \infty \right\},$$

where $g^{(m)}$ denotes the m-th derivative of g. Let further ρ_Q denote the $L_2(Q)$-norm for a probability measure Q, i.e., for square integrable functions f, g we have $\rho_Q^2(f,g) = \int (f - g)^2 dQ$. It is known that there exists a constant $A > 0$ such that for all $\epsilon > 0$,

$$H_B(\epsilon, \mathcal{G}_m, \rho_Q) \le A\, \epsilon^{-1/m}. \tag{3.9}$$

In fact, this result holds for any $L_p(Q)$-norm with $p \geq 1$. Further, it is clear that the interval $[0,1]$ in the definition of $\mathcal{G}_m$ can be replaced by any finite interval.

(iii) Another example is given by the class of bounded, monotonic functions $\mathcal{G}_{inc} = \{g : R \to [0,1],\ g \text{ increasing}\}$. For this class there exists a constant $A > 0$ such that for any probability measure Q we have for all $\epsilon > 0$,

$$H_B(\epsilon, \mathcal{G}_{inc}, \rho_Q) \leq A\,\frac{1}{\epsilon}. \tag{3.10}$$

(iv) The result in (iii) can be used to derive the metric entropy for the class of bounded total variation, where for a function g the total variation may formally be defined as $TV(g) = \int |dg|$. Since any function of bounded total variation can be written as a difference of two monotonic functions, it follows from (3.10) together with an application of the triangular inequality that for $\mathcal{G}_{bTV} = \{g : R \to [0,1], TV(g) < \infty\}$ we also have $H_B(\epsilon, \mathcal{G}_{bTV}, \rho_Q) \leq A\frac{1}{\epsilon}$ for all $\epsilon > 0$ and a constant $A > 0$.

Statements of these and other results on metric entropies can also be found in van de Geer [40]. Dudley [19] is another source for metric entropy bounds.

The classes discussed in (ii)–(iv) are "large" in the sense that their metric entropy as a function in ϵ behaves like a (positive) power of $1/\epsilon$. In (i) we presented an example for a much smaller class, where the covering numbers themselves rather than their logarithm (the entropy) behave like (positive) powers of $1/\epsilon$. This holds more generally for many so-called Vapnik–Chervonenkis (VC) subgraph classes of functions. (This fact and a definition of VC-subgraph classes can be found in all textbooks on empirical processes.)

Theorem 3.2 has several *applications in time series analysis* (cf. Dahlhaus [11]). As demonstrated in Section 2 it leads to a simple proof of the asymptotic normality of Whittle estimates. An application which goes beyond classical asymptotics is the derivation of the asymptotic distribution of test statistics for composite hypothesis (cf. Example 3.4 in Dahlhaus [11]).

4 Exponential inequalities

The crucial result in this section is a maximal inequality for $(E_n(\phi))_{\phi\in\Phi}$ presented below. This exponential inequality is the main ingredient to the proof of Theorem 3.2, the convergence in distribution of the process $(E_n(\phi))_{\phi\in\Phi}$ to a Gaussian process. In Section 5 we exploit the maximal inequality for deriving rates of convergence of nonparmetric MLEs of the spectral density. A more general theory is presented in Dahlhaus and Polonik [18].

We assume that all of the moments of the stationary distribution of our process exist (as for instance for a Gaussian process). The cumulants of a random variable X are denoted by

$$\mathrm{cum}_j(X) = j\text{-th cumulant of } X. \tag{4.1}$$

They are defined as the coefficients of the power series of the log-characteristic function, i.e.,

$$\log \mathrm{E}\exp\{isX_t\} = \sum_{j=1}^{\infty} \frac{(is)^j}{j!}\,\mathrm{cum}_j(X_t), \qquad s \in R.$$

There is a one-to-one correspondence between the cumulants and the moments of a random variable. For instance, $\mathrm{cum}_1(X_t) = \mathrm{E}X_t$ and $\mathrm{cum}_2(X_t) = \mathrm{var}(X_t)$. Utilizing the cumulants we now derive an exponential inequality for a sequence of random variables S_n, $n = 1, 2, \ldots,$ that fulfill the following:

(B1) $|\mathrm{E}(S_n)| \le C\,\rho_2$, or, alternatively $|E(S_n)| \le \frac{1}{\sqrt{n}} C\,\rho_\infty$, and

(B2) $|\,\mathrm{cum}_j(S_n)| \le C^j\, j!\, n^{-j/2} n \rho_\infty^{j-2} \rho_2^2,\ j \ge 2$,

with some constants $C, \rho_\infty, \rho_2 > 0$. In Lemma 4.2 below we state that in the Gaussian case these conditions are satisfied for $S_n = E_n(\phi)$ and the norms from (3.1) and (3.2).

Lemma 4.1. *Suppose that S_n satisfies* (B1) *and* (B2). *Then we have for $\eta > 0$ that*

$$P(|S_n| \ge \eta) \le (2+e)\exp\left(-\frac{1}{2}\frac{\eta^2}{8C^2\rho_2^2 + \frac{2C\rho_\infty\eta}{\sqrt{n}}}\right). \tag{4.2}$$

Here "e" denotes the Euler constant. This Bernstein-type inequality is utilized in the proof of Proposition 3.1 in order to derive the almost sure convergence by an application of the Borel–Cantelli lemma. The crucial maximal inequality presented below also heavily rests on (4.2).

We only present an outline of the proof. A more direct proof for the Gaussian case is given in Dahlhaus and Polonik [18], Appendix 1. Let $S_n^* = S_n - \mathrm{E}S_n$. Under (B2) the characteristic function can uniquely be extended to $s \in \mathbb{C}$, and we obtain

$$\begin{aligned}
P(S_n^* \ge \eta) &\le \exp(-t\eta)\mathrm{E}\exp(tS_n^*)\\
&= \exp(-t\eta)\exp\left(\sum_{j=1}^{\infty}\frac{t^j}{j!}\,\mathrm{cum}_j(S_n^*)\right)\\
&= \exp(-t\eta)\exp\left(\sum_{j=2}^{\infty}\frac{t^j}{j!}\,\mathrm{cum}_j(S_n)\right) && (4.3)\\
&\le \exp(-t\eta)\exp\left(C^2t^2\rho_2^2\sum_{j=2}^{\infty}\left(\frac{tC\rho_\infty}{\sqrt{n}}\right)^{j-2}\right) && (4.4)\\
&= \exp(-t\eta)\exp\left(C^2t^2\rho_2^2\frac{1}{1-\frac{Ct\rho_\infty}{\sqrt{n}}}\right). && (4.5)
\end{aligned}$$

Equality (4.3) follows from $\mathrm{cum}_1(X) = \mathrm{E}X$ and $\mathrm{cum}_j(X) = \mathrm{cum}_j(X - \mathrm{E}X)$ for $j \ge 2$. Equality (4.4) exploits assumption (B2), and for (4.5) we need $n > t^2C^2\rho_\infty^2$. The latter can be seen to hold for the choice $t = \eta/(2C^2\rho_2^2 + \frac{C\rho_\infty\eta}{\sqrt{n}})$. This choice leads one to an exponential bound for $P(S_n^* \ge \eta)$. In order to arrive at the asserted inequality for S_n instead of S_n^* we exploit the triangle inequality and (B1). □

Recall the two metrics defined above in (3.1) and (3.2).

Lemma 4.2. *If the process $(X_t, t \in \mathbb{Z})$ is Gaussian, then there exists a constant $C > 0$ such that* (B1) *and* (B2) *hold for $S_n = E_n(\phi)$ with $\rho_\infty = \rho_\infty(\phi)$ and $\rho_2 = \rho_2(\phi)$.*

The exponential inequality follows also as a special case from Theorem 4.4 in Dahlhaus and Polonik [18] where the locally stationary case has been considered. (Note that $\tilde{v}(\phi) = 0$ in the notation of that paper).

Lemma 4.2 enables us to exploit the exponential inequality presented in Lemma 4.1, which leads to the following crucial maximal inequality. Recall the constants τ_2, and τ_∞ from assumptions (A4) and (A5), respectively. Note that the bound τ_2^2 is the analog to a bound for the variance that is used in similar results for the classical empirical process.

Theorem 4.3. *Let* $K = \frac{\tau_\infty}{\pi}$. *There exists a constant* $M > 0$ *such that for all* $\eta > 0$ *and* $C_0, C_1 > 0$ *satisfying*

$$\eta \le \frac{C_1 \sqrt{n} \tau_2^2}{2CK}, \tag{4.6}$$

$$\eta \le 8\sqrt{n}\tau_2, \tag{4.7}$$

$$\eta \ge C_0 \left(\int_{a/(2^6\sqrt{n})}^{\tau_2} (H_B(u, \Phi, \rho_2))^{1/2}\, du \ \vee \tau_2 \right), \tag{4.8}$$

and

$$C_0^2 \ge M^2(8C + C_1) \tag{4.9}$$

with $C > 0$ *from Lemma 4.2, the following inequality holds:*

$$P\left(\sup_{\phi \in \Phi} |E_n(\phi)| > \eta \right) \le M \exp\left\{ -\frac{1}{2} \frac{\eta^2}{M^2(8C + C_1)\tau_2^2} \right\}. \tag{4.10}$$

This theorem is a special case of a theorem in Dahlhaus and Polonik [18] that treats the locally stationary case (see Section 6.2). It is similar in style to Theorem 5.11 of van de Geer [40], but van de Geer treats the usual empirical process based on i.i.d. observations (she also provides a martingale version of this theorem). The proof of this theorem is adapted to van de Geer's proof of her Theorem 5.11. All this is possible, because the empirical spectral process and the usual empirical process are very similar in nature, as has been mentioned above already. Moreover, the exponential inequality (4.2) is very close to the Bernstein inequality that is exploited in the empirical process case.

The crucial assumption which eventually determines rates of convergence of nonparametric MLE's (see Section 5) is (4.8). In order to obtain good (or optimal) rates, good bounds for the metric entropy with bracketing need to be available. Some examples for such bounds are provided above.

5 Nonparametric maximum likelihood estimation

In Section 2 we already discussed parametric minimum Whittle likelihood estimation. Here we show how the results from the previous sections can be utilized to arrive at consistency results and (optimal) rates of convergence for nonparametric MLE's.

Analogously to (2.17) and (2.18) we denote the nonparametric Whittle likelihood and its theoretical counterpart as

$$\mathcal{L}_n(\phi) = \frac{1}{4\pi} \int_{-\pi}^{\pi} \left\{ \log \phi(\lambda) + \frac{I_n(\lambda)}{\phi(\lambda)} \right\} d\lambda \tag{5.1}$$

and

$$\mathcal{L}(\phi) = \lim_{n\to\infty} \mathrm{E}\mathcal{L}_n(\phi) = \frac{1}{4\pi} \int_{-\pi}^{\pi} \left\{ \log \phi(\lambda) + \frac{f(\lambda)}{\phi(\lambda)} \right\} d\lambda, \tag{5.2}$$

respectively. Here $\phi \in \Phi$, where Φ is a class of positive functions that serves as a model for the true underlying spectral density f.

Further, let $\widehat{\phi}_n = \text{argmin}_{\phi\in\Phi} \mathcal{L}_n(\phi)$ be the nonparametric (Whittle) likelihood estimator in Φ, and let $\phi_{opt} = \text{argmin}_{\phi\in\Phi} \mathcal{L}(\phi)$ denote its theoretical counterpart. If the model Φ is correctly specified, i.e., if $f = \phi_0$ for some $\phi_0 \in \Phi$, then we have $\phi_{opt} = f$. This can be seen easily. In general, however, $\widehat{\phi}_n$ will converge to ϕ_{opt} rather than to f. By definition of $\widehat{\phi}_n$ and ϕ_{opt}, we have

$$\mathcal{L}_n(\widehat{\phi}_n) \leq \mathcal{L}_n(\phi_{opt}) \tag{5.3}$$

and, similarly,

$$\mathcal{L}(\phi_{opt}) \leq \mathcal{L}(\widehat{\phi}_n). \tag{5.4}$$

Combining (5.3) and (5.4) we obtain

$$\begin{aligned} 0 &\leq \mathcal{L}(\widehat{\phi}_n) - \mathcal{L}(\phi_{opt}) \leq \big(\mathcal{L}_n - \mathcal{L}\big)(\phi_{opt}) - \big(\mathcal{L}_n - \mathcal{L}\big)(\widehat{\phi}_n) && (5.5)\\ &= \frac{1}{4\pi}\frac{1}{\sqrt{n}} E_n\left(\frac{1}{\phi_{opt}} - \frac{1}{\widehat{\phi}_n}\right) = \frac{1}{4\pi}\big(F_n - F\big)\left(\frac{1}{\phi_{opt}} - \frac{1}{\widehat{\phi}_n}\right) \\ &\leq \frac{1}{4\pi}\sup_{\phi\in\Phi}\big(F_n - F\big)\left(\frac{1}{\phi_{opt}} - \frac{1}{\phi}\right). && (5.6)\end{aligned}$$

The quantity $\mathcal{L}(\widehat{\phi}_n) - \mathcal{L}(\phi_{opt})$ can be interpreted as a distance between ϕ_{opt} and $\widehat{\phi}_n$. This becomes more obvious in case the model is correctly specified, i.e., $\phi_{opt} = f$. In this case we have

$$\mathcal{L}(\widehat{\phi}_n) - \mathcal{L}(f) = \frac{1}{4\pi}\int_{-\pi}^{\pi} \log\left(\frac{\widehat{\phi}_n(\lambda)}{f(\lambda)}\right) + \left(\frac{f(\lambda)}{\widehat{\phi}_n(\lambda)} - 1\right) d\lambda \tag{5.7}$$

which is the Kullback–Leibler divergence between two spectral densities (see Parzen [29]). Another way to write this is the following. For a function $g : [-\pi, \pi] \to R$ with $g\,f < 1$ define

$$\|g\|_f := \int_{-\pi}^{\pi} \log\left(\frac{1}{1 - g(\lambda)\,f(\lambda)}\right) - g(\lambda)\,f(\lambda)\,d\lambda. \tag{5.8}$$

Since $\log(1/x) \geq 1 - x,\ x > 0$, we have $\|g\|_f \geq 0$, and hence this is a reasonable distance measure. Now we have

$$\mathcal{L}(\widehat{\phi}_n) - \mathcal{L}(f) = \left\|\frac{1}{f} - \frac{1}{\widehat{\phi}_n}\right\|_f. \tag{5.9}$$

The above reformulations show that if $f \in \Phi$, then

$$0 \leq \left\|\frac{1}{f} - \frac{1}{\widehat{\phi}_n}\right\|_f \leq \frac{1}{4\pi}\sup_{\phi\in\Phi}\left|(F_n - F)\left(\frac{1}{f} - \frac{1}{\phi}\right)\right|. \tag{5.10}$$

Let $\Phi^* = \left\{\frac{1}{f} - \frac{1}{\phi};\, \phi \in \Phi\right\}$. Combining the last inequality with Proposition 3.1 immediately gives the following consistency result.

Proposition 5.1. *If $f \in \Phi$ and $N_B(\epsilon, \Phi^*, \rho_{1,F}) < \infty\ \forall \epsilon > 0$, then*

$$\left\|\frac{1}{f} - \frac{1}{\widehat{\phi}_n}\right\|_f \to 0 \quad \textit{in probability as } n \to \infty.$$

Examples. Note that we consider classes of functions of the form $1/\phi$ so that we have to be careful if ϕ becomes small. Here we will assume that the true spectral density f is bounded away from 0 by a constant $\gamma > 0$. Hence, we also assume that the functions in the model class have the same lower bound. For instance, consider the Sobolev classes $\mathcal{G}_m$ discussed in Section 3. The idea is to truncate the functions $\phi \in \mathcal{G}_m$ at level γ, to obtain $\phi_\gamma = \phi\, 1\{\phi \geq \gamma\} + \gamma\, 1\{\phi < \gamma\}$. Hence our model class is

$$\mathcal{G}_{\gamma,m} = \{\phi_\gamma;\ \phi \in \mathcal{G}_m\}.$$

We now show that the bounds for the metric entropy of $\mathcal{G}_m$ can essentially (up to constants) be used as bounds for the metric entropy of $\mathcal{G}^*_{\gamma,m}$. We have

$$\left|\frac{1}{\phi_\gamma} - \frac{1}{\psi_\gamma}\right| = \left|\frac{\psi_\gamma - \phi_\gamma}{\psi_\gamma\, \phi_\gamma}\right| \leq \frac{1}{\gamma^2}|\psi_\gamma - \phi_\gamma|.$$

By utilizing (3.4), we obtain $N_B(\epsilon, \mathcal{G}^*_{\gamma,m}, \rho_{1,F}) \leq N_B(\gamma^2\epsilon, \mathcal{G}_m, \rho_{1,F}) \leq \exp(A\,\epsilon^{-m})$. (Combine (3.4) and (3.9) to get the upper bound).

Similar arguments of course hold for other classes like the class of bounded monotonically increasing functions $\mathcal{G}_{inc}$ defined above. Similar to the smoothness classes just discussed, we can define the corresponding truncated class $\mathcal{G}_{\gamma,inc}$, and we obtain the bound

$$H_B(\epsilon, \mathcal{G}^*_{\gamma,inc}, \rho_{1,F}) \leq H_B(\gamma^2\epsilon, \mathcal{G}_{inc}, \rho_{1,F}) < A\frac{1}{\epsilon} < \infty. \tag{5.11}$$

The distance measure $\|\cdot\|_f$ is not very tractable. We would rather like to use an L_2-metric, because bracketing numbers in terms of this distance measure are needed in order to derive the rates of convergence (see below). Fortunately, if Proposition 5.1 holds, then $\|\cdot\|_f$ can be estimated by the L_2-metric from below. This is not a surprise, because the above reformulations show (see (5.7) and (5.9)), that $\left\|\frac{1}{\widehat{\phi}_n} - \frac{1}{f}\right\|_f = \int \Psi(f/\widehat{\phi}_n)$ where $\Psi(x) = \log(1/x) + (x-1)$. Now, the function Ψ has a minimum at $x = 1$, and hence, if f is (uniformly) close to $\widehat{\phi}_n$, then Ψ behaves like $\frac{1}{2}(1-x)^2$. This heuristics can be made precise, and leads to the following lemma. Recall the definition of ρ_2 given in (3.2).

Lemma 5.2. *Suppose that $\widehat{\phi}_n$ and f are continuous with $\left\|\frac{1}{\widehat{\phi}_n} - \frac{1}{f}\right\|_f \to 0$ in probability as $n \to \infty$. Then on a set with probability tending to 1 as $n \to \infty$, we have*

$$0 \leq \frac{1}{8}\,\rho_2^2\left(\frac{f}{\widehat{\phi}_n},\ 1\right) \leq \left\|\frac{1}{\widehat{\phi}_n} - \frac{1}{f}\right\|_f \leq \sup_{\phi\in\Phi}\left|(F_n - F)\left(\frac{1}{\phi} - \frac{1}{f}\right)\right|. \tag{5.12}$$

Hence, we have an estimate of the ρ_2-distance of the MLE and the true spectral density in terms of the empirical spectral process. This estimate enables us to derive rates of convergence for the MLE depending on L_2-bracketing numbers (or entropy). Similar inequalities are exploited in van de Geer [40] to derive rates of convergence of nonparametric MLE in various models. The basic idea is as follows. On the set where (5.12) holds, we have

$$\left\{\rho_2\left(\frac{f}{\widehat{\phi}_n},\ 1\right) \geq L\delta_n\right\} \subset \left\{\sup_{\phi\in\Phi;\ \rho_2\left(\frac{f}{\phi},\,1\right)\geq L\,\delta_n} \frac{\left|(F_n - F)\left(\frac{1}{\phi} - \frac{1}{f}\right)\right|}{\frac{1}{8}\rho_2^2\left(\frac{f}{\phi},\ 1\right)} \geq 1\right\}.$$

Eventually, this leads us to

$$P\left(\rho_2\left(\frac{f}{\widehat{\phi}_n},\,1\right)\geq L\delta_n\right)\leq P\left(\sup_{\phi\in\Phi;\ \rho_2\left(\frac{f}{\phi},\,1\right)\geq L\,\delta_n}\frac{\left|(F_n-F)\left(\frac{1}{\phi}-\frac{1}{f}\right)\right|}{\frac{1}{8}\rho_2^2\left(\frac{f}{\phi},\,1\right)}\geq 1\right)+o_P(1). \tag{5.13}$$

The $o_P(1)$-term comes from the exceptional set where (5.12) does not hold. In order to derive good rates of convergence for $\rho_2(\frac{f}{\widehat{\phi}_n},1)$ we want to find the smallest sequence δ_n such that the probability on the right hand side in (5.13) tends to zero as L gets large. This can be done by utilizing the so-called peeling device (e.g., van de Geer [40]) and then applying our maximal inequality given in Theorem 4.3.

The peeling device enables us to rewrite the probability on the right-hand side of (5.13) in an appropriate form, so that we can apply our maximal inequality. Note that the supremum in the probability on the right-hand side of (5.13) is extended over functions where ρ_2 is *bigger* than some quantity, whereas our maximal inequality is for suprema over functions with norm *less than* some quantity. We have

$$\begin{aligned}
&P\left[\sup_{g,h\in\Phi^*;\ \rho_2(\frac{g}{h},1)\geq L\delta_n}\frac{|E_n(g-h)|}{\rho_2^2\left(\frac{g}{h},1\right)}\geq\frac{\sqrt{n}}{8}\right]\\
&\leq\sum_{j=0}^{\infty}P\left[\sup_{L\,2^{j+1}\delta_n\geq\rho_2(\frac{g}{h},1)\geq L\,2^j\delta_n}\frac{|E_n(g-h)|}{\rho_2^2\left(\frac{g}{h},1\right)}\geq\frac{\sqrt{n}}{8}\right]\\
&\leq\sum_{j=0}^{\infty}P\left[\sup_{\rho_2(\frac{g}{h},1)\leq L\,2^{j+1}\delta_n}|E_n(g-h)|\geq\frac{\sqrt{n}\,\left(L2^j\delta_n\right)^2}{8}\right].
\end{aligned} \tag{5.14}$$

Applying our maximal inequality (4.10) to the last expression we get a sum of exponentials. As already mentioned, we utilize our maximal inequality presented in Theorem 4.3. It turns out that the crucial quantity is the covering integral that appears on the right-hand side of (4.8).

In what follows we only consider classes with finite covering integral, so that the lower limit in (4.8) plays no role (asymptotically) and is therefore set equal to 0. We define

$$J_B(\delta,\Phi^*)=\int_0^\delta(H_B(u,\Phi^*,\rho_2))^{1/2}du\vee\delta.$$

Then the following theorem presents the connection between the behavior of J_B and the rates of convergence of the MLE. Again this theorem is an analog to results of van de Geer [40].

Theorem 5.3. *Let $M(\delta,\Phi^*)\geq J_B(\delta,\Phi^*)$ such that $M(\delta)/\delta^2$ is a nonincreasing function of δ. Then if $f\in\Phi$, there exists a constant $c>0$ such that for δ_n satisfying*

$$\sqrt{n}\,\delta_n^2\geq c\,M(\delta_n,\Phi^*), \tag{5.15}$$

we have

$$\rho_2\left(\frac{f}{\widehat{\phi}_n},\,1\right)=O_P(\delta_n).$$

It remains to calculate $M(\delta,\Phi^*)$ for the special classes Φ under consideration and to find δ_n that satisfies (5.15). We now discuss some examples.

Finite covering integral. Note that if we just know that the covering integral is finite, i.e., (3.6) for Φ^*, then we can take $M(\delta, \Phi^*) \equiv const.$ and Theorem 5.3 gives us a rate of $O_P(n^{-1/4})$. Examples for classes with finite covering integrals have been presented in Section 3. (For examples of classes with infinite covering integrals see below.)

VC-subgraph classes. We did not define VC-subgraph classes formally, but we mentioned already (see examples presented at the end of Section 3) that if Φ^* is a VC-subgraph class, then often $N_B(u, \Phi^*, \rho_2) \leq Au^{-\alpha}$, for some $\alpha > 0$. Another example was presented at the end of Section 3. This is sometimes also referred to as the finite-dimensional (or the parametric) case. Note that this is a condition on the covering numbers itself rather than on the entropy which is the log-covering number. For such classes we have

$$J_B(\delta, \Phi^*) \leq C\, \delta\sqrt{\log(1/\delta)} \quad \text{for some } C > 0. \tag{5.16}$$

It follows from Theorem 5.3 that $\delta_u = \sqrt{\frac{\log u}{n}}$, which up to the log-factor is the optimal (parametric) rate of convergence.

Smoothness classes. As we have seen in Section 3 the smoothness classes $\mathcal{G}_{\gamma,m}$ satisfy $H_B(u, \Phi^*, \rho_2) \leq Au^{-\beta}$, for some $\beta > 0$. For such classes we have

$$J_B(\delta, \Phi^*) \leq C\, \delta^{1-\frac{\beta}{2}} \quad \text{for some } C > 0. \tag{5.17}$$

If $0 < \beta < 2$ then the covering integral (3.6) is finite, and the resulting rate for the corresponding MLE is $O_P(n^{-\frac{1}{2+\alpha}})$. For instance, if we use the classes of truncated smooth functions $\mathcal{G}_{\gamma,m}$ introduced above, then we have $\alpha = 1/m$ (see example given after Proposition 5.1), and we end up with the well-known rate $O_P(n^{-\frac{m}{2m+1}})$. For the class of (truncated) monotonically increasing functions $\mathcal{G}_{\gamma,inc}$ we have $\alpha = 1$ so that we obtain the well-known rate $O_P(n^{-1/3})$.

For very large classes with infinite covering integral a more general version of J_B with nonzero lower integration limit can be used to obtain a rate of convergence, which, however, is typically nonoptimal. The method of sieves can be utilized to obtain better rates in this situation.

6 Extensions

6.1 The non-Gaussian case. In this section we indicate how the above results can be extended to the non-Gaussian case, and why the results in this case usually are not as strong as in the Gaussian case.

The crucial point for the above mathematical considerations is the availability of an exponential inequality (cf. (4.2)). This inequality is actually the only point where the Gaussian assumption enters, and it enters because assumptions (B1) and (B2) can be shown to be satisfied for an underlying Gaussian process. In the non-Gaussian case it seems possible to prove (B2) for $S_n = E_n(\phi)$ with $(2j)!$ instead of $j!$ which leads to a worse exponential inequality, and consequently to slower rates of convergence (the $(2j)!$ comes up since the product theorem for cumulants of a quadratic statistic leads up to $(2j)!$ different cumulant terms—cf. Dahlhaus [11], p. 82). We now indicate why this leads to slower rates of convergence.

First note that inequalities of type (4.2) are called sub-Gaussian, because as is well known, the tails of Gaussian random variables behave like $\exp\left(-(K/\sigma)^2\right)$ as $K \to \infty$, where σ^2 denotes the variance of the Gaussian random variable. Note that inequality (4.2) is not exactly of this type, because we have a second term in the denominator in the exp-term on the r.h.s. However, if τ_2^2 is not too small compared to η^2/n (as made precise in (4.6)) this second term in the denominator is negligible. In fact, if (4.6) and (A5) are incorporated into (4.2) we obtain an inequality

$$P(|S_n| > \eta) \leq c_1 \exp\left(-c_2\left(\frac{\eta}{\rho_2}\right)^2\right), \tag{6.1}$$

which actually is exactly of sub-Gaussian type.

In the non-Gaussian case we only have an exponential inequality where essentially the exponent "2" on the r.h.s. of (6.1) is replaced by "1," so that we have an exponential bound of the type $P(|S_n| > \eta) \leq c_1 \exp(-c_2 \frac{\eta}{\rho_2})$. It is an open question whether this can be improved upon or not, meaning whether it is possible in the non-Gaussian case to derive an exponential inequality with an exponent bigger than 1. We also want to mention here that "2" is the best exponent that we can hope to get, because the empirical spectral process becomes asymptotically normal, and hence its tails will behave like Gaussian tails.

In any case, a smaller exponent in the exponential bound has the effect that our methods lead to slower (upper bounds) for the rates of convergence of our estimators. The reason is that by means of the chaining technique there is a close connection between the exponent in the exponential bound and the covering integral which appears in (4.8) in Theorem 4.3, and the behavior of the covering integral in turn is responsible for the rate of convergence of our estimator (compare the role of $J_B(\delta, \Phi^*)$ in Theorem 5.3). Note that by our definition of the covering integral it is the integral of the *square root* of the metric entropy, in other words we integrate $(H_B(u, \mathcal{F}, \rho_2))^{1/2}$ where $\mathcal{F}$ is an appropriate class of functions. An exponential bound of the type

$$P(|S_n| > \eta) \leq c_1 \exp\left(-c_2\left(\frac{\eta}{\rho_2}\right)^\alpha\right) \tag{6.2}$$

would lead to a different covering integral of the type

$$\int_0^\delta (H_B(u, \mathcal{F}, \rho_2))^{1/\alpha}\, du.$$

Hence, the smaller α (in the exponential bound) the stronger the conditions on the metric entropy and consequently the slower the derived (upper bounds for the) rates of convergence (cf. (5.15)).

An example for the above can be found in Dahlhaus [11]. There exponential inequalities for the empirical spectral process have been derived with $\alpha = 1$ and $\alpha = 1/2$ for the Gaussian and the non-Gaussian case, respectively. Hence, meanwhile one has been able to improve upon these inequalities, because we now have $\alpha = 2$ in the Gaussian case (see above) and $\alpha = 1$ in the non-Gaussian case.

Another recent example of an exponential inequality for dependent data can be found in van de Geer [41] in the present book where exponential inequalities for martingales are derived (see also van de Geer [40]). There these exponential inequalities also lead to rates of convergence of certain estimators which often turn out to be optimal (or close to optimal) if sub-Gaussian inequalities are available.

6.2 The locally stationary case. Another extension of the above methodology is to move away from the stationarity assumption. In fact, many interesting real world phenomena are nonstationary, as in the areas of speech recognition, or seismology. There exist several notions of nonstationarity. Here we will consider the so-called local stationarity. The basic idea for constructing locally stationary processes is to take the spectral representation of a stationary process and make it time dependent. This idea goes back at least to Priestley [35], but only recently a mathematical framework has been formulated in Dahlhaus [14] which results in a framework for a rigorous asymptotic analysis. The basic underlying problem is that in a nonstationary situation future observations do not contain information about the presence or the past. Hence, a theory where time simply tends to infinity does not help. The way out of this dilemma is to rescale the time to [0, 1], say. By this rescaling we get asymptotically more and more observations of each local structure leading to a meaningful asymptotic theory. This of course is in analogy to nonparametric regression. In general, locally stationary time series can be defined via a time-varying spectral representation, and this has been formalized in Dahlhaus [14]. A statistical problem of interest then is to estimate the corresponding time-varying spectrum or functionals thereof. Dahlhaus and Neumann [17] use a wavelet approach for the estimation. Dalhaus [14] and von Sachs and Schneider [45] propose to use local periodograms. Here we discuss a likelihood based approach which utilizes the so-called preperiodogram proposed in Neumann and von Sachs [27]. For more details on the likelihood approach for nonstationary processes we refer to Dahlhaus and Polonik [18]. Applications to seismic data can also be found there. For other applications of locally stationary processes in seismology see Kitagawa et al. [23] and references therein.

In this paper we only discuss a special case of a locally stationary process, namely, the time-varying AR-models (tvAR-models) which are AR-processes with time-varying coefficients where one needs to impose some smoothness on the coefficient functions in order to maintain local stationarity. In this framework we now indicate how to utilize our likelihood approach to estimate the time-varying spectral density as well as the time-varying coefficients of the tvAR-model.

A tvAR(p)-model is given by

$$X_{t,n} = \sum_{j=1}^{p} a_j\left(\frac{t}{n}\right) X_{t-j,n} + \sigma^2\left(\frac{t}{n}\right)\epsilon_t, \tag{6.3}$$

where we assume that all functions $a_j(\cdot)$, $j = 1, \ldots, p$, and $\sigma^2(\cdot)$ are of bounded variation, and the ϵ_j are i.i.d. random variables with expectation zero and variance 1. Note that we formally deal with a triangular array of random variables which is due to the rescaling of time.

The corresponding time-varying spectral density is now given as a function of two variables—the frequency λ and the rescaled time u, and it is of the form

$$f(u,\lambda) = \frac{\sigma^2(u)}{2\pi\,|1 - \sum_{j=1}^{p} a_j(u)e^{i\lambda j}|^2}. \tag{6.4}$$

Note that although we use a parametric AR-model we are facing a nonparametric estimation problem, because the coefficients are functions which will not (all) be parametrized. The corresponding time-varying Whittle likelihood is defined in Dahlhaus and Polonik [18] as

$$\mathcal{L}_n^{(h)}(g) = \sum_{t=1}^{n} \frac{h\left(\frac{t}{n}\right)^2}{H_{2,n}} \frac{1}{4\pi} \int_{-\pi}^{\pi} \left\{ \log g\left(\frac{t}{n},\lambda\right) + \frac{J_n^{(h)}\left(\frac{t}{n},\lambda\right)}{g\left(\frac{t}{n},\lambda\right)} \right\} d\lambda, \tag{6.5}$$

where as above h denotes a taper (as, for instance, the cosine taper defined in (2.10)) and $H_{2,n} = \sum_{t=1}^{n}(h(t/n))^2$. The quantity $J_n^{(h)}$ is a generalization of the preperiodogram to tapered data, namely,

$$J_n^{(h)}\left(\frac{t}{n},\lambda\right) = \frac{1}{2\pi} h\left(\frac{t}{n}\right)^{-2} \sum_{k:1\le[t+1/2\pm k]\le n} X^{(h)}_{[t+1/2+k/2],n} X^{(h)}_{[t+1/2-k/2],n} \exp(-i\lambda k), \tag{6.6}$$

where $X_{t,n}^{(h)} = h(\frac{t}{n})X_{t,n}$. The quantity $J_n^{(h)}(\frac{t}{n}, \lambda)$ can be regarded as a preliminary estimate of $f(\frac{t}{n}, \lambda)$—however, it has to be smoothed in time and frequency direction to become consistent. It is easy to prove the relation

$$I_n^{(h)}(\lambda) = \frac{1}{2\pi H_{2,n}} \left| \sum_{t=1}^{n} X_{t,n}^{(h)} \exp(-i\lambda t) \right|^2 \tag{6.7}$$

$$= \sum_{t=1}^{n} \frac{h\left(\frac{t}{n}\right)^2}{H_{2,n}} J_n^{(h)}\left(\frac{t}{n}, \lambda\right) \tag{6.8}$$

(cf. Dahlhaus [15]), i.e., the classical tapered periodogram is a weighted average of the tapered preperiodogram.

The generalized Whittle likelihood (6.5) takes over the part of the usual Whittle likelihood in the above stationary case (see Section 5). For instance, let Φ denote a (model) class of candidate time-varying spectral densities, which now in contrast to the above are functions of two variables. Then we define the nonparametric Whittle likelihood estimator in Φ as

$$\widehat{\phi}_n = \operatorname*{argmin}_{\phi \in \Phi} \mathcal{L}_n(\phi), \tag{6.9}$$

where $\mathcal{L}_n(\phi) = \mathcal{L}_n^{(1)}(\phi)$ is the likelihood in the nontapered case. There also exists a theoretical counterpart to the Whittle likelihood, which in the present case is given by

$$\mathcal{L}(\phi) = \lim_{n \to \infty} \mathrm{E}\mathcal{L}_n(\phi) = \frac{1}{4\pi} \int_0^1 \int_{-\pi}^{\pi} \left\{ \log f(u, \lambda) + \frac{f(u, \lambda)}{\phi(u, \lambda)} \right\} d\lambda\, du \tag{6.10}$$

and f is the spectral density of the underlying locally stationary process. We denote $\phi_{opt} = \operatorname{argmin}_{\phi \in \Phi} \mathcal{L}(\phi)$. If the model is correctly specified, i.e., $f \in \Phi$, then it is easy to see that $\phi_{opt} = f$.

Analogously to the stationary case discussed earlier in this paper, the corresponding (time-varying) empirical spectral process plays a central role. It is defined as

$$E_n(\phi) = \sqrt{n}(F_n(\phi) - F(\phi)), \tag{6.11}$$

where

$$F(\phi) = \int_0^1 \int_{-\pi}^{\pi} \phi(u, \lambda) f(u, \lambda) d\lambda du \tag{6.12}$$

and

$$F_n(\phi) = \frac{1}{n} \sum_{t=1}^{n} \int_{-\pi}^{\pi} \phi\left(\frac{t}{n}, \lambda\right) J_n\left(\frac{t}{n}, \lambda\right) d\lambda. \tag{6.13}$$

Similarly to the stationary case (cf. (5.5)), we now have

$$0 \le \mathcal{L}(\widehat{\phi}_n) - \mathcal{L}(\phi_{opt}) \le \frac{1}{4\pi} \sup_{\phi \in \Phi} \left(F_n - F\right) \left(\frac{1}{\phi_{opt}} - \frac{1}{\phi}\right), \tag{6.14}$$

and again the behavior of the empirical spectral process can be utilized to derive rates of convergence for the nonparametric MLE in the locally stationary case. Note, however, that J_n is not necessarily positive, so that bracketing numbers cannot be used.

Again an exponential inequality for the empirical spectral process plays a central role. In order to derive such exponential inequalities in the present case it is very useful to observe that $F_n(\phi)$ actually is a quadratic form. (This has not been utilized above.) In fact, direct calculations show that

$$F_n(\phi) = \frac{1}{n}\underline{X}_n' U_n\left(\frac{1}{2\pi}\phi\right)\underline{X}_n, \tag{6.15}$$

where $U_n(\phi)$ is an $(n \times n)$ matrix with coefficients

$$U_n(\phi)_{jk} = \hat{\phi}\left(\frac{1}{n}\left\lfloor\frac{j+k}{2}\right\rfloor, j-k\right), \quad j,k = 1,\ldots,n.$$

Here $\hat{\phi}(u,k) = \int_{-\pi}^{\pi}\phi(u,k)e^{i\lambda k}d\lambda$ and $\lfloor x \rfloor$ denotes the largest integer less than or equal to x. Properties of $U_n(\phi)$ have been investigated in Dahlhaus [15]. Due to the representation (6.15), exponential inequalities for quadratic forms can be used for deriving exponential inequalities for $\hat{F}_n(\phi)$. Such exponential inequalities can be found in Rudzkis [36]. The following exponential inequality is presented in Dahlhaus and Polonik [18].

For a complex square matrix A with Hermite A^H we denote $\|A\|_{spec} = \max\{\sqrt{\lambda}\,|\lambda$ eigenvalue of $A^H A\}$ the spectral norm of A, and $\|A\|_2 = (\sum_{i,j}|a_{ij}|^2)^{1/2}$ the Frobenius norm of A.

Proposition 6.1. *Let* $\underline{X}_n \sim \mathcal{N}(0,\Sigma_n)$ *and* $S_n = \underline{X}_n' A_n \underline{X}_n - tr(A_n\Sigma_n)$. *If* $\|\Sigma_n^{1/2}A_n\Sigma_n^{1/2}\|_{spec} \leq \tau_\infty$ *and* $n^{-1/2}\|\Sigma_n^{1/2}A_n\Sigma_n^{1/2}\|_2 \leq \tau_2$, *then we have, for all* $\eta > 0$,

$$P(n^{-1/2}S_n \geq \eta) \leq \exp\left(-\frac{1}{4}\frac{\eta^2}{\tau_2^2 + \frac{\tau_\infty\eta}{\sqrt{n}}}\right).$$

This exponential inequality can then be used to derive maximal inequalities for the empirical spectral process which in turn leads to rates of convergence for our MLE similar to what has been outlined in the stationary case.

As an example we consider the special case of a locally stationary AR(p) model (6.4) where only the variance σ^2 is time varying, but the AR-coefficients are constant over time. In fact, we assume a shape restriction on the variance function σ^2, namely, that σ^2 is a (bounded) unimodal function. This model has applications in seismology (cf. Dahlhaus and Polonik [18]). Note that here the corresponding time-varying spectral density is of the form

$$f(u,\lambda) = \frac{\sigma^2(u)}{2\pi\,|1 - \sum_{j=1}^{p} a_j e^{i\lambda j}|^2}. \tag{6.16}$$

This means that the spectral density factorizes into $f(u,\lambda) = \sigma^2(u)\cdot h_\theta(\lambda)$ where $\theta = (a_1,\ldots,a_p)$. Note that $\{h_\theta, \theta \in \mathbf{R}^p\}$ is a continuously parametrized class (see Section 3) which has a much smaller entropy than the class of all unimodal functions (compare the two entropies given in (3.10) and (3.9), respectively). Hence, it is not difficult to see that the class of spectral densities (6.16) inherits the entropy of the larger class, the class of all bounded unimodal functions. This knowledge enables us to derive some asymptotic results for the time-varying empirical spectral distribution that are similar to the non-time-varying case discussed in Section 3. Consequently we get results about consistency and rates of convergence for the semiparametric MLE of the time-varying spectral density (6.16) in the same way as we did in Section 5 in the stationary case. This in turn allows us to derive rates of convergence for the time-varying variance. In fact, let

$$\rho_2(\phi) = \left(\int_0^1 \int_{-\pi}^{\pi} |\phi(u,\lambda)|^2 \, d\lambda \, du \right)^{1/2}, \tag{6.17}$$

be a generalization of the norm previously used for functions in λ only (note that we use the same notation; cf. (3.2)), and let $\rho_2(\phi, \psi) = \rho_2(\phi - \psi)$ be the induced metric. Then with $\widehat{\phi}_n(u,\lambda) = \frac{\widehat{\sigma}^2(u)}{2\pi\,|1 - \sum_{j=1}^p \hat{a}_j e^{i\lambda j}|^2}$ we have

$$\rho_2^2\left(\frac{f}{\widehat{\phi}_n}, 1\right) = \int_0^1 \int_{-\pi}^{\pi} \left| \frac{f(u,\lambda)}{\hat{\phi}(u,\lambda)} - 1 \right|^2 d\lambda \, du \tag{6.18}$$

$$= \int_0^1 \int_{-\pi}^{\pi} \left(\frac{\sigma^2(u)}{\widehat{\sigma}^2(u)} \frac{|1 + \sum_{j=1}^p \hat{a}_i e^{i\lambda j}|^2}{|1 + \sum_{j=1}^p a_i e^{i\lambda j}|^2} - 1 \right)^2 d\lambda \, du. \tag{6.19}$$

It is heuristically clear that in this semiparametric situation the constant AR-coefficients can be estimated at a faster rate than the time-varying variance—in fact, we can expect a $\sqrt{n}$-rate of convergence for the estimation of the AR-coefficients so that

$$\rho_2^2\left(\frac{f}{\widehat{\phi}_n}, 1\right) \approx \rho_2^2\left(\frac{\sigma^2}{\widehat{\sigma}_n^2}, 1\right) = \frac{1}{2\pi} \int_0^1 \left(\frac{\sigma^2(u)}{\widehat{\sigma}^2(u)} - 1 \right)^2 du, \tag{6.20}$$

and the nonparametric estimator $\widehat{\sigma}_n^2$ will "inherit" the rate of convergence from $\hat{\phi}_n$ which will be $O_P(n^{-1/3})$. This rate follows by similar arguments as have been used for estimating a monotonic spectral density (see the discussion at the end of Section 5).

Details can be found in Dahlhaus and Polonik [18]. There we also discuss the calculation of $\widehat{\sigma}^2(\cdot)$ and $\hat{a}_k$, $k = 1, \ldots, p$ if shape-restrictions like monotonicity or unimodality are imposed on $\sigma^2(\cdot)$ where ideas from isotonic regression can be used.

References

[1] R. Azencott and D. Dacunha-Castelle: *Series of Irregular Observations.* Springer, New York, 1986.

[2] R. Beran and L. Dümbgen: Modulation of estimators and confidence sets. *Annals of Statistics* **26** (1998), 1826–1856.

[3] L. Breiman, J.H. Friedman, R.A. Olshen and C.J. Stone: *Classification and Regression Trees.* Wadsworth, Belmont, CA, 1984.

[4] D.R. Brillinger: *Time Series: Data Analysis and Theory*, 2nd ed. Holden Day, San Francisco, 1981.

[5] P.J. Brockwell and R.A. Davis: *Time Series: Theory and Methods*, 2nd ed. Springer, New York, 1991.

[6] P. Bühlmann and B. Yu: Analysing bagging. To appear in *Annals of Statistics* (2002).

[7] N. Christiani and J. Shawe-Taylor: *An Introduction to Support Vector Machines and Other Kernel-based Learning Methods.* Cambridge University Press, 2000.

[8] J.W. Cooley and J.W. Tukey: An algorithm for the machine calculation of complex Fourier series. *Mathematical Computation* **19** (1965), 297–301.

[9] H. Cramér: On harmonic analysis in certain functional spaces. *Arkiv Math. Astr. Fysik* **28** (1942), 1–7.

[10] R. Dahlhaus: Spectral analysis with tapered data. *Journal of Time Series Analysis* **4** (1983), 163-175.

[11] R. Dahlhaus: Empirical spectral processes and their applications to time series analysis. *Stochastic Processes and Their Applications* **30** (1988), 69–83.

[12] R. Dahlhaus: Small sample effects in time series analysis: a new asymptotic theory and a new estimate. *Annals of Statistics* **16** (1988), 808–841.

[13] R. Dahlhaus: Nonparametric high resolution spectral estimation. *Probability Theory and Related Fields* **85** (1990), 147–180.

[14] R. Dahlhaus: Fitting time series models to nonstationary processes *Annals of Statistics* **25** (1997), 1–37.

[15] R. Dahlhaus: A likelihood approximation for locally stationary processes. *Annals of Statistics* **28** (2000), 1762–1794.

[16] R. Dahlhaus and D. Janas: A frequency domain bootstrap for ratio statistics in time series analysis. *Annals of Statistics* **24** (1996), 1934–1963.

[17] R. Dahlhaus and M. Neumann: Locally adaptive fitting of semiparametric models to nonstationary time series. *Stochastic Processes and Their Applications* **91** (2001), 277–308.

[18] R. Dahlhaus and W. Polonik: Empirical spectral processes and maximum likelihood estimation for locally stationary time series. Manuscript in preparation (2001).

[19] R.M. Dudley: A course in empirical processes. *Ecole d'Eté de Probabilités de Saint Flour XII-1982, Lecture Notes in Math.* **1097** 1–142, Springer, New York, 1984.

[20] G. Fay and P. Soulier: The periodogram of an i.i.d. sequence. *Stochastic Processes and Their Applications* **92** (2001), 315–343.

[21] L. Giraitis and P.M. Robinson: Parametric estimation under long range dependence. In: P. Doukhan, G. Oppenheim and M.S. Taqqu (eds.), *Theory and Applications of Long Range Dependence*. Birkhäuser, Boston, 2001.

[22] J. Hoffmann-Jørgensen: *Probability With a View Towards Statistics.* Chapman and Hall, New York, 1994.

[23] G. Kitagawa, T. Takanami, and N. Matsutomo: *Signal extraction problems in seismology. International Statistical Review* **69** (2001), 129–152.

[24] T. Mikosch: Periodogram estimates from heavy-tailed data. In: R.J. Adler, R.E. Feldman and M.S. Taqqu (eds.), *A Practical Guide to Heavy Tails.* 241–258, Birkhäuser, Boston, 1998.

[25] T. Mikosch and R. Norvaiša: Uniform convergence of the empirical spectral distribution function. *Stochastic Processes and Their Applications* **70** (1997), 85–114.

[26] E. Moulines and Ph. Soulier: Semi-parametric spectral estimation for fractional processes. In: P. Doukhan, G. Oppenheim, and M.S. Taqqu (eds.), *Theory and Applications of Long Range Dependence.* Birkhäuser, Boston, 2001.

[27] M. Neumann and R. von Sachs: Wavelet thresholding in anisotropic function classes and application to adaptive estimation of evolutionary spectra. *Annals of Statistics* **25** (1997), 38–76.

[28] E. Paparoditis: Frequency domain bootstrap for time series. In: H.G. Dehling, T. Mikosch, and M. Sørensen (eds.), *Empirical Processes for Dependent Data.* Birkhäuser, Boston, 2002.

[29] E. Parzen: Autoregressive Spectral Estimation. In: D.R. Brillinger and P.R. Krishnaiah (eds.), *Time Series in the Frequency Domain. Handbook of Statistics*, vol. 3. Elsevier, Amsterdam, 1983.

[30] D. Pollard: *Convergence of Stochastic Processes.* Springer, New York, 1984.

[31] W. Polonik: Measuring mass concentration and estimating density contour clusters: an excess mass approach. *Annals of Statistics* **23** (1995), 855–881.

[32] W. Polonik: Concentration and goodness of fit in higher dimensions. *Annals of Statistics* **27** (1999), 1210–1229.

[33] W. Polonik and Q. Yao: Asymptotics of set-indexed conditional empirical processes based on dependent data. *Journal of the American Statistical Association* **95** (2000), 509–519

[34] W. Polonik and Q. Yao: Conditional minimum volume predictive regions for stochastic processes. *Journal of Multivariate Analysis* **80** (2002), 234–255.

[35] M.B. Priestley: Evolutionary spectra and non-stationary processes. *Journal of the Royal Statistical Society, Series B* **27** (1965), 204–237.

[36] R. Rudzkis: Large deviations for estimates of spectrum of stationary series. *Lithuaian Mathematics Journal* **18** (1978), 214–226.

[37] A. Schuster: On the investigation of hidden periodicities with application to a supposed 26 day period of meteorological phenomena. *Terr. Magn.* **3** (1898), 13–41.

[38] R.H. Shumway and D.S. Stoffer: *Time Series Analysis and Its Applications.* Springer, New York, 2000.

[39] M. Taniguchi and Y. Kakizawa: *Asymptotic Theory of Statistical Inference for Time Series.* Springer, New York, 2000.

[40] S. van de Geer: *Empirical Processes in M-Estimation.* Cambridge University Press, Cambridge (UK), 2000.

[41] S. van de Geer: *On Hoeffding's inequalities for dependent random variables*, In: H.G. Dehling, T. Mikosch, and M. Sørensen (eds.), *Empirical Processes for Dependent Data.* Birkhäuser, Boston, 2002.

[42] A. van der Vaart and J.A. Wellner: *Weak Convergence and Empirical Processes With Application to Statistics.* Springer, New York, 1996.

[43] V.N. Vapnik: *Statistical Learning Theory.* Wiley, New York, 1998.

[44] V.N. Vapnik: *The Nature of Statistical Learning Theory*, 2nd ed. Springer, New York, 2000.

[45] R. von Sachs and K. Schneider: Wavelet smoothing of evolutionary spectra by non-linear thresholding. *Applications of Computational Harmonic Analysis* **3** (1996), 268–282.

[46] P. Whittle: Estimation and information in stationary time series. *Ark. Mat* **2** (1953), 423–434.

[47] P. Whittle: Some recent contributions to the theory of stationary processes. Appendix to *A Study in the Analysis of Stationary Time Series*, by H. Wold, 2nd ed. 196–228. Almqvist and Wiksell, Uppsala, 1954.

Rainer Dahlhaus
Institut für Angewandte Mathematik
Universität Heidelberg
69120 Heidelberg, Germany
dahlhaus@statlab.uni-heidelberg.de

Wolfgang Polonik
Department of Statistics
University of California, Davis
Davis, CA 95616, U.S.A.
wpolonik@ucdavis.edu

Empirical Processes Techniques for the Spectral Estimation of Fractional Processes

PHILIPPE SOULIER

ABSTRACT. Empirical processes techniques have only recently been applied to the spectral analysis of time series. Dahlhaus (1988) and Mikosch and Norvaiša (1997) proved a functional central limit theorem for the empirical spectral measure under weak dependence assumptions. This paper presents new applications of empirical processes techniques to the spectral analysis of time series. Different empirical contrast functions are considered, and applied to strongly dependent processes. Statistical applications are presented, including parametric estimation, goodness-of fit tests and adaptive estimation.

1 Introduction: spectral analysis of time series

Let $X = (X_t)_{t\in Z}$ be a second order stationary process and let $\gamma_X = \mathrm{cov}(X_0, X_t)$, $t \in \mathbb{Z}$, denote its autocovariance function. By Herglotz's Lemma (see Brockwell and Davis (1991), Theorem 4.3.1), there exists a measure G_X on $[-\pi, \pi]$ such that

$$\gamma_X(t) = \int_{-\pi}^{\pi} e^{itx} G_X(dx).$$

G_X is called the spectral measure of the process X, and for any bounded measurable function h, define

$$G_X(h) = \int_{-\pi}^{\pi} h(x) G_X(dx),$$

and the seminorm $\rho_{G_X}(h) = \{G_X(h^2)\}^{1/2}$. If G_X is absolutely continuous with respect to Lebesgue measure on $[-\pi, \pi]$, its density f_X is called the spectral density of X. It is well known that the existence of a spectral density is equivalent to the existence of a linear representation for X, i.e., there exist a sequence $(\epsilon_t)_{t\in\mathbb{Z}}$ of uncorrelated random variables with zero mean and finite variance σ^2 and a square summable series $(a_t)_{t\in\mathbb{Z}}$ such that

$$X_t = \mu + \sum\nolimits_{j\in\mathbb{Z}} a_j \epsilon_{t-j}, \tag{1.1}$$

$$a_0 = 1, \ \sum_{j\in\mathbb{Z}} a_j^2 < \infty, \ \mathbb{E}[\epsilon_j] = 0, \ \mathbb{E}[\epsilon_j \epsilon_k] = \sigma^2 \delta_{j,k}. \tag{1.2}$$

The spectral density f_X of the process X can then be expressed as

$$f_X(x) = \frac{\sigma^2}{2\pi} \left| \sum\nolimits_{j\in\mathbb{Z}} a_j e^{ijx} \right|^2.$$

When $\epsilon = (\epsilon_t)_{t\in\mathbb{Z}}$ is an i.i.d. sequence, X is called a linear process (in the strict sense). Note that by convention, the spectral density of a white noise sequence with variance σ^2 is $\sigma^2/(2\pi)$. X is a Gaussian process if and only if ϵ is a Gaussian white noise, i.e., the variables $(\epsilon_t)_{t\in\mathbb{Z}}$ are jointly Gaussian and uncorrelated. The spectral density of a Gaussian process characterizes the distribution of the process. For non-Gaussian linear processes, this is no longer true. Nevertheless, the dependence structure of a stationary process is related to the smoothness of the spectral density. Loosely speaking, a process is said to be weakly dependent if its spectral density is smooth, and strongly dependent if its spectral density has at least one singularity in $[0, \pi]$. See Section 3 for a working definition of strong dependence.

The periodogram. Spectral analysis of time series is the generic name of the set of techniques based on the spectral density of the process, in contrast to time domain methods which are based on the autocovariance function. The oldest and most natural tool of spectral analysis is the periodogram. Given observations $X_1, \dots, X_n$, the ordinary discrete Fourier transform (dft) and the periodogram are, respectively, defined as

$$d_n^X(x) = (2\pi n)^{-1/2} \sum_{t=1}^{n} X_t e^{itx}, \tag{1.3}$$

$$I_n^X(x) = |d_n^X(x)|^2 = (2\pi n)^{-1} \left| \sum_{t=1}^{n} X_t e^{itx} \right|^2. \tag{1.4}$$

Defining for $t = 1-n, \dots, n-1$, the empirical autocovariance as

$$\hat{\gamma}_X(t) = n^{-1} \sum_{k=1}^{n-|t|} X_k X_{k+|t|},$$

we obtain the following expression of the periodogram:

$$I_n^X(x) = (2\pi)^{-1} \sum_{t=1-n}^{n-1} \hat{\gamma}_X(t) e^{itx}.$$

The empirical spectral measure $\hat{G}_n$ is defined, for any bounded measurable function h, as

$$\hat{G}_n(h) = \int_{-\pi}^{\pi} h(x) I_n^X(x)dx.$$

For instance, defining $e_t(x) = e^{itx}$, we obtain, for $|t| \le n-1$,

$$\hat{G}_n(e_t) = \int_{-\pi}^{\pi} e^{itx} I_n^X(x)dx = \hat{\gamma}_X(t).$$

The empirical spectral measure $\hat{G}_n$ is the basis for the spectral analysis of time series. The most famous example of its use is the Whittle pseudo-likelihood or contrast function defined as

$$\mathcal{W}_n(h) = \int_{-\pi}^{\pi} \log(h(x))dx + \hat{G}_n(h^{-1}),$$

for a positive measurable function h. In a weak dependence context, statistical inference based on minimization of the Whittle likelihood leads to asymptotically normal and efficient estimators in the parametric context. See, for instance, Brockwell and Davis (1991).

An alternative to the Whittle contrast function for spectral domain statistical inference is the empirical logarithmic contrast function defined as

$$\mathcal{L}_n(h) = \int_0^\pi \{\log(I_n^X(x)) - \log(h(x)\}^2.$$

Taniguchi (1987) considered a version of $\mathcal{L}_n$ where I_n^X is replaced by a consistent preliminary estimator of the spectral density. Until recently, an asymptotic theory for statistics based on the logarithmic contrast was available only for weakly dependent Gaussian processes. Moreover, it was not widely used because it leads to estimates which are less efficient than those based on the Whittle likelihood, at least in the Gaussian case.

These contrast functions have discretized forms. For instance, the logarithmic contrast can be replaced by

$$\tilde{\mathcal{L}}_n(h) = \sum_{k=1}^{n} \left\{\log(I_n^X(x_k)) - h(x_k)\right\}^2,$$

where the usual choice for the frequencies $x_k, k = 1, \dots, n$ is $x_k = 2k\pi/n$, the so-called Fourier frequencies. This discretized form has a slight advantage over the integral form insofar that it is mean shift invariant and it can be considered as easier to compute.

Statistical inference based on these contrast functions involves the study of functionals of the renormalized periodogram ordinates such as

$$Z_n(h) = n^{-1/2} \sum_{k=1}^{n/2} h(x_k)\phi(I_n^X(x_k)/f_X(x_k)),$$

where f_X is the spectral density of the stationary process X. (The integral form of the contrast function yields an integral functional.) The choice $\phi(x) = x$ corresponds to Whittle's contrast, and $\phi(x) = \log(x)$ yields the logarithmic contrast. It is therefore of obvious interest when a functional central limit theorem is available for the process Z_n indexed by a suitable class of functions $\mathcal{H}$.

Such results are obtained for Gaussian processes and for non-Gaussian linear processes by using different techniques. Under the Gaussian assumption, it is possible to obtain exponential inequalities for $Z_n(h)$. These exponential inequalities and chaining arguments yield the functional central limit theorems. Dahlhaus (1988) and Dahlhaus and Polonik (2001) used this scheme to obtain a functional central limit theorem for the empirical spectral measure $\hat{G}_n$ (corresponding to $\phi(x) = x$), of a weakly dependent Gaussian process. In this paper, we apply this scheme to obtain a functional central limit theorem for the empirical logarithmic contrast function in the context of strongly dependent Gaussian processes.

For non-Gaussian linear processes which admit the linear representation (1.1), such powerful tools are (not yet) available. It is necessary to resort to the so-called Bartlett decomposition (Bartlett, 1955) which consists in relating the renormalized periodogram ordinates of the process X to those of the white noise ϵ:

$$I_n^X(x_k)/f_X(x_k) \approx 2\pi I_n^\epsilon(x_k)/\sigma^2,$$

and to approximate $Z_n(h)$ by

$$\tilde{Z}_n(h) = n^{-1/2} \sum_{k=1}^{n/2} h(x_k)\phi(2\pi I_n^\epsilon(x_k)/\sigma^2).$$

Hence it is of interest to have an asymptotic theory for $\tilde{Z}_n$. In the case $\phi(x) = x$, standard martingale techniques can be used to derive the weak convergence of the finite-dimensional distributions of $\tilde{Z}_n$. See, for instance, the proof of Theorem 2 in Robinson (1995b). Mikosch and Norvaiša (1997) obtained a functional central limit theorem for the empirical spectral measure $\hat{G}_n$ of a non-Gaussian linear process under the assumption of a finite fourth moment for the innovations.

In the case $\phi(x) = \log(x)$, and in the case of general nonlinear functionals ϕ, very little was known until recently, even though the idea of using the logarithmic contrast can already be found in Chen and Hannan (1980). In this paper, we present the recent results of Fay and Soulier (2001) who proved weak convergence of the finite-dimensional distributions of $\tilde{Z}_n$ for various functions ϕ. A functional central limit theorem has been obtained only in the special case where $\tilde{Z}_n$ is indexed by the set of the indicator functions of intervals of the real lines.

The statistical applications of empirical processes techniques to the spectral analysis of time series are the same as in the case of i.i.d. observations. We will present here applications to parametric estimation, goodness-of-fit tests and nonparametric estimation of the spectral density. Empirical processes techniques are also very important in change-point problems, but we will not consider these here. We refer the reader to the papers of Giraitis and Leipus (1990, 1992).

Organization of the paper. In Section 2, we investigate the dependence structure of the periodogram ordinates of an i.i.d. sequence. We state a general central limit theorem for weighted sums of nonlinear functionals of the periodogram ordinates of an i.i.d. sequence. An important particular case is a functional central limit theorem for the empirical distribution function of the periodogram ordinates. Some statistical applications for strict-sense linear processes are presented.

In Section 3, fractional processes are introduced. These processes are a particular case of long memory or strongly dependent processes. In Section 4, we restrict this class of processes to Gaussian processes. Under the Gaussian assumption, exponential inequalities are obtained for functionals of the periodogram of a fractional process. These inequalities are used to prove a functional central limit theorem for the empirical logarithmic contrast. Statistical applications in the parametric and nonparametric context are presented in Section 4. They should be compared to the statistical applications based on the Whittle contrast presented in the related paper of Dahlhaus and Polonik (2001).

2 The periodogram of an i.i.d. sequence at the Fourier frequencies

Recall that we have defined the so-called Fourier frequencies as

$$x_k = 2k\pi/n, \quad 0 \leq k \leq n-1.$$

The quantities $d_n^X(x_k)$ and $I_n^X(x_k)$, $0 \leq k \leq n-1$, are usually referred to as the dft and the periodogram ordinates, respectively. This choice is motivated by several reasons. The orthogonality properties of the trigonometric functions e^{it} at the Fourier frequencies imply that for $k, j = 0, \ldots, n-1$,

$$\sum_{t=1}^{n} e^{it(x_k - x_j)} = n\delta_{j,k}, \tag{2.1}$$

except for $x_k - x_j = \pm\pi$. This relation implies that the dft and periodogram computed at the Fourier frequencies (other than 0 and π) are mean-shift invariant. From a statistical point of view, this means that the unknown mean of the process need not be estimated to compute statistics based on the dft or periodogram ordinates. These properties also imply that the transformation $(X_1, \ldots, X_n) \to (d_n^X(x_0), \ldots, d_n^X(x_{n-1}))$ is one-to-one. Hence, any statistic, i.e., any measurable function of $X_1, \ldots, X_n$, is a function of the dft ordinates.

Because of the anomalous behavior at the frequencies 0 and π, and because of the symmetry of the functions considered, it is customary to define $\tilde{n} = [(n-1)/2]$ and to consider dft and periodogram ordinates computed at the Fourier frequencies $x_1, \ldots, x_{\tilde{n}}$, which exclude 0 and π.

If the random variables $X_1, \ldots, X_n$ are i.i.d. Gaussian with zero mean and variance σ^2, then the dft ordinates $d_n^X(x_1), \ldots, d_n^X(x_{\tilde{n}})$ are uncorrelated complex jointly Gaussian random variables, hence are independent. Moreover, the real and imaginary part are also uncorrelated and they are Gaussian random variables with zero mean and variance $\sigma^2/(4\pi)$. Thus the renormalized periodogram ordinates $2\pi I_n^X(x_k)/\sigma^2$ are i.i.d. standard exponential random variables, i.e., their cumulative distribution function is $F(x) = 1 - e^{-x}$. This property relies on the Gaussian assumption. If now X is a sequence of i.i.d. random variables with finite fourth moment, zero mean, variance σ^2 and fourth cumulant $\kappa_4 = \mathbb{E}[X_0^4] - 3\sigma^4$, then the dft ordinates are still uncorrelated, but the periodogram ordinates satisfy

$$4\pi^2 \mathrm{cov}\{I_n^X(x_k), I_n^X(x_l)\} = \sigma^4 \delta_{k,l} + \kappa_4/n. \tag{2.2}$$

The periodogram ordinates are correlated if the fourth cumulant κ_4 is not zero. Nevertheless, the Lindeberg–Feller central limit theorem implies that for a *fixed* number, say u, of frequencies $\{x_{k_1}, \ldots, x_{k_u}\}$, the vector of dft ordinates $d_n^X(x_{k_1}), \ldots, d_n^X(x_{k_u})$ converges in distribution to a vector of independent complex Gaussian random variables, whose real and imaginary parts are each independent Gaussian random variables with zero mean and variance $\sigma^2/(4\pi)$. Hence, a vector of fixed dimension of renormalized periodogram ordinates converges in distribution to a vector of i.i.d. standard exponential random variables, i.e., $2\pi I_n^X(x_{k_1})/\sigma^2, \ldots, 2\pi I_n^X(x_{k_u})/\sigma^2$ converge in distribution to $(Y_1, \ldots, Y_u)$ where the Y_i's are i.i.d. standard exponentials. This "asymptotic independence" holds no longer if an increasing number of Fourier frequencies is considered, as is the case with most statistics of interest. For instance, using (2.2), we get

$$\lim_{n\to\infty} \mathrm{var}\left[\frac{2\pi}{\tilde{n}} \sum_{k=1}^{\tilde{n}} I_n^X(x_k)\right] = \sigma^4 + \kappa_4/2.$$

This apparent contradiction has induced an important literature. See, for instance, Kokoszka and Mikosch (2000) for a review.

With statistical applications in view, it is important to investigate the asymptotic behavior of functionals of the periodogram of a white noise and to determine when this asymptotic behavior is the same as that of a sequence of i.i.d. exponential random variables.

Hereafter, to simplify notation, we assume that $Z = (Z_t)_{t\in\mathbb{Z}}$ is a sequence of i.i.d. mean zero, unit variance random variables with finite fourth moment and fourth cumulant $\kappa_4^Z = \mathbb{E}[Z_0^4] - 3$. Let m be a fixed integer and set $K_n = [(n-1)/2m]$. The pooled (or aggregated) periodogram ordinates of Z are defined as

$$\bar{I}_{n,k}^Z = \sum_{s=(k-1)m+1}^{km} I_n^Z(x_s), \quad 1 \le k \le K_n.$$

Pooling periodogram ordinates is used when computing nonlinear functionals. In the first place, pooling might be a necessity to ensure integrability of the functionals. For instance, if Z is a Gaussian white noise, then $\mathbb{E}[1/I_n^Z(x_k)] = \infty$, whereas $\mathbb{E}[1/(I_n^Z(x_k) + I_n^Z(x_{k+1}))] < \infty$. A second reason for pooling is to increase the asymptotic efficiency of statistics based on nonlinear functionals of the periodogram (see examples in Section 2.3 below). The choice of K_n ensures that $x_k \neq \pi$ for all $k = 1, \ldots, x_{mK_n}$. To proceed further, we introduce some notation. Let Y_m denote a generic random variable with a $\Gamma(m, 1)$ distribution. For any measurable function ϕ defined on $\mathbb{R}$ such that $\mathbb{E}[\phi^2(Y_m)] < \infty$, define

$$\gamma_m(\phi) = \mathbb{E}[\phi(Y_m)] \text{ and } \sigma_m^2(\phi) = \operatorname{var}(\phi(Y_m)). \tag{2.3}$$

Let $(\beta_{n,k})_{1 \leq k \leq K_n}$ be a triangular array of real numbers such that

$$\sum_{k=1}^{K_n} \beta_{n,k}^2 = 1, \text{ and } \lim_{n \to \infty} \max_{1 \leq k \leq K_n} |\beta_{n,k}| = 0. \tag{2.4}$$

For a function ϕ such that $\mathbb{E}[\phi^2(Y_m)] < \infty$, define

$$S_{m,n}^Z(\phi) = \sum_{k=1}^{K_n} \beta_{n,k}\{\phi(2\pi \bar{I}_{n,k}^Z) - \gamma_m(\phi)\}. \tag{2.5}$$

Define finally the empirical distribution function of the periodogram ordinates for all $x \geq 0$ as

$$\hat{F}_{n,Z}(x) = \tilde{n}^{-1} \sum_{k=1}^{\tilde{n}} \mathbf{1}_{\{2\pi I_n^Z(x_k) \leq x\}}, \tag{2.6}$$

and recall that we have set $F(x) = 1 - e^{-x}$.

If Z is Gaussian white noise, then as mentioned in the introduction, the periodogram ordinates are i.i.d. standard exponential random variables, hence the pooled periodogram ordinates are i.i.d. random variables with a $\Gamma(m, 1)$ distribution. The following properties are then straightforward consequences of the i.i.d. property.

- If the Lindeberg condition (2.4) holds, then $S_{m,n}(\phi)$ converges weakly to the Gaussian distribution with zero mean and variance $\sigma_m^2(\phi)$. This convergence can be strengthened to uniform convergence over functional classes under entropy conditions (see, for instance, Pollard (1984)).

- A particular case of the preceding property is the functional central limit theorem for the empirical distribution function of the periodogram ordinates. The process $\sqrt{n}\{\hat{F}_n - F\}$ converges in the space of $\mathcal{D}([0, \infty[)$ of left-limited right-continuous (càdlàg) functions on $[0, \infty)$ to a Gaussian process with covariance function $2F(x)(1 - F(x))$.

- The maximum of the periodogram ordinates converges weakly to the standard Gumbel distribution, i.e.,

$$\lim_{n \to \infty} \mathbb{P}(2\pi \max_{1 \leq k \leq \tilde{n}} I_n^Z(x_k) - \log(\tilde{n}) \leq x) = e^{-e^{-x}}. \tag{2.7}$$

In order to investigate the heuristic claim that the periodogram ordinates of an i.i.d. sequence behave like i.i.d. standard exponential random variables, it is natural to consider the limit behavior of the previous functionals in the non-Gaussian case. Very few results were known until recently.

Linear functionals were studied first since they are related to the Whittle pseudo-likelihood. Applying the martingale central limit theorem, if (2.4) holds and if, moreover,

$$\gamma = \lim_{n \to \infty} n^{-1} \left(\sum_{k=1}^{\tilde{n}} \beta_{n,k} \right)^2$$

exists, then $\sum_{k=1}^{\tilde{n}} \beta_{n,k}(2\pi I_n^Z(x_k) - 1)$ converges weakly to a Gaussian distribution with zero mean and variance $1+\gamma\kappa_k$. The occurrence of the fourth cumulant in the asymptotic distribution is a notable departure from the i.i.d. behavior. Martingale techniques can be adapted to the study of polynomial functionals of the periodogram ordinates. For general nonlinear functionals ϕ, earlier results gave very partial results such as bounds for the variance of $S_{m,n}^Z(\phi)$. An important contribution to the theory was made by Chen and Hannan (1980) who obtained a bound for the variance of $S_{m,n}(\log)$. More importantly, they introduced the methodology still used by now for studying such nonlinear functionals.

The behavior of the empirical distribution function of periodogram ordinates was more thoroughly investigated. Under the only assumption that Z_0 has finite variance, Freedman and Lane (1980) proved the uniform weak consistency of the empirical distribution function of periodogram ordinates, i.e.,

$$\sup_{x\geq 0} |F_n(x) - F(x)| \xrightarrow{P} 0.$$

This result was then improved on by several authors, but a central limit theorem was not obtained; see Kokoszka and Mikosch (2000) for a review.

The last mentioned problem was the first one to receive a full answer. Davis and Mikosch (1999) proved that the convergence (2.7) holds if $\mathbb{E}[|Z_0|^s] < \infty$ for some $s > 2$.

In the following subsections, we present the nearly exhaustive solutions obtained by Fay and Soulier (2001) to the first two mentioned problems and some statistical applications.

2.1 Central limit theorem for nonlinear functionals of the periodogram. As noted above, the asymptotic distribution of $S_{m,n}(\phi)$ may depend on the fourth cumulant of Z_0 and on the weight sequence. To describe this dependency precisely, we introduce some more notation and assumptions. If ϕ satisfies $\mathbb{E}[\phi^2(Y_m)] < \infty$, the Cauchy–Schwarz inequality implies that the following quantity is well defined:

$$\rho_m(\phi) = \mathbb{E}[(Y_m - m)\phi(Y_m)]. \tag{2.8}$$

The coefficient $\rho_m(\phi)$ can be interpreted in terms of the Laguerre expansion of ϕ. The Laguerre polynomials, defined as

$$L_{m,n}(x) = \frac{e^x}{n! x^{m-1}} \frac{d^n}{dx^n}(x^{n+m-1}e^{-x}),$$

form an orthonormal basis of the space of functions ϕ such that $\mathbb{E}[\phi^2(Y_m)] < \infty$, which can thus be expanded as

$$\phi = \sum_{k=0}^{\infty} C_{m,k}(\phi) L_{m,k},$$

where $C_{m,k}(\phi) = \mathbb{E}[\phi(Y_m)L_{m,k}(Y_m)]$ is the k-th Laguerre coefficient of ϕ. Hence the coefficient $\rho_m(\phi)$ is the first term in this expansion.

The condition $\mathbb{E}[\phi^2(Y_m)] < \infty$ is not sufficient to prove a central limit theorem for $S_{m,n}^Z(\phi)$ when Z is a non-Gaussian white noise. It must be strengthened by imposing either a smoothness condition or an integrability condition on ϕ. Either one of the following assumptions will be used.

Assumption 2.1. ϕ is twice differentiable and there exists an integer ν such that

$$\max_{x\in\mathbb{R}} \frac{|\phi(x)| + |\phi'(x)| + |\phi''(x)|}{1 + |x|^\nu} < \infty.$$

Assumption 2.2. There exists an integer $\alpha \geq 3$ and an integer $\beta \geq 4$ such that

$$\int_{\mathbb{R}^{2m}} \phi^2(|x|^2)(1+|x|)^{-\alpha} dx < \infty,$$

$$\int_{\mathbb{R}^{4m}} |\phi(|x|^2)\phi(|y|^2)|(1+|x|+|y|)^{-\beta} dxdy < \infty.$$

These assumptions are imposed by the method of proof which relies on Edgeworth expansions of moments of functionals which depend on a finite number of dft ordinates in both the smooth case and the nonsmooth case. In the case of smooth functionals, the results of Götze and Hipp (1978) are used and the above assumptions are sufficient. In the case of nonsmooth functionals, these results cannot be applied. Instead, Edgeworth expansions of the joint density of a finite number of dft ordinates are used. This technique was introduced by Chen and Hannan (1980). To apply this technique, an assumption on the distribution of Z_0 is needed.

Assumption 2.3. There exists a real $r \geq 1$, such that $\int_{-\infty}^{+\infty} |\mathbb{E}(e^{itZ_0})|^r dt < \infty$.

This integrability condition is a strengthening of the so-called Cramér condition (cf. Bhattacharya and Rao (1976)). This condition ensures, in particular, that for $n \geq r$, $d_n^Z(x_k)$ has a density w.r.t. Lebesgue measure, and thus that the periodogram ordinates $I_n^Z(x_k)$ are almost surely positive for $n \geq r$ and all $k = 1, \dots, \tilde{n}$. This allows us, for instance, to consider statistics based on log-periodogram ordinates. This condition might be more than needed, but up to now, no weaker assumption is known to be sufficient.

Condition (2.4) on the weight sequence $(\beta_{n,k})_{1 \leq k \leq K_n}$ must also be strengthened. We will need the following conditions;

$$\text{There exists a real } \gamma \text{ such that } \lim_{n\to\infty} n^{-1}\kappa_4^Z \left(\sum_{k=1}^{K_n} \beta_{n,k}\right)^2 = \gamma\kappa_4^Z. \tag{2.9}$$

$$\text{For all } \epsilon > 0, \text{ there exists a constant } C_\epsilon \text{ such that } \max_{1\leq k\leq K} |\beta_{n,k}| \leq C_\epsilon \mu_n^{\epsilon-1/2}, \tag{2.10}$$

where $\mu_n := \#\{k : 1 \leq k \leq K, \beta_{n,k} \neq 0\}$. Condition (2.9) automatically holds if $\kappa_4^Z = 0$. If $\kappa_4^Z \neq 0$, it holds if $\sum_{k=1}^{K_n} \beta_{n,k} = 0$ for all n, which is the case for the coefficients of some projection estimators such as the GPH estimator. It also holds with $\gamma = 0$ if $\mu_n = o(n)$ since by the Cauchy–Schwarz inequality, $\sum_{k=1}^{K_n} |\beta_{n,k}| = O(\sqrt{\mu_n})$. In view of (2.2), it is easily seen that condition (2.9) is needed to ensure the convergence of the variance of $S_{m,n}^Z(\phi)$ in the case $\phi(x) = x$ and $\kappa_4^Z \neq 0$. We will see that it suffices in the general case. Condition (2.10) is a technicality imposed by the method of proof. In practice it is not an important restriction. It holds, for instance, if $\beta_{n,k} = h(y_k)$ for some sequence of evenly spaced frequencies y_k and a smooth function h or $h(x) = \log(x)$. More generally, it holds if $\max_{1\leq k\leq K_n} |\beta_{n,k}|$ is a slowly varying function of μ_n.

We can now state the main result of this section. It is stated as Theorem 2 in Fay and Soulier (2001) and is derived from a more general result on functionals of the dft ordinates.

Theorem 2.4. *Let $(Z_t)_{t\in\mathbb{Z}}$ be a mean zero, unit variance white noise with finite moment of order $\mu \geq 4$. Assume either*

- *Assumption 2.1 holds and $\mu \geq 4\nu \vee 4$;*
- *Assumptions 2.2 and 2.3 hold and $\mu \geq \alpha \vee \beta$.*

Then $S_{m,n}^Z(\phi)$ *converges weakly to the Gaussian distribution with zero mean and variance* $\sigma_m^2(\phi) + \gamma\kappa_4^Z\rho_m^2(\phi)$. *Moreover, asymptotic normality still holds without conditions* (2.10) or (2.9) in the following cases:

- If $\mu_n = o(n^{2/3})$, then (2.10) is not needed and (2.9) holds with $\gamma = 0$.
- If $\rho_m(\phi) = 0$, then (2.9) and (2.10) are not needed. Hence, in that case the central limit theorem holds under the same assumption on the weights $\beta_{n,k}$ as in the Gaussian case and with the same asymptotic variance $\sigma_m^2(\phi)$.

Remark 2.5. We conjecture that Theorem 2.4 holds without condition (2.10) but we have not been able to prove it. The asymptotic variance $\sigma_m^2(\phi)+\gamma\kappa_4^Z\rho_m^2(\phi) = 0$ vanishes only if $\phi(y) = cy$ for some constant c and $\kappa_4^Z = -2$. This last condition is equivalent to $\mathrm{var}(Z_0^2) = 0$, i.e., $Z_0 = \pm 1$ almost surely. This case is of limited interest and can be studied directly.

We can conclude that if $\kappa_4^Z = 0$, then, apart from the restrictions on ϕ (Assumptions 2.1 or 2.2) and on the weight sequence (condition (2.10)), the periodogram ordinates of an i.i.d. sequence behave asymptotically like a sequence of i.i.d. standard exponential random variables. If $\kappa_4^Z \neq 0$, then this conclusion is no longer true, except for functionals ϕ such that $\rho_m(\phi) = 0$. In that case, the sequence $\phi(I_{n,k}^Z)$ behaves asymptotically like a sequence of i.i.d. random variables. The condition $\rho_m(\phi) = 0$ means that the Laguerre expansion of ϕ has no linear component.

Example 2.6. In statistical applications, an important case of a nonlinear functional is the logarithm. It is well known that $\gamma_m(\log) = \Psi(m)$ and $\sigma_m^2(\log) = \Psi'(m)$, where Ψ denotes the digamma function: $\Psi(z) = \Gamma'(z)/\Gamma(z)$ (cf. Johnson and Kotz (1970)), and an elementary computation yields $\rho_m(\log) = 1$. Thus, if Assumption 2.3 holds, and if $\mathbb{E}[|Z_0|^{4m+1}] < \infty$, then

$$K_n^{-1/2}\sum_{k=1}^{K_n}\{\log(2\pi \bar{I}_{n,k}^Z) - \Psi(m)\} \xrightarrow{d} \mathcal{N}\left(0, \Psi'(m) + \frac{\kappa_4^Z}{2m}\right).$$

2.2 Functional central limit theorem for the empirical spectral distribution function. Applying Theorem 2.4 to the indicator functions $y \to \mathbf{1}_{\{y\le x\}}$ yields a functional central limit theorem for the empirical spectral distribution function of the periodogram ordinates. Since the indicator functions are not smooth, it is stated under Assumption 2.3. It is not clear whether this condition is really needed.

Theorem 2.7. *If Assumption* 2.3 *holds and if* $\mathbb{E}(|Z_0|^8) < \infty$, *then* $\sqrt{n}(\hat{F}_{n,Z}(x) - F(x))$ *converges in the space* $\mathcal{D}([0,\infty[)$ *of left-limited right-continuous (càdlàg) functions on* $[0,\infty)$ *to the Gaussian process* $G(x)$ *with covariance function*

$$\mathbb{E}[G(x)G(y)] = 2F(x\wedge y)(1 - F(x\vee y)) + \kappa_4^Z e^{-x-y}xy.$$

If $q \geq 4$ *is an integer such that* $\mathbb{E}(|Z_0|^q) < \infty$, *then*

$$\lim_{n\to\infty} n^{q/2}\mathbb{E}[(\hat{F}_{n,Z}(x) - F(x))^q] = \begin{cases} 0 & \text{if } q \text{ is odd,} \\ (2F(x)(1-F(x)) + \kappa_4^Z e^{-2x}x^2)^{q/2} & \text{if } q \text{ is even.}\end{cases}$$

Remark 2.8. If $\kappa_4^Z = 0$, then the limit process is the same as if $(Z_t)_{t\in\mathbb{Z}}$ were a Gaussian white noise, or, equivalently, if the periodogram ordinates $2\pi\bar{I}_{n,k}$ were i.i.d. random variables with $\Gamma(m,1)$ distribution (i.i.d. exponentials in the case $m=1$). If enough moments of Z_0 are finite, the limiting moments are also the same as in the Gaussian case. Thus, the difference to the behavior of an i.i.d. sequence appears only through the fourth cumulant.

Remark 2.9. The convergence of the finite-dimensional distributions is an immediate consequence of Theorem 2.4 and holds under the finiteness of the fourth moment of Z_0 only. Tightness is proved using the criterion for empirical processes of Shao and Yu (1996) and requires the finiteness of the eighth moment of Z_0.

2.3 Statistical applications. An immediate application of Theorem 2.7 is the derivation of a Kolmogorov–Smirnov test of whiteness, since $\sup_{x\geq 0} |\hat{F}_{n,Z}(x) - F(x)|$ converges in distribution to $\sup_{x\geq 0} |G(x)|$. This is of limited interest since the distribution of the process G depends on the fourth cumulant κ_4^Z and the variance of Z is assumed to be known. Thus we see that in order to apply spectral methods to non-Gaussian linear processes, we need statistics which are scale invariant and whose limit distribution is independent of the fourth cumulant. Adapting an idea of Mokkadem (1997), Fay (2000) considered the following test statistic which has both these properties. Define

$$T_n = K_n^{-1} \sum_{k=1}^{K_n} \{\log(\bar{I}_{n,k}^Z) - \Psi(m)\} - \log \left\{ (mK_n)^{-1} \sum_{k=1}^{K_n} \bar{I}_{n,k}^Z \right\}. \tag{2.11}$$

It is clear that T_n is scale invariant. Its asymptotic distribution does not depend on the fourth cumulant, as can be seen from the following heuristic argument. By the central limit theorem, we have

$$\log \left\{ (mK_n)^{-1} \sum_{k=1}^{K_n} \bar{I}_{n,k}^Z \right\} = \log\{1 + K_n^{-1/2} Z_n\} \approx K_n^{-1/2} Z_n,$$

where $Z_n = K_n^{-1/2} \sum_{k=1}^{K_n} (\bar{I}_{n,k}^Z/m - 1)$. Hence T_n is asymptotically equivalent to $\tilde{T}_n = K_n^{-1} \sum_{k=1}^{K_n} \phi(\bar{I}_{n,k}^Z)$, where $\phi(z) = \log(z) - \Psi(m) - z/m + 1$. By construction, $\rho_m(\phi) = 0$, which implies that the asymptotic distribution of $\sqrt{n}\tilde{T}_n$ is independent of the fourth cumulant. More precisely, if Assumption 2.3 holds, applying Theorem 2.4, we obtain that $\sqrt{n}\tilde{T}_n$ and $\sqrt{n}T_n$ converge in distribution to $\mathcal{N}(0, 2m\Psi'(m) - 2)$.

Statistical applications of interest rather concern linear processes in the strict sense. Let X be a process which admits the linear representation (1.1) with respect to an i.i.d. sequence ϵ with zero mean and variance σ^2 and define (to be coherent with the previous notation) $Z = \epsilon/\sigma$. The spectral density of the process X is then $f_X(x) = (2\pi)^{-1}\sigma^2 |\sum_{j\in\mathbb{Z}} a_j e^{ijx}|^2$ and the Bartlett decomposition consists in writing

$$\bar{I}_{n,k}^X / f_X(y_k) \approx 2\pi \bar{I}_{n,k}^Z, \tag{2.12}$$

for some sequence of frequencies y_k, $1 \leq k \leq K_n$. The choice of the frequencies y_k is not very important. Hereafter, we will choose, for convenience, $y_k = (2K_n - 1)\pi/(2K_n)$, $1 \leq k \leq K_n$. The approximation (2.12) can be justified under miscellaneous assumptions. In this section, we will consider the following one.

Assumption 2.10.

$$\sum_{j\in\mathbb{Z}} |j|^{3/4} |a_j| < \infty.$$

This assumption is a weak dependence assumption, slightly stronger than what is classically assumed in the time series literature. It is needed when dealing with nonlinear functionals such as the log-periodogram. It implies, in particular, that the spectral density is bounded and smooth.

Estimation of the one-step linear prediction error. As mentioned above, the first attempt to derive the asymptotic distribution of $S_{m,n}(\log)$ was made by Chen and Hannan (1980) in order to prove a central limit theorem for an estimator of the one-step linear prediction error which, in the case of linear processes, coincides with the innovation variance σ^2. The so-called Kolmogorov formula (see Brockwell and Davis (1991), Theorem 5.8.1) states that

$$\log(\sigma^2) = (2\pi)^{-1} \int_{-\pi}^{\pi} \log(2\pi f_X(x))dx.$$

Hence a natural estimator of $\log(\sigma^2)$ is given by

$$\log(\hat{\sigma}_n^2) = K_n^{-1} \sum_{k=1}^{K_n} \{\log(2\pi \bar{I}_{n,k}^X) - \Psi(m)\}.$$

Chen and Hannan (1980) only proved the mean square consistency of this estimator, and they introduced the technique of Edgeworth expansions of the joint density of dft ordinates under condition 2.3, which eventually turned out to be successful. Using Bartlett's decomposition, this estimator can be decomposed as

$$\begin{aligned}\log(\hat{\sigma}_n^2) &= K_n^{-1} \sum_{k=1}^{K_n} \log(2\pi f_X(y_k)) + K_n^{-1} \sum_{k=1}^{K_n} \{\log(\bar{I}_{n,k}^X / f_X(y_k)) - \Psi(m)\} \\ &= K_n^{-1} \sum_{k=1}^{K_n} \log(2\pi f_X(y_k)) + K_n^{-1} \sum_{k=1}^{K_n} \{\log(2\pi \bar{I}_{n,k}^Z) - \Psi(m)\} + \text{ remainder term} \\ &=: \log(\tilde{\sigma}_n^2) + S_{m,n}^Z(\log) + \text{ remainder term.}\end{aligned}$$

Under Assumption 2.10, if, moreover, f_X is bounded away from zero, then $\log(\tilde{\sigma}_n^2)$ converges to $\log(\sigma^2)$. Under Assumption 2.3, Theorem 2.4 implies that $S_{m,n}^Z(\log)$ converges to a Gaussian distribution with variance $2m\Psi'(m) + \kappa_4^Z$. To prove the asymptotic normality of $\sqrt{n}(\log(\hat{\sigma}_n^2) - \log(\sigma^2))$, it only remains to prove that the remainder term vanishes asymptotically. This was done by Fay, Moulines, and Soulier (2002, Theorem 2). Altogether, we obtain

Proposition 2.11. *If Assumptions* 2.3 *and* 2.10 *hold, and if* $\mathbb{E}[|Z_0|^s] < \infty$ *for some* $s \geq 4m+1$, *then* $\sqrt{n}(\log(\hat{\sigma}_n^2) - \log(\sigma^2))$ *converges weakly to the Gaussian distribution with zero mean and variance* $2m\Psi'(m) + \kappa_4^Z$.

Goodness-of-fit test. The statistic T_n defined in (2.11) can be adapted to test goodness-of-fit of an estimated spectral density. Assume that a spectral density $f(\hat{\theta}_n, .)$ from a parametric model $\{f(\theta, .), \theta \in \Theta\}$ is fitted to the observation $X_1, \ldots, X_n$. If the model is correctly specified, and if $\hat{\theta}_n$ is a $\sqrt{n}$-consistent estimator of θ_0, define

$$\hat{T}_n = K_n^{-1} \sum_{k=1}^{K_n} \{\log(\bar{I}_{n,k}^X / f_X(\hat{\theta}_n, y_k)) - \Psi(m)\} - \log\left\{(mK_n)^{-1} \sum_{k=1}^{K_n} \bar{I}_{n,k}^X / f(\hat{\theta}_n, y_k)\right\}.$$

Under mild technical assumptions on the parameterization, Fay and Philippe (2001) proved that $\sqrt{n}\hat{T}_n$ converges in distribution to $\mathcal{N}(0, 2m\psi'(m) - 2)$.

3 Fractional processes

3.1 Definition. Let $Y = \{Y_t\}_{t\in\mathbb{Z}}$ be a mean zero covariance stationary process, with spectral density f^* bounded above and away from zero. For $d < 1/2$, a covariance stationary process X can be defined as

$$X = \mu + (I - B)^{-d} Y, \tag{3.1}$$

where B is the backshift operator, the fractional differencing operator $(I - B)^{-d}$ is defined by the infinite series

$$(I - B)^{-d} = \sum_{j=0}^{\infty} \frac{\Gamma(d + j)}{\Gamma(d)\Gamma(j + 1)} B^j,$$

and Γ denotes the Gamma function. If d is a negative integer, this expansion is actually finite (with the convention $\Gamma(z) = 0$ if z is a negative integer). The series $\sum_{j=0}^{\infty} \Gamma(d + j)/\Gamma(j + 1)$ is summable if $d < 0$ but it is only square summable if $d \in (0, 1/2)$. The process $\{X_t\}_{t\in\mathbb{Z}}$ is said to be a fractional process, and d is called the fractional differencing parameter, or the memory parameter when $d > 0$. The spectral density of the process X is given by

$$f_X(x) = |1 - e^{ix}|^{-2d} f^*(x). \tag{3.2}$$

When $0 < d < 1/2$, the spectral density is unbounded at zero, and the autocovariance function γ_X of the process X is not summable. More precisely, it decays hyperbolically to zero as $|t|$ tends to infinity:

$$\gamma_X(t) \approx c|t|^{2d-1}. \tag{3.3}$$

When $-1/2 < d < 0$, the spectral density of X vanishes at zero, but if Y is invertible then X is also invertible. The asymptotic behavior (3.3) is still valid. If $d \leq -1/2$, then X is no longer invertible.

For $d \in (0, 1/2)$, (3.3) implies that fractional processes are a particular case of long range dependent processes. These processes are now widely used and their main properties are well known, an important one being convergence of the renormalized partial sums to the fractional Brownian motion of index $H = 1/2 + d$. See Beran (1994) for a review of long range dependent processes.

By far the most familiar fractional processes are the FARIMA processes, where the process Y in (3.1) is an ARMA process (cf. Granger and Joyeux (1980).

Several generalizations of fractional processes have been introduced. They include nonstationary fractional processes (see, for instance, Hurvich and Chen (2000)), fractional processes with a fractional pole outside zero (Gray, Zhang, and Woodward (1989), Oppenheim, Ould-Haye, and Viano (2000), Giraitis and Leipus (1995)), long memory bilinear models (Giraitis, Robinson, and Surgailis (2000)).

In the sequel, we will only consider stationary and invertible processes with a fractional pole at zero, i.e., processes satisfying (3.1) with $d \in (-1/2, 1/2)$.

3.2 The periodogram of a fractional process. A characteristic feature of the periodogram ordinates of a fractional process was first observed by Künsch (1986): the "asymptotic independence" property of the renormalized periodogram ordinates does not hold. Indeed, for a fractional process with spectral density satisfying (3.2), if the function f^* is bounded above and away from zero and continuous at zero, for fixed $1 \leq k < j \leq \tilde{n}$, it holds that

$$\lim_{n\to\infty} |\mathbb{E}[I_n(x_k)/f(x_k)] - 1| \neq 0, \tag{3.4}$$

$$\lim_{n\to\infty} |\mathrm{cov}(I_n(x_k)/f(x_k), I_n(x_j)/f(x_j))| \neq 0. \tag{3.5}$$

Hurvich and Beltrao (1993) later provided an explicit expression of these limits.

In order to reduce this correlation between dft and periodogram ordinates, a simple technique is tapering, which we present now. Let $(h_{n,t})_{1\leq t\leq n}$ be a sequence of complex numbers and define $H_n^2 = \sum_{t=1}^n |h_{n,t}|^2$. Since $h_{n,t} = 1$ for all $t = 1, \ldots, n$, is a particular case of a data-taper, we will use the same notation as before and define the tapered Fourier transform and periodogram of a process X as

$$d_n^X(x) = (2\pi H_n^2)^{-1/2} \sum_{t=1}^n h_{n,t} e^{itx} \quad \text{and } I_n^X(x) = |d_n(x)|^2.$$

There are many ways to define a data-taper. The main requirement is that they attain their maximum at $n/2$ and decay to zero smoothly at the edges of the observation domain in order to avoid edge effects. The original motivation for the use of a data-taper was to reduce leakage due to spectral peaks; see, for instance, Tukey (1967), Brillinger (1981) or Dahlhaus (1990). In the context of fractional processes, an additional motivation is to reduce the asymptotic correlation between the dft and periodogram ordinates. Giraitis, Robinson, and Samarov (2000) and Velasco (2000) used the cosine bell taper, Velasco (1999) used the Kolmogorov taper and Hurvich and Chen (2000) introduced a new taper, related to the cosine bell taper which we will use here. Hereafter, we set

$$h_{n,t} = 1 - e^{ix_t}, \quad 1 \leq t \leq n. \tag{3.6}$$

With this choice, it holds that $H_n^2 = 2n$, and the shift invariance property is not lost, i.e., it holds that $\sum_{t=1}^n h_{t,n} \exp(itx_j) = 0$ for $1 \leq j \leq \tilde{n}$. This is an advantage of this data-taper with respect to Kolmogorov's taper used by Velasco (1999) which does not share this property.

Tapering has one drawback, which is that adjacent tapered dft ordinates are correlated, even if the process under consideration is a white noise. With the data-taper considered here, if Z is a unit variance white noise, then $\mathbb{E}[d_n^Z(x_k)\overline{d_n^Z(x_{k+1})}] = -1/(2\pi)$, and $\mathbb{E}[d_n^Z(x_k)\overline{d_n^Z(x_j)}] = 0$ if $k < j - 1$.

A simple solution to this problem consists in dropping one out of two Fourier frequencies in the definition of the statistics considered. This results in a loss of efficiency which can be partially compensated by pooling of periodogram ordinates. The pooled tapered periodogram of a process X is defined as

$$\bar{I}_{n,k}^X = \sum_{t=m(k-1)+1}^{mk} I_n^X(x_{2t}), \quad 1 \leq k \leq [(n-1)/4m]. \tag{3.7}$$

For the sake of simplicity, we will also denote $K_n = [(n-1)/4m]$ when the taper periodogram is used. In the rest of this paper, all the results are stated for the tapered periodogram, using the data-taper defined in (3.6).

In order to obtain covariance bounds for the dft ordinates, whether tapered or not, it is necessary to specify an assumption on the function f^* appearing in (3.2). The one we consider here is convenient since it is satisfied by most commonly used models of fractional processes, in particular, by FARIMA processes. For $M > 0$, let $\mathcal{L}^*(M)$ be the class of even continuous functions h on $[-\pi, \pi]$ such that

$$\sup_{s\in[0,\pi]} |h(s)| \leq M \text{ and } \sup_{x,y\in[0,\pi]} |xh(x) - yh(y)| \leq M|x - y|.$$

If f^* is bounded away from zero and $\log(f^*) \in \mathcal{L}^*(M)$, then the correlation between dft ordinates can be conveniently bounded, see Robinson (1995a, b) and Moulines and Soulier (1999) in the nontapered case, and Hurvich and Chen (2000) and Hurvich, Moulines, and Soulier (2002) in the tapered case.

4 Empirical processes techniques for fractional Gaussian processes

4.1 Exponential inequalities. Exponential inequalities constitute essential tools of empirical processes techniques. In the present context, we need to obtain inequalities for weighted sums of functionals of the renormalized pooled tapered periodogram

$$\sum_{k=1}^{K_n} \beta_{n,k}\phi(\bar{I}_{n,k}^X/f_X(y_k)).$$

Having statistical applications in view, it suffices to consider the cases $\phi(x) = x$ and $\phi(x) = \log(x)$. The former corresponds to the Whittle contrast, and the latter to the logarithmic contrast.

For any positive integer K, and $v \in \mathbb{R}^K$, denote $|v|_2^2 = \sum_{k=1}^K v_k^2$ and $|v|_\infty = \max_{1\leq k\leq K} |v_k|$.

Proposition 4.1. *Let $\delta \in [0, 1/2)$ and $M > 0$. Let $\mathcal{F}^*$ be a compact subclass of $\mathcal{L}^*(M)$. For any real ρ, there exists a real $\eta(\phi, \rho)$, which depends only on ρ and on the function ϕ, such that, for any real α, any $v \in \mathbb{R}^{K_n}$ satisfying $|v|_2 \leq 1$ and $|\alpha||v|_\infty \leq \eta(\phi, \rho)$, and for any Gaussian process X with spectral density $f_X = e^{dg+l^*}$ with $|d| \leq \delta$ and $l^* \in \mathcal{F}^*$, it holds that*

$$\mathbb{E}\left[\exp\left\{\alpha \sum_{k=1}^{K_n} v_k(\phi(\bar{I}_{n,k}^X/f_X(y_k)) - \gamma_m(\phi))\right\}\right] \leq C(\delta, \mathcal{F}^*, \rho)e^{(1+\rho)\sigma_m^2(\phi)\alpha^2/2}. \tag{4.1}$$

Proposition 4.1 easily yields a Bernstein-type inequality.

Proposition 4.2. *Let $\delta \in [0, 1/2)$, and $M > 0$. Let $\mathcal{F}^*$ be a compact subclass of $\mathcal{L}^*(M)$. There exists a constant $\tau(\phi)$ which depends only on the function ϕ such that for any positive real x, any $v \in \mathbb{R}^{K_n}$ satisfying $|v|_2 \leq 1$, for any $d \in [-\delta, \delta]$ and $l^* \in \mathcal{F}^*$, and for any Gaussian process X with spectral density $f_X = e^{dg+l^*}$, it holds that*

$$\mathbb{P}\left(\sum_{k=1}^{K_n} v_k\{\phi(\bar{I}_{n,k}^X/f(y_k)) - \gamma_m(\phi)\} > \tau(\phi)\sqrt{x} + 2|v|_\infty x\right) \leq C(\delta, \mathcal{F}^*)e^{-x}. \tag{4.2}$$

Remark 4.3. The constants $\eta(\rho, \phi)$ and $\tau(\phi)$ in the previous results are numerical constants which can be evaluated. $\gamma_m(\phi)$ and $\sigma_m^2(\phi)$ are defined in Section 2. In the case $\phi(x) = \log(x)$, $\gamma_m(\phi) = \Psi(m)$ and $\sigma_m^2(\phi) = \Psi'(m)$ and in the case $\phi(x) = x$, $\gamma_m(\phi) = \sigma_m^2(\phi) = m$. Propositions 4.1 and 4.2 are proved in the case $\phi(x) = \log(x)$ in, respectively, Iouditsky, Moulines, and Soulier (2001) and Soulier (2001b). The case $\phi(x) = x$ can be treated similarly and is actually easier.

Remark 4.4. Uniform bounds cannot be obtained over the whole class $\mathcal{L}^*$ because it is not equicontinuous at zero. Hence attention must be restricted to arbitrary compact subclasses. In the context of parametric estimation, the class $\mathcal{F}^*$ is restricted to the true spectral density, while in the context of nonparametric estimation, $\mathcal{F}^*$ is a compact subset of a classical functional space, such as a Sobolev ellipsoid (see Section 4.3).

4.2 Functional central limit theorem. In the sequel, we restrict our attention to the case $\phi(x) = \log(x)$. Recall that $y_k = (2k-1)\pi/(2K_n)$ and define

$$\epsilon_{n,k} - \log(\bar{I}^X_{n,k}/f(xy_k)) - \Psi(m).$$

Let $\mathcal{H}$ be a given class of functions and consider the process Z_n indexed by $h \in \mathcal{H}$,

$$Z_n(h) = K_n^{-1/2} \sum_{k=1}^{K_n} h(y_k)\epsilon_{n,k}.$$

A functional central limit theorem is usually proved in two steps. The first one is to prove convergence of the finite-dimensional distributions and the second is to prove asymptotic equicontinuity. In the present context, the convergence of the finite-dimensional distributions of the process Z_n is obtained by an application of Theorem 3 in Soulier (2001a). Define

$$\|h\|_n^2 = K_n^{-1} \sum_{k=1}^{K_n} h^2(y_k), \quad \text{and} \quad \|h\|_{n,\infty} = \max_{1 \le k \le K_n} |h(y_k)|.$$

Proposition 4.5. *Let X be a Gaussian process with spectral density $f = e^{dg+l^*}$ with $d \in (-1/2, 1/52)$ and $l^* \in \mathcal{L}^*(M)$ for some $M > 0$. Let $h \in L^2([0, \pi])$ be such that*

$$\|h\|_{n,\infty} \le C(h)\log(n), \quad \textit{and} \quad \lim_{n\to\infty} \|h\|_n^2 = \pi^{-1} \int_0^\pi h^2(y)dy.$$

Then $K_n^{-1} \sum_{k=1}^n h(y_k)\epsilon_{n,k}$ converges in distribution to $\mathcal{N}(0, \Psi'(m)\pi^{-1} \int_0^\pi h^2(y)dy)$.

The Bernstein inequality (4.2) implies that there exist positive constants a and b such that

$$\mathbb{P}(Z_n(h) > x) \le C(M, \delta)e^{-x^2/(a\|h\|_n^2 + b\|h\|_n x)}.$$

This inequality implies asymptotic equicontinuity of the process $(Z_n(h))_{h\in\mathcal{H}}$ under a metric entropy condition. See, for instance, Theorems 11.1 and 11.7 in Ledoux and Talagrand (1991).

Proposition 4.6. *Let $\mathcal{H}$ be a subset of $L^2([0, \pi], dx)$ such that for all n, the radius of $\mathcal{H}$ with respect to the norm $\|.\|_n$, i.e. $R_n := \max_{h\in\mathcal{H}} \|h\|_n$, is finite and the following conditions on the entropy integrals hold:*

$$\sup_n \int_0^{R_n} H(u, \mathcal{H}, \|.\|_n)du < \infty, \tag{4.3}$$

$$\lim_{\eta\to 0} \limsup_{n\to\infty} \int_0^\eta H(s, \mathcal{H}, \|.\|_n)ds = 0. \tag{4.4}$$

Then the process $(Z_n(h))_{h\in\mathcal{H}}$ is asymptotically equicontinuous.

Altogether, we obtain the following.

Theorem 4.7. *Under the assumptions of Propositions* 4.5 *and* 4.6, *the process* $(Z_n(h))_{h\in\mathcal{H}}$ *weakly converges to the centered Gaussian process* $(Z^*(h))_{h\in\mathcal{H}}$ *with almost surely continuous and bounded paths and covariance*

$$\mathbb{E}[Z^*(h_1)Z^*(h_2)] = \Psi'(m)\pi^{-1}\int_0^\pi h_1(x)h_2(x)dx.$$

Remark 4.8. The conditions of Propositions 4.5 and 4.6 are *ad hoc*. They are meant to be satisfied by classes of functions which are smooth except at zero and behave like $\log(x)$ at zero. This is what is needed for applications to fractional processes.

Remark 4.9. Proposition 4.5 also holds if the ordinary (nontapered) periodogram is considered. Tapering is needed here only to prove the exponential inequality (4.2). It is not known if such an inequality can be proved for the ordinary periodogram.

Remark 4.10. In the standard theory of empirical processes, when a Bernstein inequality is available, and entropy with bracketing is used rather than metric entropy, then asymptotic equicontinuity holds with the square root of the entropy instead of the entropy as assumed here in (4.3). See, for instance, Dedecker and Louhichi (2001). For weighted sums of i.i.d. random variables with subexponential tails, Corollary 8.8 in van de Geer (2000) implies that asymptotic equicontinuity holds if the integral of the square root of the entropy with bracketing is finite and if the class $\mathcal{H}$ is uniformly bounded. This last condition is unacceptable with applications to fractional processes in view. For the statistical applications presented here, the exponential inequalities of Section 4.1 and Theorem 4.7 are sufficient.

4.3 Statistical applications.

Parametric estimation. Let s be a positive integer and Θ be a compact subset of $\mathbb{R}^s$, $\theta \to d(\theta)$ be a continuous map from Θ into $(-1/2, 1/2)$ and $\theta \to l^*(\theta, .)$ be a continuous map from Θ into $\mathcal{C}([0,\pi])$, the space of real-valued continuous functions on $[0,\pi]$, endowed with the supremum norm. Assume that $l^*(\theta, .)$ and $d(\theta)$ are continuously differentiable with respect to θ and $\theta \to \nabla_\theta l^*(\theta, .)$ is also a continuous map from Θ into $\mathcal{C}([0,\pi])$, where $\nabla_\theta h(\theta, .)$ denotes the gradient of the function h with respect to θ. Define $l(\theta) = d(\theta)g + l^*(\theta, .)$, $\mathcal{H} = \{l(\theta, .), \theta \in \Theta\}$ and $\dot{\mathcal{H}} = \{\nabla_\theta l(\theta, .), \theta \in \Theta\}$.

The conditions stated above on the parameterization imply that the classes $\mathcal{H}$ and $\dot{\mathcal{H}}$ satisfy the entropy condition (4.3) and (4.4). These conditions also imply that the following limits hold uniformly with respect to $\theta, \theta' \in \Theta$:

$$\lim_{n\to\infty} K_n^{-1}\sum_{k=1}^{K_n} l(\theta, y_k)l(\theta', y_k) = \pi^{-1}\int_0^\pi l(\theta, y_k)l(\theta', y)dy,$$

$$\lim_{n\to\infty} K_n^{-1}\sum_{k=1}^{K_n} \nabla_\theta l(\theta, y_k)\nabla_\theta l(\theta', y_k)^T = \pi^{-1}\int_0^\pi \nabla_\theta l(\theta, y_k)\nabla_\theta l(\theta', y)^T dy.$$

Finally it must be assumed that the model is identifiable. For instance, we assume that if $\theta \neq \theta'$, then $\int_0^\pi (l(\theta, y) - l(\theta', y))^2 dy > 0$.

Let now X be a Gaussian process with positive spectral density f_0 such that $l_0 = \log(f_0)$ can be expressed as

$$l_0 = d(\theta_0)g + l^*(\theta_0, .)$$

for an unknown value of the parameter θ_0 which lies in the interior of Θ, and with $l^*(\theta_0, .) \in \mathcal{L}^*(M)$ for some $M > 0$.

The previous considerations imply that the conditions of Theorem 4.7 hold, thus the processes $Z_n(l(\theta,.))_{\theta\in\Theta}$ and $Z_n(\nabla_\theta l(\theta,.))_{\theta\in\Theta}$ are asymptotically equicontinuous and converge, respectively, to the Gaussian processes $Z^*(l(\theta,.))_{\theta\in\Theta}$ and $Z^*(l(\nabla_\theta\theta,.))_{\theta\in\Theta}$ which have almost surely continuous and bounded paths.

Define now the log-periodogram regression estimator of $\hat{\theta}_n$ of θ_0 as

$$\hat{\theta}_n = \arg\min_{\theta\in\Theta} \sum_{k=1}^{K_n} \{\log(\bar{I}_{n,k}) - \Psi(m) - l(\theta, y_k)\}^2. \tag{4.5}$$

The uniform central limit theorem for the process $(Z_n(l(\theta,.))_{\theta\in\Theta}$ and the identifiability condition imply that $\hat{\theta}_n$ is consistent.

The proof of such a result is classical, so we just briefly outline it. See, for instance, the proof of a similar result for the Whittle minimum contrast estimator in Dahlhaus (1988) and Dahlhaus and Polonik (2001). Note that $\hat{\theta}_n$ also minimizes

$$\tilde{L}_n(\theta) = 2K_n^{-1/2} Z_n(l(\theta_0,.) - l(\theta,.)) + \mathcal{K}_n(\theta),$$

where

$$\mathcal{K}_n(\theta) = K_n^{-1} \sum_{k=1}^{K_n} \{l(\theta, y_k) - l(\theta_0, y_k)\}^2.$$

$\mathcal{K}_n(\theta)$ is obviously minimized by $l(\theta_0,.)$ and under the assumptions, converges uniformly over Θ to $\mathcal{K}(\theta) = \int_0^\pi (l(\theta,y) - l(\theta_0,y))^2 dy$. Since $\hat{\theta}_n$ also minimizes $\tilde{L}_n$, then it obviously holds that $\tilde{L}_n(\hat{\theta}_n) - \tilde{L}_n(\theta_0) \le 0$, and

$$\begin{aligned}
0 \le \mathcal{K}_n(\hat{\theta}_n) &= \mathcal{K}_n(\hat{\theta}_n) - \mathcal{K}_n(\theta_0) \\
&= \mathcal{K}_n(\hat{\theta}_n) - \tilde{L}_n(\hat{\theta}_n) + \tilde{L}_n(\hat{\theta}_n) - \tilde{L}_n(\theta_0) + \mathcal{K}_n(\theta_0) - \mathcal{K}_n(\theta) \\
&\le 2K_n^{-1/2} Z_n(l(\theta_0,.) - l(\theta,.)) + 2K_n^{-1/2} Z_n(l(\theta_0,.) - l(\theta,.)) \\
&\le 8K_n^{-1/2} \sup_{\theta\in\Theta} |Z_n(l(\theta,.))| = O_P(K_n^{-1/2}),
\end{aligned}$$

the last equality being a consequence of the asymptotic equicontinuity. Thus $\mathcal{K}_n(\hat{\theta}_n)$ tends to zero in probability, which implies that $\hat{\theta}_n$ is consistent because of the identifiability assumption.

If, moreover, the matrix

$$J(\theta_0) := (2\pi)^{-1} \int_0^\pi \nabla_\theta l(\theta_0, y_k) \nabla_\theta l(\theta_0, y)^T dy$$

is invertible, then $\sqrt{n}(\hat{\theta}_n - \theta_0)$ is asymptotically Gaussian with covariance matrix $m\Psi'(m)J^{-1}(\theta_0)$. We prove this in the case of a one-dimensional parameter for notational simplicity.

Since $\hat{\theta}_n$ is consistent and minimizes $\tilde{L}_n$, and since θ_0 lies in the interior of Θ, for large enough n, it holds that $\partial_\theta \tilde{L}_n(\hat{\theta}_n) = 0$. Hence

$$\begin{aligned}
0 = \partial_\theta \tilde{L}_n(\hat{\theta}_n) &= 2K_n^{-1/2} Z_n(\partial_\theta l(\hat{\theta}_n,.)) + 2K_n^{-1} \sum_{k=1}^{K_n} \{l(\hat{\theta}_n, y_k) - l(\theta_0, y_k)\} \partial_\theta l(\hat{\theta}, y_k) \\
&= 2K_n^{-1/2} Z_n(\partial_\theta l(\hat{\theta}_n,.)) + 2(\hat{\theta}_n - \theta_0) K_n^{-1} \sum_{k=1}^{K_n} \partial_\theta l(\tilde{\theta}_{n,k}, y_k) \partial_\theta l(\hat{\theta}_n, y_k).
\end{aligned}$$

Hence, for large enough n,

$$\sqrt{K_n}(\hat{\theta}_n - \theta_0) = \frac{-Z_n(\partial_\theta l(\hat{\theta}_n, .))}{K_n^{-1} \sum_{k=1}^{K_n} \partial_\theta l(\tilde{\theta}_{n,k}, y_k) \partial_\theta l(\hat{\theta}_n, y_k)}.$$

By uniform convergence of Z_n to the Gaussian process Z^* and of the denominator of the last expression to $J(\theta)$ over the class $\dot{\mathcal{H}}$, we get that $\sqrt{K_n}(\hat{\theta}_n - \theta_0)$ converges in distribution to a mean zero Gaussian variable with variance $\Psi'(m) J^{-1}(\partial_\theta l(\theta_0, .))$.

Nonparametric estimation. We consider now the problem of nonparametric estimation of the spectral density of a fractional process. It is customary in nonparametric functional estimation to assess the performance of an estimator through the L^2 risk. Since the spectral density of a fractional process is not necessarily square integrable, this choice is not possible here. Hence an alternative measure of the estimation accuracy is needed. We consider the log-spectrum (often referred to as the *cepstrum* in the signal processing literature) instead of the spectral density, and use the logarithmic risk function, defined for positive functions f and h by

$$R(f, h) = \int_{-\pi}^{\pi} \{\log(f(x) - \log(h(x))\}^2 dx = \| \log(f) - \log(h) \|^2.$$

It is thus natural to use a nonparametric version of the log-periodogram regression estimator (4.5). Instead of minimizing the contrast over infinite-dimensional sieves, which is numerically difficult, we will use finite-dimensional sieves, and a penalization to select the dimension of the sieve. This method is the model selection approach to adaptive estimation due to Birgé and Massard (1997). A very simple method is to use the FEXP estimator, originally suggested by Robinson (1994), and later studied by Moulines and Soulier (1999) and Soulier (2000, 2001b). The FEXP estimation procedure consists in estimating d and the p first Fourier coefficients of l^*, by ordinary least squares:

$$(\hat{d}, \hat{\theta}_0, \ldots, \hat{\theta}_{p-1}) = \arg \min_{\bar{d}, \bar{\theta}_0, \ldots, \bar{\theta}_{p-1}} \sum_{k=1}^{K_n} \left(\log(\bar{I}_{n,k}) - \Psi(m) - \bar{d} g(y_k) - \sum_{j=0}^{p-1} \bar{\theta}_j \cos(j y_k) \right)^2 .$$

For a given p, an estimator of the log-spectrum is then defined as

$$\hat{l}_p(x) = \hat{d}_p g(x) + \sum_{j=0}^{p-1} \hat{\theta}_j \cos(jx).$$

The dimension of the estimated model is selected by conservative penalization. For a given nondecreasing sequence of integers $p_n^* \leq K_n$ and a positive real κ, define

$$S_{p,n}(\kappa) = \|Y_n - \hat{l}_p\|_n^2 + \kappa p K_n^{-1}, \tag{4.6}$$

$$\hat{p}(\kappa) = \arg \min_{1 \leq k \leq p_n^*} S_{p,n}(\kappa). \tag{4.7}$$

The choice $\kappa = 4\Psi'(m)$ yields the well-known Mallows C_L criterion, which is an "asymptotically optimal" criterion in a very weak sense (cf. Polyak and Tsybakov (1989) in the classical case of a regression with i.i.d. noise and Moulines and Soulier (2000) in the present context). The model selection approach takes advantage of the powerful tools of empirical processes theory to obtain risk bounds for the penalized least squares estimators, for a suitable choice of κ.

Theorem 4.11. *Let $\delta \in [0, 1/2)$ and $M > 0$. Let $\mathcal{F}^*$ be a compact subset of $\mathcal{L}^*(M)$. Let X be a stationary Gaussian process with spectral density $f_X = e^{dg+l^*}$ with $|d| \le \delta$ and $e^{l^*} \in \mathcal{F}^*$. Define $p_n^* = \sqrt{K_n}/\log(K_n)$ and for $\kappa > 0$, let $\hat{p}(\kappa)$ be defined as in* (4.7). Then there exists a universal choice of the constant $\kappa > 0$ and a constant $C(\delta, \mathcal{F}^*, \kappa)$, such that

$$\mathbb{E}[\|\hat{l}_{\hat{p}(\kappa)} - l\|_n^2] \le \inf_{1 \le p \le p_n^*} \left\{ \sum_{j=p}^{\infty} \theta^2 + \kappa p K_n^{-1} \right\} + C(\delta, \mathcal{F}^*, \kappa) K_n^{-1}.$$

This result (proved in Soulier (2001b), Theorem 3) expresses that the FEXP estimator is adaptive to the target function. For instance, if the true l^* has only finitely many nonvanishing Fourier coefficients, then the mean square error of the estimator is $O(n^{-1})$, i.e., the parametric rate is achieved.

The main tool in its proof is a moment bound for the supremum of the empirical process Z_n over a finite-dimensional sieve. More precisely, for $q \le K_n^{1/2}/\log(K_n)$, define $\Phi_q = \{h : h(x) = \bar{d}g(x) + \sum_{j=0}^{p-1} \bar{\theta}_j \cos(jx),\ \|h\|_n = 1\}$. The Bernstein inequality (4.2) and a standard chaining argument yield

$$\mathbb{P}(\sup_{h \in \Phi_q} Z_n(h) \ge K_n^{-1/2}\{\alpha_1\sqrt{q} + \alpha_2\sqrt{\xi} + \alpha_3\xi\}) \le C(\delta, \mathcal{F}^*)e^{-q-\xi}, \tag{4.8}$$

for some numerical constants α_1, α_2 and α_3. Integrating with respect to ξ and summing over $q \le K_n^{1/2}/\log(K_n)$ yield the moment bounds used in the proof of Theorem 4.11.

The performance of an adaptive estimator is usually also assessed by its minimax properties over functional classes. In the present context, as in classical nonparametric regression, the so-called analytic and Sobolev classes are of interest (cf. Iouditsky, Moulines, and Soulier (2001) or Soulier (2001b) for details). For $\gamma > 0$, $\beta > 1/2$ and $L > 0$, define

$$\mathcal{A}(\gamma, L) = \left\{ h : h = \sum_{j=0}^{\infty} \theta_j h_j,\ \sum_{j=0}^{\infty} e^{2\beta j}\theta_j^2 \le L^2 \right\}, \tag{4.9}$$

$$\mathcal{S}(\beta, L) = \left\{ h : h = \sum_{j=0}^{\infty} \theta_j h_j,\ \sum_{j=0}^{\infty} (1+j)^{2\beta}\theta_j^2 \le L^2 \right\}. \tag{4.10}$$

Theorem 4.11 yields the minimax properties of the FEXP estimator.

Corollary 4.12. *Let $0 < \gamma_* < \gamma^*$, $1/2 < \beta_* < \beta^*$, $L^* > 0$, $M > 0$ and $\delta \in [0, 1/2[$. Then*

$$\limsup_{n\to\infty} \sup_{\beta_*<\beta<\beta^*} \sup_{0<L<L^*} \sup_{|d|\le\delta} \sup_{l^*\in\mathcal{A}(\gamma,L)} \frac{n}{\log(n)} \mathbb{E}[\|\hat{l}_{\hat{p}(\kappa)} - l\|_n^2] < \infty,$$

$$\limsup_{n\to\infty} \sup_{\gamma_*<\beta<\gamma^*} \sup_{0<L<L^*} \sup_{|d|\le\delta} \sup_{l^*\in\mathcal{S}(\beta,L)\cap\mathcal{L}^*(M)} n^{2\beta/(2\beta+1)} \mathbb{E}[\|\hat{l}_{\hat{p}(\kappa)} - l\|_n^2] < \infty.$$

We conclude by giving the corresponding lower bounds.

Theorem 4.13. *Let $\beta > 1/2$, $\gamma > 0$, $L > 0$, and $\delta \in [0, 1/2)$. Then*

$$\liminf_n \inf_{\hat{l}_n} \sup_{|d|\le\delta} \sup_{l^*\in\mathcal{S}(\beta,L)} n^{2\beta/(2\beta+1)} \mathbb{E}_{d,l^*}[\|\hat{l}_n - l\|^2] > 0, \tag{4.11}$$

$$\liminf_n \inf_{\hat{l}_n} \sup_{|d|\le\delta} \sup_{l^*\in\mathcal{A}(\gamma,L)} n\log^{-1}(n) \mathbb{E}_{d,l^*}[\|\hat{l}_n - l\|^2] \ge 1/\gamma, \tag{4.12}$$

where the infimum $\inf_{\hat{l}_n}$ is taken over all possible estimators of $l = \log(f)$ based on a sample $X_1, \dots, X_n$ of a covariance stationary process with spectral density $f = e^{dg+l^*}$.

Thus, the adaptive FEXP estimator achieves the minimax rates of convergence, in the long range dependence as well as in the short range dependence cases, and these rates are the same in both cases. The log-periodogram regression and the empirical processes techniques are crucial to obtain these results.

REFERENCES

[1] M.S. Bartlett. *An Introduction to Stochastic Processes.* Cambridge University Press, Cambridge, 1955.

[2] J. Beran. *Statistics for Long Memory Processes.* Chapman Hall, London, 1994.

[3] R.N. Bhattacharya and R.R. Rao. *Normal Approximation and Asymptotic Expansions.* Wiley Series in Probability and Mathematical Statistics. Wiley, New York, 1976.

[4] L. Birgé and P. Massart. From model selection to adaptive estimation. In D. Pollard, E. Torgersen, and G. Yang (Eds.) *Festschrift for Lucien Le Cam: Research in Probability and Statistics*, 55–87, Springer, New York, 1997.

[5] D.R. Brillinger. *Time Series. Data Analysis and Theory.* Holden-Day Series in Time Series Analysis, Holden-Day, San Frasncisco, 1981.

[6] P.J. Brockwell and R.A. Davis. *Time Series: Theory and Methods.* Springer Series in Statistics. Springer-Verlag, New York, 1991.

[7] Z.-G. Chen and E.J. Hannan. The distribution of periodogram ordinates. *Journal of Time Series Analysis* **1** (1980), 73–82.

[8] R. Dahlhaus. Empirical spectral processes and their applications to time series analysis. *Stochastic Processes and Their Applications* **30** (1988), 69–83.

[9] R. Dahlhaus. Nonparametric high resolution spectral estimation. *Probability Theory and Related Fields* **85** (1990), 147–180.

[10] R. Dahlhaus and W. Polonik. Empirical spectral processes and nonparametric maximum likelihood estimation for time series. In H. Dehling, T. Mikosch and M. Sorensen (Eds.) *Empirical Processes for Dependent Data*, Birkhäuser, Boston, 2002.

[11] R.A. Davis and T. Mikosch. The maximum of the periodogram of a non-Gaussian sequence. *Annals of Probability* **27** (1999), 522–536.

[12] J. Dedecker and S. Louhichi. Chaining from Rosenthal inequality. In H. Dehling, T. Mikosch, and M. Sorensen (Eds.) *Empirical Processeses for Dependent Data* Birkhäuser, Boston, 2002.

[13] G. Fay. Théorèmes limites pour les processus linéaires en courte et en longue portée. Ph. D. Thesis, Ecole Nationale Supérieures des Télécommunications, Paris, 2000.

[14] G. Fay, E. Moulines, and Ph. Soulier. Central limit theorems for nonlinear functionals of the periodogram of a non-Gaussian time series. To appear in *Journal of Time Series Analysis*, 2002.

[15] G. Fay and A. Philippe. A goodness-of-fit test for linear time series. Preprint, 2001.

[16] G. Fay and Ph. Soulier. The periodogram of an i.i.d. sequence. *Stochastic Processes and Their Applications* **92** (2001), 315–343.

[17] D. Freedman and D. Lane. The empirical distribution of fourier coefficients. *Annals of Statistics* **8** (1980), 1244–1251.

[18] S.A. van de Geer. *Applications of Empirical Process Theory.* Cambridge Series on Statistical and Probabilistic Mathematics. Cambridge University Press, Cambridge, 2000.

[19] L. Giraitis and R. Leipus. Functional CLT for nonparametric estimates of the spectrum and change-point problem for a spectral function. *Lithuanian Mathematical Journal* **30** (1990), 302–322.

[20] L. Giraitis and R. Leipus. Testing and estimating in the change-point problem of the spectral function. *Lithuanian Mathematical Journal* **32** (1992), 15–29.

[21] L. Giraitis and R. Leipus. A generalized fractional differencing approach in long-memory modeling. *Lithuanian Mathematical Journal* **35** (1995), 53–65.

[22] L. Giraitis, J. Hidalgo, and P.M. Robinson. Gaussian estimation of parametric spectral density with unknown pole. Preprint, 1999.

[23] L. Giraitis, P.M. Robinson, and A. Samarov. Adaptive rate optimal estimation of the long memory parameter. *Journal of Multivariate Analysis* **72** (2000), 183–207.

[24] L. Giraitis, P.M. Robinson, and D. Surgailis. A model for long memory conditional heteroscedasticity. *Annals of Applied Probability* **10** (2000), 1002–1024.

[25] F. Götze and C. Hipp. Asymptotic expansions in the central limit theorem under moment conditions. *Zeitschrift für Wahrscheinlichkeitstheorie und verwandte Gebiete* **42** (1978), 67–87.

[26] C.W.J. Granger and R. Joyeux. An introduction to long memory time series and fractional differencing. *Journal of Time Series Analysis* **1** (1980), 15–30.

[27] H.L. Gray, N.F. Zhang, and W.A. Woodward. On generalized fractional processes. *Journal of Time Series Analysis* **10** (1989), 233–257.

[28] J. Hidalgo and Ph. Soulier. Estimation of the location and exponent of the spectral singularity of a long memory process. Prépublication 141 de l'Université d'Evry Val d'Essonne, 2001.

[29] C.M. Hurvich and K. Beltrao. Asymptotics of the low-frequency ordinates of the periodogram of a long-memory time series. *Journal of Time Series Analysis* **14** (1993), 455–472.

[30] C.M. Hurvich and W.W. Chen. An efficient taper for potentially overdifferenced long-memory time series. *Journal of Time Series Analysis* **21** (2000), 155–180.

[31] C.M. Hurvich, E. Moulines, and Ph. Soulier. The FEXP estimator for non-Gaussian, potentially nonstationary processes. *Stochastic Processes and Their Applications* **97** (2002), 307–340.

[32] A. Iouditsky, E. Moulines, and Ph. Soulier. Adaptive estimation of the fractional differencing coefficient. *Bernoulli* **7** (2001), 1–33.

[33] N.L. Johnson and S. Kotz. *Continuous Univariate Distributions I.* New York, Wiley, 1970.

[34] P. Kokoszka and T. Mikosch. The periodogram at the Fourier frequencies. *Stochastic Processes and Their Applications* **86** (2000), 49–79.

[35] H.R. Künsch. Discrimination between monotonic trends and long-range dependence. *Journal of Applied Probability* **23** (1986), 1025–1030.

[36] M. Ledoux and M. Talagrand. *Probability in Banach Spaces.* Number 23 in Ergebnisse der Mathematik und ihrer Grenzgebiete. Springer, Berlin, 1991.

[37] T. Mikosch and R. Norvaiša. Uniform convergence of the empirical spectral distribution function. *Stochastic Processes and Their Applications* **70** (1997), 85–114.

[38] A. Mokkadem. A measure of information and its applications to test for randomness against ARMA alternatives and to goodness-of-fit test. *Stochastic Processes and Their Applications* **72** (1997), 145–159.

[39] E. Moulines and Ph. Soulier. Data driven order selection for projection estimator of the spectral density of time series with long range dependence. *Journal of Time Series Analysis* **21** (2000), 193–218.

[40] E. Moulines and Ph. Soulier. Semiparametric estimation for fractional processes. In P. Doukhan, G. Oppenheim, and M.S. Taqqu (Eds.) *Theory and Applications of Long Range Dependence*. Birkhäuser, Boston, 2001.

[41] G. Oppenheim, M. Ould Haye, and M.C. Viano. Long memory with seasonal effects. *Statistical Inference for Stochastic Processes* **3** (2000), 53–68.

[42] D. Pollard *Convergence of Stochastic Processes.* Springer Series in Statistics. Springer, New York, 1984.

[43] B.S. Polyak and A.B. Tsybakov. Asymptotic optimality of the C_p test for the orthogonal series estimation of regression. *Theory of Probability and Its Applications* **35** (1989), 293–306.

[44] P.M. Robinson. Time series with long range dependence. In A.C. Sims (Ed.) *Advances in Econometrics. Proceedings of the Sixth World Congress*, 47–95, Cambridge University Press, Cambridge, 1994.

[45] P.M. Robinson. Log-periodogram regression of time series with long range dependence. *Annals of Statistics* **23** (1995a), 1043–1072.

[46] P.M. Robinson. Gaussian semiparametric estimation of long range dependence. *Annals of Statistics* **24** (1995b), 1630–1661.

[47] Q.M. Shao and Y.X. Yu. Weak convergence for weighted empirical processes of dependent sequences. *Annals of Probability* **24** (1996), 2098–2127.

[48] Ph. Soulier. Estimation adaptative de la densité spectrale d'un processus gaussien faiblement ou fortement dépendant. *Comptes Rendus de l'Académie des Sciences de Paris, Série I* **330** (2000), 733–736.

[49] Ph. Soulier. Moments bounds and central limit theorem for functions of Gaussian vectors. *Statistics and Probability Letters* **54** (2001), 193–203.

[50] Ph. Soulier. Adaptive estimation of the spectral density of weakly or strongly dependent Gaussian process. *Mathematical Methods of Statistics* **10** (2001), 331–354.

[51] M. Taniguchi. Minimum constrast estimation for spectral densities of stationary processes. *Journal of the Royal Statistical Society* **49** (1987), 315–325.

[52] J.W. Tukey. An introduction to the calculation of numerical spectrum analysis. In B. Harris (Ed.) *Advanced Seminar on Spectral Analysis of Time Series*, 25–46, New York, Wiley, 1967.

[53] C. Velasco. Nonstationary log-periodogram regression. *Journal of Econometrics* **91** (1999), 325–371.

[54] C. Velasco. Non-Gaussian log-periodogram regression. *Econometric Theory* **16** (2000), 44–79.

Philippe Soulier
Département de Mathématiques
Université d'Evry Val d'Essonne
Boulevard François Mitterand
91025 Evry cedex, France
philippe.soulier@maths.univ-evry.fr

Part V

The Tail Empirical Process in Extreme Value Theory

Tail Empirical Processes Under Mixing Conditions

HOLGER DREES[1]

ABSTRACT. Limit theorems for tail processes of absolutely regular time series are surveyed. First we discuss Rootzén's [29] result on the uniform tail empirical process and generalizations thereof. Then similar results on the uniform tail quantile process are derived. If the observations come from a distribution function that belongs to the domain of attraction of an extreme value distribution, then one can prove weighted approximations of the linearly standardized tail quantile function. Moreover, asymptotic normality can be deduced for many estimators of interest in extreme value statistics. Finally, we apply the limit theorems to particular linear and nonlinear time series models.

1 Introduction

If one is interested in the probabilistic behavior of extreme observations or the performance of statistical procedures based on these data, it is reasonable to examine not the whole empirical process but only the tail that corresponds to the extreme observations under consideration. Throughout this article we focus on the large observations in a univariate time series. Of course, one can deal with the small observations in the same way by analyzing the large values after multiplying the time series by -1.

Let X_i, $i \in \mathbb{N}$, be a stationary real-valued time series with common distribution function (d.f.) F. We study the stochastic behavior of large observations, i.e., exceedances over a high threshold u_n or a deterministic number k_n of the largest observations in the sample X_i, $1 \le i \le n$. Depending on the asymptotic behavior of the (expected) number of extreme values, i.e., of $n(1-F(u_n))$ and k_n, respectively, as the sample size n increases two different types of empirical processes are appropriate for the asymptotic analysis. If the (expected) number of extremes remains bounded, then the empirical point process $\sum_{i=1}^n \varepsilon_{(i/n, X_i)} 1_{\{X_i > u_n\}}$ (and its projection onto the first and second coordinates) is a very useful tool for studying the probabilistic behavior of the largest observations. For example, Resnick [27, Chapter 4] employed empirical point process theory to establish the asymptotics of extremes and records of independent, identically distributed (i.i.d.) sequences and infinite moving averages. In [23] similar point processes based on more general time series are examined under suitable mixing conditions. Since then, the literature on point processes has grown considerably and their asymptotics is well understood; see, e.g., [21], [20], [5] and [6] for general results and [7] for an analysis of a specific time series model. It turned out that, while for i.i.d. data the empirical point process can be approximated by a Poisson process, in the case of dependent observations typically a Poisson cluster process occurs as the limiting process.

[1] Supported by the Heisenberg program of the DFG.

In contrast to the analysis of the probabilistic behavior of extremes, the statistical inference about the tail parameters of F, extreme quantiles and exceedance probabilities, must be based on an increasing number of extreme observations as the sample size increases. In that case, tail versions of the standard empirical processes are more appropriate for the asymptotic analysis. In the sequel we will concentrate on these processes, since the literature on this topic is less easily accessible than that on empirical point processes.

To start with, we consider a time series U_i, $i \in \mathbb{N}$, of uniformly on $(0, 1)$ distributed random variables. Recall the definition of the pertaining empirical process for sample size n:

$$E_n(x) = \sqrt{n}(F_n(x) - x), \qquad x \in [0, 1],$$

where $F_n(x) = n^{-1} \sum_{i=1}^{n} 1_{\{U_i \le x\}}$ defines the empirical d.f.

If one is only interested in the exceedances over the threshold $1 - v_n$ with $v_n \to 0$ but the expected number of exceedances nv_n tends to ∞, then the tail empirical process e_n defined by

$$\begin{aligned} e_n(x) &= \frac{1}{\sqrt{nv_n}} \sum_{i=1}^{n} (1_{\{U_i > 1 - v_n x\}} - v_n x) \\ &= -\sqrt{\frac{n}{v_n}}(F_n(1 - v_n x) - (1 - v_n x)) = -\frac{1}{\sqrt{v_n}} E_n(1 - v_n x) \end{aligned} \tag{1.1}$$

suggests itself. Note that the standardization is chosen such that $E(e_n(x)) = 0$ and, in case of independent observations, $\mathrm{Var}(e_n(x)) = x(1 - v_n x) \to x$ as $n \to \infty$. It is well known that in the latter case e_n converges in distribution to a Brownian motion. Moreover, sharp results about weighted approximations of this process can be found in the monograph [3]. However, much less is known about the asymptotics of e_n for dependent observations. Section 2 reviews the main results for absolutely regular time series.

For applications in extreme value statistics, it is often more convenient to examine the tail quantile process. For a time series of uniform random variables, it is defined as

$$g_n(t) = \frac{n}{\sqrt{k_n}} \left(1 - U_{n-[k_n t]:n} - \frac{k_n}{n} t \right), \qquad t \in [0, 1],$$

where $U_{i:n}$ denotes the ith order statistic among U_j, $1 \le j \le n$, (i.e., $\min_{1 \le i \le n} U_i = U_{1:n} \le U_{2:n} \le \cdots \le U_{n:n} = \max_{1 \le i \le n} U_i$) and $(k_n)_{n \in \mathbb{N}}$ is an intermediate sequence, i.e., $k_n \to \infty$ and $k_n/n \to 0$. Since $(g_n(t))_{0 \le t \le 1}$ depends only on the $k_n + 1$ largest observations, k_n/n corresponds to v_n in the definition of the uniform tail empirical process. In fact, with F_n^{-1} denoting the empirical quantile function, i.e.,

$$F_n^{-1}(t) = U_{i:n} \quad \text{for } t \in \left(\frac{i-1}{n}, \frac{i}{n} \right],$$

the following analog to (1.1) shows the close relationship between the asymptotic theory for e_n and g_n with $v_n = k_n/n$:

$$\begin{aligned} g_n(t) &= -\frac{n}{\sqrt{k_n}} \left(F_n^{-1}\left(1 - \frac{k_n}{n} t\right) - \left(1 - \frac{k_n}{n} t\right)\right) \\ &= -\sqrt{\frac{n}{v_n}} \left(F_n^{-1}(1 - v_n t) - (1 - v_n t)\right). \end{aligned}$$

In Section 3 we will learn that a limit theorem for e_n together with stochastic bounds for $(e_n(t)/t)_{0 < t \le 1}$ implies a similar limit theorem for g_n.

Though the crucial step in the asymptotic analysis of classical tail processes is to study the case of uniform random variables, for practical applications we need more general limit theorems. Typically these can be derived from the versions for uniform random variables by a combination of quantile transformation techniques and the standard δ-method. For example, the *tail empirical quantile function*

$$Q_n(t) := Q_{n,k_n}(t) := F_n^{-1}\left(1 - \frac{k_n}{n}t\right) = X_{n-[k_n t]:n}$$

based on a sample of size n taken from a stationary time series X_i, $i \in \mathbb{N}$, with common continuous d.f. F can be represented as

$$Q_n(t) = F^{-1}(U_{n-[k_n t]:n}) = F^{-1}\left(1 - \frac{k_n}{n}t - \frac{\sqrt{k_n}}{n}g_n(t)\right),$$

where $U_i = F(X_i)$ are uniformly distributed. If F has a Lebesgue density f, then one may apply a Taylor expansion of F^{-1} to check that, for all arguments t, the tail quantile process

$$g_{n,F}(t) = \frac{n}{\sqrt{k_n}} f\left(F^{-1}\left(1 - \frac{k_n}{n}t\right)\right)\left(F_n^{-1}\left(1 - \frac{k_n}{n}t\right) - F^{-1}\left(1 - \frac{k_n}{n}t\right)\right). \tag{1.2}$$

behaves asymptotically as $g_n(t)$. Drees and de Haan [14] obtained weighted approximations for $g_{n,F}$ from approximations of g_n under almost minimal conditions on F in the case of i.i.d. data, and the methods used there easily carry over to more general time series.

However, the process (1.2) is not suitable for applications in extreme value statistics, because the normalization factor depends on the unknown density which is particularly difficult to estimate accurately in the tail where little data is available. In [10, 11] it is advised instead to use weighted approximations of the tail empirical quantile function Q_n after a suitable affine standardization with normalizing constants not depending on t. It was shown that the asymptotic behavior of a large class of important estimators in extreme value statistics can be easily derived from these approximations. This advantage, however, comes at the cost of weight functions depending on the unknown underlying d.f. via its extreme value index.

In Section 3 limit theorems for the uniform tail empirical quantile function g_n are given under suitable mixing conditions, while Section 4 deals with weighted approximations of the affinely standardized process Q_n. In Section 5 we apply these results to establish the asymptotics of well-known tail estimators. Finally, we describe some well-known time series models that satisfy the assumptions of the main theorems.

It is worth mentioning that the above classical tail empirical processes, while a powerful tool to analyze the performance of statistical procedures to infer the tail of F, unlike the empirical point processes, do not contain any information about the temporal order of the extreme observations. Hence they cannot be used directly to establish the asymptotics of estimators for the extremal index and other quantities describing the clustering of extremes. Unfortunately, general limit theorems for empirical processes that take into account an increasing number of large observations as the sample size increases, while capturing the full temporal structure of these extremes, are not yet available in the literature.

2 The uniform tail empirical process

Limit theorems for the uniform tail empirical process e_n have been almost exclusively studied for absolutely regular (i.e., β-mixing) time series. Recall that a time series X_i, $i \in \mathbb{N}$, is called *absolutely regular* if

$$\beta_X(l) = \beta(l) := \sup_{m\in\mathbb{N}} E\left(\sup_{A\in\mathcal{B}_{m+l+1}^{\infty}} \left|P(A \mid \mathcal{B}_1^m) - P(A)\right|\right) \longrightarrow 0 \tag{2.1}$$

as $l \to \infty$. Here $\mathcal{B}_1^m$ and $\mathcal{B}_{m+l+1}^{\infty}$ denote the σ-fields generated by $(X_i)_{1\le i\le m}$ and $(X_i)_{i\ge m+l+1}$, respectively. Hence $\beta(l)$ is the maximal expected variational distance between the unconditional and the conditional distribution of the future of the time series given the past when future and past time periods are separated by l time points. In particular, the β-mixing property rules out long range dependence. It is, though, satisfied by recurrent Markov chains under mild extra conditions and, more specifically, by ARMA-, ARCH- and GARCH-models. In these examples the mixing coefficients decrease geometrically, i.e., $\beta(l) = O(\tau^l)$ for some $\tau \in (0,1)$, as $l \to \infty$ (see [9, Section 2.4] for details).

The first limit theorem for e_n is due to Rootzén [29]. Its proof is based on the blocks approach, that is, the stationary sequence U_i, $1 \le i \le n$, of uniform random variables is split into blocks of length $l_n \to \infty$ such that asymptotically the dependence between nonadjacent blocks vanishes. For this, one assumes that the β-coefficients pertaining to the U_i satisfy

$$\frac{\beta(l_n)}{l_n} n \longrightarrow 0 \qquad \text{for some sequence } l_n = o\left((nv_n)^{1/2}/\log^2(nv_n)\right). \tag{2.2}$$

(In fact, in [29] a slightly stronger condition is imposed, but a minor modification of the proof given there shows that (2.2) is sufficient.)

To establish convergence of the finite-dimensional marginals of e_n, it is natural to assume that the covariance of the number of exceedances over two different thresholds in one block converges after a suitable standardization:

$$\frac{1}{l_n v_n} \operatorname{Cov}\left(\sum_{i=1}^{l_n} 1_{\{U_i>1-v_n x\}}, \sum_{i=1}^{l_n} 1_{\{U_i>1-v_n y\}}\right) \longrightarrow c(x,y) \tag{2.3}$$

for all $0 < x, y \le 1$. Indeed, if the sequence U_i, $i \in \mathbb{N}$, is geometrically β-mixing then, under very weak extra conditions, (2.3) is necessary and sufficient for the convergence of the finite-dimensional marginals.

To see this, assume $\log^2 n = o(l_n)$ and $n^{-1}\log^2 n \log^4(\log n) = o(v_n)$. Split the sequence U_i, $1 \le i \le n$, into alternating $m_n = [n/(l_n + s_n)]$ large blocks of length l_n and m_n small blocks of length $s_n = [c \log n]$ (and a negligible remainder), where c is chosen such that $\beta(s_n) m_n \to 0$. According to a result by Eberlein [15], the variational distance between the distribution of the large blocks and the distribution of m_n i.i.d. copies of the first large block vanishes asymptotically, and the same holds true for the small blocks. Moreover, using the fact that U_i is geometrically β-mixing and Remark 3 on page 10 of [9], one can show that

$$E\left(\frac{1}{\sqrt{nv_n}} \sum_{i=1}^{s_n} 1_{\{U_i>1-v_n\}}\right)^2 = O\left(\frac{\log^2 n}{n}\right).$$

It follows that the sum of m_n i.i.d. copies of $(nv_n)^{-1/2}\sum_{i=1}^{s_n} 1_{\{U_i>1-v_n\}}$ converges to 0 in probability. Therefore, $e_n(x)$ converges if and only if the sum of m_n copies of

$$B_1 = (nv_n)^{-1/2} \sum_{i=1}^{l_n} 1_{\{U_i>1-v_n x\}}$$

converges, and an analogous conclusion holds true for all linear combinations of any finite-dimensional marginals of e_n. Since B_1 is bounded by $l_n(nv_n)^{-1/2}$, which tends to 0, the Normal Convergence Criterion and its Corollary given in [24, p. 328] show that a limit of these marginals, if it exists, must be normal, and convergence holds if and only if (2.3) is satisfied.

So up to now we know that, under (2.2) and weak extra conditions, (2.3) is necessary and sufficient for the weak convergence of the finite-dimensional marginals, and that the limiting process is Gaussian, if it exists. However, Rootzén [30] demonstrated by an example that (2.2) and (2.3) are not sufficient to prove weak convergence of e_n, even if the β-coefficients decrease geometrically. Therefore, he assumed in addition that for all $\theta \in (0, 1)$ there exists D_θ such that for $N_n(x, y) := \sum_{i=1}^{l_n} 1_{\{1-v_n y<U_i\le 1-v_n x\}}$

$$E\left(N_n^2(x, y) \mid N_n(x, y) \neq 0\right) \le D_\theta \qquad \text{for all } \theta \le x < y \le 1. \tag{2.4}$$

That is, the second moment of any cluster size is supposed to be bounded, where a cluster is defined as a nonempty set of exceedances in one block belonging to a fixed interval $(1-v_n y, 1-v_n x]$. In fact, condition (2.4) is only needed to establish the following bound:

$$\frac{1}{l_n v_n} E\left(\sum_{i=1}^{l_n} 1_{\{1-v_n y<U_i\le 1-v_n x\}}\right)^2 \le D_\theta\ (y-x) \quad \text{for all } \theta \le x < y \le 1. \tag{2.5}$$

Though (2.5) lacks a nice interpretation in terms of cluster sizes, in view of (2.3) this condition seems more natural than (2.4).

To judge how stringent condition (2.5) is recall that, for geometrically β-mixing time series, condition (2.3) for all l_n of order larger than $\log^2 n$ but smaller than $(nv_n)^{1/2}$ is indeed necessary for the convergence of e_n. If v_n is a regularly varying function of n, i.e., $v_{[\lambda n]} \sim \lambda^\alpha v_n$, $\lambda > 0$, for some $\alpha \in [-1, 0)$, then we may choose l_n slowly varying such that $l_n v_n \to 0$, so that by monotonicity

$$\frac{1}{l_n v_n} E\left(\sum_{i=1}^{l_n} 1_{\{U_i>1-v_n x\}} \cdot \sum_{i=1}^{l_n} 1_{\{U_i>1-v_n y\}}\right) \longrightarrow c(x, y)$$

uniformly in x and y. Hence

$$\begin{aligned} c(\lambda^\alpha x, \lambda^\alpha y) \quad &\longleftarrow \quad \frac{1}{l_n v_n} E\left(\sum_{i=1}^{l_n} 1_{\{U_i>1-v_{[\lambda n]}x\}} \cdot \sum_{i=1}^{l_n} 1_{\{U_i>1-v_{[\lambda n]}y\}}\right) \\ &\sim \quad \frac{\lambda^\alpha}{l_{[\lambda n]} v_{[\lambda n]}} E\left(\sum_{i=1}^{l_{[\lambda n]}} 1_{\{U_i>1-v_{[\lambda n]}x\}} \cdot \sum_{i=1}^{l_{[\lambda n]}} 1_{\{U_i>1-v_{[\lambda n]}y\}}\right) \\ &\longrightarrow \quad \lambda^\alpha c(x, y), \end{aligned}$$

i.e., the limiting covariance function is homogeneous of order 1. Moreover, it defines a positive, bounded measure on the unit square via $[0, x] \times [0, y] \mapsto c(x, y)$ which is also homogeneous. It follows that $c(y, y) - 2c(x, y) + c(x, x)$ is of the order $y - x$, that is, the limit of the left-hand side of (2.5) is bounded by $D_\theta\ (y - x)$, provided the aforementioned mild extra conditions hold. This shows that (2.5) is a reasonable condition which does not restrict the class of potential limiting processes.

Now we are ready to state Rootzén's [29] main result.

Theorem 2.1. *Under the conditions* (2.2), (2.3), *and* (2.5), *for some* $\theta \in [0, 1]$, *the uniform tail empirical process* e_n *converges weakly in the Skorohod space* $D[\theta, 1]$ *to a pathwise continuous centered Gaussian process* e *with covariance function* c.

Since convergence of the finite-dimensional marginals has already been established, it remains to prove tightness of $(e_n)_{n\in\mathbb{N}}$. For this, one uses the aforementioned approximation of blocks separated in time by independent blocks and the classical tightness criterion as given by Billingsley [1, Theorem 15.5].

The convergence of e_n describes the asymptotic behavior of observations of the order $1 - v_n$, but even in case of $\theta = 0$ it contains little information about the most extreme observations, as both $e_n(x)$ and $e(x)$ are close to 0 for x nearby the origin.

Stronger assertions about the asymptotic behavior of extreme observations may be provided by limit theorems for the weighted tail empirical process e_n/q, where $q > 0$ is a suitable weight function in $D[0, 1]$ which tends to 0 at a suitable rate as the argument tends to 0. In this way differences between $e_n(x)$ and its limit $e(x)$ are inflated, so that convergence of $e_n(x)/q(x)$ is more informative for x close to 0.

Of course, one may expect convergence only if $q(x)$ does not tend to 0 too fast as x tends to 0. It is well known that in case of i.i.d. random variables $e_n/q \to e/q$ weakly in $D[0, 1]$ if

$$\inf_{\vartheta \le x \le 1} q(x) > 0 \quad \text{for all } \vartheta > 0 \tag{2.6}$$

and

$$(x \log|\log x|)^{1/2} = o\,(q(x)) \quad \text{as } x \downarrow 0. \tag{2.7}$$

In fact, if $q(x)x^{-1/2}$ is decreasing in a neighborhood of 0, then (2.7) is also necessary for the convergence of e_n/q.

It turns out that, in the present setting of β-mixing time series, the conditions of Theorem 2.1 with $\theta = 0$ are sufficient to prove convergence of e_n/q, but $q(x)$ must tend to 0 more slowly than in (2.7). For simplicity, we confine ourselves to weight functions q satisfying (2.6) and

$$x^{\nu}|\log x|^{\mu} = O\,(q(x)) \quad \text{as } x \downarrow 0. \tag{2.8}$$

Theorem 2.2. *Under the conditions of Theorem* 2.1 *with* $\theta = 0$, *we have*

$$\frac{e_n}{q} \longrightarrow \frac{e}{q}$$

weakly in $D[0, 1]$ *for all* q *satisfying* (2.6) *and* (2.8) *with* $\nu < 1/4$, *or* $\nu = 1/4$ *and* $\mu > 1/4$.

To prove convergence of e_n/q for weight functions that converge to 0 nearly as fast as it is allowed in (2.7), we must strengthen either condition (2.2) on the relationship between l_n and nv_n, or the moment condition (2.5).

Theorem 2.3. *Suppose* (2.2) *and* (2.3) *hold, and* q *meets the conditions* (2.6) *and* (2.8) *with* $1/4 < \nu \le 1/2$. *Then*

$$\frac{e_n}{q} \longrightarrow \frac{e}{q}$$

weakly in $D[0, 1]$, *provided one of the following conditions is satisfied:*

(i) $l_n = o((nv_n)^{-3(1-2\nu)/(4(1-\nu))} \log^{-3\mu/(2(1-\nu))}(nv_n))$ *and* (2.5) *holds true for* $\theta = 0$,

(ii) $\frac{1}{l_n v_n} E(\sum_{i=1}^{l_n} 1_{\{1-v_n y<U_i\le 1-v_n x\}})^4 \le D(y-x)$ *for some* $D>0$ *and all* $0\le x<y\le 1$, *and* $\mu > 1/4$ *if* $\nu = 1/2$.

Note that for geometrically β-mixing time series condition (i) implies $\nu < 1/2$, or $\nu = 1/2$ and $\mu > 1/3$, which is slightly stronger than what is required in (ii).

To prove Theorems 2.2 and 2.3, one has to verify that both $\sup_{0<x\le\vartheta} |e(x)/q(x)|$ and $\sup_{0<x\le\vartheta} |e_n(x)/q(x)|$ tend to 0 in probability as $\vartheta \downarrow 0$ and that this holds uniformly in n for the latter. For both expressions one splits the interval $(0,\vartheta]$ into subintervals $(\vartheta e^{-(i+1)}, \vartheta e^{-i}]$, on which, e.g., $e(x)/q(x)$ can be bounded by $\sup |e(x)|/q(\vartheta e^{-(i+1)})$ if q is increasing in a neighborhood of 0. In addition, one uses a bound on the exceedance probabilities of the supremum of a Gaussian process established by Fernique [16], and again the aforementioned blocks approach for e_n. For details about the proof and further refinements of the assertions we refer to [12].

By a standard probability integral transform one can derive limit theorems for the tail empirical process of time series X_i, $i\in\mathbb{N}$, whose marginal d.f. F is continuous and belongs to the domain of attraction of an extreme value distribution G_γ. Recall (e.g., from [23, Section 1.1]) that by definition $F\in D(G_\gamma)$ for some $\gamma\in\mathbb{R}$ if and only if

$$\lim_{n\to\infty} F^n(a_n x + b_n) = G_\gamma(x) := \exp\left(-(1+\gamma x)^{-1/\gamma}\right), \qquad x\in\mathbb{R}, \tag{2.9}$$

for some $a_n > 0$ and $b_n \in \mathbb{R}$, where $G_0(x)$ is interpreted as the limit of $G_\gamma(x)$ for $\gamma \to 0$, i.e., $\exp(-e^{-x})$. Then the parameter γ is called *extreme value index*. Convergence (2.9) is equivalent to

$$\lim_{t\to\infty} t(1 - F(a(t)x + b(t))) = -\log G_\gamma(x), \qquad x\in\mathbb{R}, \tag{2.10}$$

for some functions $a:\mathbb{R}\to(0,\infty)$ and $b:\mathbb{R}\to\mathbb{R}$. The convergence for $x=0$ indicates that one may assume $b(t) = F^{-1}(1-1/t)$. Now choose a sequence u_n converging to the right endpoint $F^{-1}(1)$ of the support of F such that $n\,(1-F(u_n))\to\infty$. Hence with $\sigma_n := a(1/(1-F(u_n)))$, one has

$$T_n(x) := \frac{1-F(\sigma_n x + u_n)}{1-F(u_n)} \longrightarrow -\log G_\gamma(x) = (1+\gamma x)^{-1/\gamma}, \qquad x\ge 0,$$

as $n\to\infty$.

Define the tail empirical d.f. by

$$e_{n,F}(x) := \frac{1}{\sqrt{n(1-F(u_n))}} \sum_{i=1}^{n} \left(1_{\{X_i > u_n + \sigma_n x\}} - (1 - F(u_n + \sigma_n x))\right).$$

For the time series of uniformly distributed random variables $U_i := F(X_i)$ and $v_n = 1 - F(u_n)$, it follows that

$$e_{n,F} = (e_n\,(T_n(x)))_{0\le x\le 1} \qquad \text{almost surely.}$$

Hence Theorem 2.1 yields (see [29, Theorem 2.1])

Corollary 2.4. *Suppose that* (2.2) *holds for the β-mixing coefficients of the time series* X_i, $i\in\mathbb{N}$, *and* $v_n := 1 - F(u_n)$. *If*

$$\frac{1}{l_n(1-F(u_n))} \operatorname{Cov}\left(\sum_{i=1}^{l_n} 1_{\{X_i > u_n + \sigma_n x\}}, \sum_{i=1}^{l_n} 1_{\{X_i > u_n + \sigma_n y\}}\right) \longrightarrow c_F(x,y)$$

for all $0 < x, y \leq 1$, *and if for all* $0 < \theta < x_\gamma := 1/\max(-\gamma, 0)$ *(with the convention* $1/0 := \infty$*) there is a constant* D_θ *such that*

$$\frac{1}{l_n(1-F(u_n))} E\left(\sum_{i=1}^{l_n} 1_{\{u_n+\sigma_n x < X_i \leq u_n+\sigma_n y\}}\right)^2 \leq D_\theta\left((1+\gamma x)^{-1/\gamma} - (1+\gamma y)^{-1/\gamma}\right)$$

for $0 \leq x < y \leq \theta$, *then*

$$e_{n,F} \longrightarrow e_F$$

weakly in $D[0, x_\gamma)$, *where* e_F *is a continuous centered Gaussian process with covariance function* c_F.

It is more difficult to carry over the limit theorems for the weighted tail empirical process, because for that purpose one needs conditions on the rate at which convergence holds in (2.10). Since the weighted approximations of tail quantile processes we will establish in Section 4 are more useful for applications in extreme value statistics, we do not give any generalizations of Theorems 2.2 and 2.3 here.

3 The uniform tail quantile process

The following modification of a lemma by Vervaat [34] is an important tool to derive limit theorems for quantile processes from approximations of the pertaining empirical process.

Lemma 3.1. *If, for some* $\varepsilon > 0$, *we have nondecreasing functions* $z_n \in D[0, 1+\varepsilon]$ *and positive constants* $\lambda_n \to 0$ *such that*

$$\left(\frac{z_n(x) - x}{\lambda_n}\right)_{0 \leq x \leq 1+\varepsilon} \longrightarrow z$$

for some continuous function z, *then for generalized inverse functions* $z_n^{-1}(t) = \inf\{x \in [0, 1+\epsilon] \mid z_n(x) \geq t\}$, *restricted to* $t \in [0, 1]$, *one has*

$$\left(\frac{z_n^{-1}(t) - t}{\lambda_n}\right)_{0 \leq t \leq 1} \longrightarrow -z.$$

Since the limiting functions are continuous, in both cases the functions converge uniformly. Note that, for sufficiently large n, the set considered in the definition of the generalized inverse is nonempty, because $z_n(1+\varepsilon)$ converges to $1+\varepsilon$.

By Skorohod's representation theorem (see [27, p. 6]), in the setting of Theorem 2.1 with $\theta = 0$ there exist versions $\tilde{e}_n$ and $\tilde{e}$ of e_n and e, respectively (i.e., processes with the same distribution), such that $\tilde{e}_n \to \tilde{e}$ a.s. Obviously, the convergence even holds in $D[0, 1+\varepsilon]$ if (2.3) and (2.5) hold for all $0 \leq x \leq y \leq 1+\varepsilon$ and some $\epsilon > 0$. Now let $v_n = k_n/n$ and $\lambda_n = k_n^{-1/2}$ for some intermediate sequence $(k_n)_{n\in\mathbb{N}}$, and apply Vervaat's lemma pathwise to

$$z_n = (x + \lambda_n \tilde{e}_n(x))_{0 \leq x \leq 1+\varepsilon} =^d \left(\frac{1 - F_n(1 - v_n x)}{v_n}\right)_{0 \leq x \leq 1+\varepsilon}$$

(with $=^d$ denoting identity in distribution). Since

$$z_n^{-1} =^d \left(\frac{1 - F_n^{-1}(1 - v_n t)}{v_n} \right)_{0 \le t \le 1} = (t + \lambda_n g_n(t))_{0 \le t \le 1}$$

and $-\tilde{e} =^d e$, we obtain the following corollary to Theorem 2.1.

Corollary 3.2. *If for some* $\varepsilon > 0$, (2.2), (2.3) *and* (2.5) *hold for all* $0 \le x \le y \le 1 + \varepsilon$, *and* $v_n = k_n/n$, *then*

$$g_n \longrightarrow e$$

weakly in $D[0, 1]$.

Again one is interested in weighted approximations for g_n which bear more information about the asymptotic behavior of the largest observations. It is less trivial to deduce these from Theorems 2.2 and 2.3, since one needs stochastic upper and lower bounds for $V_n(t)/t$, where

$$V_n(t) := \frac{n}{k_n}(1 - U_{n-[k_n t]:n}), \quad 0 \le t \le 1,$$

denotes the *uniform tail empirical quantile function*. Note that

$$g_n(t) = \sqrt{k_n}(V_n(t) - t), \quad 0 \le t \le 1,$$

so that bounds for $g_n(t)/t$ follow immediately.

Lemma 3.3. *Suppose that for some* $\varepsilon > 0$ *the conditions* (2.2) *and* (2.5) *are satisfied for* $0 \le x \le y \le 1 + \varepsilon$ *and* $v_n = k_n/n$. *Then*

$$\sup_{1/(2k_n) \le t \le 1} \frac{V_n(t)}{t} = O_p(1) \quad \textit{and} \quad \sup_{0 \le t \le 1} \frac{t}{V_n(t)} = O_p(1).$$

The proof is based on the same ideas used to establish Theorems 2.2 and 2.3: divide the n observations into blocks of length l_n and split the interval $(0, 1]$ into subintervals of type $(\lambda e^{-(i+1)}, \lambda e^{-i}]$, $i \ge 0$ and $(\lambda, 1]$, $\lambda \in (0, 1)$ (see [12]).

For i.i.d. random variables U_i such in probability linear bounds are known for the whole empirical quantile function (and hence for V_n, too); see, e.g., [31, Section 10.4]. For φ-mixing time series similar results were established by Puri and Tran [26, Theorem 1.1]. However, φ-mixing implies β-mixing and we will see that φ-mixing with a suitable rate implies condition (2.5) as well. Hence the conditions of Lemma 3.3 are somewhat weaker than the ones used by Puri and Tran. (Of course, the results are not fully comparable as Puri and Tran did not confine themselves to the tail of the empirical quantile function.)

Using Lemma 3.3 it is easy to verify that $\sup_{1/(2k_n) \le t \le \vartheta} |g_n(t)/q(t)|$ converges to 0 as $\vartheta \downarrow 0$ if and only if $\sup_{1/(2k_n) \le t \le \vartheta} |e_n(t)/q(t)|$ tends to 0, which has been verified in the proofs of Theorems 2.2 and 2.3. Hence we obtain the following.

Corollary 3.4. *Suppose that for* $v_n = k_n/n$ *and some fixed* $\varepsilon > 0$ *the conditions of Theorems* 2.2 *or* 2.3 *are satisfied for* $0 < x \le y \le 1 + \varepsilon$. *Then*

$$\frac{g_n}{q} 1_{[1/(2k_n), 1]} \longrightarrow \frac{e}{q}$$

weakly in $D[0, 1]$.

Notice that the restriction to the interval $[1/(2k_n), 1]$ is necessary, because $g_n(t)/q(t) \sim n k_n^{-1/2}(1 - U_{n:n})/q(t) \longrightarrow \infty$ a.s. as $t \downarrow 0$.

4 The tail empirical quantile function

In this section we study the asymptotics of the linearly standardized tail empirical quantile function

$$\frac{Q_n(t) - D_n}{c_n} = \frac{X_{n-[k_n t]:n} - D_n}{c_n}, \qquad 0 \le t \le 1,$$

for β-mixing stationary time series X_i, $i \in \mathbb{N}$, with d.f. $F \in D(G_\gamma)$ (see (2.9)). Here $c_n > 0$ denotes a constant and D_n a suitable real-valued random variable. We assume that F is continuous (at least, in a neighborhood of the right endpoint of its support), though with minor modifications the results can be proved for more general d.f.'s. Then we may apply the limit theorems stated in Section 3 to the time series $U_i := F(X_i)$ which is β-mixing with coefficients bounded by those of the original time series.

To see what kind of results are to be expected, note that by an inversion of (2.10) one obtains

$$\lim_{\lambda \downarrow 0} \frac{F^{-1}(1-\lambda x) - F^{-1}(1-\lambda)}{a(1/\lambda)} = (-\log G_\gamma)^{-1}(x) = \frac{x^{-\gamma} - 1}{\gamma}, \qquad x > 0,$$

where the right-hand side is interpreted as $-\log x$ if $\gamma = 0$. In fact, this convergence is equivalent to $F \in D(G_\gamma)$ (see [17]).

Thus one may argue heuristically as follows:

$$\begin{aligned}
\frac{Q_n(t) - F^{-1}(1-k_n/n)}{a(n/k_n)} &= \frac{F^{-1}\left(1 - V_n(t)k_n/n\right) - F^{-1}(1-k_n/n)}{a(n/k_n)} \\
&\approx \frac{(V_n(t))^{-\gamma} - 1}{\gamma} \\
&= \frac{t^{-\gamma}\left(1 + k_n^{-1/2} g_n(t)/t\right)^{-\gamma} - 1}{\gamma} \\
&\approx \frac{t^{-\gamma} - 1}{\gamma} - k_n^{-1/2} t^{-(\gamma+1)} g_n(t),
\end{aligned} \tag{4.1}$$

and it is plausible that

$$k_n^{1/2}\left(\frac{Q_n(t) - F^{-1}(1-k_n/n)}{a(n/k_n)} - \frac{t^{-\gamma}-1}{\gamma}\right)_{0<t\le 1} \longrightarrow (t^{-(\gamma+1)} e(t))_{0<t\le 1} \tag{4.2}$$

weakly in $D(0,1]$ under conditions analogous to those used in Section 2.

To extend this result to convergence on the closed interval $[0,1]$, one must again introduce a weight function, which depends on the underlying d.f. F via its extreme value index. If q denotes an admissible weight function in the sense of Theorem 2.2 or 2.3, then obviously $t \longmapsto t^{-(\gamma+1)} q(t)$ is appropriate in the present context. If one uses the standardization of Q_n with $F^{-1}(1-k_n/n)$, though, a problem arises in the immediate neighborhood of $t = 0$ if $t^{-(\gamma+1)} q(t) \longrightarrow 0$ as $t \downarrow 0$, which may be true for $\gamma < -1/2$. For, in that case, the left-hand side of (4.2) divided by the weight function converges to ∞ as $t \downarrow 0$. One can avoid this problem by a restriction to the interval $[1/(2k_n), 1]$ as we did in Section 3, yet this approach renders the application of the limit theorem less convenient. Alternatively, one may replace $F^{-1}(1-k_n/n)$ in the numerator of (4.2) by the random variable

$$D_n = X_{n:n} + \frac{a(n/k_n)}{\gamma} \tag{4.3}$$

if $\gamma < -1/2$. Observe that, under the conditions stated below, $D_n - F^{-1}(1-k_n/n) = o_p(k_n^{-1/2})$, so in the unweighted limit theorem the assertion remains essentially the same if one replaces $F^{-1}(1-k_n/n)$ with D_n.

Next, we state the exact conditions under which a weighted version of (4.2) can be proved. First we need analogs to (2.2), (2.3), and (2.5), namely

$$\frac{\beta(l_n)}{l_n} n \to 0 \text{ for some sequence } l_n = o(k_n^{1/2}/\log^2 k_n), \tag{4.4}$$

with $\beta = \beta_X$ defined in (2.1),

$$\frac{n}{k_n l_n} \operatorname{Cov}\left(\sum_{i=1}^{l_n} 1_{\{X_i > F^{-1}(1-\frac{k_n}{n}x)\}}, \sum_{i=1}^{l_n} 1_{\{X_i > F^{-1}(1-\frac{k_n}{n}y)\}}\right) \longrightarrow c(x, y) \tag{4.5}$$

and

$$\frac{n}{k_n l_n} E\left(\sum_{i=1}^{l_n} 1_{\{F^{-1}(1-\frac{k_n}{n}y) < X_i \le F^{-1}(1-\frac{k_n}{n}x)\}}\right)^2 \le D\,(y - x) \tag{4.6}$$

for all $0 < x \le y \le 1 + \varepsilon$ and some $D, \varepsilon > 0$.

In addition, we need a condition to ensure that the approximation (4.1) is sufficiently accurate, i.e., that the approximation error is of smaller order than $k_n^{-1/2}$. To this end, define

$$R(\lambda, x) := \frac{F^{-1}(1-\lambda x) - F^{-1}(1-\lambda)}{a(1/\lambda)} - \frac{x^{-\gamma} - 1}{\gamma},$$

so that $R(\lambda, x) \to 0$ as $\lambda \downarrow 0$. In fact, it can be proved that, for all $\eta > 0$,

$$\sup_{0 < x \le 1+\varepsilon} x^{\gamma+\eta} |R(\lambda, x)| \longrightarrow 0. \tag{4.7}$$

Hence, for k_n increasing not too fast, it follows that

$$k_n^{1/2} \sup_{0 < x \le 1+\varepsilon} \frac{|R(k_n/n, x)|}{x^{-(\gamma+1)} q(x)} \longrightarrow 0 \tag{4.8}$$

as $n \to \infty$.

We get the following analog to Theorem 2.2.

Corollary 4.1. *If the time series X_i, $i \in \mathbb{N}$, has a continuous marginal d.f. $F \in D(G_\gamma)$ and satisfies* (4.4)–(4.6) *for some $\varepsilon > 0$ and some sequence $k_n \to \infty$ with $k_n/n \to 0$ such that* (4.8) *holds, then*

$$\left(\frac{t^{\gamma+1}}{q(t)} \sqrt{k_n} \left(\frac{Q_n(t) - D_n}{a(n/k_n)} - \frac{t^{-\gamma} - 1}{\gamma}\right)\right)_{t \in [0,1]} \longrightarrow \left(\frac{t^{\gamma+1}}{q(t)} (t^{-(\gamma+1)} e(t))\right)_{t \in [0,1]}$$

weakly in $D[0, 1]$ (with the convention $0/0 := 0$). Here $D_n = F^{-1}(1 - k_n/n)$ if $\gamma \ge -1/2$, and D_n is defined in (4.3) *if $\gamma < -1/2$, and q is a weight function of the type considered in Theorem* 2.2.

Likewise, one can establish a counterpart to Theorem 2.3 if

$$l_n = o(k_n^{-3(1-2\nu)/(4(1-\nu))} \log^{-3\mu/(2(1-\nu))} k_n)$$

or (4.6) is replaced with the corresponding condition on the fourth moment. Note that, in contrast to Section 3, here we require not only a lower bound on the rate at which k_n tends to infinity (implicitly given by (4.4)), but also an upper bound is necessary for (4.8) to hold. If the rate of convergence in (4.7) is particularly slow, equivalently, if the extreme value approximation is poor, then both conditions may contradict each other. However, if, as usual, the time series is geometrically β-mixing, then (4.4) merely implies $\log^2 n \log^4(\log n) = o(k_n)$ (cf. Section 2). In practice, this does not constitute a severe restriction unless the sample size is huge. Moreover, it does not contradict (4.8) if the rate in (4.7) is polynomial.

If one imposes a second order strengthening of the basic condition $F \in D(G_\gamma)$, then more order statistics can be taken into account, yet in general the limiting process will not be centered any longer. To be more precise, assume

$$\frac{F^{-1}(1-\lambda x) - F^{-1}(1-\lambda)}{a(1/\lambda)} = \frac{x^{-\gamma}-1}{\gamma} + \Phi(\lambda)\Psi(x) + o(\Phi(\lambda))$$

for some function $\Phi(\lambda) \to 0$ as $\lambda \downarrow 0$ and a function $\Psi(x)$ that is not a multiple of $(x^{-\gamma}-1)/\gamma$. De Haan and Stadtmüller [18, Theorem 1] proved that Φ is regularly varying at 0 with some exponent $\varrho \geq 0$. If ϱ is strictly positive (which essentially corresponds to a polynomial rate in (4.7)), then for an appropriate choice of a and Φ one has $\Psi(x) = (x^{\varrho-\gamma}-1)/(\gamma-\varrho)$. Now assume that

$$\lim_{n\to\infty} \sqrt{k_n}\,\Phi(k_n/n) = \mu \in \mathbb{R}; \tag{4.9}$$

notice that (4.8) holds if and only if $\mu = 0$, while for $|\mu| > 0$ the tail empirical quantile function is based on more order statistics. In that case, under the conditions (4.4)–(4.6),

$$\left(\frac{t^{\gamma+1}}{q(t)}\sqrt{k_n}\left(\frac{Q_n(t)-\overline{D}_n}{a(n/k_n)} - \frac{t^{-\gamma}-1}{\gamma}\right)\right)_{t\in[0,1]} \longrightarrow \left(\frac{t^{\gamma+1}}{q(t)}\left(t^{-(\gamma+1)}e(t) + \mu\overline{\Psi}(t)\right)\right)_{t\in[0,1]}$$

weakly in $D[0,1]$, where $\overline{D}_n = D_n - a(n/k_n)\Phi(k_n/n)\, 1_{\{\varrho\neq\gamma\}}/(\gamma-\varrho)$ and $\overline{\Psi} = \Psi + 1_{\{\varrho\neq\gamma\}}/(\gamma-\varrho)$. In the important special case $\Phi(t) = ct^{\varrho}$ for some $c \in \mathbb{R}$ and $\varrho > 0$, (4.9) is satisfied with $\mu > 0$ if $k_n \sim (\mu/c)^{2/(2\varrho+1)} n^{2\varrho/(2\varrho+1)}$. If k_n is of larger order, then the appropriate standardizing factor is $1/\Phi(k_n/n)$ instead of $\sqrt{k_n}$, and the limit is deterministic yet different from 0.

5 Applications: Statistical tail functionals

One important advantage of limit theorems for empirical processes is that one can easily derive the asymptotics of statistics that can be represented as a smooth functional of the process. In the classical setting, the so-called statistical functionals are functionals of the empirical distribution function. In [10, 11] a large class of estimators of the extreme value index γ were introduced that can be represented as a functional of the tail empirical quantile function Q_n.

More precisely, let T be a Borel-measurable real-valued functional on the set of functions $z \in D(0,1]$ such that $t^{\gamma+1}(z(t)-c)/q(t) \longrightarrow 0$ as $t \downarrow 0$, for some $c \in \mathbb{R}$. Here q denotes a weight function of the type considered in Theorem 2.2. In addition, we assume the following three regularity conditions.

- T is location- and scale-invariant:

$$T(az+b) = T(z) \tag{5.1}$$

for all $a > 0$, $b \in \mathbb{R}$ and all functions as described above;

- $$T\left(\left(\frac{t^{-\gamma}-1}{\gamma}\right)_{t\in(0,1]}\right)=\gamma; \tag{5.2}$$

- T is differentiable at $\left((t^{-\gamma}-1)/\gamma\right)_{t\in(0,1]}$ in the following sense: there exists a signed measure $\nu_{T,\gamma}$ on (0, 1] with $\int t^{-(\gamma+1)}q(t)|\nu_{T,\gamma}|(dt)<\infty$ such that

$$\frac{1}{\varepsilon_n}\cdot\left(T\left(\left(\frac{t^{-\gamma}-1}{\gamma}+\varepsilon_n z_n(t)\right)_{t\in(0,1]}\right)-T\left(\left(\frac{t^{-\gamma}-1}{\gamma}\right)_{t\in(0,1]}\right)\right)\longrightarrow\int z(t)\,\nu_{T,\gamma}(dt) \tag{5.3}$$

for all $\varepsilon_n\downarrow 0$ and z_n satisfying

$$\sup_{0<t\le 1}\frac{t^{\gamma+1}}{q(t)}|z_n(t)-z(t)|\longrightarrow 0 \tag{5.4}$$

for some continuous function z of the aforementioned type. (This is a special kind of Hadamard differentiability of T.)

Then $T(Q_n)$ is called a smooth statistical tail functional. Since (5.4) is exactly the type of convergence we have established in Corollary 4.1, using the δ-method one can derive the asymptotic normality of the estimator $T(Q_n)$ for γ as follows. By Corollary 4.1 and the Skorohod representation theorem there exist versions of Q_n and e such that (5.4) holds almost surely for $\varepsilon=k_n^{-1/2}$,

$$z_n(t):=\sqrt{k_n}\left(\frac{Q_n(t)-D_n}{a(n/k_n)}-\frac{t^{-\gamma}-1}{\gamma}\right) \tag{5.5}$$

and

$$z(t)=t^{-(\gamma+1)}e(t).$$

Hence by (5.1)–(5.3)

$$\begin{aligned}\sqrt{k_n}\,(T(Q_n)-\gamma) &= \sqrt{k_n}\left(T\left(\frac{Q_n-D_n}{a(n/k_n)}\right)-T\left(\left(\frac{t^{-\gamma}-1}{\sigma}\right)_{t\in(0,1]}\right)\right)\\ &\longrightarrow \int t^{-(\gamma+1)}e(t)\,\nu_{T,\gamma}(dt)\quad\text{a.s.,}\end{aligned}$$

the latter being a centered normal random variable with variance

$$\sigma^2_{T,\gamma}:=\int_{(0,1]}\int_{(0,1]}(st)^{-(\gamma+1)}c(s,t)\,\nu_{T,\gamma}(ds)\,\nu_{T,\gamma}(dt).$$

For details see [12].

An important example is the so-called maximum likelihood estimator in a generalized Pareto model (see [32]) with pertaining signed measure

$$\nu_{ML,\gamma}(dt)=\frac{(\gamma+1)^2}{\gamma^2}\left(t^{\gamma}-(2\gamma+1)t^{2\gamma}\right)dt+\frac{\gamma+1}{\gamma}\,\varepsilon_1(dt).$$

It can be shown that for this estimator the asymptotic variance equals $\sigma^2_{ML,\gamma} = (\gamma+1)^2 c(1,1)$, i.e., in comparison with the classical i.i.d. setting the variance is increased by the factor $c(1,1)$. In [10, 11] many other functionals of Q_n are examined in the i.i.d. setting. Their asymptotic normality also follows for the mixing time series considered here. Likewise, one can study estimators of extreme quantiles; see [13].

As another potential field of application of the results described in the preceding sections, Rootzén [29] suggested studying the asymptotic behavior of the tail quantile plot and the mean and median excess plots, which are popular graphical tools for judging the tail heaviness of the underlying distribution function.

6 Examples

In this section we demonstrate that well-known time series models satisfy the conditions of Corollary 4.1, so that the tail quantile function can be approximated by the Gaussian limiting process $(t^{-(\gamma+1)}e(t))_{t\in[0,1]}$.

Because Doukhan [9] gives a comprehensive discussion of β-mixing time series, we refrain from giving general sufficient conditions for β-mixing. However, it is worth mentioning that under weak extra conditions Harris recurrent Markov chains are geometrically β-mixing; see [9, Section 2.4].

Conditions of type (4.7) or (4.8) are common in extreme value theory; hence they are not specific for the situation when dependent random variables are observed. For example, if

$$F^{-1}(1-x) = c\frac{x^{-\gamma}-1}{\gamma} + d + O(x^{\varrho-\gamma}),$$

then (4.8) holds with $a(\lambda) = c\lambda^\gamma$ if

$$k_n = o(n^{2\varrho/(2\varrho+1)}).$$

See also the discussion at the end of Section 4.

It remains to check the convergence (4.5) of the covariance function and the moment condition (4.6). While in general the verification may be difficult particularly for the latter, the task becomes much easier if an additional mixing condition holds. The following result is taken from [13].

Lemma 6.1. *Suppose that $l_n = o(n/k_n)$ and*

$$\frac{n}{k_n}P\Big\{X_1 > F^{-1}\left(1-\frac{k_n}{n}x\right), X_{1+m} > F^{-1}\left(1-\frac{k_n}{n}y\right)\Big\} \longrightarrow c_m(x,y) \tag{6.1}$$

for all $m \geq 1$, $0 < x, y \leq 1+\varepsilon$.

(i) *If, in addition, there exist $D_1 \geq 0$ and a sequence $\tilde\rho(m)$ satisfying $\sum_{m=1}^\infty \tilde\rho(m) < \infty$ such that for all $m \geq 1$ and all $0 < x < y \leq 1+\varepsilon$,*

$$\frac{n}{k_n}P\{X_1 \in I_n(x,y), X_{1+m} \in I_n(x,y)\} \leq (y-x)\left(\tilde\rho(m) + D_1\frac{k_n}{n}\right) \tag{6.2}$$

with $I_n(x,y) = (F^{-1}(1-yk_n/n), F^{-1}(1-xk_n/n)]$, then the conditions (4.5) *and* (4.6) *hold with*

$$c(x,y) := \min(x,y) + \sum_{m=1}^\infty (c_m(x,y) + c_m(y,x)).$$

(ii) *In particular, if the time series is ρ-mixing, i.e.,*

$$\rho(m) := \sup_{l\in\mathbb{N}} \sup_{U\in L_2(\mathcal{B}_1^l), V\in L_2(\mathcal{B}_{l+m+1}^\infty)} \frac{|\operatorname{Cov}(U,V)|}{(\operatorname{Var}(U)\operatorname{Var}(V))^{1/2}} \longrightarrow 0$$

(with $L_2(\mathcal{A})$ denoting the space of square integrable $\mathcal{A}$-measurable functions) and $\sum_{m=1}^\infty \rho(m) < \infty$, then the assertions stated in (i) *hold true.*

If the vector (X_1, X_{1+m}) belongs to the domain of attraction of an extreme value distribution G, that is, for i.i.d. copies (Y_i, Z_i) of (X_1, X_{1+m}), one has

$$\mathcal{L}\left(\frac{\max_{1\le i\le n} Y_i - d(n)}{c(n)}, \frac{\max_{1\le i\le n} Z_i - d(n)}{c(n)}\right) \longrightarrow G$$

weakly for suitable normalizing constants, then

$$\begin{aligned}
&\lim_{t\downarrow 0} t^{-1} P\{1 - F(X_1) < tx \quad \text{or} \quad 1 - F(X_{1+m}) < ty\} \\
&\quad = -\log G\left((-\log G_1)^{-1}(x), (-\log G_2)^{-1}(y)\right) =: \ell(x,y)
\end{aligned}$$

for all $x, y > 0$, where G_1 and G_2 denote the marginal d.f.'s of G. Hence (6.1) holds with $c_m(x,y) = \ell(x,0) + \ell(0,y) - \ell(x,y) = x + y - \ell(x,y)$.

Next, we turn to specific time series models. First we examine classical causal linear time series

$$X_i = \sum_{j=0}^\infty \Psi_j Z_{i-j}, \qquad i \in \mathbb{N},$$

where the innovations Z_i are i.i.d. with balanced heavy tails, i.e., the d.f. F_Z of Z_1 satisfies $F_Z \in D(G_\gamma)$ for some $\gamma > 0$ and

$$\lim_{x\to\infty} \frac{1 - F_Z(x)}{F_Z(-x)} = \frac{p}{q} \qquad \text{for some } p = 1 - q \in (0,1).$$

This is equivalent to the conditions

$$1 - F_Z(x) = pL(x)x^{-1/\gamma}, \quad F_Z(-x) = qL(x)x^{-1/\gamma}$$

for some slowly varying function L, that is, $L(\lambda x)/L(\lambda) \to 1$ as $\lambda \to \infty$ for all $x > 0$ (see, e.g., [23, Theorem 1.6.2])

For simplicity we assume that the coefficients Ψ_j decrease at a geometric rate:

$$|\Psi_j| = O(\tau^j) \quad \text{for some } \tau \in (0,1). \tag{6.3}$$

Then the d.f. F of X_1 belongs to the domain of attraction of G_γ. More precisely, $1 - F(x) \sim c_\Psi(1 - F_Z(x))$ as $x \to \infty$ with

$$c_\Psi := \sum_{j=0}^\infty \left(p\Psi_j^{1/\gamma} 1_{\{\Psi_j>0\}} + q|\Psi_j|^{1/\gamma} 1_{\{\Psi_j<0\}}\right).$$

If F_Z has a Lebesgue density f_Z which is L_1-Lipschitz continuous:

$$\int |f_Z(z+u) - f_Z(z)|\, dz = O(u) \qquad \text{as } u \downarrow 0,$$

then X_i, $i \in \mathbb{N}$, is geometrically β-mixing; see [9, Theorem 2.3.2]. Thus condition (4.4) is fulfilled for $l_n = [c \log n]$ and suitable large $c > 0$ if k_n is of order larger than $\log^2 n \log^4(\log n)$. Moreover, by standard methods one can show that (6.1) holds with

$$c_m(x, y) \;=\; c_\Psi^{-1} \sum_{j=1}^{\infty} \min\left(x|\Psi_j|^{1/\gamma}, y|\Psi_{j+m}|^{1/\gamma}\right)\left(1_{\{\Psi_j>0, \Psi_{j+m}>0\}} + \frac{q}{p} 1_{\{\Psi_j<0, \Psi_{j+m}<0\}}\right);$$

see, e.g., [4] where similar convergence results are proved under conditions on Ψ_j much weaker than (6.3).

If X_i, $i \in \mathbb{N}$, is a finite order moving average, i.e., $\Psi_j = 0$ for all $j \geq j_0$, or if $\gamma < 1/2$, then the time series is geometrically ρ-mixing (see [2, p. 18]), and Lemma 6.1 yields the conditions (4.5) and (4.6). Another example is a first order autoregressive model, i.e., $X_i = \theta X_{i-1} + Z_i$, for which (6.2) is verified directly in [13]. Hence for all these models the tail empirical quantile function can be approximated by a Gaussian process using Corollary 4.1, provided k_n does not grow too fast.

From this we can derive the asymptotic normality not only for the Hill estimator considered in several papers (e.g., [19] and [28]), but for the much larger class of statistical tail functionals, including the maximum likelihood estimator, the moment estimator by Dekkers et al. [8] and Pickands's [25] estimator for the extreme value index.

As a second example, consider the solutions of the stochastic recursion

$$X_i = A_i X_{i-1} + B_i, \qquad i \in \mathbb{N}, \tag{6.4}$$

with i.i.d. $[0, \infty)^2$-valued random vectors (A_i, B_i). Kesten [22] proved that under some moment conditions on A_i and B_i a unique stationary solution exists and the d.f. F of X_i belongs to $D(G_\gamma)$ where γ is given by the relation $EA_1^{1/\gamma} = 1$. It was shown by Stărică [33] and Drees [12] that under mild extra conditions on the moments and the rate of convergence of k_n the conditions of Corollary 4.1 are met and that the limiting covariance function is given by

$$c(x, y) = \min(x, y) + \sum_{j=1}^{\infty}\left(x\int_0^{y/x} P\left\{\prod_{i=1}^{j} A_i > t^\gamma\right\} dt + y\int_0^{x/y} P\left\{\prod_{i=1}^{j} A_i > t^\gamma\right\} dt\right). \tag{6.5}$$

Finally, consider a first order autoregressive time series with conditional heteroscedasticity (ARCH(1)):

$$Y_i = (\alpha_0 + \alpha_1 Y_{i-1}^2)^{1/2} Z_i, \qquad i \in \mathbb{N},$$

with i.i.d. innovations Z_i. This is an unsophisticated model for the returns of a risky investment. Since $X_i = Y_i^2$ satisfies (6.4) with $A_i = \alpha_1 Z_i^2$ and $B_i = \alpha_0 Z_i^2$, it is easily seen that, under suitable moment conditions on Z_i, Corollary 4.1 applies to Y_i as well but now the limiting covariance function equals $c_Y = c/2$ with c given in (6.5).

References

[1] P. Billingsley: *Convergence of Probability Measures.* Wiley, New York, 1968.

[2] D. Bosq: *Nonparametric Statistics for Stochastic Processes (2nd ed.).* Springer, Berlin, 1998.

[3] M. Csörgő, and L. Horváth: *Weighted Approximations in Probability and Statistics*. Wiley, Chichester, 1993.

[4] S. Datta, and W.P. McCormick: Inference for the tail parameters of a linear process with heavy tail innovations. *Annals of the Institute of Statistical Mathematics* **50** (1998), 337–359.

[5] R.A. Davis, and T. Hsing: Point processes and partial sum convergence for weakly dependent random variables with infinite variance. *Annals of Prabability* **23** (1995), 879–917.

[6] R.A. Davis, and T. Mikosch: The sample autocorrelations of heavy-tailed processes with applications to ARCH. *Annals of Statistics* **26** (1998), 2049–2080.

[7] R.A. Davis, and S.I. Resnick: Limit theory for bilinear processes with heavy-tailed noise. *Annals of Applied Probability* **6**, 1191–1210.

[8] A.L.M. Dekkers, J.H.J. Einmahl, and L. de Haan: A moment estimator for the index of an extreme value distribution. *Annals of Statistics* **17** (1989), 1833–1855.

[9] P. Doukhan: *Mixing, Properties and Examples*. Springer, New York, 1994.

[10] H. Drees: On smooth statistical tail functionals. *Scandinavian Journal of Statistics* **25** (1998), 187–210.

[11] H. Drees: A general class of estimators of the extreme value index. *Journal of Statistical Planning and Inference* **66** (1998), 95–112.

[12] H. Drees: Weighted approximations of tail processes for β-mixing random variables. *Annals of Applied Probability* **10** (2000), 1274–1301.

[13] H. Drees: Extreme quantile estimation for dependent data with application to finance. Submitted to *Bernoulli*.

[14] H. Drees, and L. de Haan: Conditions for quantile process approximations. *Stochastic Models* **15** (1999), 485–502.

[15] E. Eberlein: Weak convergence of partial sums of absolutely regular sequences. *Statistics and Probability Letters* **2** (1984), 291–293.

[16] X.M. Fernique: Regularite des trajectoires des fonctions aleatoires gaussiennes. In: *Ecole d'Eté de Probabilités des Saint-Flour IV-1974* (P.-L. Hennequin, ed.). Springer, Berlin, 1975, 1–96.

[17] L. de Haan: Slow variation and characterization of domains of attraction. In: *Statistical Extremes and Applications* (J. Tiago de Oliveira, ed.). Reidel Publishing, Dordrecht, 1984, 31–48.

[18] L. de Haan, and U. Stadtmüller: Generalized regular variation of second order. *Journal of the Australian Mathematical Society Series A* **61** (1996), 381–395.

[19] T. Hsing: On tail estimation using dependent data. *Annals of Statistics* **19** (1991), 1547–1569.

[20] T. Hsing: On some estimates based on sample behavior near high level excursions. *Probability Theory and Related Fields* **95** (1993), 331–356.

[21] T. Hsing, J. Hüsler, and M.R. Leadbetter: On the exceedance point process for a stationary sequence. *Probability Theory and Related Fields* **78** (1988), 97–112.

[22] H. Kesten: Random difference equations and renewal theory for products of random matrices. *Acta Mathematica* **131** (1973), 207–248.

[23] M.R. Leadbetter, G. Lindgren, and H. Rootzén: *Extremes and Related Properties of Random Sequences and Processes.* Springer, Berlin, 1983.

[24] M. Loève: *Probability Theory I (4th ed.).* Springer, New York, 1974.

[25] J. Pickands III: Statistical inference using extreme order statistics. *Annals of Statistics* **3** (1975), 119–131.

[26] M.L. Puri, and L.T. Tran: Empirical distribution functions and functions of order statistics for mixing random variables. *Journal of Multivariate Analysis* **10** (1980), 402–425.

[27] S.I. Resnick: *Extreme Values, Regular Variation, and Point Processes.* Springer, New York, 1987.

[28] S.I. Resnick, and C. Stărică: Asymptotic behavior of Hill' estimator for autoregressive data. *Communications in Statistics: Stochastic Models* **13** (1997), 703–721.

[29] H. Rootzén: The tail empirical process for stationary sequences. Preprint, Chalmers University Gothenburg, 1995.

[30] H. Rootzén: Example which satisfies (1.1), (2.2), and (2.4) of "The tail empirical process," but not the conclusion of Theorem 2.1. Unpublished manuscript, Chalmers University Gothenburg, 1998.

[31] G.R. Shorack, and J.A. Wellner: *Empirical Processes with Applications to Statistics.* Wiley, New York, 1986.

[32] R.L. Smith: Estimating tails of probability distributions. *Annals of Statististics* **15** (1987), 1174–1207.

[33] C. Stărică: On the tail empirical process of solutions of stochastic difference equations. Preprint, Chalmers University Gothenburg, 1999. (Available as gzipped postscript file at www.math.chalmers.se/~starica/resume/aarch.ps.gz.)

[34] W. Vervaat: Functional central limit theorems for processes with positive drift and their inverses. Zeitschrift für Wahrscheinlichkeitstheorie und verwandte Gebiete **23** (1972), 245–253.

Holger Drees
Faculty of Mathematics and Computer Science
Saarland University
P.O. Box 15 11 50
66041 Saarbruecken, Germany
drees@num.uni-sb.de

Part VI

Bootstrap Techniques

On the Bootstrap and Empirical Processes for Dependent Sequences

DRAGAN RADULOVIĆ

ABSTRACT. The goal of this paper is to describe recent advances in empirical processes theory and several bootstraps for dependent data. We will address both theory and applications. The first three sections deal with the heuristics, motivation, statement of results and applications. The last section is devoted to mathematical techniques behind the theory. Although some results presented here are new (bootstrap for Markov chains), this is not a research paper, and the presented proofs do not contain all the details.

1 I.i.d. case

Although the resampling idea can be traced back earlier, it is generally accepted that the bootstrap was introduced by Efron [18]. This rather simple and intuitive method has been proven a powerful tool as well as a source for some very interesting and nontrivial mathematics. The idea is as follows:

Let $X_i : i = 1, \ldots, n$ be an i.i.d. P sequence (independent and identically distributed sequence with common distribution P). Let $H_n : \mathbf{R}^n \to \mathbf{R}$, be a statistic of interest (the extension from $\mathbf{R}$ to $\mathbf{R}^d$ is straightforward). Throughout this paper we will work with the following two characteristic examples:

$$H_n(X_1, \ldots, X_n) = \sqrt{n}(\mathrm{Med} - \mathrm{Med}_n(X_1, \ldots, X_n)), \tag{1.1}$$

where Med stands for the median and $\mathrm{Med}_n(X_1, \ldots, X_n)$ for the sample median, and

$$H_n(X_1, \ldots, X_n) = \sqrt{n} \sup_{t \in \mathbf{R}} |F(t) - F_n(t)|, \tag{1.2}$$

where F and F_n are the cumulative distribution function and its sample counterpart. Suppose now that we are interested in estimating a confidence interval for Med or $F(t)$. Classically, this is solved by applying weak limit theorems, that is,

$$H_n(X_1, \ldots, X_n) \stackrel{W}{\to} Z \Rightarrow \mathcal{L}(H_n(X_1, \ldots, X_n)) \approx \mathcal{L}(Z). \tag{1.3}$$

Since the theory often gives a closed form of the distribution $\mathcal{L}(Z)$, one can then compute an approximate confidence interval by referring to tabulated values for the limit distribution. This approach has the following restrictions.

Restrictions:

(A) It relies on difficult mathematical theorems which narrow the applications. (It is very easy to construct interesting statistics H_n for which we do not have limiting theorems.)

(B) A closed form is sometimes hard to obtain. (Often, it is much easier to just prove that the limiting distribution exists.)

(C) In almost all cases of interest the limiting distribution $\mathcal{L}(Z)$ depends on some unknown parameters.

The bootstrap approach bypasses these problems (at least for some important classes of statistics). Let X_i^* be an i.i.d. P_n sequence of random variables. Here, $P_n = 1/n \sum_{i=1}^n \delta_{X_i}$ is the empirical distribution, putting mass $1/n$ at each observation. The way to obtain the X_i^*'s is to "randomly pick" one observation at a time (with replacement) from $\{X_1, \dots, X_n\}$. In other words,

$$X_i^* = X_j : j = 1, \dots, n \text{ with probability } 1/n.$$

Now, since the empirical measure $P_n \approx P$, (which is equivalent to $\mathcal{L}(X_i^*) \approx \mathcal{L}(X_i)$), one can expect that

$$\mathcal{L}(H_n(X_1, \dots, X_n)) \approx \mathcal{L}(H_n(X_1^*, \dots, X_n^*)). \tag{1.4}$$

The latter distribution obviously depends on $\{X_i\}_{i=1}^n$, and the proper way to write it would be $\mathcal{L}(H_n(X_1^*, \dots, X_n^*)| \{X_1, \dots, X_n\})$. In order to simplify notation we will skip the "condition on the sample part" and write H_n^* instead of $H_n(X_1^*, \dots, X_n^*)| \{X_1, \dots, X_n\}$.

Now, in order to bypass the "weak limit theorems" approach, all we need to observe is that the right-hand side of (1.4) *does not depend on any unknown quantity*, and is fully computable. Thus, in order to justify the bootstrap one has to demonstrate that (1.4) holds, or, more formally,

$$d(\mathcal{L}(H_n), \mathcal{L}(H_n^*)) \to 0 \text{ a.s. or in probability.} \tag{1.5}$$

Here, d stands for some (any) distance that metrizes weak convergence. How do our two examples fit here? The median case is well studied and the limiting distribution $\mathcal{L}(Z)$ is a centered normal random variable. The variance is unknown but can be estimated from the sample. Thus, bootstrapping will not hurt us, but it is not necessary. The main advantage here is that it can be done automatically, without referring to tables and parameter estimations.

The cumulative distribution case is more complex. If the X's are supported on $\mathbf{R}$, then the classical approach goes flawlessly. Actually, the limiting distribution of (1.2) (after the transformation $Y_i := F^{-1}(X_i)$) is the supremum of the Brownian bridge and it is one of the very few cases where we do not have to worry about any constant. If the X's are supported on $\mathbf{R}^d$, then the transformation "trick" does not apply and the limiting distribution heavily depends on $F(t)$. Since it is the supremum of a Gaussian process, it is very hard to compute in closed form. This is one of the classical examples where we have a "weak limit result" but we can seldom use it. The bootstrap on the other hand, trivially bypasses this difficulty, for the distribution $\mathcal{L}(H_n^*)$ can be easily numerically estimated.

The bootstrap (i.e., (1.5)) has been justified very early (see Efron and Tibshirani [19], Bickel and freedman [4], and Singh [41]) for a variety of statistics. Around the same time, it was observed that for some statistics of interest the bootstrap does not work (see Bickel and Freedman [4]). Through the eighties it was evident that essentially all the cases for which the bootstrap works are related to Gaussian limits (i.e., statistics H_n have $\sqrt{n}$ imbedded) and all the counterexamples were not related to Gaussian limits and H_n's did not have $\sqrt{n}$.

The role of empirical processes theory was twofold.

First, via the delta method and M-estimation, the uniform central limit theorems provided a justification of (1.3) for many statistics H_n. In other words, having

$$\sqrt{n}(P_n f - Pf)_{f\in\mathcal{F}} \overset{W}{\to} G_P(f)_{f\in\mathcal{F}}, \tag{1.6}$$

for large classes $\mathcal{F}$, one can prove weak convergence: $\mathcal{L}(H_n) \to \mathcal{L}(Z)$ for a wealth of interesting statistics H_n. Thus, we have a positive solution to the problem of Restriction (A).

Second, it solved the "$\sqrt{n}$" problem. Namely, Giné and Zinn [21], proved that (1.6) holds if and only if

$$\sqrt{n}(P_n^* f - P_n f)_{f\in\mathcal{F}} \xrightarrow{W} G_P(f)_{f\in\mathcal{F}} \text{ in probability.} \tag{1.7}$$

This result does not only explain the mystery behind the normalization $\sqrt{n}$ but also justifies the bootstrap for all statistics H_n, with asymptotic behavior explained in (1.6), and essentially solves Restrictions (B) and (C). It is worth noticing that although empirical processes had a pivotal role in the acceptance and understanding of the bootstrap, they were not studied with this application in mind. In fact, both theories started to develop at around the same time, initiated by Efron in 1979 and by Dudley [17] in 1978, but it took a full decade before their synthesis. We finish this section with two perhaps surprising results.

First, As we mentioned before the bootstrap works if $d(\mathcal{L}(H_n), \mathcal{L}(H_n^*)) \to 0$. However, it is desirable to estimate the rate of this convergence, in particular, since the method is in competition with the classical approach where we exploit $d(\mathcal{L}(H_n), \mathcal{L}(\mathcal{Z})) \to 0$. One of the biggest surprises was that in the classical case, where $H_n = \sqrt{n}(\overline{X}_n - EX)/\sigma$, the bootstrap works better. Namely, Hall [22] showed that under sufficient moment assumptions

$$d(\mathcal{L}(H_n), \mathcal{L}(H_n^*)) = o_P(n^{-1/2})$$

while, in general,

$$d(\mathcal{L}(H_n), \mathcal{L}(Z)) = O(n^{-1/2}).$$

So, we can use the bootstrap not only in the cases when the limiting distribution $\mathcal{L}(Z)$ is intractable, but also in simple well understood cases.

Second, the Giné–Zinn result confirmed the "folklore" by proving that (1.6) converges if and only if (1.7) does. They, like most authors, justified the bootstrap indirectly by showing that both the original statistic H_n and the H_n^*, converge to the same limit. However, we defined that the bootstrap works if

$$d(\mathcal{L}(H_n), \mathcal{L}(H_n^*)) \to 0 \tag{1.8}$$

without invoking any limiting distribution. Thus, it is conceivable that for some statistics H_n we could have that (1.8) holds while both $\mathcal{L}(H_n)$ and $\mathcal{L}(H_n^*)$ diverge. In other words, it would be possible to infer on a statistic H_n even though its distribution does not converge. This was confirmed by Radulović [36].

2 Dependent case

Let us consider a concrete example related to Internet traffic. Let X_i be the traffic in kilobytes for the i-th second, observed on some Internet link such as the optical cable coming out of a large university. It is by now generally confirmed, both via experiments and theory, that the X_i's are heavily dependent, and, if properly collected (avoiding intra-day cycles) constitute a stationary sequence. There are numerous statistical aspects that have been studied in relation to this sequence. We will focus on two of them: the estimation of the location parameter and the marginal distribution. It has been observed that the traffic is very "bursty" and "spiky," so the mean is generally replaced by more robust location estimators like the median, trimmed mean, or Huber [23] type of location estimators. Also, it appears that the marginal distributions of this process are not Gaussian and that they seem to lack Gaussian (or exponential) tails. In order to construct confidence intervals for the "true" marginal, it is natural to consider the following statistics,

$$\sqrt{n} \sup_{t\in\mathbf{R}} |F_n(t) - F(t)| \tag{2.1}$$

or, if we are more concerned with the (positive) tail behavior,

$$\sqrt{n} \sup_{t\in\mathbf{R}} W(t)|F_n(t) - F(t)|, \tag{2.2}$$

where the weight function $W(t) \ : \mathbf{R} \to [0, 1]$, $W(t)$ increasing, is designed to put more mass into the tail. As in the case of i.i.d. observations, in order to proceed one needs "weak limit results"; for each of these statistics we need to prove the existence of a random variable Z such that

$$H_n(X_1, \ldots, X_n) \stackrel{W}{\to} Z.$$

However, the mentioned examples, in particular, the trimmed mean and the weighted cumulative distribution, are not "textbook cases." Only recently the necessary results were proved under reasonably general assumptions. Empirical processes played a pivotal role. Similar to the i.i.d. case, weak limit theorems for various statistics of interest can be deduced from the empirical CLT. We would just mention a few important contributions: Arcones and Yu [1], and Doukhan et al. [16].

These results, although immensely important, suffer from Restrictions (B) and (C) even more than in the i.i.d. case. For example, even the simplest form of the statistic (2.1), with X_i's uniformly distributed on [0, 1], has limiting statistic $Z = \sup_{t\in[0,1]} |G(t)|$, where $G(t)$ is a Gaussian process, with the (computationally intractable) covariance structure

$$\mathrm{Cov}(G(s), G(t)) = s(1-t) + \lim_{n\to\infty} \sum_{i\neq j}^{n} (E1_{X_i\le s}1_{X_j\le t} - st).$$

Things become even more complicated if we consider X_i's supported on $\mathbf{R}^d$, and/or a weighted version of the same statistic (2.2). It is not hard to see that essentially all of the statistics of interest, in this setting (dependent, stationary data) will suffer from similar problems.

So, how does one construct a confidence interval for the marginal distributions of a stationary process? An appropriate bootstrap would be very helpful here. Namely, if one can construct a bootstrap sequence $\{X_i^*\}_{i=1}^n$, such that

$$\sqrt{n}(F_n^*(t) - F_n(t)) \to G(t),$$

one could approximate the distribution of Z, (or even better, the distribution of $\sqrt{n}\sup|F_n(t) - F(t)|$) by simply resampling X_i^*'s.

Therefore, we face the following two questions:

- How do we construct the bootstrap sequence in such a way that we capture all the dependency without imposing any particular model (regression, Markov, etc.)?
- How do we prove the empirical CLT's which would then establish bootstrap for a wealth of statistics (in particular, (1.1) and (2.2))?

Before answering these questions, let us see why the original (naive) bootstrap fails in this setting. Let $\{X_i\}$ be a sequence of weakly dependent stationary random variables with nontrivial covariance (i.e., $|\operatorname{Cov}(X_i, X_j)| > 0$ for at least one pair $(i, j), i \neq j$). Consider the central limit theorem for the mean

$$H_n := n^{-1/2} \sum_{i=1}^{n} (X_i - EX_i).$$

Since H_n is designed with inference on EX_i in mind, which clearly depends only on the marginal distribution P, one might expect that the bootstrap random variables X_i^* (which are i.i.d. $P_n \approx P$) would work in this case.

It is well known that under some moment and weak dependency conditions

$$H_n \xrightarrow{W} N(0, \sigma^2), \tag{2.3}$$

where

$$\sigma^2 = \lim_{n\to\infty} \operatorname{Var}(H_n) = \lim_{n\to\infty} (\operatorname{Var} X_1 + \sum_{i \neq j} \operatorname{Cov}(X_i, X_j))$$

$$= \operatorname{Var}(X_1) + 2 \sum_{i=2}^{\infty} \operatorname{Cov}(X_1, X_i).$$

On the other hand, for $H_n^* := n^{-1/2} \sum_{i=1}^{n} (X_i^* - EX_i^*)$,

$$\operatorname{Var}^*(H_n^*) = \operatorname{Var}^*(X_1^*) + \sum_{i \neq j} \operatorname{Cov}^*(X_i^*, X_j^*)$$

(since the X_i^* are (conditionally) i.i.d.)

$$= E^*(X_1^*)^2 - (E^* X_1^*)^2 = \frac{1}{n} \sum_{i=1}^{n} X_i^2 - (\frac{1}{n} \sum_{i=1}^{n} X_i)^2 \to \operatorname{Var}(X_1)$$

if (X_i) is ergodic with finite variance.

Thus, even if H_n^* converged, it would converge to the wrong limit. One way to fix this problem would be to assume a particular structure of the underlying dependency, fit the parameters and then resample from the approximate model. Obviously, this approach could work only if we were lucky enough to correctly guess the model.

We believe that the true nature of the bootstrap is to completely rely on the empirical information, without imposing models. The above counterexample gives us a clue of how to fix the problem. Namely, what if we fix $b \in N$, $b \ll n$, and then consider overlapping blocks of size b,

$$B_1 := \{X_1, \ldots, X_b\},\ B_2 := \{X_2, \ldots, X_{b+1}\}, \ldots, B_{n-b+1} := \{X_{n-b+1}, \ldots, X_n\}.$$

Let us now resample, with replacement from the set $\{B_1, \ldots, B_{n-b+1}\}$. Formally, for $m = n/b$ (we assume for simplicity that n is an integer multiple of b) and $\{I_i\}_{i=1}^{m}$ i.i.d. uniform on $\{1, \ldots, n - b + 1\}$, we construct

$$B_{I_1}, B_{I_2}, \ldots, B_{I_m} \text{ and let } B_i^* = B_{I_i}.$$

The sample X_i^* consists of the entries from the blocks $B_1^*, \ldots, B_m^*$. This procedure is called the moving blocks bootstrap, or MBB. With this modification, the above example leads to the following: Let $Y_k^* := X_{(k-1)b+1}^* + \cdots + X_{kb}^*$, then clearly

$$n^{-1/2} \sum_{k=1}^{n} (X_k^* - E^*(X_k^*)) = n^{-1/2} \sum_{k=1}^{m} (Y_k^* - E^*(Y_k^*))$$

and since the Y_k^* are i.i.d. (by construction of X_i^*'s),

$$\mathrm{Var}^*(H_n^*) = \frac{m}{n} \mathrm{Var}^*(Y_1^*)$$

(for $Y_k := X_{(k-1)b+1} + \cdots + X_{kb}$)

$$= \frac{1}{b} \left(\frac{1}{n-b+1} \sum_{k=1}^{n-b+1} Y_k^2 - \left(\frac{1}{n-b+1} \sum_{k=1}^{n-b+1} Y_k \right)^2 \right) \tag{2.4}$$

which under some moment and weak dependence conditions converges to

$$\frac{1}{b} \mathrm{Var}(X_1 + \cdots + X_b).$$

A few comments are in order.

(a) If b is sufficiently large the expression above will be close to σ^2. The next logical step would be to let $b := b(n)$ such that $b \to \infty$, as $n \to \infty$, and $b/n \to 0$. With appropriate rates on $b(n)$ and with some assumptions on the dependence, one could hope that the weak limits of H_n^*, could be established. This is exactly the line of thought which the authors of this method Liu and Singh [26] and Künsch [25] took.

(b) Through the above example we have encountered "moments and weak dependence conditions" several times. We needed them to justify the CLT of the mean for the dependent data (2.3) and the LLN of the squares for dependent data (2.4). The first was related to the original statistics H_n and the second to its bootstrap version H_n^*. It is worth investigating the relation between these two requirements.

(c) The presented method, if extendable, would allow for inference on large classes of interesting statistics *without imposing particular dependence conditions*. It is thus imperative to generalize the method to a wider class of statistics H_n.

From a theoretical point of view, we see two directions: establishing the MBB CLT of the mean under minimal conditions (comment (b)) and deriving the MBB for the empirical CLT (comment (c)).

Blockwise bootstrap of the mean. First, we need to introduce Rosenblatt's [39] α-mixing as a measure of dependence. Let $\mathcal{A}$ and $\mathcal{B}$ be two σ fields in $(\Omega, \mathcal{T}, \mathbf{P})$, then

$$\alpha(\mathcal{A}, \mathcal{B}) = \sup_{(A,B) \in \mathcal{A} \times \mathcal{B}} |P(A \cap B) - P(A)P(B)|.$$

Let $\{X_i\}_{i=-\infty}^{\infty}$ be a strictly stationary sequence of random variables. Then we define

$$\alpha_n := \alpha(\mathcal{F}_1, \mathcal{G}_n),$$

where $\mathcal{F}_1 := \sigma(X_i : i \leq 1)$ and $\mathcal{G}_n := \sigma(X_i : i \geq n+1)$. The sequence X_i is α-mixing if $\alpha_n \to 0$. In the original MBB papers the authors assumed the following conditions: *Künsch*:

$$E|X_1|^{6+\delta} < \infty \text{ and } \sum_{i=1}^{\infty}(1+i)^2\alpha_i^{\delta/(6+\delta)} < \infty \text{ for some } \delta > 0. \tag{2.5}$$

Liu and Singh:

$$E|X_1|^{6+\delta} < \infty \text{ for some } \delta > 0 \text{ and the } X_i\text{'s are } m\text{-dependent.} \tag{2.6}$$

Doukhan et al. [15] showed the following: let $\alpha^{-1}(t)$ be the cadlag inverse of the monotone function $\alpha(t) := \alpha_{[t]}$, and let $Q_{X_1}(t)$ be the quantile function of $|X_1|$, which is the cadlag inverse of the distribution function of $|X_1|$. If

$$\int_0^1 \alpha^{-1}(t)Q_{X_1}^2(t)dt < \infty,$$

then

$$\mathrm{Var}(H_n) \to \sigma^2 < \infty \text{ and } H_n \xrightarrow{W} N(0, \sigma^2).$$

They also showed that these conditions are (slightly) weaker than the classical condition

$$E|X_1|^{2+\delta} < \infty \text{ and } \sum_{i=1}^{\infty}\alpha_i^{\delta/(2+\delta)} < \infty \text{ for some } \delta > 0, \tag{2.7}$$

given in Ibragimov and Linnik [24]. There is an obvious discrepancy between conditions for the bootstrap (2.5) and (2.6) on the one hand and for the CLT (2.7) on the other hand. This matter was settled with the following result of Radulović [33].

Theorem 2.1. *Let X_i be an α-mixing sequence, $\sigma_n^2 := \mathrm{Var}(X_1 + \cdots + X_n)$ and $\sigma_n^{*2} := \mathrm{Var}^*(X_1^* + \cdots + X_n^*)$; then*

$$\frac{1}{\sigma_n}\sum_{i=1}^{n}(X_i - EX_1) \xrightarrow{W} N(0,1), \ b \to \infty \text{ and } b/n \to 0$$

implies

$$\frac{1}{\sigma_n^*}\sum_{i=1}^{n}(X_i^* - E^*X_1^*) \xrightarrow{W} N(0,1) \text{ in probability.}$$

By "in probability" we refer to definition (1.5). This result resembles the i.i.d. case for which Giné and Zinn [20] proved the "if and only if" version of the same result. The final twist to this story is due to Peligrad [30]. She demonstrated that the converse of the result above is not true.

Theorem 2.2. *Let X_i be an α-mixing sequence of bounded random variables. Let $\sigma_n^2 := \mathrm{Var}(X_1+\cdots+X_n)$, $\sigma_n^{*2} := \mathrm{Var}^*(X_1^*+\cdots+X_n^*)$ and $\sum_{i=1}^{\infty} i^{-1}\alpha_i < \infty$. Then if $\lim_{n\to\infty}\sigma_n^2/n = \sigma^2 < \infty$, we have*

$$\sigma_n^{*2}/n \to \sigma^2 \text{ a.s.}$$

and

$$\frac{1}{\sqrt{n}}\sum_{i=1}^{n}(X_i^* - E^*X_1^*) \xrightarrow{W} N(0,\sigma^2) \text{ a.s.}$$

Bradley [6] constructed a stationary sequence of bounded, pairwise independent, α-mixing random variables with rate $\alpha_i = O(1/i)$, such that

$$\frac{1}{\sqrt{n}}\sum_{i=1}^{n}(X_i - EX_1) \text{ does not converge weakly.}$$

The "bizarre" consequence is that, due to pairwise independence, $\sigma_n^2/n = \text{Var}(X_1 + \cdots + X_n)/n = \text{Var}(X_1) < \infty$. Thus the MBB captures the variance correctly but "overshoots" the weak limit.

The blockwise bootstrap empirical CLT. The above results, though adding a great deal to the understanding of the mechanism and of the restrictions to the MBB, have only limited practical value. We are still not able to tackle the two examples posed at the beginning of this section: the location and the marginal distributions problem. For this we will need to prove the bootstrap empirical CLT, i.e., to identify the classes of functions $\mathcal{F}$ for which

$$\sqrt{n}(P_n^* f - P_n f)_{f\in\mathcal{F}} \to G_P(f)_{f\in\mathcal{F}},$$

where $P_n = n^{-1}\sum_{i=1}^{n}\delta_{X_i}$ and $P_n^* = n^{-1}\sum_{i=1}^{n}\delta_{X_i^*}$ for X_i^* constructed by the MBB procedure. The motivation is obvious. As in the i.i.d. case, the empirical CLT's are the essential ingredients for deriving the CLT for a wide class of interesting statistics H_n.

The results are split into two groups: the classical case $\mathcal{F} := \{f_t(x) = 1_{x\le t} : t \in R\}$ and the more complex classes $\mathcal{F}$, (typically VC (Vapnik–Červonenkis) or bracketing classes).

For the first case, pioneering results are due to Bühlmann [7] and Naik-Nimbalkar and Rajarshi [27]. They prove that

$$\sqrt{n}(P_n^* 1_{x\le t} - P_n 1_{x\le t})_{t\in\mathbf{R}} \to G_P(t)_{t\in\mathbf{R}}, \text{ in probability} \tag{2.8}$$

under the following conditions on the block size b and α-mixing coefficients α_i: *Naik-Nimbalkar and Rajarsh (for $X_i \in \mathbf{R}$ only)*:

$$\sum_{i=1}^{\infty} i^7\alpha_i^{1/2-\tau} < \infty \text{ and } b = O(n^{1/2-\varepsilon}) \text{ for some/any } \tau, \varepsilon \in (0, 1/2).$$

Bühlmann (for $X_i \in \mathbf{R}^d$):

$$\sum_{i=1}^{\infty} i^{32}\alpha_i^{1/2} < \infty \text{ and } b = O(n^{1/2-\varepsilon}) \text{ for some/any } \varepsilon \in (0, 1/2).$$

These results, although obtained under much stronger conditions then the weakest known sufficient conditions for their nonbootstrap version (i.e., $\alpha_n = O(n^{-\eta})$, $\eta > 3$ see Yoshihara [44]) represent an important step. This was the first time that the MBB was extended to the infinite dimensional setting of empirical processes.

The final results in this setting are due to Radulović [35] and Peligrad [30].

Theorem 2.3 (Peligrad). *Let X_i be an α-mixing sequence such that*

$$\sum_{i\ge n}^{\infty}\alpha_i = O(n^{-\eta}) \textit{ and } n^{\eta} \ll b \ll n^{1/3-\varepsilon} \textit{ for } 0 < \eta < \frac{1}{3} - \varepsilon,\ \varepsilon \in (0, 1/3).$$

Then

$$\sqrt{n}(P_n^* 1_{x\le t} - P_n 1_{x\le t})_{t\in\mathbf{R}} \to G_P(t)_{t\in\mathbf{R}} \textit{ a.s.}$$

Theorem 2.4 (Radulović). *Let X_i be an α-mixing sequence such that*

$$\alpha_n = O(n^{-\eta}),\ \eta > 1 \text{ and } b = O(n^{\varepsilon}) \text{ for } 0 < \varepsilon < \min\left(\frac{1}{3}, \frac{\eta}{4} - \frac{1}{6}\right).$$

Then

$$\sqrt{n}(P_n^* 1_{x \le t} - P_n 1_{x \le t})_{t \in \mathbf{R}} \to G_P(t)_{t \in \mathbf{R}}, \text{ in probability.}$$

It has been observed (Künsch [25]) that the optimal block size is $b = cn^{1/3}$, which can nearly be obtained in the above results (provided that the α-mixing coefficients decay fast enough). It is worth observing the surprising fact that the above conditions on α-mixing are weaker than those (Yoshihara's) for the CLT. Thus, the bootstrap works even if the strongest (to date) sufficient conditions for the CLT are not satisfied.

With these results, we are now able to solve the second problem, related to the marginal distribution. In order to construct a confidence interval for $\sqrt{n}\sup_t |F_n(t) - F(t)|$, one does not have to compute the distribution of $\sup_{t\in[0,1]} |G(t)|$, where

$$\mathrm{Cov}(G(s), G(t)) = s(1-t) + \lim_{n\to\infty} \sum_{i \ne j}^{n} (E 1_{X_i \le s} 1_{X_j \le t} - st).$$

It is sufficient to resample the quantity $\sqrt{n}\sup_t |F_n^*(t) - F_n(t)|$. Now we are left with the first problem of location estimation. Here we will need the full power of empirical processes. The extension to larger classes $\mathcal{F}$ of functions requires an additional assumption on the weak dependence. We will use β-mixing as defined by Volkonskii and Rozanov [43]. For two σ fields $\mathcal{A}$ and $\mathcal{B}$, we let

$$\beta(\mathcal{A}, \mathcal{B}) = \frac{1}{2} \sup \sum_{(i,j)\in I\times J} |P(A_i \cap B_j) - P(A_i)P(B_j)|,$$

where the supremum is taken over all finite measurable partitions $(A_i)_{i\in I}$ and $(B_j)_{j\in J}$, with $A_i \in \mathcal{A}$ and $B_j \in \mathcal{B}$.

The β-mixing condition is more restrictive than α-mixing but it allows one to use exponential inequalities (see Doukhan [14]) which are essential in this context. There are two types of classes that are generalizations of indicators of half lines: VC-type classes and bracketing-type classes (see van der Vaart and Wellner [42]). The strongest results obtained so far for VC classes, for the usual (nonbootstrap) empirical CLT are due to Arcones and Yu [1]. They established

$$\sqrt{n}(P_n f - Pf)_{f\in\mathcal{F}} \to G_P(f)_{f\in\mathcal{F}}$$

(where G_P is a centered, tight Gaussian process indexed by the class $\mathcal{F}$) under the following assumptions:

$$E|F|^p < \infty \text{ and } \sum_{i=1}^{\infty} (\ln i)^{2(p-1)/p} \beta_i^{(p-2)/p} < \infty \text{ for some } p > 2 \tag{2.9}$$

(where F is the envelope function for the class $\mathcal{F}$).

There are very few results for the MBB under these settings. A result by Radulović [34] establishes the MBB for VC classes.

Theorem 2.5. *Let $\mathcal{F}$ be a VC class of functions with envelope function F satisfying $EF^p < \infty$ for some $p > 2$. Let X_i^* be the MBB sequence with block size $b = O(n^\rho)$ and $0 < \rho < \frac{p-2}{2(p-1)}$. If the original sequence X_i is β-mixing with $\beta_i = O(i^{-q})$ and $q > p/(p-2)$, then*

$$\sqrt{n}(P_n^* f - P_n f)_{f\in\mathcal{F}} \to G_P(f)_{f\in\mathcal{F}} \text{ in probability.}$$

These conditions are only marginally stronger than required for the nonbootstrap version (2.9). With this theorem we are in a position to bootstrap a large class of interesting statistics, in particular, location estimators, (see the Applications in Radulović [34])

The bracketing classes (in the nonbootstrap setting) were considered by Doukhan et al. [16]. They established the empirical CLT under conditions close to necessity. Unfortunately, we do not have similar results for the MBB. The strongest result (to the best of our knowledge) is due to Bühlmann [8]. He treated bracketing classes under rather restrictive conditions (exponential β-mixing decay and polynomial growth of bracketing numbers).

It is worth mentioning that the Giné–Zinn approach which relates stochastic equicontinuity of the bootstrap process to that of the original one, fails here. It is interesting though, that if we considered a different blockwise scheme (the one which *would resample from nonoverlapping blocks)*, it would be easy to recover the Giné–Zinn result. Unfortunately, resampling from nonoverlapping blocks has some statistical limitations and has not been seriously considered in the literature.

3 Markov chains

There are at least two different ways to bootstrap countable (regarding space and time) Markov chains. The first, naive, way is to form a transition probability matrix $\{p_{i,j}^n\}$ estimated from a sample $\{X_1, \dots, X_n\}$. The bootstrap random variables are now formed by using $\{p_{i,j}^n\}$ as if it was the original matrix. The second way to bootstrap explores the regenerative property of Markov chains. The original sample $\{X_1, \dots, X_n\}$ is broken into blocks of random size starting with a designated atom. These blocks are independent (by the regenerative property) and one can now form a bootstrap sample by resampling from the blocks. It is not clear which of the methods is better and in what sense. After an extensive search in the literature, we could not find any research devoted to this problem.

The first way to bootstrap has been studied by numerous authors (Datta and McCormick [11], Athreya and Fuh [3]). To the best of our knowladge there are no results related to empirical processes for this type of bootstrap.

The second method, the renewal bootstrap, is more general, since it easily extends to continuous time Markov chains with an atom. Also, it is plausible that one could mimic the ideas of Athreya and Ney [2] and Nummelin [28] by introducing an artificial atom and derive the bootstrap procedure that applies to nonatomic Markov chains. Some work along this line has been reported by Paparoditis and Politis [29] as well as Rajarshi [38].

Formally we define the renewal bootstrap as follows. Under reasonable conditions on a Markov chain with atom Δ, (say $\Delta = 1$), we can define $\mathcal{T}_k : \Omega \to \mathbf{N}_0 \cup \{\infty\}$ by

$$\begin{aligned} \mathcal{T}_0 &:= \inf\{n \geq 0 : X_n = 1\}, \\ \mathcal{T}_k &:= \inf\{n > \mathcal{T}_{k-1} : X_n = 1\}, \ k \geq 1. \end{aligned} \tag{3.1}$$

It is well known (see Chung [9], Chapter I.14) that for any initial probability measure ν and all $k \in \mathbf{N}_0$, $\mathcal{T}_k < \infty$, P_ν a.s. *The renewal times* are defined by

$$\tau_0 := \mathcal{T}_0 + 1, \quad \tau_k := \mathcal{T}_k - \mathcal{T}_{k-1}, \ k \geq 1. \tag{3.2}$$

It is a classical fact that $\{\tau_k\}_{k=1}^{\infty}$ is i.i.d. and independent of the choice of the initial probability measure ν. Throughout the remainder of the paper we set $\tau = \tau_1$. Finally, we let

$$k_n(\omega) := \max\left\{k : \sum_{i=0}^{k} \tau_i \le n\right\} \text{ if defined and } -1 \text{ otherwise.} \tag{3.3}$$

Given the sample $\{X_1, \ldots, X_n\}$, we define the blocks

$$B_i = (X_{\mathcal{T}_{i-1}+1}, X_{\mathcal{T}_{i-1}+2}, \ldots, X_{\mathcal{T}_i}) \text{ for } i = 1, \ldots, k_n(\omega).$$

The bootstrap blocks B_i^*, $i = 1, \ldots, k_n(\omega)$, are obtained by resampling with replacement from $\{B_i\}$. More precisely: let $\{I_j\}$, $j = 1, \ldots, k_n(\omega)$ be an i.i.d. sequence of random variables uniformly distributed on the set $\{1, \ldots, k_n(\omega)\}$, then

$$B_j^* := B_{I_j} \text{ for } j = 1, \ldots, k_n(\omega).$$

Clearly, the length of the i-th bootstrap block is τ_{I_i}. In order to simplify notation we will denote it by τ_i^*. Finally, lining up the blocks $(B_1^*, \ldots, B_{k_n}^*)$, we form a bootstrap sample $(X_1^*, \ldots, X_{n^*}^*)$, where $n^* = \sum_{i=1}^{k_n} \tau_i^*$ and $(X_1^*, \ldots, X_{\tau_1^*}^*) = B_1^*$, $(X_{\tau_1^*+1}^*, \ldots, X_{\tau_1^*+\tau_2^*}^*) = B_2^*, \ldots$, etc.

As in the previous two sections the theoretical work related to the bootstrap will be divided into two categories: *Establishing a bootstrap procedure for a simple case under minimal assumptions and extending the procedure to empirical processes setings*. The results here are scattered. Datta and McCormick [10] established the bootstrap CLT for the mean under a third moment assumption. Athreya and Fuh treated a somewhat different problem related to empirical estimates of a stationary probability measure $\pi(i)$ and of the transition matrix $\{p_{i,j}\}$. Radulović [37] mimicked the results for the MBB and stationary sequences and proved the following.

Theorem 3.1. *Let X_i be an aperiodic positive recurrent Markov chain with an atom. Then $E\tau^2 < \infty$ and*

$$W_n := \frac{1}{\sqrt{n}} \sum_{i=1}^{n} (f(X_i) - \mu) \xrightarrow{W} N(0, \sigma^2) \tag{3.4}$$

imply

$$W_n^* := \frac{1}{\sqrt{n^*}} \sum_{i=1}^{n^*} (f(X_i^*) - E^*(f(X_1^*)) \xrightarrow{W} N(0, \sigma^2) \textit{ in probability,}$$

where $n^* := \sum_{i=1}^{k_n} \tau_i^*$, is the bootstrap sample size.

Under the same conditions we obtain the empirical CLT.

Theorem 3.2. *If $E\tau^2 < \infty$, then*

$$\sqrt{n}(P_n(f) - P(f))_{f\in\mathcal{F}} \xrightarrow{W} G_P(f)_{f\in\mathcal{F}} \tag{3.5}$$

implies

$$\sqrt{n^*}(P_{n^*}^*(f) - P_n(f))_{f\in\mathcal{F}} \xrightarrow{W} G_P(f)_{f\in\mathcal{F}} \textit{ in probability,} \tag{3.6}$$

where $P_n = \frac{1}{n}\sum_{i=1}^{n} \delta_{X_i}$, $P_{n^*}^* = \frac{1}{n^*}\sum_{i=1}^{n^*} \delta_{X_i^*}$.

Remark 3.3. Due to the results by Bolthausen [5] (see Theorem 2 in [5]) there is a connection between the α-mixing coefficients and the renewal times for countable Markov chains:

$$\sum_{i=1}^{\infty} i^{\lambda}\alpha_i < \infty \Leftrightarrow E(\tau^{2+\lambda}) < \infty, \text{ for every } \lambda \geq 0. \tag{3.7}$$

The condition $E\tau^2 < \infty$ is equivalent to $\sum_{i=1}^{\infty} \alpha_i < \infty$, and it is close to necessity for the CLT of the mean, even if f is uniformly bounded. At this point it is not known if it is a necessary condition for the CLT. The usual sufficient and necessary conditions $E(\sum_{i=\mathcal{T}_1+1}^{\mathcal{T}_2} f(X_i))^2 < \infty$ and $E(\sum_{i=\mathcal{T}_1+1}^{\mathcal{T}_2}(f(X_i) - \mu_f)) = 0$, do not imply $E\tau^2 < \infty$. The fact that there are stationary sequences (not Markov chains) for which we have the CLT even if $\alpha_i \to 0$ arbitrarily slowly, strongly suggests that the condition above is not necessary.

Remark 3.4. It is easy to show that (3.5) follows from

$$\mathcal{F} = \{f : |f| \leq F\}, \ EF^{2+\delta} < \infty \text{ and } E\tau^{2+2/\delta} < \infty. \tag{3.8}$$

This class is very rich and it does not presuppose any geometric conditions (bracketing, VC, etc.) usually associated with the uniform CLT. The conditions (3.8) are sharp but not necessary. Interestingly, the result above establishes a bootstrap version of (3.5) even if condition (3.8) is not satisfied.

Applications. Clearly, Theorem 3.1 justifies the use of the renewal Markov bootstrap for a variety of situations and under minimal conditions. For example: letting $f(x) = x$, $f(x) = 1_{x=\Delta}$ and $f(x) = x^2$ we can construct bootstrap confidence intervals related to EX, $E\tau$ and EX^2.

The most interesting applications are related to Theorem 3.2. A number of applications dealing with countable state Markov chains are related to estimates of the stochastic matrices $\{p(i, j)\}$, the stationary probabilities $\pi(i)$ and the renewal times τ. The classical results (related to the uniform CLT) are due to Derman [13]. He considered the usual (maximum likelihood) estimators

$$\pi_n(i) = \frac{1}{n}\sum_{k=1}^{n} 1_{(X_k=i)}$$

and

$$p_n(i, j) = \frac{\sum_{k=1}^{n-1} 1_{(X_k=i, X_{k+1}=j)}}{\sum_{k=1}^{n} 1_{(X_k=i)}} \text{ with the convention } 0/0 = 0.$$

Assuming $E\tau^2 < \infty$, he proved that

$$\sqrt{n}(\pi_n(i) - \pi(i))_{i\in A} \overset{W}{\to} N(0, \Sigma_{\pi,A}) \tag{3.9}$$

and

$$\sqrt{n}(p_n(i, j) - p(i, j))_{i,j\in B} \overset{W}{\to} N(0, \Sigma_{p,B}),$$

where A and B are any fixed finite subsets of $\mathbf{N}$ and $\mathbf{N} \times \mathbf{N}$, respectively. A serious constraint here is that A and B are finite. So, for example one cannot use these results to construct any test related to "the most frequent state" (i.e., i_o and $\pi(i_o)$ such that $\pi(i_o) \geq \pi(i)$ for every $i \in \mathbf{N}$)

since we do not know *a priori* which finite set A to choose. Applying (3.5) and (3.8) we get (3.9) for possibly infinite A's and consequently the following result:

$$\sup_{i\in\mathbf{N}} \sqrt{n}\ |\pi_n(i) - \pi(i)| \xrightarrow{W} \sup_{i\in\mathbf{N}} |G_\Pi(i)|, \tag{3.10}$$

where G_Π is a centered, tight Gaussian process indexed by i and depending on the stationary probability Π. As already commented by Athreya and Fuh [3], the result (3.9) is very hard to use since the covariance matrix $\Sigma_{\pi,A}$ is extremely complicated. Consequently, we have even more problems with (3.10) and the covariance matrix of the Gaussian process $G_\Pi(i)$. Theorem 3.2 shows that one could bootstrap the statistics $\pi_n(i)$ in (3.9) and (3.10) (for finite as well as for infinite A's) and circumvent the cumbersome estimation of covariance structures.

One cannot get an analogous result for $p(i, j)$ and $p_n(i, j)$. Namely, since $p_n(i, j)$ has at most n nonzero entries, it is very easy to construct a "nice" Markov chain for which $\sup_{i,j\in\mathbf{N}} \sqrt{n}|p_n(i, j) - p(i, j)| = \sqrt{n}$ for all n and almost all ω. If we consider the possibly more relevant case of weighted $p(i, j)$, we get weak convergence for

$$\sup_{i,j\in\mathbf{N}} \sqrt{n}|\pi_n(i)p_n(i, j) - \pi(i)p(i, j)|. \tag{3.11}$$

In this case $P^{(2)}(i, j) := \pi(i)p(i, j)$ is a two-dimensional marginal distribution for the original Markov chain $\{X_i\}$. Let $Y_i = (X_{i-1}, X_i)$, $i > 1$. Clearly, $\{Y_i\}$ is a countable Markov chain with stationary probability $\Pi_Y(i, j) = P^{(2)}(i, j)$. Therefore, we can apply (3.8). We get (3.9) and establish weak convergence for (3.11).

4 Technical details

It is well known that most of the bootstrap proofs reduce to an application of the LLN. For example, the typical requirement $EX^2 < \infty$, in the bootstrap setting becomes $E^*(X^*)^2 = n^{-1}\sum_{i=1}^n X_i^2 < \infty$. For the bootstrap empirical CLT, we need a uniform LLN. Here, one has to compare the covering numbers (i.e., the minimum number of $L_2(P)$ balls needed to cover $\mathcal{F}$ with resolution x)

$$N(x, \mathcal{F}, L_2(P))$$

with its bootstrap counterpart

$$N(x, \mathcal{F}, L_2(P_n))$$

for which we need

$$\sup_{f\in\mathcal{F}} |Ef(X)^2 - E^* f(X^*)^2| \to 0. \tag{4.1}$$

In both examples, we needed an LLN for the "squares." In our setting, this induces an extra difficulty.

First the X_i's are not independent, and because of the MBB's blocking structure we are dealing with triangular arrays: for Y_i^* and Y_i defined earlier in computation (2.4), we have

$$E^*(Y_1^*/b)^2 = \frac{1}{n}\sum_{i=1}^n (Y_i/b)^2 = \frac{1}{n}\sum_{i=1}^n \left(\frac{X_i + \cdots + X_{i+b}}{b}\right)^2.$$

Through this section we will use the "circular" version of the MBB. The original sample is extended with the first b observation, i.e., we resample blocks from $\{X_1, \dots, X_n, X_1, \dots, X_b\}$. With this construction, we do not have to worry about Y_i's for $i > n - b$.

MBB for the mean.

Comment on the proof for Theorem 2.1. First, it is fairly easy to replace σ_n^* with $\sqrt{m}\sigma_b$. In order to prove weak convergence for

$$\frac{1}{\sqrt{m}\sigma_b}\sum_{i=1}^{n}(X_i^* - E^*X_1^*) = \frac{1}{\sqrt{m}\sigma_b}\sum_{i=1}^{m}(Y_i^* - E^*Y_1^*)$$

by the classical conditions for the CLT of an i.i.d. sequence, it is sufficient to prove that, for any $\delta > 0$,

$$\sum_{i=1}^{m} P^*\left(\left|Y_i^*/\sqrt{m}\sigma_b\right| > \delta\right) \to 0 \text{ in probability,}$$

$$\sum_{i=1}^{m} E^*\left(\left|Y_i^*/\sqrt{m}\sigma_b\right| 1_{|Y_i^*/\sqrt{m}\sigma_b|>\delta}\right) \to 0 \text{ in probability,}$$

and

$$\sum_{i=1}^{m} \mathrm{Var}^*\left(\left|Y_i^*/\sqrt{m}\sigma_b\right| 1_{|Y_i^*/\sqrt{m}\sigma_b|<\delta}\right) \to 1 \text{ in probability.}$$

Using the definition of the MBB and the notation $Y_{i,\delta} := Y_i 1_{|Y_i/\sqrt{m}\sigma_b|<\delta}$, the above is equivalent to

$$\frac{m}{n}\sum_{i=1}^{n} 1_{|Y_i/\sqrt{m}\sigma_b|>\delta} \to 0 \text{ in probability,}$$

$$\frac{m}{n}\sum_{i=1}^{n} |Y_i/\sqrt{m}\sigma_b| 1_{|Y_i/\sqrt{m}\sigma_b|>\delta} \to 0 \text{ in probability,}$$

and

$$\frac{1}{n}\sum_{i=1}^{n}\left(Y_{i,\delta}/\sigma_b\right)^2 - \left(\frac{1}{n}\sum_{i=1}^{n}\left(Y_{i,\delta}/\sigma_b\right)\right)^2 \to 1 \text{ in probability.}$$

The essential ingredient for the proof is the following LLN for weakly dependent triangular arrays.

Lemma 4.1 (Lemma 1 in Radulović [33]). *Let $\{X_i^n\}$ be a triangular array such that, for every fixed n, the "row" $\{X_i^n : i = 1, \dots, n\}$ is strictly stationary. Let us also assume that $\{X_i^n\}$ is uniformly integrable and that $L := \lim E(X_1^n)$ exists and is finite. Finally suppose that there exists a triangular array of positive real numbers $\{a_i^n\}$ such that for every $M \in R$,*

$$|\mathrm{Cov}(Y1_{|Y|<M}, Z1_{|Z|<M})| \le M^2\alpha_i^n \text{ and } \frac{1}{n}\sum_{i=1}^{n}\alpha_i^n \to 0,$$

where Y and Z are any real random variables measurable with respect to $\sigma(X_1^n)$ and $\sigma(X_i^n)$, respectively. Then

$$\frac{1}{n}\sum_{i=1}^{n} X_i^n \to L \textit{ in probability.}$$

Finally, we need to observe that for any $Y_1/\sigma_b = \sigma_b^{-1}\sum_{i=1}^{b} X_i \to N(0,1)$ together with $E(Y_1/\sigma_b)^2 = 1$ implies that $\{(Y_1/\sigma_b)^2\}_{b=1}^{\infty}$ is uniformly integrable.

This, and the above lemma allows one to bypass the usual moment requirements commonly associated with the CLT and the LLN. The proof of Theorem 2.1 now easily follows.

Comment on the proof for Theorem 2.2. Similar to Theorem (2.1), the Lindeberg condition coupled with the assumption

$$\frac{1}{m}\sum_{i=1}^{n}(Y_i/b - \overline{X}_n)^2 \to \sigma^2 \text{ a.s.} \tag{4.2}$$

implies

$$\frac{1}{\sqrt{n}}\sum_{i=1}^{n}(Y_i^* - EY^*) \overset{W}{\to} N(o,\sigma^2) \text{ a.s.}$$

The hard part is to verify Assumption (4.2), which is an a.s. LLN for weakly dependent triangular arrays. This follows easily after applying the following two powerful lemmas.

Lemma 4.2 (Lemma 3.1 in Shao and Yu [40]). *Let $\{\xi_n\}$ be a sequence of random variables with $E\xi_n = 0$ and $\sup_{n\geq 1} E\xi_n^2 < \infty$. Assume there is a constant $C > 0$ such that for any $n \geq 1$,*

$$\sup_{k>0} E\left(\sum_{i=k+1}^{k+n}\xi_i\right)^2 \leq Cn.$$

Then

$$\lim_{n\to\infty}\frac{|\sum_{i=1}^{n}\xi_i|}{\sqrt{n}\log^2(n)} = 0 \textit{ a.s.}$$

Lemma 4.3 (Theorem 2.2 in Peligrad [31]). *Assume $\{X_n\}_{n\geq 1}$ is an α-mixing sequence of centered random variables such that $|X_i| < C$ a.s. Then there is a universal constant K such that for every $x > 0$ and every $n \in N$,*

$$P\left(\max_{1\leq i\leq n}\left|\sum_{k=1}^{i} X_i\right| > x\right) \leq Kx^{-2}\left(\sum_{i=1}^{n} EX_i^2 + C^2 n\sum_{i=1}^{n}\alpha_i\right).$$

MBB for the empirical CLT. In order to prove that for a certain class $\mathcal{F}$ of functions

$$Z_n^*(f) := \sqrt{n}(P_n^* f - P_n f)_{f\in\mathcal{F}} \to G_P(f)_{f\in\mathcal{F}}, \text{ “in probability”}$$

it is sufficient (and necessary) to establish the following:

(a) Finite-dimensional convergence (FIDI): That is,

$$(Z_n^*(f_1), \ldots, Z_n^*(f_k)) \to (G_P(f_1), \ldots, G_P(f_k)) \text{ in probability}$$

for any finite subset $\{f_1, \ldots, f_k\} \subset \mathcal{F}$.

(b) Stochastic equicontinuity:

$$\lim_{\delta \to 0} \limsup_{n \to \infty} P^*(\sup_{d(f,g) \le \delta} |Z_n^*(f)| > \varepsilon) = 0 \text{ “in probability”}$$

for some distance d (typically $L_2(P)$ distance). By “in probability,” we mean a.s. along subsequences.

Theorems 2.1 and 2.2 settle the (a) part easily. The (b) part is much more difficult and proofs are very technical. We will only outline the main idea and pinpoint the difficulties. It is by now well established that in order to bound the quantity

$$\sup_{d(f,g) \le \delta} |Z_n^*(f)|,$$

one has to use some kind of “chaining” argument and good exponential inequalities. Let us cover the class $\mathcal{F}$ with subclasses $\mathcal{H}_k$ such that for every $f \in \mathcal{F}$ there exist $h_f^k \in \mathcal{H}_k$ with $d(h_f^k, f) \le \delta/2^k$. A chaining argument (see Pollard [32]) by cleverly adding and subtracting appropriate h_f^k's and h_g^k's produces

$$P^*(\sup_{d(f,g) \le \delta} |Z_n^*(f)| > \varepsilon)$$
$$\le C_1 \int_{\phi_n}^{\delta} N(x, \mathcal{F}, L_2(P_n))dx + C_2 P^*(\sup_{d(f,g) \le \phi_n} |Z_n^*(f)| > \varepsilon/2), \quad (4.3)$$

for any $\phi_n \ge C_3/n^{1/2}$. Since in our cases the class $\mathcal{F}$ is VC, one could try to invoke a well-known property of VC classes:

$$N(x, \mathcal{F}, L_2(P_n)) \le K_1 \left(\frac{(P_n F^2)^{1/2}}{x} \right)^{K_2}, \text{ where } F \text{ is an envelope function for } \mathcal{F}. \quad (4.4)$$

Unfortunately, the chaining argument above heavily relies on an exponential inequality which in turn requires independence. In order to use chaining in the MBB setting one has to explore the “blocking structure,” and

$$Z_n^*(f) = \frac{1}{\sqrt{n}} \sum_{i=1}^{n} (f(X_i^*) - E^* f(X_1^*)) = \frac{1}{\sqrt{m}} \sum_{i=1}^{m} (Y_{i,f}^* - E^* Y_{i,f}^*),$$

where $Y_{i,f}^* := f(X_{(i-1)b+1}^*) + \cdots + f(X_{ib}^*)$. We could formally define a class

$$\widetilde{\mathcal{F}} = \{\widetilde{f}(x_1, \ldots, x_b) := \sum_{i=1}^{b} f(x_i)/\sqrt{b} : f \in \mathcal{F}\}$$

and a process

$$\widetilde{Z}_n^*(\widetilde{f})_{\widetilde{f} \in \widetilde{\mathcal{F}}} = \frac{1}{\sqrt{m}} \sum_{i=1}^{m} (\widetilde{f}(\overrightarrow{X}_i^*) - E^* \widetilde{f}(\overrightarrow{X}_1^*))_{\widetilde{f} \in \widetilde{\mathcal{F}}},$$

where $\overrightarrow{X}_i^* = (X^*_{(i-1)b+1}, \ldots, X^*_{ib})$. Now, the $\widetilde{f}(\overrightarrow{X}_i^*)$ are i.i.d. and the class $\widetilde{\mathcal{F}}$ is VC. However, this class depends on n (through b) and although we can still formally use the bound (4.4) the envelope $\widetilde{F} := \sup |\widetilde{f}| \to \infty$ and the bound is useless.

Thus in order to control the integral part of (4.3) one has to consider a more painful approach of relating the quantities $N(x, \mathcal{F}, L_2(P_n))$ and $N(x, \mathcal{F}, L_2(P))$. In words, we need the uniform law of large numbers (4.1).

As mentioned before we will not go into the details of these proofs, since they are rather technical. We will only emphasize the key ideas.

First, an important observation (due to Bühlmann) played a pivotal role. Namely, a careful inspection of the chaining argument reveals that to obtain (4.3) one needs Bernstein's exponential inequality, *but only for finitely many functions*, (i.e., only for $h_f^k \in \bigcup_{k=1}^{\phi_n} \mathcal{H}_k$). This in turn means that the uniform LLN has to be established only for finitely many (depending on n) functions.

Second, the uniform LLN (for the case $f(x) = 1_{x\le t}$) can be established directly using the classical Davydov [12] covariance inequality:

$$|\operatorname{Cov}(X, Y)| \le 8\alpha(X, Y)^r \, \|X\|_p \, \|Y\|_q$$

for positive r, p, q such that $r + p^{-1} + q^{-1} = 1$. This is mainly due to the fact that the covering number $N(x, \mathcal{F}, L_2(P))$ in this setting is of the order x^{-2}. For the more complicated case ($\mathcal{F}$ is VC), one has to use the following powerful lemma.

Lemma 4.4 (Proposition 2 in Doukhan et al. [16]). *Let $\{X_i\}_{i\ge1}$ be a sequence of random variables in a Polish space. For any integer $j > 0$, let $b_j = \beta(\sigma(X_j), \sigma(X_i : i > j)$. Then there exists a sequence $\{X_i'\}_{i\ge1}$ of* independent random variables *such that, for any positive integer j, X_j' has the same distribution as X_j and $P(X_j \ne X_j') \le b_j$.*

Now we can use the "blocking trick." Namely,

$$\begin{aligned} b^{-1/2}\sum_{j=1}^{n} f(X_j) &= \sum_{i=1:i \text{ even}}^{n/b} \widetilde{f}(\overrightarrow{X}_i) + \sum_{i=1:i \text{ odd}}^{n/b} \widetilde{f}(\overrightarrow{X}_i) \\ &= \sum_{i=1:i \text{ even}}^{n/b} (\widetilde{f}(\overrightarrow{X}_i) - \widetilde{f}(\overrightarrow{X_i'})) + \sum_{i=1:i \text{ odd}}^{n/b} (\widetilde{f}(\overrightarrow{X}_i) - \widetilde{f}(\overrightarrow{X_i'})) + \sum_{i=1}^{n/b} \widetilde{f}(\overrightarrow{X_i'}) \end{aligned}$$

(here we revoke the above notation with $\widetilde{f}(\overrightarrow{X_i'}) = \sum_{j=(i-1)b+1}^{ib} f(X_i')/\sqrt{b})$. By the above lemma, the $\widetilde{f}(\overrightarrow{X_i'})$ are independent and $P(|\widetilde{f}(\overrightarrow{X}_i) - \widetilde{f}(\overrightarrow{X_i'})| > 0) \le \beta_b$. Since the X_i are β-mixing we can reduce our problem to the i.i.d. case. To obtain optimality (for moment and β-mixing conditions) one has to carefully balance two opposing forces. Namely, we would like to increase b in order to speed up the β_b's convergence to 0, while on the other hand, we want n/b as large as possible in order to have better exponential inequalities.

Third, in both cases (indicators and VC) the reminder term $P^*(\sup_{d(f,g)\le\phi_n} |Z_n^*(f)| > \varepsilon/2)$ is fairly easy to bound using the VC properties.

We briefly comment on the Markov chain results. Since the renewal property ensures that even the nonbootstrapped blocks B_i are independent, most of the proofs reduce to the i.i.d. case. Theorem 3.1 is obtained by mimicking the proof of Theorem 2.1 for the MBB bootstrap, and Theorem 3.2 by essentially repeating the Giné–Zinn argument.The main difficulties in proving Theorems 3.1 and 3.2 are related to the random size of the resampled blocks. This creates problems with random stopping times (in the real and Banach space settings). One could replace a random stopping time with its expectation. However, the bootstrap random variables are formed by resampling from a random number of blocks. Conditioning on a sample, the number of blocks is not random but one problem still remains: the proof of Theorem 3.2 is closely related to Giné and Zinn's [21] proof for the i.i.d. case. There, the authors verified the bootstrap stochastic equicontinuity by comparing it with the original process. This was done by applying a chain of very carefully crafted inequalities in a Banach space setting, some of which do not hold if the bootstrap is resampled from a random number of variables. The solution for this problem is rather technical but straightforward.

References

[1] Arcones, M.A. and Yu, B. Central limit theorems for empirical and U-processes of stationary mixing sequences. *J. Theor. Probab.* **7** (1994), 47–71.

[2] Athreya, K.B. and Ney, P. A new approach to the limit theory of recurrent Markov chains. *Trans. Amer. Math. Soc.* **245** (1978), 493–501.

[3] Athreya, K.B. and Fuh, C.D. Bootstrapping Markov chains: countable case. *J. Statist. Plan. Infer.* **33** (1992), 311–331.

[4] Bickel, P. and Freedman, D. Some asymptotic theory for the bootstrap. *Ann. Statist.* **9** (1981), 1196–1216.

[5] Bolthausen, E. The Berry-Esséen theorem for functionals of discrete Markov chains. *Z. Wahrscheinlichkeitstheorie verw. Geb.* **54** (1980), 59–73.

[6] Bradley, R.C. A stationary, pairwise independent, absolutely regular sequence for which the central limit theorem fails. *Probab. Th. Rel. Fields* **81** (1989), 1–10.

[7] Bühlmann, P. Blockwise bootstrapped empirical process for stationary sequences. *Ann. Statist.* **22** (1994), 995–1012.

[8] Bühlmann, P. The blockwise bootstrap for general empirical processes of stationary sequences. *Stoch. Proc. Appl.* **58** (1995), 247–265.

[9] Chung, K. L. *Markov Chains with Stationary Transition Probabilities.* Springer, Berlin, 1967.

[10] Datta, S. and McCormick, W.P. Regeneration-based bootstrap for Markov chains. *Canad. J. Statist.* **21** (1993), 181–193.

[11] Datta, S. and McCormick, W.P. Some continuous Edgeworth expansion for Markov chains with applications to bootstrap. *J. Multivar. Anal.* **52** (1995), 83–106.

[12] Davydov, Y.A. Convergence of distributions generated by stationary stochastic processes. *Th. Probab. Appl.* (1968), 691–696.

[13] Derman, C. Some asymptotic distribution theory for Markov chains with a denumerable number of states. *Biometrika* **43** (1956), 285–294.

[14] Doukhan, P. *Mixing: Properties and Examples.* Lecture Notes in Statistics, Springer, Berlin, 1994.

[15] Doukhan, P., Massart, P. and Rio, E. The functional central limit theorem for strongly mixing processes. *Ann. Inst. H. Poincare, Probab. Stat.* **30** (1994), 63–82.

[16] Doukhan, P. Massart, P. and Rio, E. Invariance principles for absolutely regular empirical processes. *Ann. Inst. H. Poincare, Probab. Stat.* **31** (1995), 393–427.

[17] Dudley, R. M. Central limit theorems for empirical measures. *Ann. Probab.* **6** (1978), 899–929.

[18] Efron, B. Bootstrap methods: another look at the jackknife. *Ann. Statist.* **7** (1979), 1–26.

[19] Efron, B. and Tibshirani, R.J. *An Introduction to the Bootstrap.* Chapman and Hall, New York, 1993.

[20] Giné, E. and Zinn, J. Necessary conditions for the bootstrap of the mean. *Ann. Statist.* **17** (1989), 684–691.

[21] Giné, E. and Zinn, J. Bootstrapping general empirical measures. *Ann. Probab.* **18** (1990), 851–869.

[22] Hall, P. *The Bootstrap and Edgeworth Expansions.* Springer, Berlin, 1992.

[23] Huber, P.J. Robust estimation of a location parameter, *Ann. Math. Statist.* **35** (1964), 73–101.

[24] Ibragimov, I.A. and Linnik, Y.V. *Independent and Stationary Sequences of Random Variable.* Wolters-Noordhoff, Groningen, 1971.

[25] Künsch, H.R. The jackknife and the bootstrap for general stationary observations. *Ann. Statist.* **17** (1989), 1217–1241.

[26] Liu, R. Singh, K. Moving blocks jackknife and bootstrap capture weak dependence. In: *Exploring the Limits of Bootstrap.* Wiley, New York, pp. 225–248, 1992.

[27] Naik-Nimbalkar U.V. and Rajarshi, M.B. Validity of blockwise bootstrap for empirical processes with stationary observations. *Ann. Statist.* **22** (1994), 980–994.

[28] Nummelin, E. A Splitting technique for Harris recurrent chains. *Z. Wahrscheinlichkeitstheorie verw. Geb.* **43** (1978), 309–318.

[29] Paparoditis, E. and Politis, D.N. The local bootstrap for Markov processes. *J. Statist. Plan. Inf.* **108** (2002).

[30] Peligrad, M. On the blockwise bootstrap for empirical processes for stationary sequences. *Ann. Probab.* **26** (1998), 887–901.

[31] Peligrad, M. Convergence of stopped sums of weakly dependent random variables. *Electr. J. Probab.* **4** (1999), number 13.

[32] Pollard, D. *Convergence of Stochastic Processes.* Springer, New York, 1984.

[33] Radulović, D. The bootstrap of the mean for strong mixing sequences under minimal conditions. *Stat. Prob. Letters* **28** (1996), 65–72.

[34] Radulović, D. The bootstrap for empirical processes based on stationary observations. *Stoch. Proc. Appl.* **65** (1996), 259–279.

[35] Radulović, D. The bootstrap of empirical processes for α-mixing sequences. Progress in Probability, Vol 43, Birkhäuser, Basel, 1998.

[36] Radulović, D. Can we bootstrap even if CLT fails?, *J. Theor. Probab.* **11** (1998), 813–830.

[37] Radulović, D. Renewal type bootstrap for Markov chains. Preprint (2001).

[38] Rajarshi, M.B. Bootstrap in Markov-sequences based on estimates of transition density, *Ann. Inst. Statist. Math.* **42** (1990), 253–268.

[39] Rosenblatt, M. A central limit theorem and a strong mixing condition, *Proc. Nat. Acad. Sci. U.S.A.* **42** (1956), 43–47.

[40] Shao, Q. and Yu, H. Bootstrapping the sample mean for stationary mixing sequences, *Stoch. Proc. Appl.* **48** (1995), 175–190.

[41] Singh, K. On the asymptotic accuracy of Efron's bootstrap. *Ann. Statist.* **9** (1981), 1187–1195.

[42] Vaart, A.W. van der and Wellner J.A. *Weak Convergence and Empirical Processes.* Springer, New York, 1996.

[43] Volkonskii, V.A. and Rozanov, Y.A. Some limit theorems for random functions, Part I, *Th. Probab. Appl.* **4** (1959), 178–197.

[44] Yoshihara, K. Note on an almost sure invariance principle for some empirical processes, *Yokohama Math. J.*, **27** (1979), 105–110.

Dragan Radulović
Program in Applied and Computational Mathematics
Department of Mathematics
Fine Hall 213
Princeton University, NJ 08648, U.S.A.
dradulov@math.princeton.edu

Frequency Domain Bootstrap for Time Series

EFSTATHIOS PAPARODITIS

ABSTRACT. The paper discusses frequency domain bootstrap methods for time series including some recent developments. Attention is focused on nonparametric resampling methods of the periodogram and their application to statistical inference in the frequency domain.

1 Introduction

It is common in time series analysis that the (limiting) distribution of some statistics of interest depends in a complicated way on certain characteristics of the underlying stochastic process making statistical inference based on large sample approximations difficult to implement. In such situations bootstrap based methods offer a potentially useful alternative.

Working, for instance, in the time domain, one can use bootstrap methods by resampling whole blocks of consecutive observations (Künsch (1989), Liu and Singh (1992)) or by applying parametric models together with i.i.d. resampling of estimated residuals (Freedman (1984), Swanepoel and van Wyk (1986), Kreiss and Franke (1992)). See also Radulović (2002) in this volume.

On the other hand, bootstrap methods for time series have been also used successfully in the frequency domain and in particular in resampling periodogram ordinates. The advantage of such an approach is that in order for a bootstrap procedure to generate replicates of the periodogram it is not necessary first to generate replicates of the observed series. Such a bootstrap procedure can operate directly in the frequency domain bypassing; therefore, the difficult problem of mimicking the (potentially very complicated) dependence structure of the underlying process in the time domain. This possibility is due to a basic nonparametric property of the periodogram according to which for quite a general class of stochastic processes, the periodogram ordinates behave for large sample sizes and for any fixed set of frequencies, like independent exponentially distributed random variables, the parameters of which equal the value of the spectral density function at the corresponding frequencies.

Several approaches have been proposed for bootstrapping time series in the frequency domain. For Gaussian processes frequency domain resampling methods have been investigated by Nordgaard (1992), Theiler et al. (1994), and Braun and Kulperger (1997). Hartigan (1990) proposes perturbed periodogram estimates of variance. In this paper, however, we are interested in nonparametric, frequency domain methods of bootstrapping the periodogram. The periodogram is a basic tool in spectral analysis and several statistics of interest can be expressed as functionals of the periodogram; cf. the paper by Dahlhaus and Polonik (2002) and Soulier (2002) in this volume. Apart from a discussion of the different resampling methods proposed, our interest is focused on those classes of periodogram-based statistics to which they can be successfully applied.

In the following we consider a stochastic process $\mathbf{X} = \{X_t, t \in \mathbb{Z}\}$ that is generated by

(A1) $X_t = \sum_{j=-\infty}^{\infty} \psi_j \varepsilon_{t-j}, \sum_{j\in\mathbb{Z}} j^2|\psi_j| < \infty,$

where

(A2) $\{\varepsilon_t, t \in \mathbb{Z}\}$ is a sequence of independent, identically distributed random variables satisfying $E(\varepsilon_t) = 0$ and $E(\varepsilon_t^8) < \infty$.

Denote by f the spectral density of $\mathbf{X}$ which can be expressed as

$$f(\lambda) = (2\pi)^{-1}\sigma_\varepsilon^2|\Psi(e^{i\lambda})|^2,$$

where $\Psi(z) = \sum_{j\in\mathbb{Z}} \psi_j z^j$ and $\sigma_\varepsilon^2 = E[\varepsilon_t^2]$. We assume that

(A3) $\inf_{\lambda\in[0,\pi]} f(\lambda) > 0$.

Let $X_1, X_2, \ldots, X_n$ be observations from $\mathbf{X}$ and denote by $I_n(\lambda)$ the periodogram defined by

$$I_n(\lambda) = \frac{1}{2\pi n}\left|\sum_{t=1}^{n} X_t e^{-i\lambda t}\right|^2. \tag{1.1}$$

The periodogram $I_n(\lambda)$ is usually calculated at the so-called Fourier frequencies

$$\lambda_j = 2\pi j/n, \quad j = 0, 1, \ldots, N, \quad \text{where } N = [n/2].$$

Under assumptions (A1) and (A2) we have $\mathrm{E}[I_n(\lambda_j)] = f(\lambda_j) + O(n^{-1})$, $\mathrm{Cov}[I_n(\lambda_j), I_n(\lambda_k)] = f^2(\lambda_j) + O(n^{-1})$ for $0 < \lambda_j = \lambda_k < \pi$ and $\mathrm{Cov}[I_n(\lambda_j), I_n(\lambda_k)] = n^{-1}\eta_4 f(\lambda_j) f(\lambda_k) + o(n^{-1})$ if $\lambda_j \neq \lambda_k$, where $\eta_4 = \kappa_4/\sigma_\varepsilon^4$ and $\kappa_4 = E[\varepsilon_t^4] - 3\sigma_\varepsilon^4$ is the fourth cumulant of ε_t; cf. Anderson (1971), p. 476.

In the sequel, we denote quantities (expectation, covariance, etc.) taken with respect to the corresponding bootstrap probability measures P^* or P^+ with an asterisk * or with a cross $^+$, respectively. Furthermore, in order to metrize the distance between probability measures we use Mallow's d_2 metric on the space $\{\mathcal{P} : \mathcal{P}$ is a probability measure on $(\mathbb{R}, \mathcal{B}), \int |x|^2 d\mathcal{P} < \infty\}$. This metric is defined by $d_2(\mathcal{P}_1, \mathcal{P}_2) = \inf\{\mathrm{E}|Y_1 - Y_2|^2\}^{1/2}$ where the infimum is taken over all real-valued random variables (Y_1, Y_2) which have marginal distributions $\mathcal{P}_1$ and $\mathcal{P}_2$, respectively. Note that convergence in the d_2 metric implies weak convergence and convergence of the first two moments; cf. Bickel and Freedman (1981).

2 Nonparametric residual-based bootstrap

The first nonparametric approach to bootstrapping the periodogram which we consider is based on the expression

$$I_n(\lambda_j) = f(\lambda_j)\frac{I_{n,\varepsilon}(\lambda_j)}{2\pi\sigma_\varepsilon^2} + R_n(\lambda_j), \tag{2.1}$$

where $I_{n,\varepsilon}(\lambda) = (2\pi n)^{-1}|\sum_{t=1}^{n} \varepsilon_t \exp\{-i\lambda t\}|^2$ and the remainder term $R_n(\lambda_j)$ satisfies under assumptions (A1) and (A2), $\max_{\lambda_j\in[0,\pi]} E[R_n^2(\lambda_j)] = O(n^{-1})$, cf. Brockwell and Davis (1991), Proposition 10.3.1. For bootstrap purposes, the interesting aspect of expression (2.1) is that the relation between the periodogram of the observed series on the one hand and of the unknown spectral density and the (rescaled) periodogram of the unobserved i.i.d. sequence on the other hand, can be approximately described by means of a multiplicative regression model.

Taking into account the asymptotic independence of the periodogram ordinates and ignoring the asymptotically negligible term $R_n(\lambda_j)$, this expression suggests that bootstrap replicates of the periodogram can be obtained by replacing the unknown spectral density f by an estimator $\hat{f}$ and mimicking the sampling behavior of $I_{n,\varepsilon}(\lambda_j)/\sigma_\varepsilon^2$ by means of an i.i.d. resampling of appropriately defined frequency domain residuals.

To implement this idea let $\hat{f}$ be an estimator of f satisfying

(B1) $\sup_{\lambda\in[0,\pi]} |\hat{f}(\lambda) - f(\lambda)| \to 0$ a.s.

The algorithm used to generate bootstrap periodogram replicates $I_n^*(\lambda_j)$, $j = -N, -N + 1, \ldots, N$ can then be described by the following two steps.

Step 1. Let $\widehat{U}_j = I_n(\lambda_j)/\hat{f}(\lambda_j)$ and define the rescaled residuals

$$\widetilde{U}_j = \frac{\widehat{U}_j}{N^{-1}\sum_{s=1}^{N}\widehat{U}_s}, \quad j = 1, 2, \ldots, N.$$

Denote by $F_{n,U}$ the empirical distribution function of $\widetilde{U}_1, \widetilde{U}_2, \ldots, \widetilde{U}_N$.

Step 2. Let U_j^*, $j = 1, 2, \ldots, N$ be i.i.d. random variables with cumulative distribution function $F_{n,U}$ (conditionally on the given observations $X_1, X_2, \ldots, X_n$). Define bootstrap periodogram values as

$$I_n^*(\lambda_j) = I_n^*(-\lambda_j) = \hat{f}(\lambda_j)U_j^*, \quad j = 1, 2, \ldots, N,$$

and $I_n^*(0) = 0$.

Like centering in an additive regression model, the rescaling in Step 1 ensures that $E^*[U_j^*] = 1$ which avoids an unnecessary bias. Elementary calculations yield $\mathrm{E}^*[I_n^*(\lambda_j)] = \hat{f}(\lambda_j)$ and $\mathrm{Cov}^*[I_n^*(\lambda_j), I_n^*(\lambda_k)] = \delta_{j,k}\hat{f}(\lambda_j)\hat{f}(\lambda_k)$, where $\delta_{j,k}$ denotes Kronecker's delta.

The above procedure is called a nonparametric residual-based frequency bootstrap procedure because it resamples the periodogram by means of an i.i.d. resampling of the (nonparametrically estimated) frequency domain "residuals" $\widetilde{U}_j$, $j = 1, 2, \ldots, N$. Such an approach for bootstrapping the periodogram was proposed by Hurvich and Zeger (1987) under more restrictive (Gaussian) assumptions on the underlying process and by Franke and Härdle (1992) for inferring properties of kernel estimators of the spectral density. Note that exploiting our knowledge about the asymptotic distribution of the rescaled periodogram $I_{n,\varepsilon}(\lambda)/\sigma_\varepsilon^2$, we may modify the above bootstrap procedure by replacing the U_j^*, $j = 1, 2, \ldots, N$ by a a set of N independent standard exponentially distributed random variables, V_j^*. However, for the class of statistics considered in the sequel, all results obtained hold if the bootstrap periodogram $I_n^*(\lambda_j)$ is replaced by the modified bootstrap statistic $\tilde{I}_n^*(\lambda_j) = \hat{f}(\lambda_j)V_j^*$; cf. Franke and Härdle (1992) and Dahlhaus and Janas (1996).

2.1 Spectral density estimators. The first application of the above bootstrap procedure we discuss is in inferring properties of nonparametric (kernel) estimators of the spectral density. For this consider the estimator

$$\hat{f}_h(\lambda) = \frac{1}{n}\sum_{j=-N}^{N} K_h(\lambda - \lambda_j)I_n(\lambda_j) \tag{2.2}$$

of $f(\lambda)$, where $K_h(\cdot) = h^{-1}K(\cdot/h)$, $h > 0$, and the real-valued function K satisfies

(A4) K is a symmetric, nonnegative, and Lipschitz continuous kernel with compact support $[-\pi, \pi]$ which satisfies $(2\pi)^{-1} \int K(u)du = 1$. Furthermore, the Fourier transform $\kappa(u)$ of $K(u)$ is locally quadratic around $u = 0$, i.e., $\lim_{u\to 0}\{\kappa(0) - \kappa(u)\}/u^2$ exists, is finite, and is different from zero.

To ensure consistency of $\hat{f}_h$ we require that the smoothing bandwidth h satisfies

(A5) $h \to 0$ and $nh \to \infty$ as $n \to \infty$.

Interest is now focused on estimating the distribution of the statistic $L_n(\lambda) = \sqrt{nh}\,(\hat{f}_h(\lambda) - f(\lambda))/f(\lambda)$. For this, the residual-based bootstrap algorithm can be adapted appropriately by replacing the estimator $\hat{f}$ used to resample the periodogram by a kernel estimator $\hat{f}_b(\lambda)$. For the smoothing bandwidth b, $b > 0$, we require that

(B2) $b \to 0$ and $nb \to \infty$ as $n \to \infty$.

To estimate the distribution of $L_n(\lambda)$ the bootstrap statistic $L_n^*(\lambda) = \sqrt{nh}(\hat{f}_h^*(\lambda) - \hat{f}_b(\lambda))/\hat{f}_b(\lambda)$ is used, where $\hat{f}_h^*(\lambda)$ is defined by

$$\hat{f}_h^*(\lambda) = \frac{1}{n} \sum_{j=-N}^{N} K_h(\lambda - \lambda_j) I_n^*(\lambda_j).$$

Writing $L_n(\lambda)$ as the sum of a stochastic and a bias term, i.e.,

$$L_n(\lambda) = \sqrt{nh}\left(\frac{\hat{f}_h(\lambda) - E[\hat{f}_h(\lambda)]}{f(\lambda)}\right) + \sqrt{nh}\left(\frac{E[\hat{f}_h(\lambda)] - f(\lambda)}{f(\lambda)}\right) =: D_n(\lambda) + B_n(\lambda),$$

with an obvious notation for $D_n(\lambda)$ and $B_n(\lambda)$, we have, under assumptions (A1)–(A5), that the stochastic term $D_n(\lambda) \Rightarrow N(0, \sigma_f^2(\lambda))$ as $n \to \infty$, where $\sigma_f^2(\lambda) = (2\pi)^{-1} \int K^2(u)du$. Furthermore, if $nh^5 \to 0$, then the bias term vanishes, i.e., $B_n(\lambda) \to 0$, while for $h \sim n^{-1/5}$, which is the optimal rate from the viewpoint of minimizing an integrated mean square error criterion, the bias term $B_n(\lambda)$ is asymptotically nonnegligible. In this case, we get $B_n(\lambda) \to (4\pi)^{-1} f''(\lambda) f^{-1}(\lambda) \int u^2 K(u)du$ as $n \to \infty$. From the previous discussion it is clear that in order for the bootstrap to provide a valid approximation of the distribution of $L_n(\lambda)$, the statistic $L_n^*(\lambda)$ has to be able to estimate correctly the asymptotic behavior of the stochastic term $D_n(\lambda)$ and of the bias term $B_n(\lambda)$. To achieve the second goal in case of an asymptotically nonvanishing bias, it is essential that the smoothing bandwidth b used in the resampling stage behaves in a particular way compared to the smoothing bandwidth h used to estimate the spectral density f. The following theorem summarizes the asymptotic behavior of the nonparametric residual-based bootstrap proposal in estimating the distribution of $L_n(\lambda)$; cf. Theorem 1 of Franke and Härdle (1992).

Theorem 2.1. *Suppose that assumptions* (A1)–(A5) *and* (B2) *are satisfied.*

(i) *If* $nh^5 \to 0$ *as* $n \to \infty$, *then*

$$d_2\left(\mathcal{L}\{L_n(\lambda)\}, \mathcal{L}\{L_n^*(\lambda)|X_1, X_2, \ldots, X_n\}\right) \to 0$$

in probability.

(ii) *If* $h \sim n^{-1/5}$ *and* $hb^{-1} \to 0$ *as* $n \to \infty$, *then the same result as in* (i) *holds true.*

The condition that h tends to zero faster than b stated in the second part of the above theorem, implies that the kernel estimator $\hat{f}_b(\lambda)$ used for resampling purposes should be somewhat smoother compared to the kernel estimator $\hat{f}_h(\lambda)$ used for estimation purposes. This kind of "oversmoothing" condition is common in applications of the bootstrap to estimate the bias in nonparametric estimation; cf., for instance, Romano (1988).

2.2 Spectral means and ratio statistics. An interesting question concerning the applicability of the nonparametric residual-based bootstrap is whether apart from the class of nonparametric estimators considered, there are other classes of periodogram statistics to which this procedure can be successfully applied. To answer this question it is essential to understand the extent to which certain sampling properties of the periodogram $I_n(\lambda)$ which are not appropriately captured by its bootstrapped version $I_n^*(\lambda)$, affect the sampling properties of the statistics of interest.

To understand this point, consider, for instance, the important class of so-called spectral means. Spectral means are integrated periodogram statistics defined by

$$M(I_n, \phi) = \int_0^{\pi} \phi(\lambda) I_n(\lambda) d\lambda, \tag{2.3}$$

where

(A6) $\phi : [0, \pi] \to \mathbb{R}$ is a bounded function of bounded variation. We assume that ϕ is extended to the real line with $\phi(-\lambda) = \phi(\lambda)$ and $\phi(\lambda + 2\pi) = \phi(\lambda)$.

Several statistics useful in time series analysis are obtained as special cases of $M_n(I_n, \phi)$. For instance, for $\phi(\lambda) = 2\cos(\lambda h)$, $h \in \mathbb{N}_0$, we get $M(I_n, 2\cos(\cdot h)) = 2\int_0^{\pi} \cos(\lambda h)\, I_n(\lambda) d\lambda =: \hat{\gamma}(h)$ which is a $\sqrt{n}$-consistent estimator of the true underlying autocovariance $\gamma(h) = E(X_t X_{t+h})$. For $\phi(\lambda) = 1_{[0,x]}(\lambda)$, $x \in [0, \pi]$, we get $M(I_n, 1_{[0,x]}) = \int_0^x I_n(\lambda) d\lambda =: F_n(x)$ which is a $\sqrt{n}$-consistent estimator of the spectral distribution function $F(x) = \int_0^x f(\lambda) d\lambda$. Discretizing the integral in (2.3) along the Fourier frequencies leads to a computationally more tractable version of $M(I_n, \phi)$ which is given by

$$M_D(I_n, \phi) = \frac{2\pi}{n} \sum_{j=1}^{N} \phi(\lambda_j) I_n(\lambda_j). \tag{2.4}$$

Under some smoothness assumptions on ϕ the difference between the above statistic and (2.3) is asymptotically negligible.

From Dahlhaus (1985) it is known that

$$\sqrt{n}\,(M(I_n, \phi) - M(f, \phi)) \Rightarrow N(0, \sigma_M^2(\phi)), \tag{2.5}$$

where

$$\sigma_M^2(\phi) = 2\pi \int_0^{\pi} \phi^2(\lambda) f^2(\lambda) d\lambda + \eta_4 \left(\int_0^{\pi} \phi(\lambda) f(\lambda) d\lambda \right)^2 \tag{2.6}$$

and $\eta_4 = \kappa_4/\sigma_\varepsilon^4$. The interesting fact here is that the second term on the right-hand side of (2.6) is due to the (asymptotically vanishing) covariance of the periodogram ordinates, i.e., for the class of spectral means (2.3), the slight dependence of the periodogram ordinates sums up to a nonvanishing contribution to the asymptotic variance $\sigma_M^2(\phi)$.

Now, let $M_D(I_n^*, \phi)$ be the bootstrap analogue of $M_D(I_n, \phi)$ obtained by replacing I_n by the bootstrap periodogram I_n^*. By the independence of the periodogram ordinates and the uniform convergence (B1) it easily follows that

$$n\mathrm{Var}^*[M_D(I_n^*, \phi)] \to 2\pi \int_0^{\pi} \phi^2(\lambda) f^2(\lambda) d\lambda,$$

in probability. Comparing this with (2.6) shows that apart from two specific cases where

(i) $\kappa_4 = 0$, which is, for instance, true if ε_t is Gaussian,

or

(ii) $\int_0^\pi \phi(\lambda) f(\lambda) d\lambda = 0$,

the nonparametric residual-based periodogram bootstrap fails to estimate correctly the limiting variance of this class of statistics.

Since the above failure of the nonparametric residual-based bootstrap is due to the fact that the bootstrap periodogram ordinates are independent while the ordinary periodogram ordinates are not, we expect that this bootstrap proposal will work for statistics for which the weak and asymptotically vanishing dependence of the periodogram ordinates does not affect characteristics of their (limiting) distribution. There are some different classes of statistics for which this is true. One such class is that of nonparametric (kernel) spectral density estimators considered in the previous section. Here the $O(n^{-1})$-vanishing covariance of the periodogram does not affect the limiting variance $\sigma_f^2(\lambda)$ because of the slower, namely $n^{-1/2}h^{-1/2}$, convergence rate of these estimators. We expect that the same will be true for other linear statistics based on the periodogram provided their convergence rate is less than $n^{-1/2}$. An interesting question is, however, if there exist classes of $\sqrt{n}$-consistent statististics to which the nonparametric residual-based periodogram bootstrap can be successfully applied also in the non-Gaussian case.

Dahlhaus and Janas (1996) showed that one important class of such statistics is that of the so-called ratio statistics. Ratio statistics which are defined as the ratio of two spectral means, i.e.,

$$R(I_n, \phi) = \frac{M(I_n, \phi)}{M(I_n, 1)}, \tag{2.7}$$

have the desired property that their asymptotic distribution is not affected by the weak dependence of the periodogram. In particular, since the statistic $\sqrt{n}(R(I_n, \phi) - R(f, \phi))$ can be written as a normalized, integrated periodogram type statistic, i.e.,

$$\sqrt{n}\,(R(I_n, \phi) - R(f, \phi)) = \sqrt{n}\frac{1}{\int_0^\pi f(\lambda)d\lambda \int_0^\pi I_n(\lambda)d\lambda} \int_0^\pi w(\lambda) I_n(\lambda) d\lambda,$$

with $w := \phi \int_0^\pi f - \int_0^\pi \phi f$, we have because $\int_0^\pi w(\lambda) f(\lambda) d\lambda = 0$ that the asymptotic variance of (2.7) does not depend on the fourth cumulant of the error process. In fact,

$$\sqrt{n}\,(R(I_n, \phi) - R(f, \phi)) \Rightarrow N(0, \sigma_R^2(\phi)) \tag{2.8}$$

as $n \to \infty$, where

$$\sigma_R^2(\phi) = 2\pi \int_0^\pi w^2(\lambda) f^2(\lambda) d\lambda \Big/ \left(\int_0^\pi f^2(\lambda) d\lambda\right)^4;$$

cf. Dahlhaus and Janas (1996).

Note that some statistics useful in time series analysis are obtained as special cases of (2.7). For instance, $R(I_n, 2\cos(\cdot h)) = \hat{\gamma}(h)/\hat{\gamma}(0) =: \hat{\rho}(h)$ is the sample autocorrelation at lag h, which is a $\sqrt{n}$-consistent estimator of the lag-h autocorrelation $\rho(h) = \gamma(h)/\gamma(0)$. Furthermore, $R(I_n, 1_{[0,x]}) =: G_n(x)$ is a $\sqrt{n}$-consistent estimator of the normalized spectral distribution function $G(x) = F(x)/F(\pi)$.

By comparing Edgeworth expansions for the ratio statistics and their bootstrap versions, it was shown by Dahlhaus and Janas (1996) that for this class of statistics the nonparametric residual-based periodogram bootstrap outperforms the classical approach based on the large sample Gaussian approximation (2.8). To make this statement precise, the following assumptions are imposed:

(A7) $\mathrm{E}[\varepsilon_t^3] = 0$.

(A8) The filter coefficients $\{\psi_j\}$ and the Fourier coefficients $\{\widehat{\phi}(u)\}$ of ϕ satisfy $|\psi_j| \le \tau^{|j|}$ and for large u, $|\widehat{\phi}(u)| \le \tau^{|u|}$, where $\tau \in (0,1)$ fixed.

(A9) $(\varepsilon_t, \varepsilon_t^2)$ satisfies Cramér's condition, i.e., $\delta > 0$ and $d > 0$ exist such that for all $\|t\| > d$, $|\mathrm{E}[\exp\{it'(\varepsilon_t, \varepsilon_t^2)'\}]| \le 1 - \delta$.

(A10) Let $d_n(\lambda) := \sum_{t=1}^n X_t \exp\{-i\lambda t\}$ be the finite Fourier transform of $X_1, X_2, \dots, X_n$. For the finite Fourier transform

$$Y_n := \frac{1}{\sqrt{n}}\left(d_n\left(\frac{2\pi}{n}j(1)\right), d_n\left(\frac{2\pi}{n}j(2)\right), \dots, d_n\left(\frac{2\pi}{n}j(8)\right)\right)',$$
$$j(s) \in \{1, 2, \dots, n/2 - 1\}$$

and for $V_n := \int (\phi(\lambda), 1)' I_n(\lambda) d\lambda$, the limits of the corresponding dispersion matrices exist and are positive definite. Furthermore, $\int (\phi(\lambda), 1)\,(\phi(\lambda), 1)'\, f^2(\lambda) d\lambda$ is positive definite.

The following theorem summarizes the asymptotic properties of the bootstrap statistic $T_n^* = \sqrt{n}(R_D(I_n^*, \phi) - R_D(\hat{f}, \phi))$ in estimating the distribution of $T_n = \sqrt{n}(R(I_n, \phi) - R(f, \phi))$. Here $R_D(I_n^*, \phi) := M_D(I_n^*, \phi)/M_D(I_n^*, 1)$ and $R_D(\hat{f}, \phi) := M_D(\hat{f}, \phi)/M_D(\hat{f}, 1)$, where $M_D(\cdot, \cdot)$ is defined in (2.4).

Theorem 2.2. *Suppose that the assumptions* (A1)–(A3), (A6)–(A10), *and* (B1) *hold. Then for almost all samples* $X_1, X_2, \dots, X_n$,

$$\sup_{C \in \mathcal{C}} \left| P(D_n T_n \in C) - P^*(\hat{D}_n T_n^* \in C) \right| = o(n^{-1/2}),$$

where $\mathcal{C}$ *is the class of convex measurable sets* $C \subseteq \mathbb{R}$, $D_n^2 = S_n^{-1}$, $\hat{D}_n^2 = \hat{S}_n^{-1}$ *and* S_n *and* $\hat{S}_n$ *are the variances of* $\sqrt{n}R(I_n, \phi)$, *and* $\sqrt{n}R_D(I_n^*, \phi)$, *respectively.*

2.3 Whittle estimators. Whittle estimators are obtained by minimizing the distance

$$L_n(\theta) = \frac{1}{2\pi}\int_0^\pi \left\{\log f(\lambda, \theta) + \frac{I_n(\lambda)}{f(\lambda, \theta)}\right\} d\lambda$$

between the periodogram I_n and the parametric spectral density $f(\cdot, \theta)$. We assume that $f(\lambda, \theta) = \tau^2 g(\lambda, \vartheta)$ where $\theta = (\tau^2, \vartheta)$, $f(\cdot, \theta) \in \mathcal{F}$ and $\mathcal{F} = \{f(\cdot, \theta) : \theta \in \Theta\}$. Furthermore, we require that Kolmogorov's formula holds, that is, $\int_{-\pi}^{\pi} \log f(\lambda, \theta) d\lambda = 2\pi \log(\tau^2/2\pi)$; cf. Brockwell and Davis (1991), p. 191. In the following we do not assume that $f \in \mathcal{F}$, i.e., we allow the model to be misspecified. In this case the Whittle estimator $\hat{\theta} = (\hat{\tau}^2, \hat{\vartheta})$ converges to $\theta_0 = (\tau_0^2, \vartheta_0)$ where the later is the minimum of

$$L(\theta) = \frac{1}{2\pi}\int_0^\pi \left\{\log f(\lambda, \theta) + \frac{f(\lambda)}{f(\lambda, \theta)}\right\} d\lambda.$$

The estimator $\hat{\theta}$ fulfills the equations

$$\int_0^\pi I_n(\lambda) \frac{\partial}{\partial \vartheta} f^{-1}(\lambda, \theta)|_{\theta = \hat{\theta}} d\lambda = 0 \quad \text{and} \quad \pi^{-1}\int_0^\pi I_n(\lambda) f^{-1}(\lambda, \hat{\theta}) d\lambda = 1$$

while its asymptotic limit θ_0 satisfies the equations

$$\int_0^{\pi} f(\lambda)\frac{\partial}{\partial\vartheta} f^{-1}(\lambda,\theta)|_{\theta=\theta_0} d\lambda = 0 \quad \text{and} \quad \pi^{-1}\int_0^{\pi} f(\lambda) f^{-1}(\lambda,\theta_0)d\lambda = 1.$$

Now, apart from the class of ratio statistics, the innovation free part of the Whittle estimator $\hat{\theta}$, that is, $\hat{\vartheta}$, forms another $\sqrt{n}$-consistent statistic to which the residual-based periodogram bootstrap can be successfully applied. Note that this is not the case for the variance estimator $\hat{\tau}^2$ since the limiting variance of this estimator depends on the fourth cumulant κ_4. To approximate the distribution of $\sqrt{n}(\hat{\vartheta}-\vartheta)$ the bootstrap statistic $\sqrt{n}(\hat{\vartheta}^*-\tilde{\vartheta})$ is used, where $\hat{\theta}^* = (\hat{\tau}^{*^2}, \hat{\vartheta}^*)$ is the minimizer of

$$L_n^*(\theta) = \frac{1}{2}\log(\tau^2/2\pi) + \frac{1}{2n}\sum_{j=1}^{N} I_n^*(\lambda_j) f^{-1}(\lambda_j,\theta)$$

and $\tilde{\theta} = (\tilde{\tau}^2, \tilde{\vartheta})$ the minimizer of $\tilde{L}(\theta)$, which is defined as $L_n^*(\theta)$ after replacing the bootstrap periodogram $I_n^*(\lambda_j)$ by the spectral density estimator $\hat{f}(\lambda_j)$. To deal with the class of Whittle estimators the following assumptions are needed.

(A11) Θ is a compact subset of $\mathbb{R}^+ \times \mathbb{R}^p$ and the parameter ϑ satisfies the identifiability condition that if $\vartheta_1 \neq \vartheta_2$, then $f(\cdot,\vartheta_1) \neq f(\cdot,\vartheta_2)$ on a set with positive Lebesgue measure.

(A12) The function $g(\lambda,\vartheta)$ is three times continuously differentiable with respect to ϑ and every component of its derivative is continuous in λ.

(A13) Let $c_\vartheta(\cdot) = ((\partial/\partial\vartheta)g^{-1}(\cdot,\vartheta),\ (\partial/\partial\vartheta)^2 g^{-1}(\cdot,\vartheta))'$. The limit of the dispersion matrix $\int (c_\vartheta'(\lambda),1)' I_n(\lambda)d\lambda$ exists and is positive definite. Furthermore, $\int (c_\vartheta'(\lambda),1)'(c_\vartheta'(\lambda),1) f^2(\lambda)d\lambda$ is positive definite.

Using a Taylor series expansion of $L_n^*(\theta)$ around $\theta = \tilde{\theta}$, the following result can then be established; cf. Dahlhaus and Janas (1996), Theorem 6.

Theorem 2.3. *If* (A1)–(A3), (A11)–(A13), *and* (B1) *are satisfied, then*

$$d_2\left(\mathcal{L}\{\sqrt{n}(\hat{\vartheta}-\vartheta)\}, \mathcal{L}\{\sqrt{n}(\hat{\vartheta}^*-\tilde{\vartheta})|X_1,X_2,\ldots,X_n\}\right) \to 0$$

in probability.

Improving the above result for the case where the dimension of ϑ equals 1 (i.e., $p=1$), it was shown by Dahlhaus and Janas (1996) under some additional assumptions that the bootstrap approximation $\sqrt{n}(\hat{\vartheta}^*-\tilde{\vartheta})$ of the distribution of $\sqrt{n}(\hat{\vartheta}-\vartheta)$ is second order correct, i.e., that the nonparametric residual-based periodogram bootstrap outperforms the asymptotic Gaussian approximation.

3 Local bootstrap

An alternative way to bootstrap the periodogram and which does not require initial estimation of the unknown spectral density f and i.i.d. resampling of (estimated) innovations, can be motivated by the following observation. Recall that for large n, $I_n(\lambda)$ is approximately exponentially distributed with parameter $f(\lambda)$. Assuming that the spectral density $f(\lambda)$ is a smooth function of λ, we expect that the sampling behavior of the periodogram at any particular frequency λ_s will be very similar for periodogram ordinates corresponding to frequencies in a small neighborhood of λ_s. This implies that periodogram replicates at a frequency λ_s can be obtained by locally resampling the periodogram, i.e., by choosing with replacement between periodogram ordinates corresponding to frequencies 'close' to the frequency λ_s of interest. Borrowing ideas from nonparametric estimation, quantification of closeness in such a resampling scheme can be done by means of a resampling bandwidth. Consistency of this resampling procedure for interesting classes of periodogram statistics can then be established by allowing the neighborhood around λ_s (from which replicates are obtained) to get narrower as the sample size increases but in such a way that the number of periodogram ordinates falling in this neighborhood increases to infinity.

The algorithm used to implement the above idea of generating bootstrap replicates of the periodogram, which are denoted by $I_n^+(\lambda_j)$, $j = -N, -N+1, \ldots, N$, can be described by the following three steps.

Step 1. Select a resampling width B_n where $B_n = B(n) \in \mathbb{N}$ and $B_n \leq [N/2]$.

Step 2. Define i.i.d. discrete random variables $J_1, J_2, \ldots, J_N$ taking values in the set $\{-B_n, -B_n+1, \ldots, B_n\}$ with probability mass function $p_{n,s}$, i.e., $P(J_i = s) = p_{n,s}$ for $s = 0, \pm 1, \ldots, \pm B_n$.

Step 3. The bootstrap periodogram is then defined by

$$I_n^+(\lambda_j) = I_n^+(-\lambda_j) = I_n(\lambda_{J_j+j}) \quad \text{for } j = 1, 2, \ldots, N$$

and $I_n^+(0) = 0$.

Such an approach for bootstrapping the periodogram was proposed by Paparoditis and Politis (1999). The idea underlying the above "local" resampling scheme can be considered as a modification of the ordinary bootstrap idea of "global" resampling, i.e., resampling from the whole set of observations, and can be applied when the distribution of the sequence of random variables to be resampled changes smoothly with respect to their indexing; see Shi (1991), Falk and Reiss (1992), and Paparoditis and Politis (2000).

The following condition states our requirements on the limiting behavior of the resampling width B_n and of the resampling probabilities $p_{n,s}$.

(L1) $B_n \to \infty$ as $n \to \infty$ such that $B_n/n \to 0$. Furthermore, the resampling probabilities $\{p_{n,s} : s = -B_n, -B_n+1, \ldots, B_n\}$ satisfy

$$\sum_{s=-B_n}^{B_n} p_{n,s} = 1, \quad p_{n,s} = p_{n,-s}, \quad \text{and} \quad \sum_{s=-B_n}^{B_n} p_{n,s}^2 \to 0, \quad \text{as } n \to \infty.$$

Note that one way to obtain the resampling probabilities $p_{n,s}$ of the above resampling scheme is by $p_{n,s} = K(\pi s/B_n)/\sum_{s=-B_n}^{B_n} K(\pi s/B_n)$, where K is any commonly used kernel satisfying assumption (A4).

If assumptions (A1)–(A3) and (L1) are fulfilled, then we get as $n \to \infty$ that

$$\sup_{\lambda_j \in (0,\pi)} d_2(\mathcal{L}\{I_n^+(\lambda_j)|X_1, X_2, \ldots, X_n\}, \mathcal{L}\{I_n(\lambda_j)\}) \to 0$$

in probability. This suggests that sampling properties of statistics based on the periodogram I_n can be approximated by the corresponding properties of the same statistic based on the local bootstrapped periodogram I_n^+. Note, however, that the idea of local resampling relies on the asymptotic independence of the periodogram ordinates. In fact the bootstrap periodogram ordinates

$I_n^+(\lambda_j)$ are independent. Simple calculations yield $\mathrm{E}^+[I_n^+(\lambda_j)] = \tilde{f}(\lambda_j)$ and $\mathrm{Cov}^+[I_n^+(\lambda_j), I_n^+(\lambda_k)] = \delta_{j,k}\tilde{f}(\lambda_j)\tilde{f}(\lambda_k)$, where $\tilde{f}(\lambda) = \sum_{s=-B_n}^{B_n} p_{n,s} I_n^+(c(n,\lambda)+\lambda_s)$ and $c(n,\lambda)$ denotes the multiple of $2\pi/n$ closest to λ (the smallest one if there are two). If $\lambda \in [-\pi, 0)$ we set $c(n,\lambda) = c(n,-\lambda)$. Because of the independence of the ordinates $I_n^+(\lambda_j)$ and as in the case of the nonparametric residual-based bootstrap, we expect that inference based on the local resampling procedure can lead to correct answers only as long as the existing weak dependence between the periodogram ordinates $I_n(\lambda_j)$ does not affect the sampling characteristics of the class of statistics considered. Consistency properties for such classes of periodogram-based statistics are briefly discussed in the next sections.

3.1 Spectral density estimators. Using the local bootstrap approach of the periodogram, the distribution of the statistic $L_n(\lambda)$ considered in Section 2.1 can be approximated by means of the distribution of the bootstrap statistic $L_n^+(\lambda) = \sqrt{nh}(\hat{f}_h^+(\lambda) - \tilde{f}(\lambda))/\tilde{f}(\lambda)$, where

$$\hat{f}_h^+(\lambda) = \frac{1}{n}\sum_{j=-N}^{N} K_h(\lambda-\lambda_j) I_n^+(\lambda_j).$$

We then have the following result; cf. Paparoditis and Politis (1999), Theorem 3.2.

Theorem 3.1. *Suppose that* (A1)–(A5) *and* (L1) *are satisfied.*

(i) *If* $nh^5 \to 0$ *as* $n \to \infty$, *then*

$$d_2\left(\mathcal{L}\{L_n(\lambda)\}, \mathcal{L}\{L_n^+(\lambda)|X_1, X_2, \ldots, X_n\}\right) \to 0$$

in probability.

(ii) *If* $h \sim n^{-1/5}$ *and* $nh\sum_{s=-B_n}^{B_n} p_{n,s}^2 \to 0$ *as* $n \to \infty$, *then the same result as in* (i) *holds true.*

Note that the assumption $nh\sum_{s=-B_n}^{B_n} p_{n,s}^2 \to 0$ imposed in the second part of the above theorem is an oversmoothing assumption similar to the one imposed in the second part of Theorem 2.1. To see this, consider the simple case of uniform resampling probabilities, that is, $p_{n,s} = 1/(2B_n+1)$. In this case the above condition simplifies to $nh/(2B_n+1) \to 0$. Taking into account the fact that nh are approximately the number of periodogram ordinates effectively used in $\hat{f}_h(\lambda)$ to estimate $f(\lambda)$, this implies that $2B_n+1$ increases faster than nh, i.e., $\tilde{f}(\lambda)$ is somewhat smoother than $\hat{f}_h(\lambda)$.

3.2 Ratio statistics. For the class of ratio statistics the distribution of $T_n = \sqrt{n}(R(I_n,\phi) - R(f,\phi))$ is approximated by that of $T_n^+ = \sqrt{n}(R_D(I_n^+,\phi) - R_D(\tilde{f},\phi))$ where $R_D(I_n^+,\phi) = M_D(I_n^+,\phi)/M_D(I_n^+,1)$ and $M_D(\cdot,\cdot)$ is defined in (2.4). The following result can then be established; cf. Paparoditis and Politis (1999), Theorem 3.1.

Theorem 3.2. *Suppose that assumptions* (A1)–(A3), (A6), (A10), *and* (L1) *hold true. As* $n \to \infty$,

$$d_2\left(\mathcal{L}\{T_n\}, \mathcal{L}\{T_n^+|X_1, X_2, \ldots, X_n\}\right) \to 0$$

in probability.

3.3 Parametric fits in the frequency domain. To approximate the distribution of the innovation free part of the Whittle estimator, i.e., of $\sqrt{n}(\hat{\vartheta} - \vartheta)$, the local bootstrap estimator $\sqrt{n}(\hat{\vartheta}^+ - \tilde{\vartheta}^+)$ can be used, where $\hat{\vartheta}^+$ is obtained by minimizing the function

$$L_n^+(\theta) = \frac{1}{2}\log(\tau^2/2\pi) + \frac{1}{2n}\sum_{j=1}^{N} I_n^+(\lambda_j) f^{-1}(\lambda_j, \theta)$$

and $\tilde{\vartheta}$ by minimizing $\tilde{L}^+(\theta)$, where $\tilde{L}^+(\theta)$ is defined as $L_n^+(\theta)$ but by replacing the local bootstrap periodogram $I_n^+(\lambda_j)$ by $\tilde{f}(\lambda_j)$. We then have the following result; cf. Paparoditis and Politis (1999), Theorem 3.3.

Theorem 3.3. *If* (A1)–(A3), (A11)–(A13), *and* (L1) *are satisfied, then*

$$d_2\left(\mathcal{L}\{\sqrt{n}(\hat{\vartheta} - \vartheta)\}, \mathcal{L}\{\sqrt{n}(\hat{\vartheta}^+ - \tilde{\vartheta}^+)|X_1, X_2, \ldots, X_n\}\right) \to 0$$

in probability.

We mention here that apart from the class of Whittle estimators, the local bootstrap can be successfully applied to approximate the distribution of the innovation free part of more general classes of frequency domain parameter estimators, a result which can be also established for the nonparametric residual-based periodogram bootstrap. Such estimators are obtained by fitting a model from the parametric class $\mathcal{F} = \{f(\lambda, \theta) : \theta \in \Theta\}$ to the process $\mathbf{X}$ via minimizing the objective function

$$L(f(\cdot, \theta), \overline{f}_n) = \int_{[-\pi,\pi]} D(f(\lambda, \theta)/\overline{f}_n(\lambda))d\lambda.$$

Here $\overline{f}_n$ denotes an estimator of f and $D(x)$ is a three times continuously differentiable function on $(0, \infty)$ which has a unique minimum at $x = 1$. Note that for $\overline{f}_n(\lambda) = I_n(\lambda)$ and $D(x) = \log(x) + x^{-1}$, we are in the case of the Whittle estimators. Taniguchi (1987) investigated properties of such parameter estimators for different choices of the function $D(\cdot)$ and by allowing $\overline{f}_n$ to be a nonparametric estimator of f.

4 Capturing periodogram dependence

From the previous discussion it is clear that in order to extend the class of periodogram-based statistics for which frequency domain bootstrap methods work, it is necessary to take into account the weak dependence structure of the periodogram ordinates. In fact, the statistics considered so far and for which validity of the nonparametric periodogram bootstrap procedures has been established, either obey a convergence rate of order less than $n^{-1/2}$ or they can be expressed as normalized, integrated periodogram statistics, $\sqrt{n}\ \omega(\int f, \int I_n) \int_0^\pi \psi(\lambda) I_n(\lambda)d\lambda$ for some function $\psi(\cdot)$ which satisfies the condition $\int_0^\pi \psi(\lambda) f(\lambda)d\lambda = 0$. This condition ensures that the asymptotic variance of the corresponding statistic does not depend on the fourth cumulant of the error process. Thus, the independence of the bootstrap periodogram ordinates I_n^* and I_n^+ which was justified by the asymptotic independence of the ordinary periodogram, turns out to provide a major limitation in terms of classes of statistics to which these methods can be successfully applied.

One approach to extend the class of statistics for which frequency domain bootstrap methods work, is by modifying appropriately the bootstrap procedures considered in the previous sections.

For instance, and for the residual-based bootstrap, such an approach was proposed by Janas and Dahlhaus (1994). Using a nonparametric estimator of the quantity η_4, the bootstrap periodogram I_n^* is modified in a way that emulates also the covariance of I_n. In particular, for $j = 1, 2, \ldots, N$, a modified bootstrap periodogram is defined by

$$\widetilde{I}_n^*(\lambda_j) = \hat{f}(\lambda_j)\left(U_j^* + \{\sqrt{1+\hat{\eta}_4/2} - 1\}n^{-1}\sum_{s=1}^{N}(U_s^* - 1)\right),$$

where

$$\hat{\eta}_4 = \left(2\pi \hat{f}_2(0) - 4\pi \int_{-\pi}^{\pi} \hat{f}^2(\lambda)d\lambda\right) / \left(\int_{-\pi}^{\pi} \hat{f}(\lambda)d\lambda\right)^2.$$

Here $f_2(\cdot)$ denotes the spectral density of the squared process $\{X_t^2, t \in \mathbb{Z}\}$, while $\hat{f}$ and $\hat{f}_2$ are consistent estimators of f and f_2, respectively.

Note that apart from the problems associated with estimating nonparametrically the complicated functional η_4, such a procedure has some additional drawbacks in that it does not lead to correct estimates of the variance if one wants to apply an asymptotically nonvanishing data taper and of the skewness; see Janas and Dahlhaus (1994) and Dahlhaus and Janas (1996).

A different approach to the bootstrap of the periodogram, called the autoregressive aided periodogram bootstrap, has been recently proposed by Kreiss and Paparoditis (2001). This method combines a time domain and a frequency domain resampling approach, where a parametric fit in the time domain is used to generate periodogram ordinates that imitate the essential features of the data and the weak dependence structure of the periodogram, while a nonparametric (kernel based) correction in the frequency domain is applied in order to catch features not represented by the parametric fit. The bootstrap procedure used to generate pseudo-replicates of the periodogram can be described along the following five steps.

Step 1. Fit a linear autoregressive process of order P to the observations $X_1, X_2, \ldots, X_n$, where P may depend on the sample. Let $\hat{\mathbf{a}}(P) = (\hat{a}_1(P), \hat{a}_2(P), \ldots, \hat{a}_P(P))'$ and $\hat{\sigma}^2(P)$ be the estimators of the autoregressive coefficients and of the error variance, respectively. Consider the estimated residuals

$$\hat{\varepsilon}_t = X_t - \sum_{j=1}^{P} \hat{a}_j(P)X_{t-j}, \quad t = P+1, P+2, \ldots, n,$$

and let $\hat{F}_n$ be the empirical distribution function of the centered quantities $\tilde{\varepsilon}_t$, $t = P+1, P+2, \ldots, n$, i.e., $\tilde{\varepsilon}_t := \hat{\varepsilon}_t - \overline{\varepsilon}$ where $\overline{\varepsilon} = (n-P)^{-1}\sum_{t=P+1}^{n} \hat{\varepsilon}_t$.

Step 2. Generate bootstrap observations $X_1^\dagger, X_2^\dagger, \ldots, X_n^\dagger$, according to the autoregressive model

$$X_t^\dagger = \sum_{j=1}^{P} \hat{a}_j(P)X_{t-j}^\dagger + \varepsilon_t^\dagger,$$

where $\{\varepsilon_t^\dagger\}$ is a sequence of i.i.d. random variables with distribution function $\hat{F}_n$ (conditionally on the observations $X_1, X_2, \ldots, X_n$). Note that the bootstrap process $\mathbf{X}^\dagger = \{X_t^\dagger, t \in \mathbb{Z}\}$ possesses the spectral density $\hat{f}_{AR}(\lambda) = (2\pi)^{-1}\hat{\sigma}^2(P)|1 - \sum_{j=1}^{P} \hat{a}_j(P)\exp\{-ij\lambda\}|^{-2}$ for $\lambda \in [0, \pi]$, which is always well defined if, for instance, $(\hat{a}_1(P), \hat{a}_2(P), \ldots, \hat{a}_P(P))'$ are the Yule–Walker estimators.

Step 3. Compute the periodogram of the bootstrap observations, i.e.,

$$I_n^{\dagger}(\lambda) = \frac{1}{2\pi}\left|\sum_{t=1}^{n} X_t^{\dagger} e^{-i\lambda t}\right|^2, \quad \lambda \in [0, \pi].$$

Step 4. Let $q(\lambda) := f(\lambda)/f_{AR}(\lambda)$, where

$$f_{AR}(\lambda) = (2\pi)^{-1}\sigma^2(P)|1 - \sum_{j=1}^{P} a_j(P)\exp\{-ij\lambda\}|^{-2}.$$

Here $a_1(P), a_2(P), \ldots, a_P(P)$ are the coefficients of $X_{t-1}, X_{t-2}, \ldots X_{t-p}$ in the orthogonal projection $\hat{X}_t$ of X_t on the closed span $\overline{sp}\{X_{t-1}, X_{t-2}, \ldots X_{t-p}\}$ and $\sigma^2(P) := \mathrm{E}[\hat{X}_t - X_t]^2$. Let $\hat{q}(\lambda)$ be an estimator of $q(\lambda)$ satisfying $\sup_{\lambda\in[0,\pi]} |\hat{q}(\lambda) - q(\lambda)| \to 0$ a.s., as $n \to \infty$. For instance, $\hat{q}(\lambda)$ could be the kernel estimator

$$\hat{q}(\lambda) = \frac{1}{n}\sum_{j=-N}^{N} K_h(\lambda - \lambda_j)\frac{I_n(\lambda_j)}{\hat{f}_{AR}(\lambda_j)}, \quad \lambda \in [0, \pi].$$

Step 5. Using $I_n^{\dagger}(\lambda)$ and $\hat{q}(\lambda)$, the bootstrap periodogram is defined by

$$I_n^{\star}(\lambda_j) = \hat{q}(\lambda_j)I_n^{\dagger}(\lambda_j), \quad \text{for } j = 1, 2, \ldots, N.$$

To see the differences between the nonparametric bootstrap procedures considered so far and the new proposal, recall that using $f(\lambda) = q(\lambda)f_{AR}(\lambda)$, we have under the assumptions made that

$$I_n(\lambda_j) = q(\lambda_j)f_{AR}(\lambda_j)\frac{I_{n,\varepsilon}(\lambda_j)}{2\pi\sigma_\varepsilon^2} + R_n(\lambda_j),$$

see (2.1). A similar expression is also valid for $I_n^{\star}(\lambda_j)$, i.e.,

$$I_n^{\star}(\lambda_j) = \hat{q}(\lambda_j)\hat{f}_{AR}(\lambda_j)\frac{I_{n,\varepsilon^{\dagger}}(\lambda_j)}{2\pi\hat{\sigma}^2(P)} + R_n^{\star}(\lambda_j),$$

where $I_{n,\varepsilon^{\dagger}}(\lambda_j) = (2\pi n)^{-1}|\sum_{t=1}^{n}\varepsilon_t^{\dagger}\exp\{-i\lambda t\}|^2$ is the periodogram of the i.i.d. series $\varepsilon_1^{\dagger}, \varepsilon_2^{\dagger}, \ldots, \varepsilon_n^{\dagger}$ and $R_n^{\star}(\lambda_j)$ satisfies $\max_j \mathrm{E}^{\star}[R_n^{\star}(\lambda_j)]^2 = O_P(n^{-1})$; cf. Lemma 7.3 of Kreiss and Paparoditis (2001). Furthermore, since the periodogram ordinates $I_n^{\dagger}(\lambda_j)$ are not independent, the same is also true for $I_n^{\star}(\lambda_j)$. In fact it can be shown that (conditional on the observed series), $\mathrm{E}^{\star}[I_n^{\star}(\lambda_j)] = \breve{f}(\lambda_j) + O_P(n^{-1})$, $\mathrm{Cov}^{\star}[I_n^{\star}(\lambda_j), I_n^{\star}(\lambda_k)] = \breve{f}^2(\lambda_j) + O_P(n^{-1})$ for $0 < \lambda_j = \lambda_k < \pi$ and $\mathrm{Cov}^{\star}[I_n^{\star}(\lambda_j), I_n^{\star}(\lambda_k)] = n^{-1}\hat{\eta}_4(P)\breve{f}(\lambda_j)\breve{f}(\lambda_k) + o_P(n^{-1})$ for $\lambda_j \neq \lambda_k$, where $\breve{f}(\lambda) = \hat{q}(\lambda)\hat{f}_{AR}(\lambda)$ and $\hat{\eta}_4(P) = (\mathrm{E}^{\dagger}[\tilde{\varepsilon}_t]^4 - 3\hat{\sigma}^4(P))/\hat{\sigma}^4(P)$. Note that $\breve{f}(\lambda)$ is a uniformly consistent estimator of $f(\lambda)$ and $E^{\dagger}[\tilde{\varepsilon}_t]^4 - 3\hat{\sigma}^4(P) = (n-P)^{-1}\sum_{t=P+1}^{n}\tilde{\varepsilon}_t^4 - 3\hat{\sigma}^4(P)$ is the fourth cumulant of $\tilde{\varepsilon}_t$.

The essential feature of this new periodogram bootstrap proposal is that for periodogram statistics for which the dependence of the periodogram ordinates does not affect characteristics of their asymptotic distribution (like nonparametric spectral density estimators, ratio statistics, parametric fits in the frequency domain), the procedure works under the same set of process assumptions as those required for the nonparametric bootstrap methods considered in this paper. Furthermore, for important classes of stochastic processes the procedure leads to asymptotically valid approximations for more general classes of periodogram statistics including spectral means; see Kreiss and Paparoditis (2001) for details.

5 Goodness-of-fit testing

Frequency domain bootstrap procedures have also enjoyed success in testing the fit of a parametric time series model. The basic problem considered here is that of testing the composite hypothesis

$$H_0 : f \in \mathcal{F}_\Theta \quad \text{against} \quad H_1 : f \notin \mathcal{F}_\Theta,$$

where $\mathcal{F}_\Theta = \{f(\cdot, \theta), \theta \in \Theta\}$ denotes a class of parametric spectral density models and Θ is a parameter set. Interesting classes of test statistics can be obtained by comparing nonparametric estimators of certain spectral characteristics of the observed process with those of the parametric class postulated under the null hypothesis. For instance, in time series analysis, goodness-of-fit tests based on comparisons between hypothesized and postulated spectral distribution functions have a long history; cf. Priestley (1981) and Anderson (1993). Although a well developed asymptotic theory exist for such tests, the asymptotic distribution obtained is hard to implement. Bootstrap methods offer an alternative and potentially more powerful way to estimate the sampling behavior of the corresponding test statistics under the null.

In conjunction with the bootstrap, several goodness-of-fit statistics have been considered in the literature which can be expressed as functionals of the periodogram and in particular of the empirical process

$$D_n(\lambda) = \sqrt{n}\left\{\frac{F_n(\lambda)}{F_n(\pi)} - \frac{F(\lambda, \hat{\theta})}{F(\pi, \hat{\theta})}\right\}, \quad \lambda \in [0, \pi]$$

or of the (kernel) smoothed deviations

$$d_n(\lambda) = \frac{1}{n}\sum_{j=-N}^{N} K_h(\lambda - \lambda_j)\left(\frac{I_n(\lambda_j)}{f(\lambda_j, \hat{\theta})} - 1\right), \quad \lambda \in [0, \pi].$$

Here $f(\lambda, \hat{\theta})$ denotes the estimated parametric spectral density obtained by replacing θ by a $\sqrt{n}$-consistent estimator $\hat{\theta}$ and $F(x, \hat{\theta}) = \int_0^x f(\lambda, \hat{\theta})d\lambda$ is the corresponding estimated parametric spectral distribution function. Test statistics based on $D_n(\cdot)$ and using a parametric time domain bootstrap with resampling of estimated residuals, was considered by Chen and Romano (1999). Note that such a time domain bootstrap requires that the parametric model for the spectral density under the null can be appropriately translated to the time domain. Alternatively, one can use directly a frequency domain bootstrap approach. Hainz and Dahlhaus (1999) proposed a frequency domain bootstrap for test statistics which are smooth functionals of $D_n(\cdot)$ when $\hat{\theta}$ is the Whittle estimator. Paparoditis (2000) uses a frequency domain bootstrap in order to approximate the distribution of L_2-type statistics based on $d_n(\cdot)$. Since $D_n(\cdot)$ and $d_n(\cdot)$ are scale free, i.e., they are not affected by the innovation variance, and $d_n(\lambda) = O_P(n^{-1/2}h^{-1/2})$, an appropriately modified frequency domain periodogram bootstrap, like the nonparametric residual-based bootstrap, which generates independent periodogram replicates can be successfully applied. The modification uses the estimated parametric spectral density $f(\lambda, \hat{\theta})$ together with standard exponentially distributed innovations to generate the pseudo-replicates of the periodogram instead of using the nonparametric estimated spectral density $\hat{f}(\lambda)$ and the rescaled residuals $\widetilde{U}_j$ as in Section 2. This is important for a good power performance of the methods; cf. Hainz and Dahlhaus (1999) and Paparoditis (2000).

Such a frequency domain periodogram bootstrap leading to independent periodogram ordinates is, however, not valid if the test statistic used depends on the innovation variance. This is, for instance, true if instead of the normalized process $D_n(\cdot)$ the test statistic used is a functional of the empirical process $\sqrt{n}(F_n(\lambda) - F(\lambda, \hat{\theta}))$. An appropriate modification for this case has been proposed by Hainz and Dahlhaus (1999) along the lines suggested in Janas and Dahlhaus (1994).

References

[1] T. W. Anderson: *The Statistical Analysis of Time Series*. John Wiley, New York, 1971.

[2] T. W. Anderson: Goodness of fit tests for spectral distributions. *Annals of Statistics* **21** (1993), 830–847.

[3] P. J. Bickel and D. A. Freedman: Some asymptotic theory for the bootstrap. *The Annals of Statistics* **9** (1981), 1196–1217.

[4] J. W. Braun and R. J. Kulperger: Properties of a Fourier bootstrap method for time series. *Communications in Statistics, Theory and Methods* **276** (1997), 1329–1336.

[5] P. Brockwell and R. Davis: *Time Series: Theory and Methods*. Springer-Verlag, New York, 1991.

[6] H. Chen and J. Romano: Bootstrap-assisted goodness-of-fit tests in the frequency domain. *Journal of Time Series Analysis* **20** (1999), 619–654.

[7] R. Dahlhaus: Asymptotic normality of spectral estimates. *Journal of Multivariate Analysis* **16** (1985), 412–431.

[8] R. Dahlhaus and D. Janas: A frequency domain bootstrap for ratio statistics in time series analysis. *The Annals of Statistics* **24** (1996), 1934–1963.

[9] R. Dahlhaus and W. Polonik: Empirical spectral processes and nonparametric maximum likelihood estimation for time series. In: *Empirical Processes for Dependent Data*. (H. Dehling, T. Mikosch, M. Søresen, eds.) Birkhäuser: Boston, 2002.

[10] M. Falk and R.-D. Reiss: Bootstrapping conditional curves. In: *Bootstrapping and Related Techniques*, (K. H. Jöckel, G. Rothe and W. Sendler, eds.). Lecture Notes in Economics and Mathematical Systems, **376**, Springer Verlag: Berlin, (1992), 173–180.

[11] J. Franke and W. Härdle: On bootstrapping kernel spectral estimates. *The Annals of Statistics* **20** (1992), 121–145.

[12] D. A. Freedman: On bootstrapping two-stage least-squares estimates in stationary linear models. *The Annals of Statistics* **12** (1984), 827–842

[13] G. Hainz and R. Dahlhaus: Spectral domain bootstrap tests for stationary time series. *Beiträge zur Statistik 61, Universität Heidelberg*, (1999).

[14] J. A. Hartigan: Perturbed periodogram estimates of variance. *International Statistical Review* **58** (1990), 1–7.

[15] C. M. Hurvich and S. L. Zeger: Frequency domain bootstrap methods for time series. *Technical Report # 87-115*, (1987), Graduate School of Business Administration, New York University.

[16] D. Janas and R. Dahlhaus: A frequency domain bootstrap for time series. In: *Computationally Intensive Statistical Methods*. Proceedings of the $26th$ Symposium on the Interface (J. Sall and A. Lehman, eds.). Interface Foundation of North America, Fairfax Station, VA, (1994), 423–425.

[17] J.-P. Kreiss and J. Franke: Bootstrapping stationary autoregressive moving-average models. *Journal of Time Series Analysis* **13** (1992), 297–313.

[18] J.-P. Kreiss and E. Paparoditis: Autoregressive aided periodogram bootstrap. *Discussion Paper No 60, SFB 373, Humboldt Universität Berlin*, (2001).

[19] H. Künsch: The jackknife and the bootstrap for general stationary observations. *The Annals of Statistics* **17** (1989), 1217–1241.

[20] R. Liu and C. Singh: Moving blocks jackknife and bootstrap capture weak dependence. In: *Exploring the Limits of the Bootstrap* (R. LePage and L. Billard, eds.), Wiley: New York, (1992), 225–248.

[21] A. Nordgaard: Resampling stochastic processes using a bootstrap approach. In: *Bootstrapping and Related Techniques*. (K. H. Jöckel, G. Rothe and W. Sendler, eds.). Lecture Note in Economics and Mathematical Systems, **376**, Springer, Berlin, (1992), 181–185.

[22] E. Paparoditis and D. N. Politis: The local bootstrap for periodogram statistics. *Journal of Time Series Analysis* **20** (1999), 193–222.

[23] E. Paparoditis: Spectral density based goodness-of-fit tests for time series models. *Scandinavian Journal of Statistics* **27** (2000), 143–176.

[24] E. Paparoditis and D. N. Politis: The local bootstrap for kernel estimators under general dependence conditions. *Annals of the Institute of Statistical Mathematics* **52** (2000), 139–159.

[25] M. B. Priestley: *Spectral Analysis and Time Series*. Academic Press: New York, 1981.

[26] D. Radulović: On the bootstrap and empirical processes for dependent sequences. In: *Empirical Processes for Dependent Data*. (H. Dehling, T. Mikosch, M. Søresen, eds.) Birkhäuser: Boston, 2002.

[27] J. P. Romano: Bootstrapping the mode. *Annals of the Institute of Statistical Mathematics* **40** (1988), 565–586.

[28] S. G. Shi: Local bootstrap. *Annals of the Institute of Statistical Mathematics* **43** (1991), 667–676.

[29] P. Soulier: Empirical process techniques for spectral estimation of fractional processes. In: *Empirical Processes for Dependent Data*. (H. Dehling, T. Mikosch, M. Søresen, eds.) Birkhäuser: Boston, 2002.

[30] J. W. H. Swanepoel and J. W. Y. van Wyk: The bootstrap applied to power spectral density function. *Biometrika* **73** (1986), 135–141.

[31] M. Taniguchi: Minimum contrast estimation for spectral densities of stationary processes. *Journal of the Royal Statistical Society, Series B* **49** (1987), 315–325.

[32] J. Theiler, L. S. Paul and D. M. Rubin: Detecting nonlinearity in data with long coherence times. In: *Time Series Prediction*. (A. Weigend and N. Gershenfeld, eds.). Addison-Welsey: Reading, MA, 1994.

Efstathios Paparoditis
Department of Mathematics and Statistics
University of Cyprus
P.O. Box 20537
CY-1678 Nicosia, Cyprus
stathisp@ucy.ac.cy

Related Titles

Barndorff-Nielsen/Mikosch/Resnick, *Lévy Processes: Theory and Applications*, 0-8176-4167-X, 2001

Adler/Feldman/Taqqu, *A Practical Guide to Heavy Tails: Statistical Techniques for Analyzing Heavy-Tailed Distributions*, 0-8176-3951-9, 1998, 2nd printing, 2000

Doukhan/Oppenheim/Taqqu, *Long-Range Dependence: Theory and Applications*, 0-8176-4168-8, 2002

Kotz/Kozubowski/Podgórski, *The Laplace Distribution and Generalizations: A Revisit with Applications to Communications, Economics, Engineering, and Finance*, 0-8176-4166-1, 2001

Nolan, *Stable Distributions*, 0-8176-4159-9, forthcoming 2003

Good, *Resampling Methods: A Practical Guide to Data Analysis*, 2nd Edition, 0-8176-4243-9, 2001

Sahai/Ageel, *The Analysis of Variance*, 0-8176-4012-6, 2000

Chen/Gupta, *Parametric Statistical Change Point Analysis*, 0-8176-4169-6, 2000

Borodin/Salminen, *Handbook of Brownian Motion: Facts and Formulae*, 2nd Edition, 3-7643-6705-9, 2002

Falk/Marohn/Tewes, *Foundations of Statistical Analyses and Applications with SAS*, 3-7643-6893-4, 2002

Reiss/Thomas, *Statistical Analysis of Extreme Values: with Applications to Insurance, Finance, Hydrology and Other Fields*, 2nd Edition, 3-7643-6487-4, 2001

Groechenig, *Foundations of Time-Frequency Analysis*, 0-8176-4022-3, 2000